TB 43-0209

Color, Marking, and Camouflage Painting of Military Vehicles, Construction Equipment And Materials Handling Equipment

This manual is published to provide information and guidance to personnel charged with painting and marking equipment for which the U.S. Army has responsibility. It contains instructions for treating surfaces to remove corrosion, and procedures for preventing corrosion by applying protective coatings. Although many paint systems are covered, special emphasis is placed on the Chemical Agent Resistant Coatings (CARC), because they are particularly effective in resisting corrosion and chemical penetration, and are also decontaminated more easily than are other coatings.

Should you have suggestions or feedback on ways to improve this book please send email to Books@OcotilloPress.com

---

Edited 2021 Ocotillo Press
ISBN 978-1-954285-55-2

Ocotillo Press
Houston, TX 77017
Books@OcotilloPress.com

DEPARTMENT OF THE ARMY TECHNICAL BULLETIN

COLOR, MARKING, AND CAMOUFLAGE

PAINTING OF MILITARY VEHICLES,

CONSTRUCTION EQUIPMENT

AND

MATERIALS HANDLING EQUIPMENT

This copy is a reprint which includes current pages from Change 1.

HEADQUARTERS. DEPARTMENT OF THE ARMY
31 OCTOBER 1990

SPECIFIC HAZARD

WARNING

The local safety office and preventive medicine support activity must be consulted prior to using Chemical Agent Resistant Coatings (CARC). This also applies to all spray painting operations, regardless of the material used.

Spray Painting.

- All personnel who work in or near a spray painting booth must wear a respirator when spray painting operations are underway.

- Only one person will spray paint at a time. This is to eliminate the hazard of accidentally spraying paint on another person.

- No spray painting will be conducted in an area where unprotected personnel are within a 50-foot radius.

- No unprotected personnel can enter an area of spray painting without protection until 30 minutes after all painting/cleaning is completed.

Cleaning.

- Do not use gasoline or other hydrocarbons for cleaning.

- Avoid skin contact with cleaning solvents. Wear synthetic rubber gloves.

Fire Hazards.

- Observe fire regulations when using paint and lacquer thinners, as they are highly flammable

- Post "NO SMOKING" signs in paint warehouses.

- Listed below are the various specific hazards associated with spray painting, cleaning, fires, safety equipment, hazardous materials, temperatures and equipment.

- The dried spray dust of lusterless enamels is an extreme fire hazard. Remove this dust daily. The danger of fire can be materially reduced by the use of a water-wash or water fall type of spray booth.

- ... highly flammable. Keep away from fl... s.

- ... sanders in a paint shop or near a sp...

- ... ked with linseed oil must either be ... ad out to dry in a well ventilated area ... e spontaneous combustion.

Safe...

- ... ent must be worn during painting, ... blasting, grinding, buffing, or where ... being used. Approved respirators ... en required.

- ... CARC, use silicon rubber (... cloth) gloves for protection.

Haz... ls.

- ... ylene, toluene, naphtha, methyl ethyl ke... butyl ketone, and 1,1,1 tr... y be hazardous materials and must be ... accordance with AR 420-47, Solid ar... aste Management.

- ... carcinogens. Pay careful attention ... utions, notes, and safety procedures ... aints, primers, solvents, and ...

- ... t be observed in the handling of pa... mercury or other fungicides to pr... or skin irritation.

- ... t with lacquer thinner. It can cause a sk...

- ... be allergic to isocyanites shall not pa... thane's.

- ... to surfaces reaching 400 °F (204 °C ... produce toxic fumes and/or ... e.

- ... nent must be properly grounded be... y painting procedures.

CHANGE
NO. 1

HEADQUARTERS
DEPARTMENT OF THE ARMY
Washington D.C., *7 May 1991*

COLOR, MARKING, AND CAMOUFLAGE PAINTING OF MILITARY VEHICLES, CONSTRUCTION EQUIPMENT, AND MATERIALS HANDLING EQUIPMENT

TB 43-0209, 31 October, 1990, is changed as follows:

1. Remove old pages and insert new pages as indicated below.
2. New or changed information is indicated by a vertical bar in the margin of the page.

Remove Pages	Insert Pages
D-1 and D-2	D-1 and D-2

File this change sheet in front of the publication for reference purposes.

By Order of the Secretary of the Army:

CARL E. VUONO
General, United States Army
Chief of Staff

Official:

PATRICIA P. HICKERSON
Colonel, United States Army
The Adjutant General

Distribution:
To be distributed in accordance with DA Form 12-34-E, block 0528, requirements for TB 43-0209.

TECHNICAL BULLETIN

*No. 43-0209

HEADQUARTERS
DEPARTMENT OF THE ARMY
Washington D.C., *31 October 1990*

COLOR, MARKING, AND CAMOUFLAGE PAINTING OF MILITARY VEHICLES, CONSTRUCTION EQUIPMENT, AND MATERIALS HANDLING EQUIPMENT

REPORTING ERRORS AND RECOMMENDING IMPROVEMENTS

You can help improve this manual. If you find any mistakes or if you know of a way to improve the procedures, please let us know. Mail your letter, DA Form 2028, Recommended Changes to Publications and Blank Forms, or DA Form 2028-2 located in back of this manual direct to: U.S. Army Tank-Automotive Command, ATTN: AMSTA-MBT, Warren, Michigan 48397-5000. A reply will be furnished to you.

Approved for Public Release; Distribution is Unlimited

TABLE OF CONTENTS

***This manual supersedes TB 43-0209, 29 October 1976.**

TABLE OF CONTENTS

SECTION I. GENERAL

1. PURPOSE.

a. This bulletin is published to implement AR 750-1 in defining the color marking and camouflage pattern painting requirements for Department of the Army military vehicles (except vehicles in administrative use), construction and highway maintenance equipment, and self-propelled materials handling equipment.

b. Special color and marking requirements for vehicles in administrative use are contained in AR 58-1.

2. SCOPE.

a. These instructions apply to all Department of the Army activities and installations using, maintaining or procuring such vehicles or equipment. Specific exceptions to normal color and marking requirements are identified in later sections of this bulletin.

b. Instructions for application of special technical markings not shown in these instructions are located in specific Department of the Army publications.

c. Vehicle and equipment painting instructions are contained in TM 43-0139.

3. EXCEPTIONS.

a. Vehicles used for intelligence, criminal investigation, and similar purposes requiring concealment of the true identity of the activity involved, may be painted and marked as deemed appropriate by the responsible commander. Identification of such vehicles will be marked in inconspicuous locations as prescribed in applicable Department of the Army publications.

b. Reporting Equipment Improvement Recommendations (EIR's): You can help improve this equipment by recommending design improvements and/or better service and repair procedures. EIR's will be prepared using SF 368, Quality Deficiency Report. Instructions for preparing EIR's are provided in DA Pam 738-750. Mail directly to: Commander, U.S. Army Tank-Automotive Command, ATTN: AMSTA-QRT, Warren, Michigan 48397-5000. A reply will be furnished to you.

4. GENERAL.

a. Markings that are not authorized, are not of the authorized style, or are not in approved location per this bulletin, will be changed or relocated at the time of vehicle repainting, or when the condition of the existing marking becomes unsuitable due to deterioration or damage.

b. Vehicles and equipment will be repainted only when required for preservation purposes, or at time of vehicle

ov[illegible]horized by this bulletin. Repainting w[illegible]rely to change color unless such is con[illegible] for safety or security purposes by th[illegible]mmander. Spot painting for pres[illegible]ill be the rule rather than the exce[illegible]moval from storage to be issued or plac[illegible]icle and equipment will be marked as p[illegible]e instructions.

[illegible]markings will conform to the requ[illegible]bulletin.

[illegible]n addition to those prescribed or auth[illegible]lletin, will be applied only on specific app[illegible]rtment of the Army (paragraph 3a) or [illegible]fficial Department of the Army pub[illegible]

[illegible]culiar to any vehicle, or series of vehic[illegible]ied as specified in this bulletin. Des[illegible]ation, and location of vehicle peculiar mark[illegible]found in the appropriate vehicle tech[illegible]

[illegible]ical conditions, when it becomes nece[illegible]l the markings, all such prominent mark[illegible]vered with temporary paint (Paint, tem[illegible]s gasoline removable MIL-P-13983) on a[illegible]esponsible commander. Such mark[illegible]e temporarily camouflaged by the us[illegible]ud, or other readily removable sub[illegible]

SECT[illegible]PTION

5. [illegible] MARKINGS.

[illegible] numerals used in the marking of vehic[illegible]ment will be of the gothic style illus[illegible]. Thickness of the characters (stro[illegible] between 1/6 and 1/7In of their heig[illegible]

[illegible]s will be of the size shown and spec[illegible]in.

[illegible]arkings may be applied in the form of a[illegible]markers of the prescribed color, may be p[illegible] markers are unavailable, or when app[illegible] made on canvas or other porous surfa[illegible] markers will be of material conf[illegible]43719, Type II, Class 1, and will con[illegible] detailed specification sheets. Refl[illegible]ill be of material conforming to Fede[illegible] L-S-300, Type I or II, Class 3, refle[illegible]

[illegible]il and paint. Paper and metal sten[illegible] as individual letters, numerals, and leger[illegible]es, and may be located by National Stock[illegible] FSC 7520, Federal Supply Catalog C75[illegible]e 2-1 lists several different size meta[illegible]

TABLE 2-1

BRASS STENCIL OUTFIT		
1/2"	SE	7520-00-205-1760
1"	SE	7520-00-298-7043
1 1/2"	SE	7520-00-272-9680
2"	SE	7520-00-298-7044
3"	SE	7520-00-272-9683
5"	SE	7520-00-272-9684
METAL STENCIL OUTFIT		
4"	SE	7520-00-272-9684

(2) Black-pressure sensitive adhesive vinyl: Markings made of this material are available as die cut letters, numerals and legends of various sizes, and are listed in Table 2-2 through 2-4.

TABLE 2-2
LUSTERLESS BLACK PRESSURE-SENSITIVE DECALS NSN 7690-01

LETTER	1-IN	1-1/2 IN	2-IN	3-IN	4-IN
A	030-7126	031-4660	031-9543	032-0732	032-0741
B	030-5778	031-4661	031-5335	032-0733	032-0742
C	030-5779	031-5521	031-5536	032-2428	032-0743
D	030-5780	031-5522	031-6394	032-2430	032-2434
E	030-5781	031-5523	031-9544	032-2432	032-2436
F	030-5782	031-5524	031-5537	032-0734	032-2437
G	030-5783	031-4662	031-5538	032-0735	032-0744
H	030-5784	031-4663	031-5539	032-0736	032-0745
I	030-8542	031-4664	031-5540	032-0737	032-1402
J	030-5785	031-4665	031-5541	032-0738	032-1403
K	030-5786	031-4666	031-5542	032-0739	032-1404
L	030-8543	031-4667	031-5543	032-0740	032-1405
M	030-7816	031-4668	031-5544	031-9545	032-2438
N	030-8544	031-4669	031-5545	031-9546	032-1406
O	030-8545	031-4670	031-5546	031-9547	032-3712
P	031-6395	031-5525	031-5547	031-9548	032-2439
Q	030-8546	031-5526	031-5548	031-9549	032-2441
R	030-8547	031-5527	031-5549	031-9550	032-1407
S	030-8548	031-5528	031-5550	031-9551	032-2443
T	030-8549	031-5529	031-5551	031-9552	032-2445
U	030-8550	031-5530	031-5552	031-9553	032-2447
V	030-8551	031-5531	031-5553	031-9554	032-2449
W	030-8552	031-5532	031-5554	031-9555	032-1408
X	030-8553	031-5333	031-5555	031-9556	032-2451
Y	030-8554	031-4671	031-1400	031-9557	032-4593
Z	030-8555	031-4672	032-0727	031-9558	032-4594

d. Temporary markings will be applied with a temporary paint or removable vinyl markings in accordance with instructions contained herein.

e. Unless otherwise indicated in Section VIII, markings on vehicles and equipment will be applied in contrasting lusterless black 37030 or green 383 (34094).

f. When pressure-sensitive adhesive markers are used, and gloss white markers are unavailable, lusterless white 37875 may be used on vehicles requiring gloss white.

6. DESCRIPTION OF COLORS.

a. Refer to TM 43-0139 for color requirements for specific usage's other than those shown in these instructions.

b. Vehicles assigned to Explosive Ordnance Disposal Units will have all fenders painted lusterless red 31136 on vehicles painted lusterless green 383 or forest green.

TABLE 2-3
LUSTERLESS NUMERALS -- NSN 7690-01

NUMBER	1-IN	1-1/2 IN	2-IN	3-IN	4-IN
1	030-8556	031-4673	032-0728	031-9559	032-4595
2	030-8557	031-4674	032-0729	031-9560	032-4596
3	030-8558	031-4675	032-2421	031-9561	032-4597
4	030-8559	031-4676	032-2422	031-9562	032-4598
5	030-8560	031-4677	032-1401	031-9563	032-4599
6	030-8561	031-5534	032-2423	031-9564	032-4600
7	030-8562	031-4678	032-2424	031-9565	032-4601
8	030 8563	031 4670	032 2425	031-9566	032-2452
9	030-8564	031-4680	032-0730	031-9567	032-2453
0	030-8565	031-4681	032-0731	031-9568	032-2454

TABLE 2-4
MISCELLANEOUS DECALS -- NSN 7690-01

	1-IN	1 1/2-IN	2-IN	3-IN	4-IN	6-IN
				042-0671		
MAX SPEED	032-3713	032-3715				
NO SMOKING	032-3714	032-4602	032-4604	036-4777		
US ARMY		032-4603	032-4605	032-4607	032-4609	
MIL POLICE			032-4606	032-4608		
FLAM-MABLE						047-6457*

* with white background

SECTION III.
IDENTIFICATION MARKINGS

7. AGENCY IDENTIFICATION AND REGISTRATION NUMBER.

There is no further requirement to mark vehicles and equipment with the identification legend "US Army." When vehicles and equipment are camouflaged, both the "US Army" and the registration number will be removed from vehicle/equipment exterior. The preferred location of the registration number is any appropriate interior area which is visible from outside a locked or secured item.

8. NATIONAL SYMBOL.

a. The national symbol is a 5-point star in contrasting lusterless black 37030 or green 383 (34094) in a size that fits inside a 3 inch diameter circle. It will be applied to all tactical and combat vehicles and appropriate equipment in CONUS and overseas. The symbol will be applied to vehicles and equipment only when withdrawn for issue to US Army units. It will not be applied to ambulance or other medical service vehicles affected by international agreement concerning such vehicles.

b. The star will be applied to front and rear of vehicles/equipment as specified in Appendix C. On the rear of wheeled vehicles, the star may be centered on the tailgate, Figure 2.

9. UNIT IDENTIFICATION.

a. These markings will be applied to vehicles and equipment by units to which they are assigned. The markings will be maintained at all times but will be removed when the vehicle or equipment is permanently transferred.

b. The unit identification markings will normally appear on both the front and rear of each vehicle. Suitable surfaces (bumperette, tailgates, fenders, rear and front) may be used provided the unit identification marking does not conflict with the location and legibility of other markings. Where no suitable surface is available at the front or rear, the markings will be applied in an appropriate location on each side of the vehicle. If necessary, the unit identification may be applied to the vehicle mounted equipment. Refer to Figures 2 through 13.

c. CARC markings. Equipment painted with CARC with data plates shall have the word "CARC" painted in black or green contrasting color in a conspicuous area as near the data plate as possible. Painting shall be in block letters, as large as possible, not to exceed one inch. All major items having a log book shall have the Equipment Control Record, DA Form 2408-9, annotated in the lower left corner of Block 21, Remarks, to reflect the word "CARC" and the date applied.

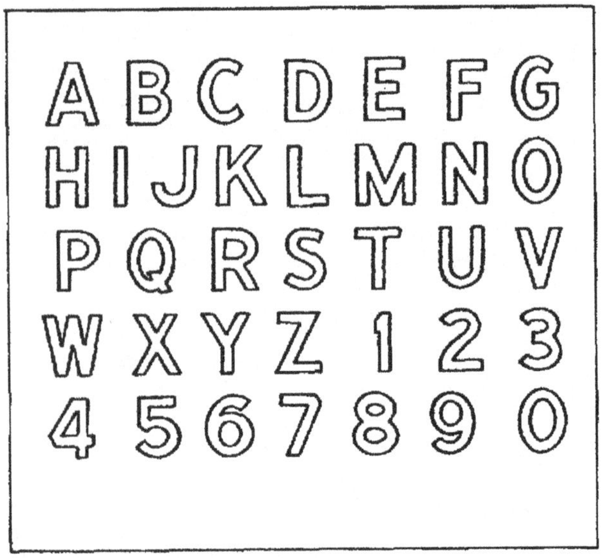

Figure 1. Gothic style alphabet and numerals.

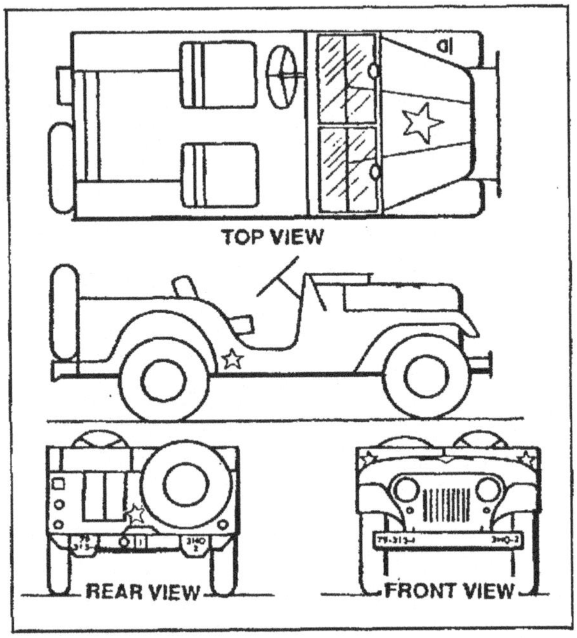

Figure 2. Marking location for 1/4-ton trucks.

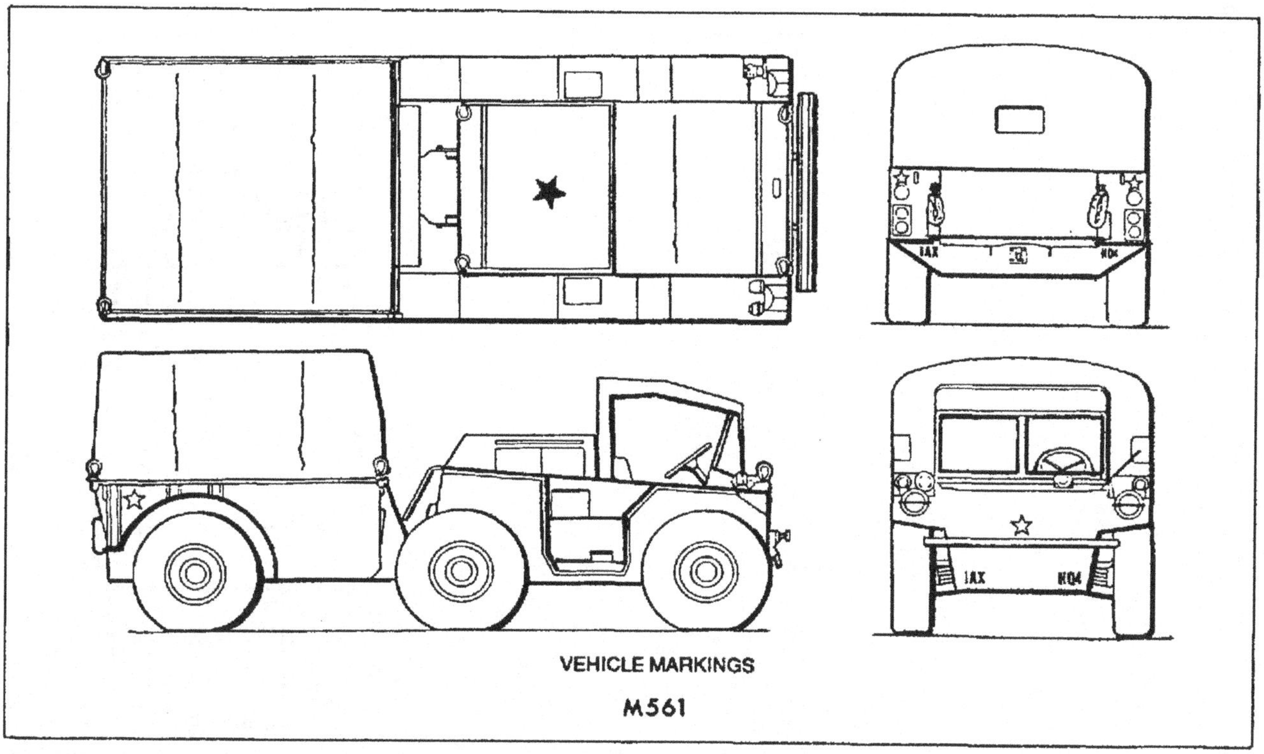

Figure 3. Marking locations for 1-1/4-ton trucks.

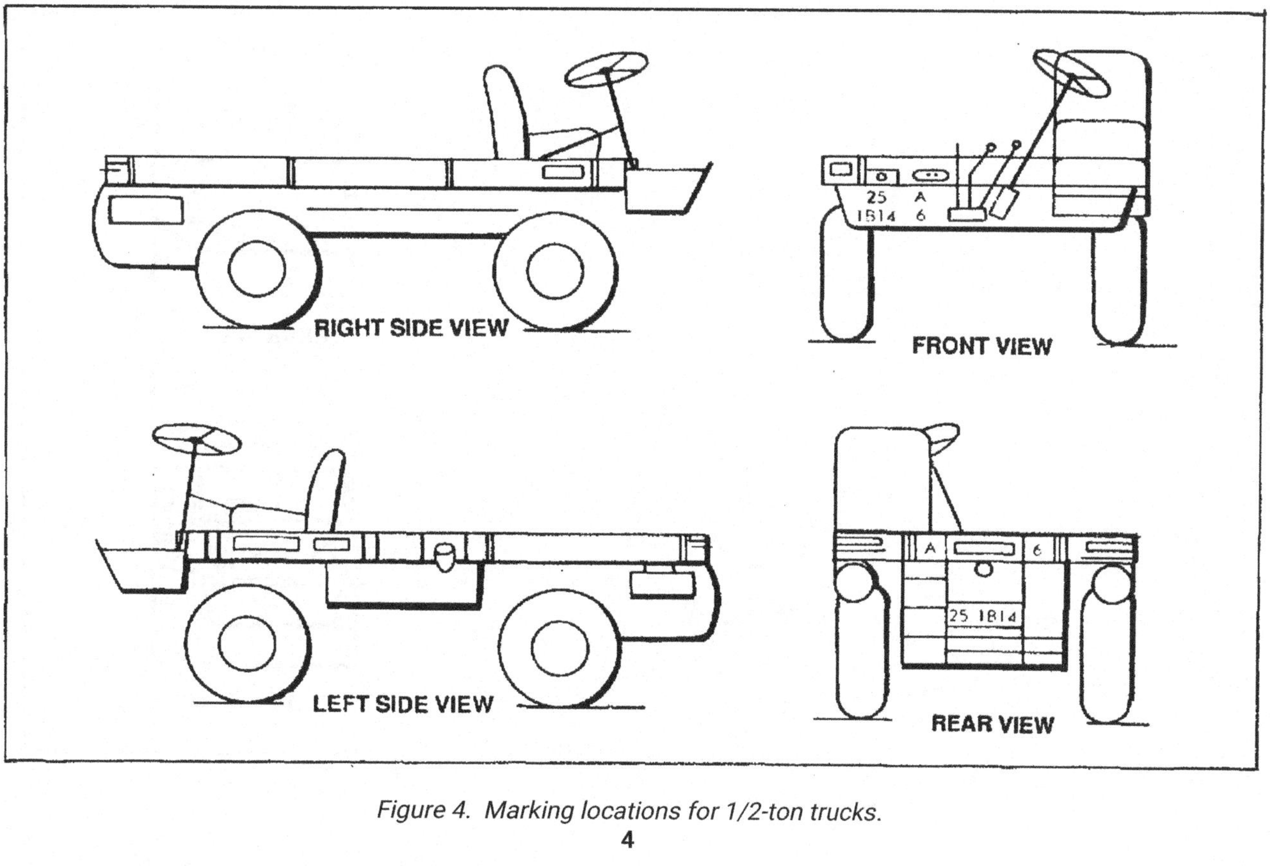

Figure 4. Marking locations for 1/2-ton trucks.

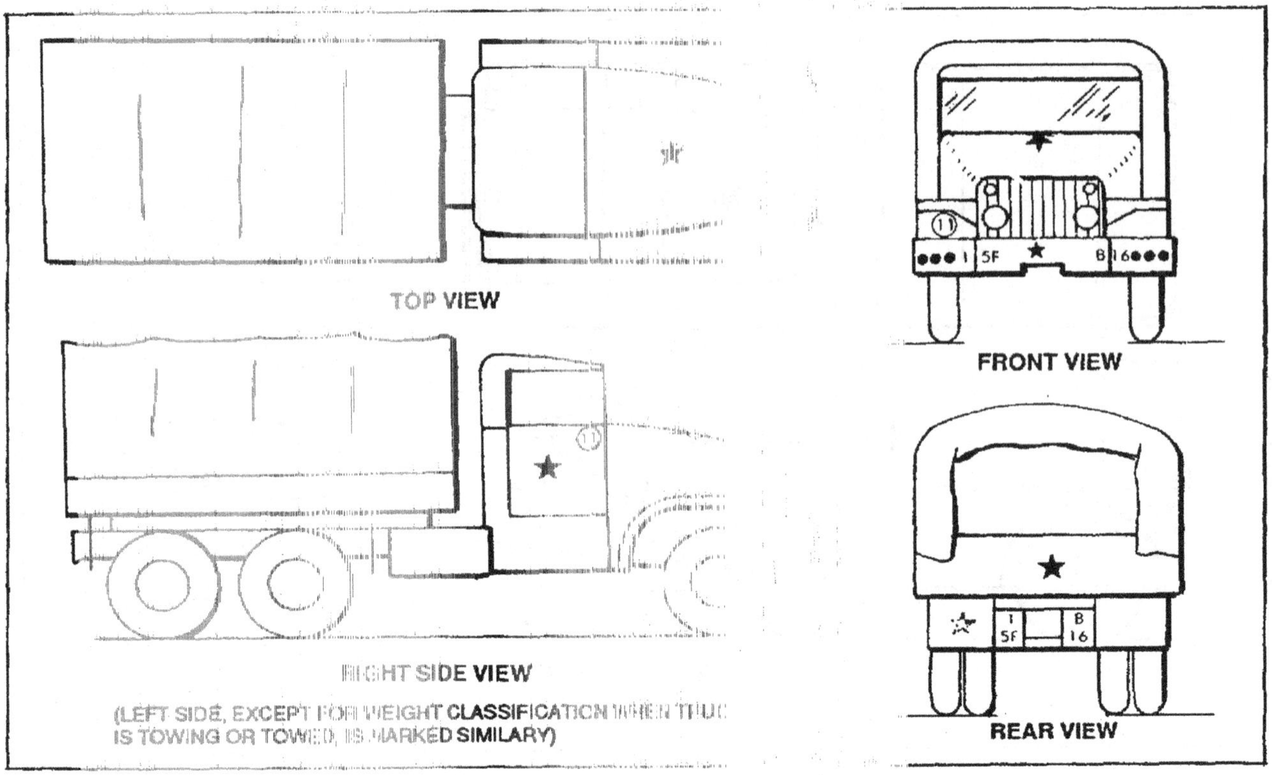

Figure 5. Marking locations—(typical) 2-1/ [illegible] rucks.

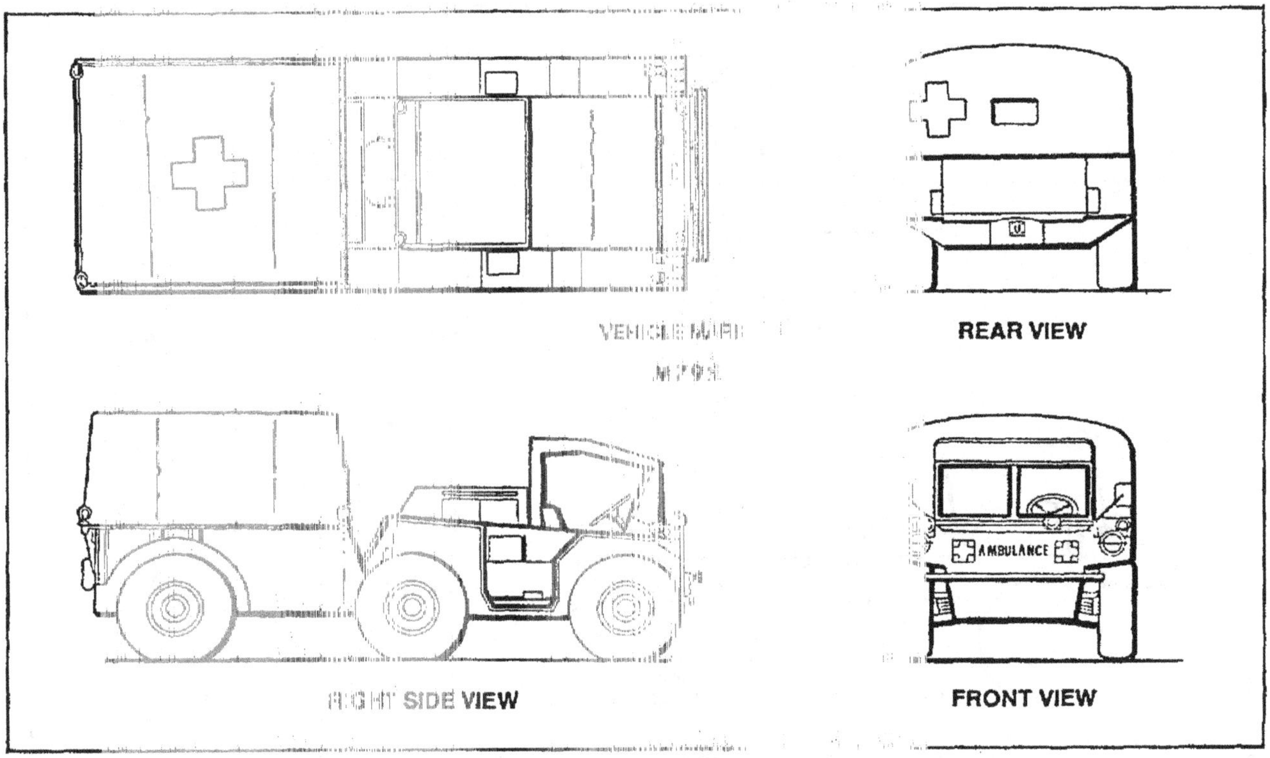

Figure 6. Marking locations – 1-1/4 [illegible]s.

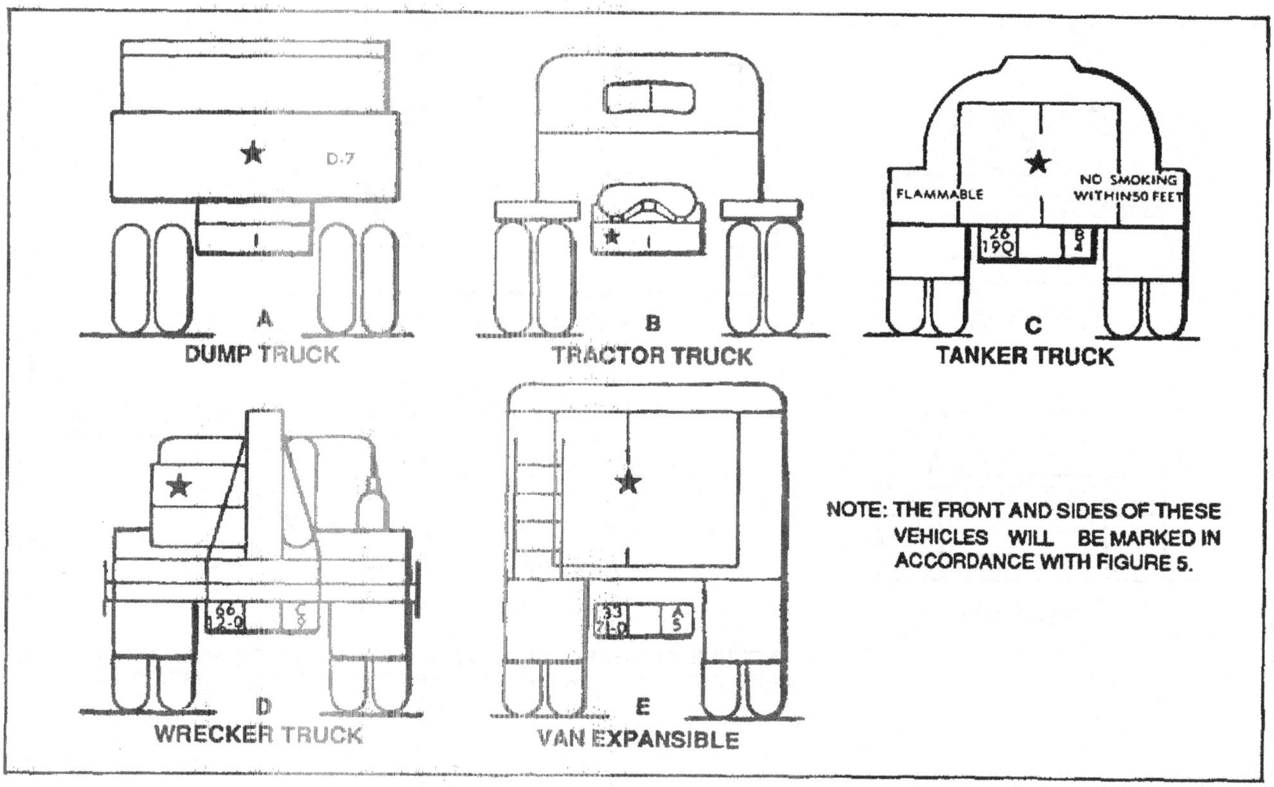

Figure 7. Marking locations -- (typical) 2-1/2-, 5- and 10-ton trucks.)

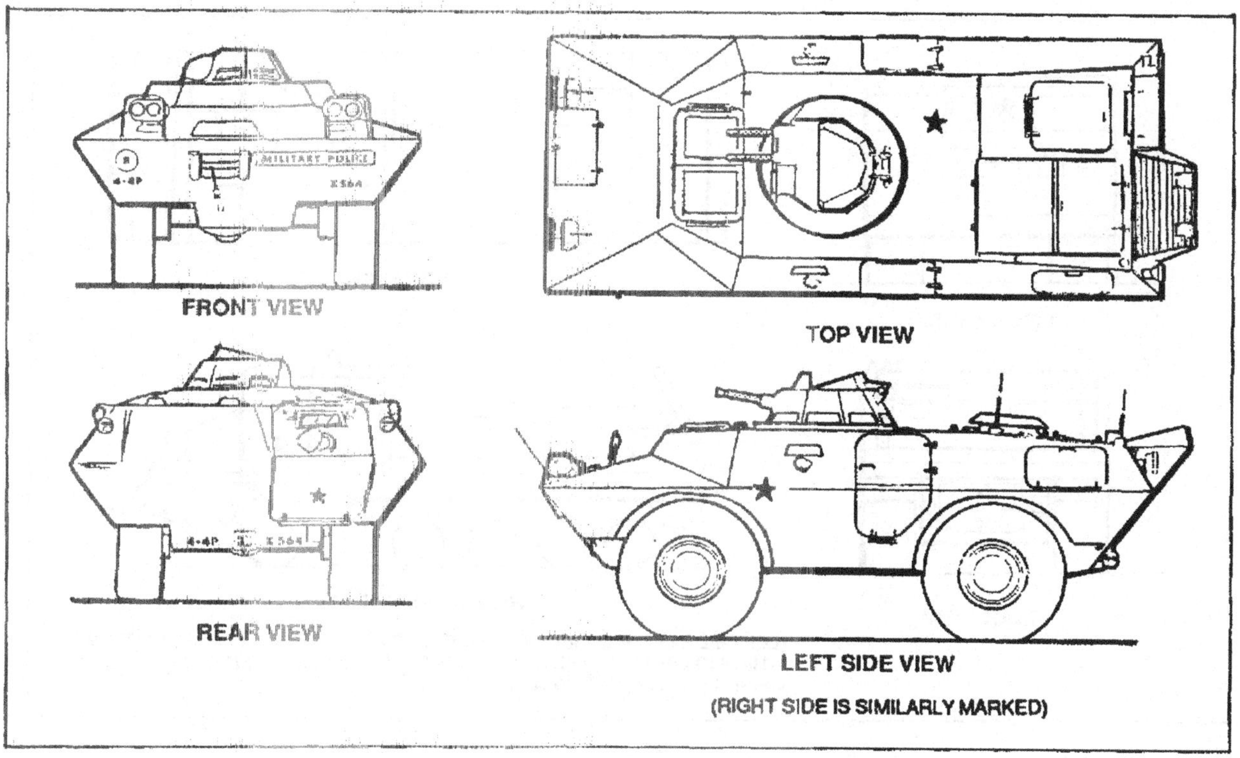

Figure 8. Marking locations -- combat vehicles. (1 of 5)

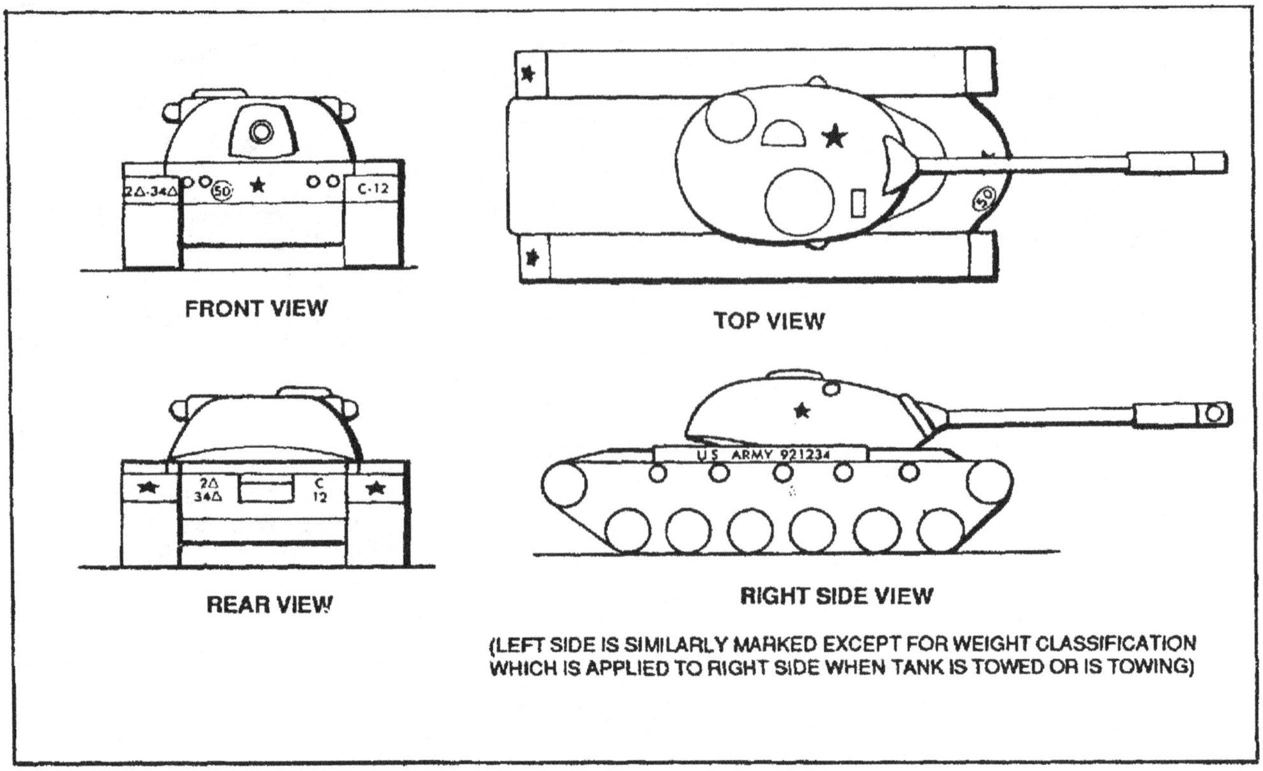

Figure 8. Marking locations -- combat vehicles. (2 of 5)

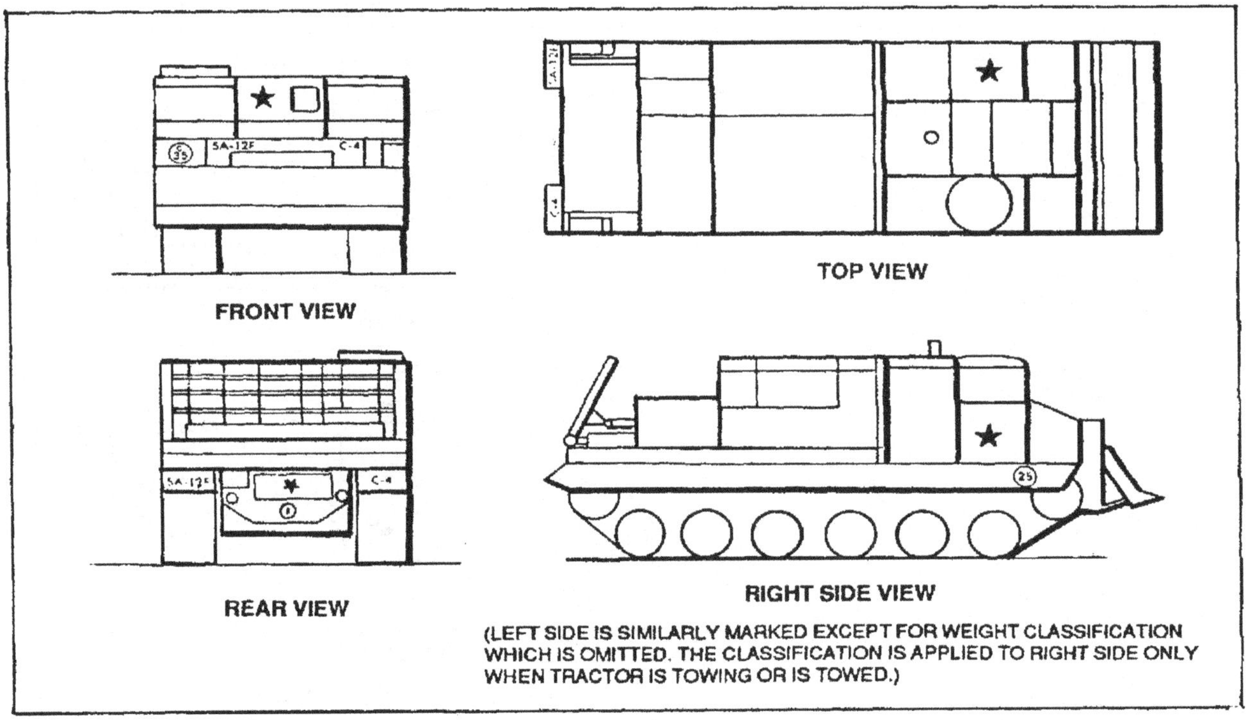

Figure 8. Marking locations--combat vehicles.(3 of 5)

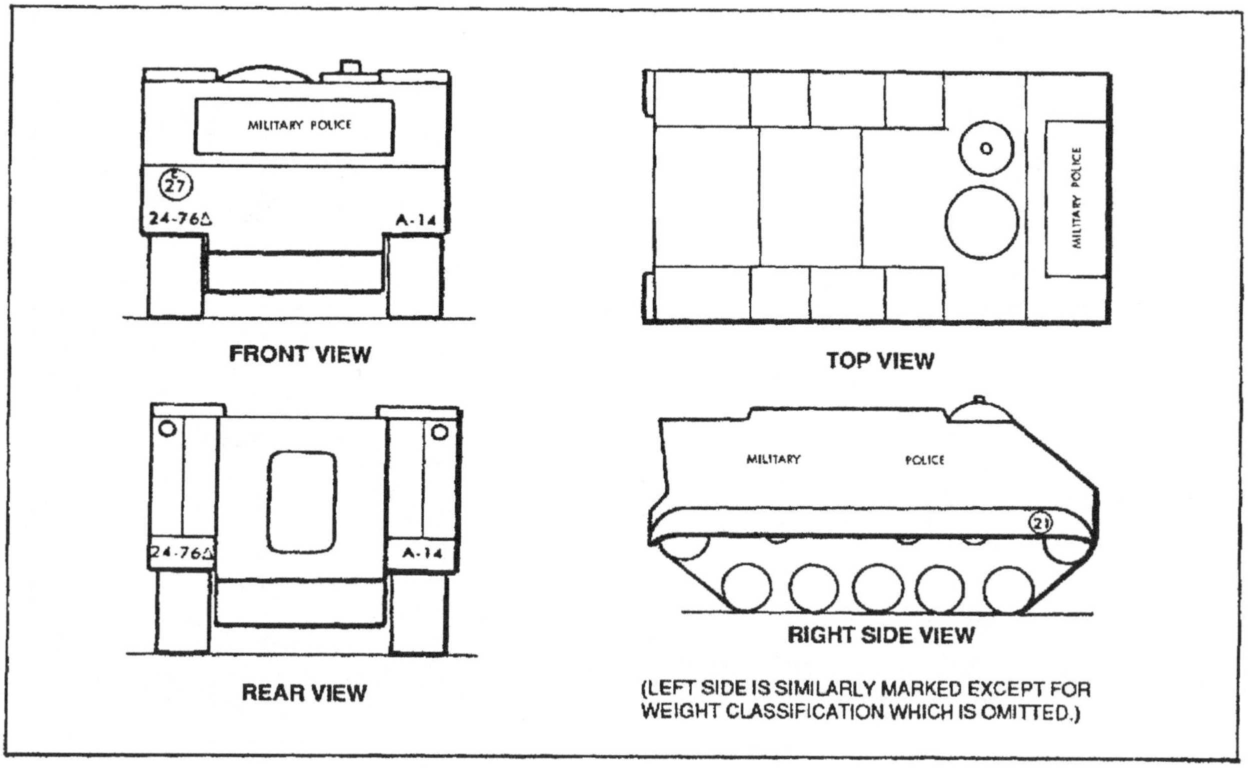

Figure 8. Marking locations -- combat vehicles. (4 of 5)

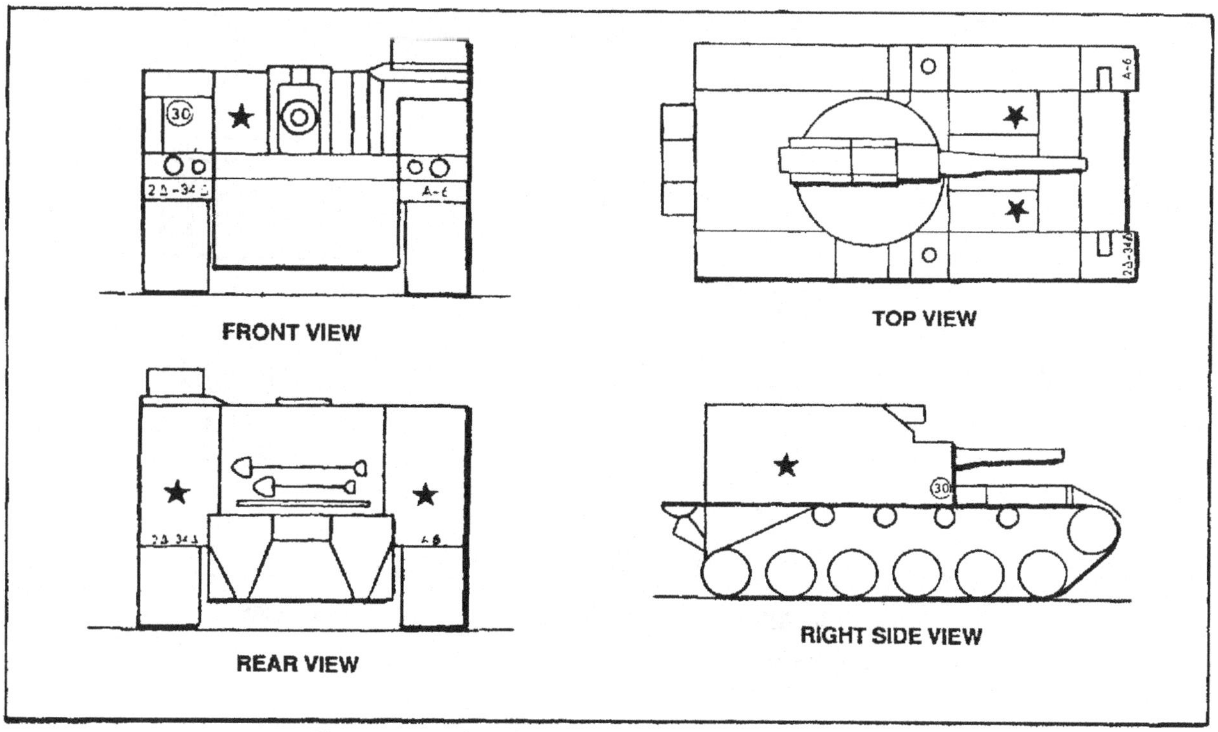

Figure 8. Marking locations -- combat vehicles. (5 of 5)

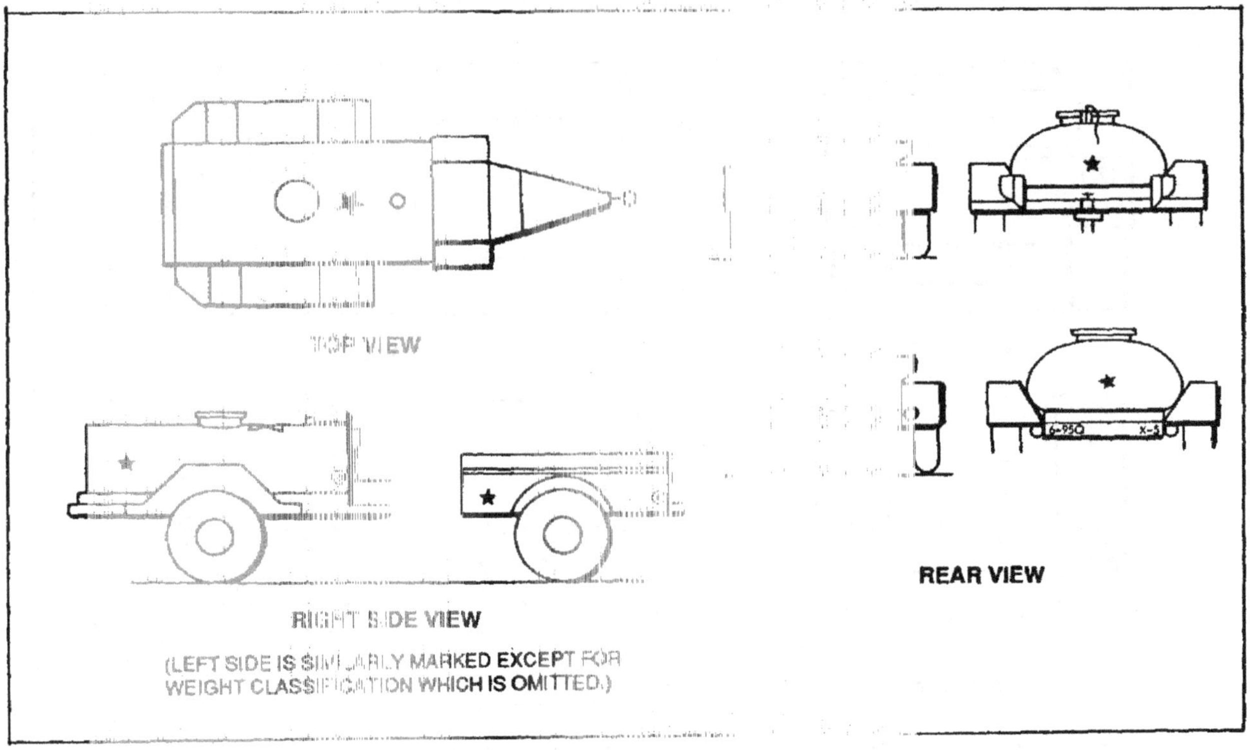

Figure 9. Marking locations – cargo ... ers.

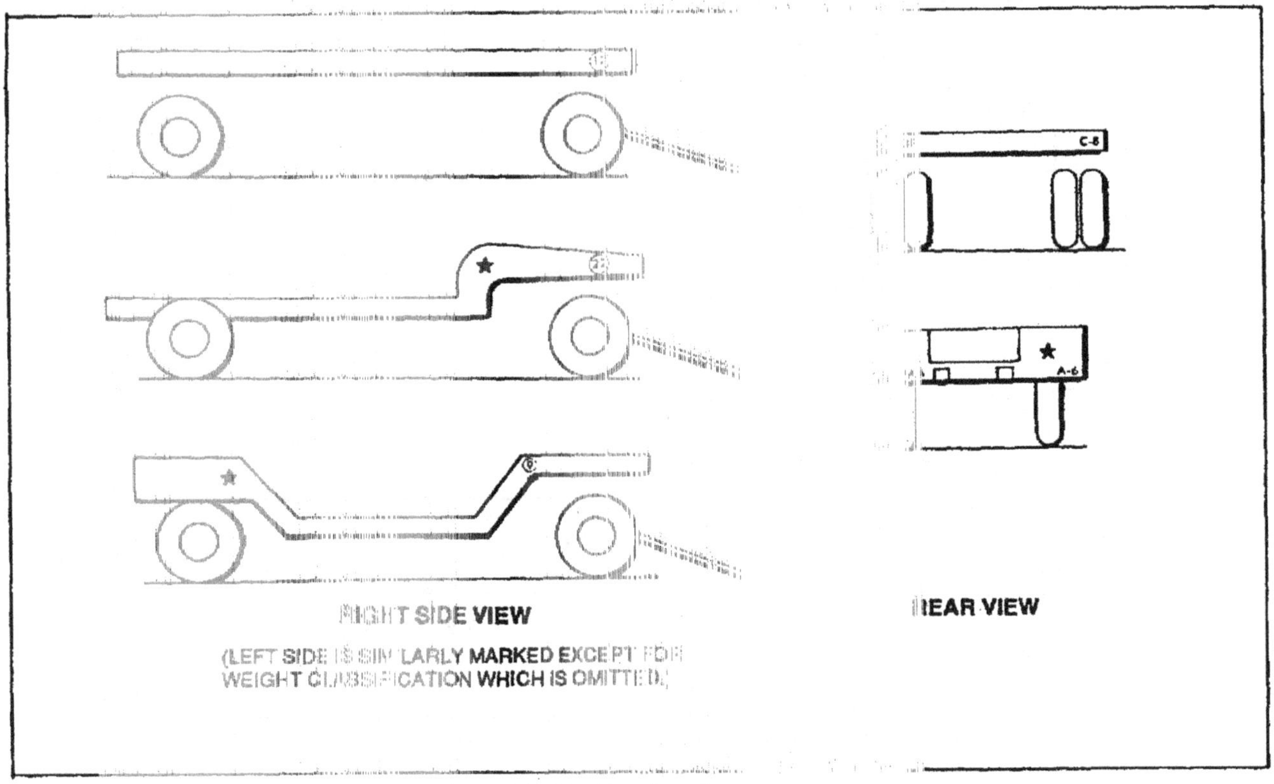

Figure 10. Marking locations ...

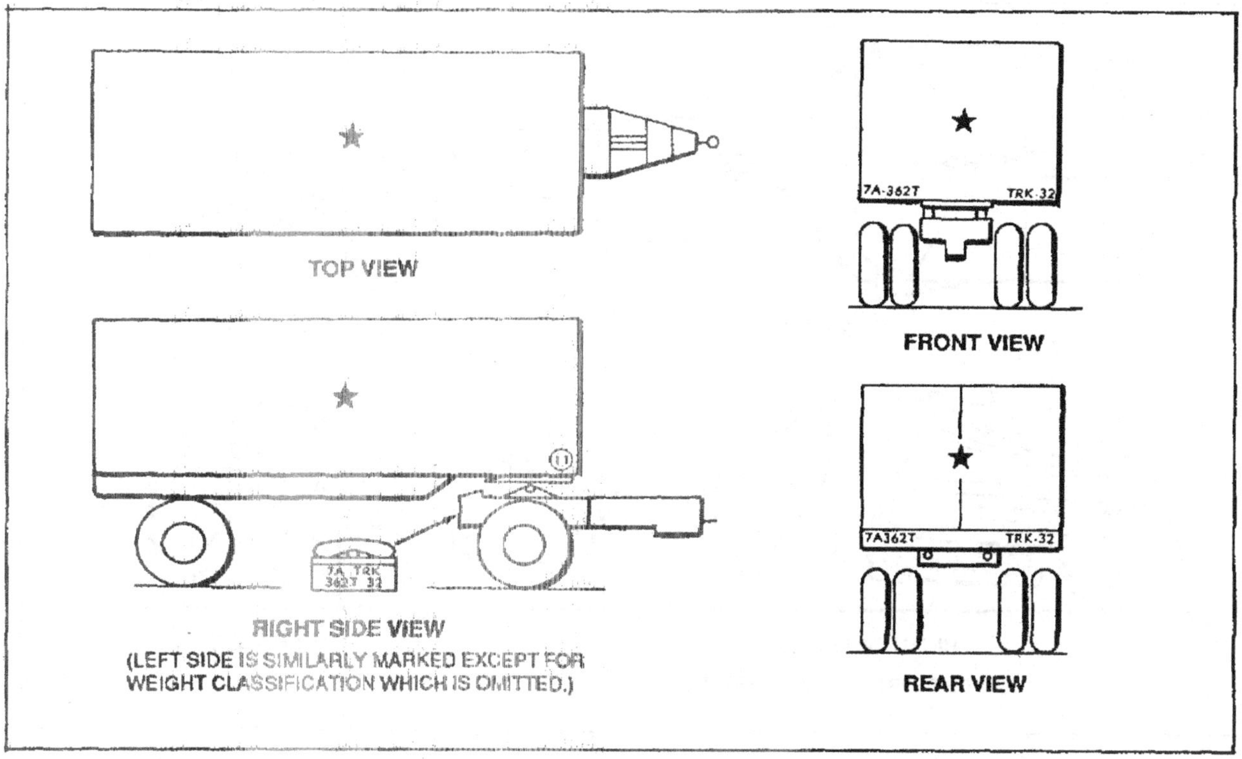

Figure 11. Marking locations -- trailer and Semitrailer van.

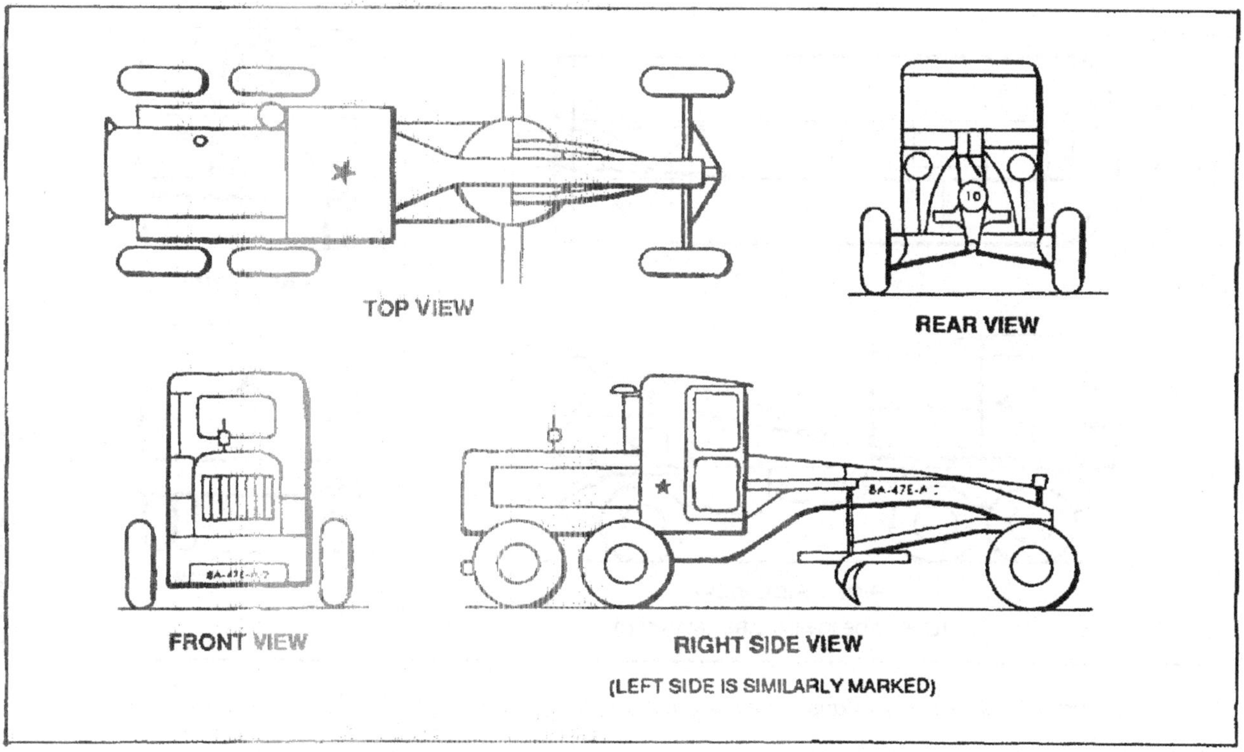

Figure 12. Marking locations -- construction and highway construction equipment. (1 of 6)

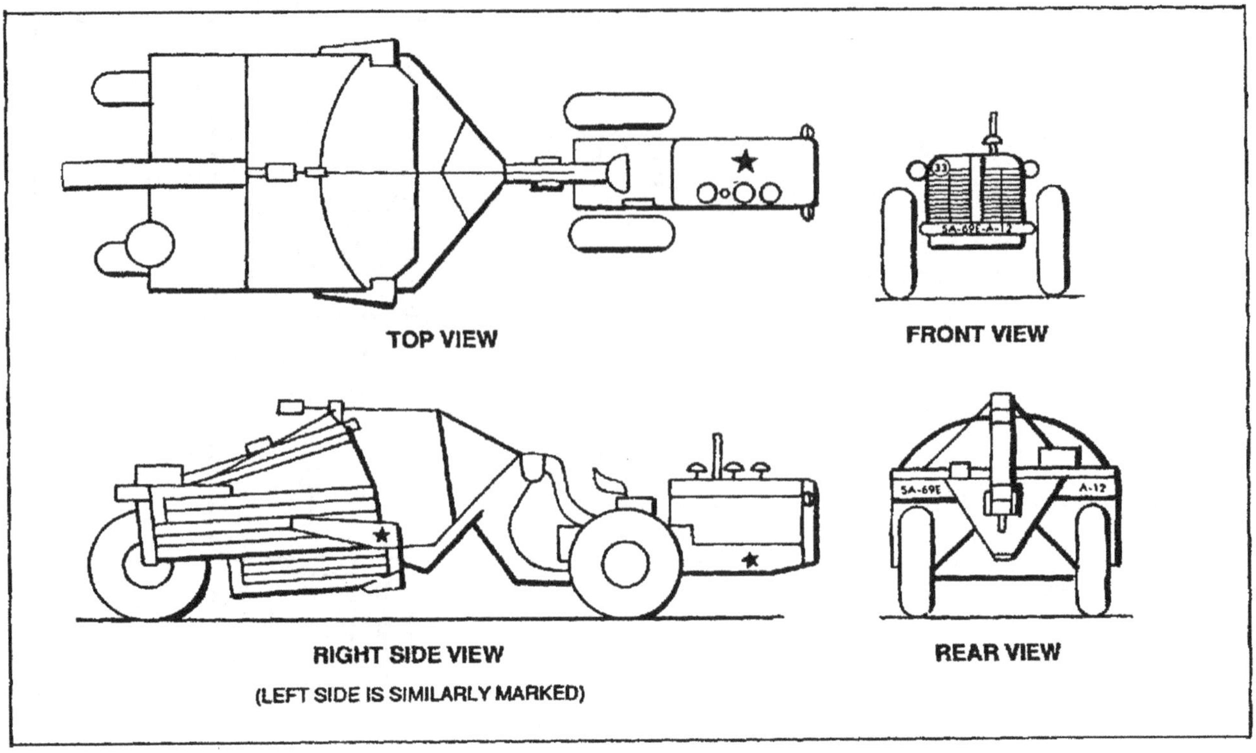

Figure 12. Marking locations -- construction and highway construction equipment. (2 of 6)

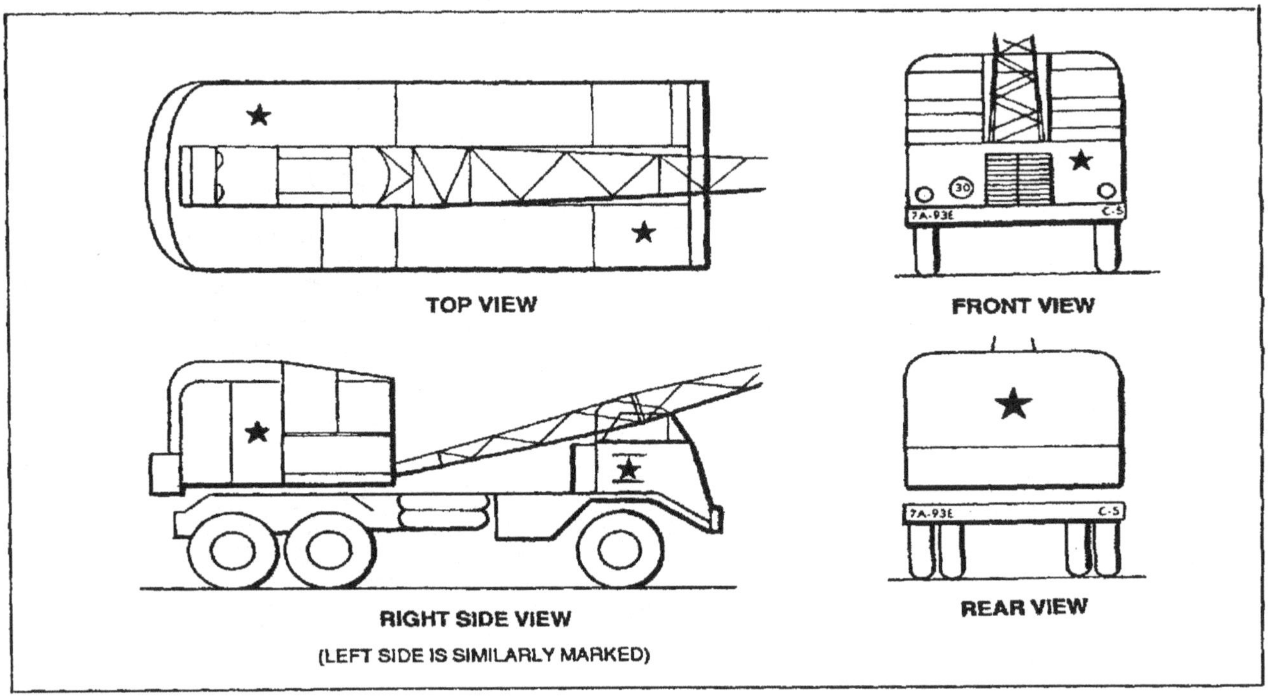

Figure 12. Marking locations -- construction and highway construction equipment. (3 of 6)

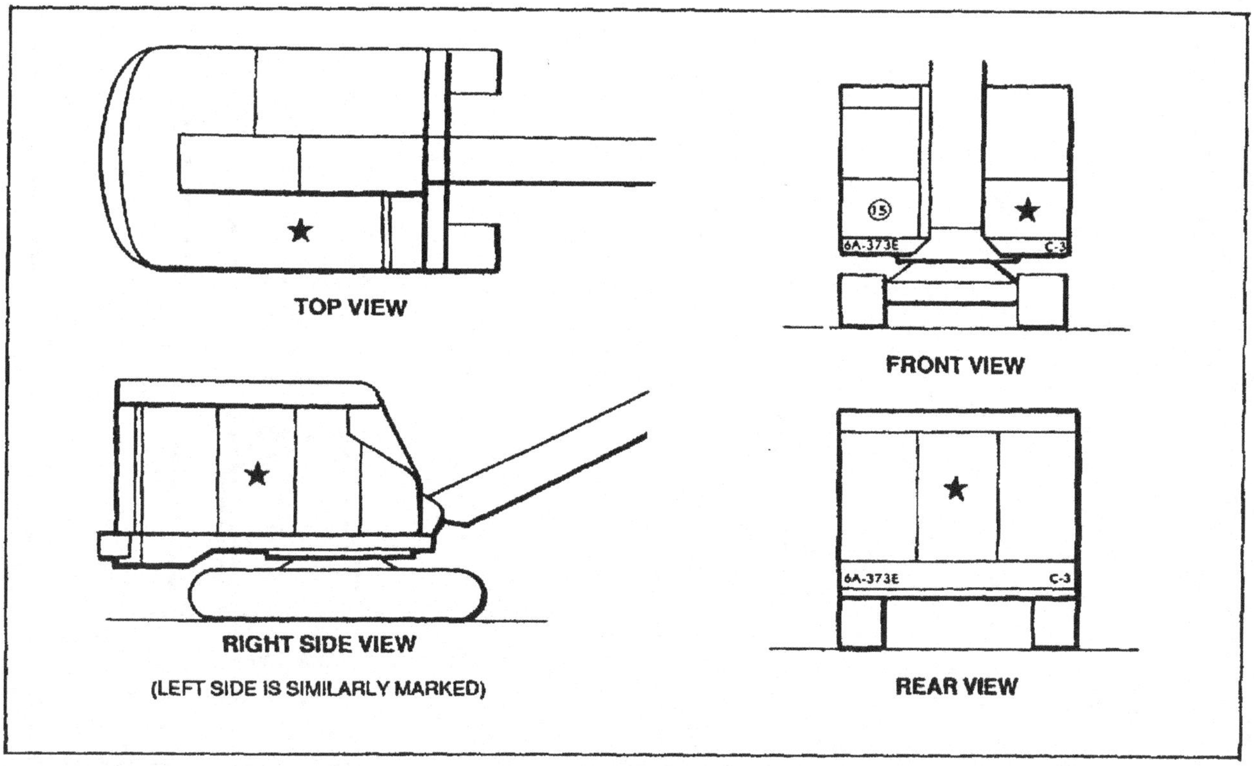

Figure 12. Marking locations -- construction and highway construction equipment. (4 of 6)

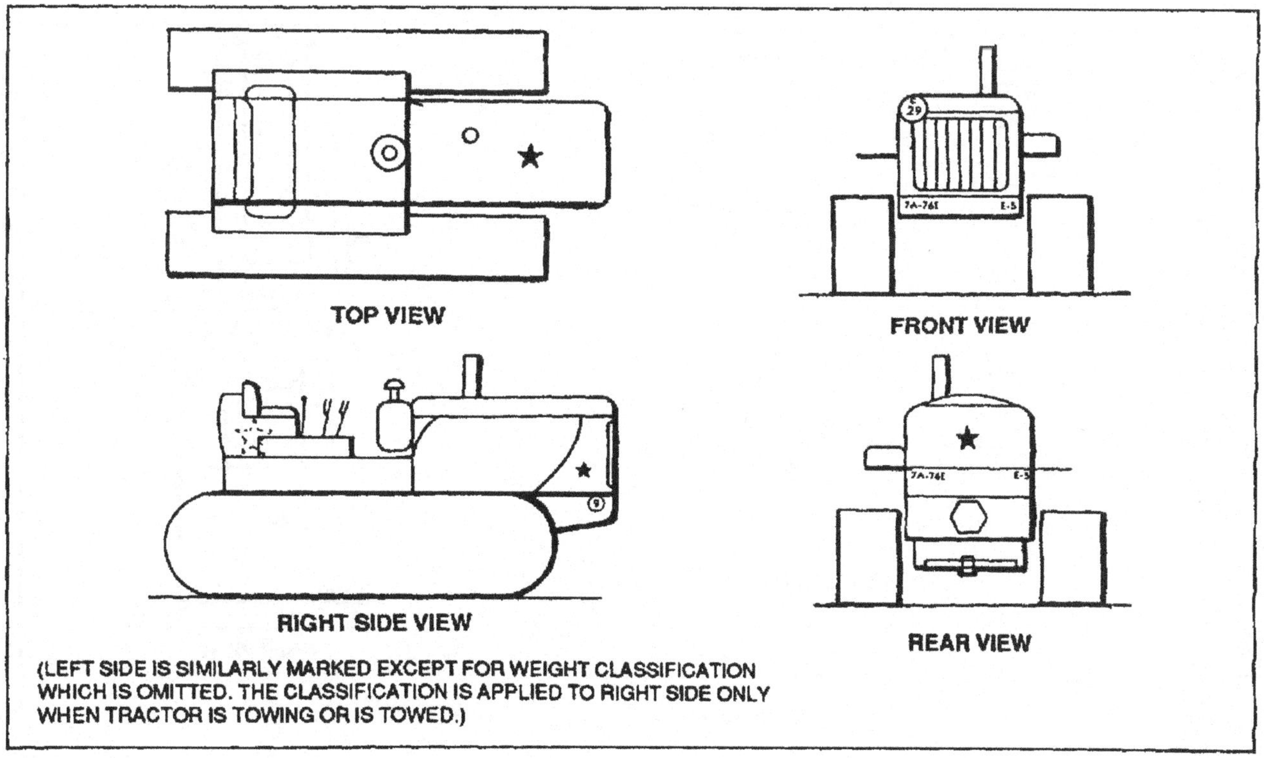

Figure 12. Marking locations -- construction and highway construction equipment. (5 of 6)

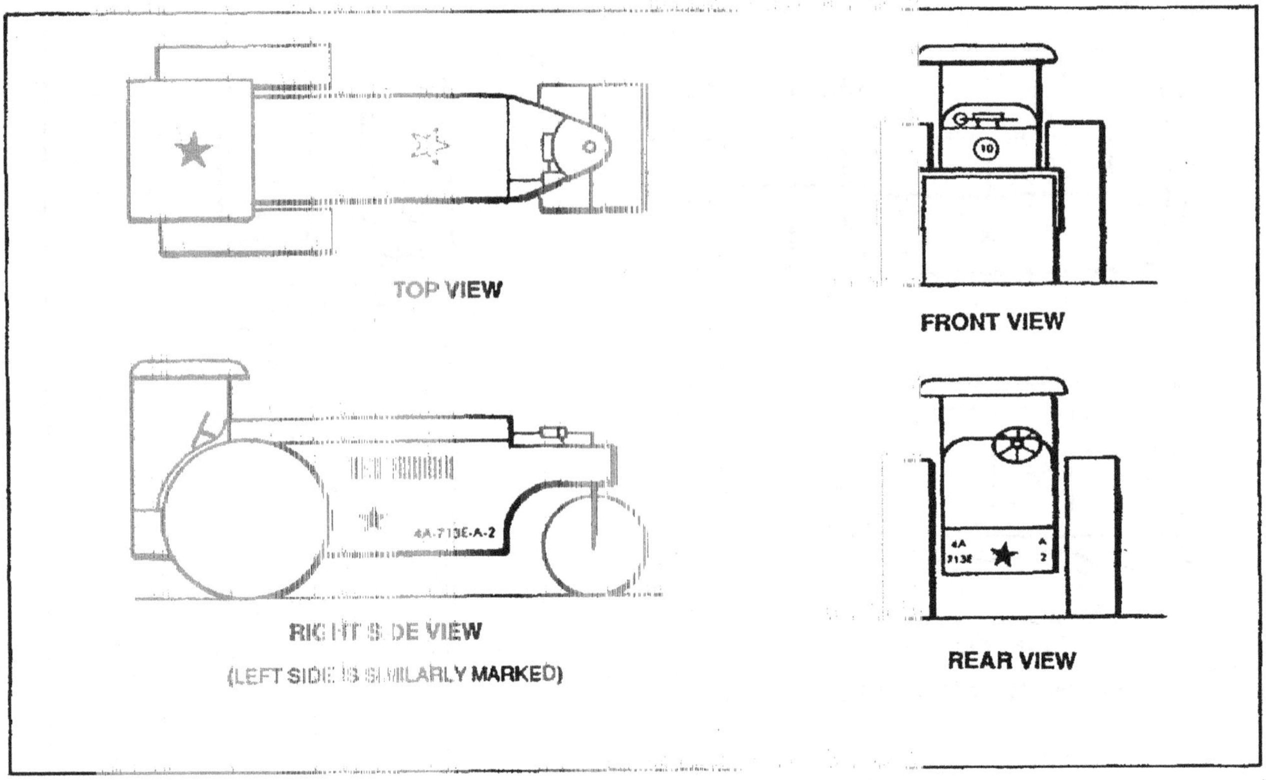

Figure 12. Marking locations -- construction and highway maintenance equipment. (6 of 6)

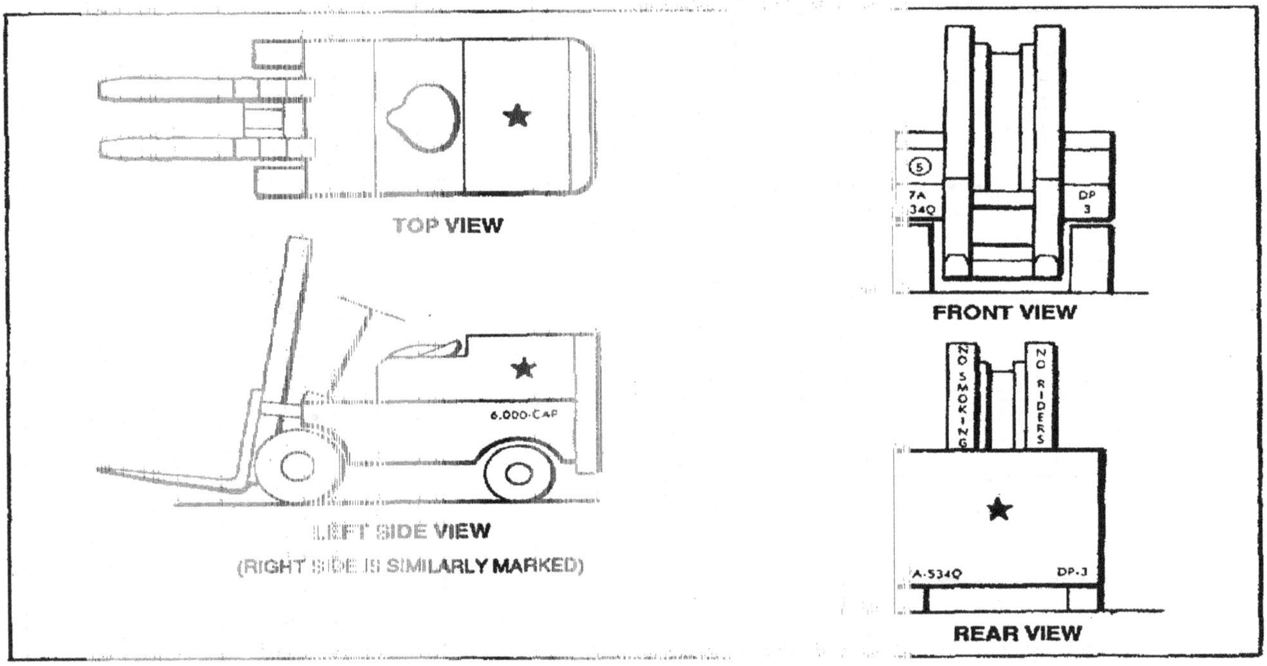

Figure 13. Marking locations -- materials handling equipment. (1 of 4)

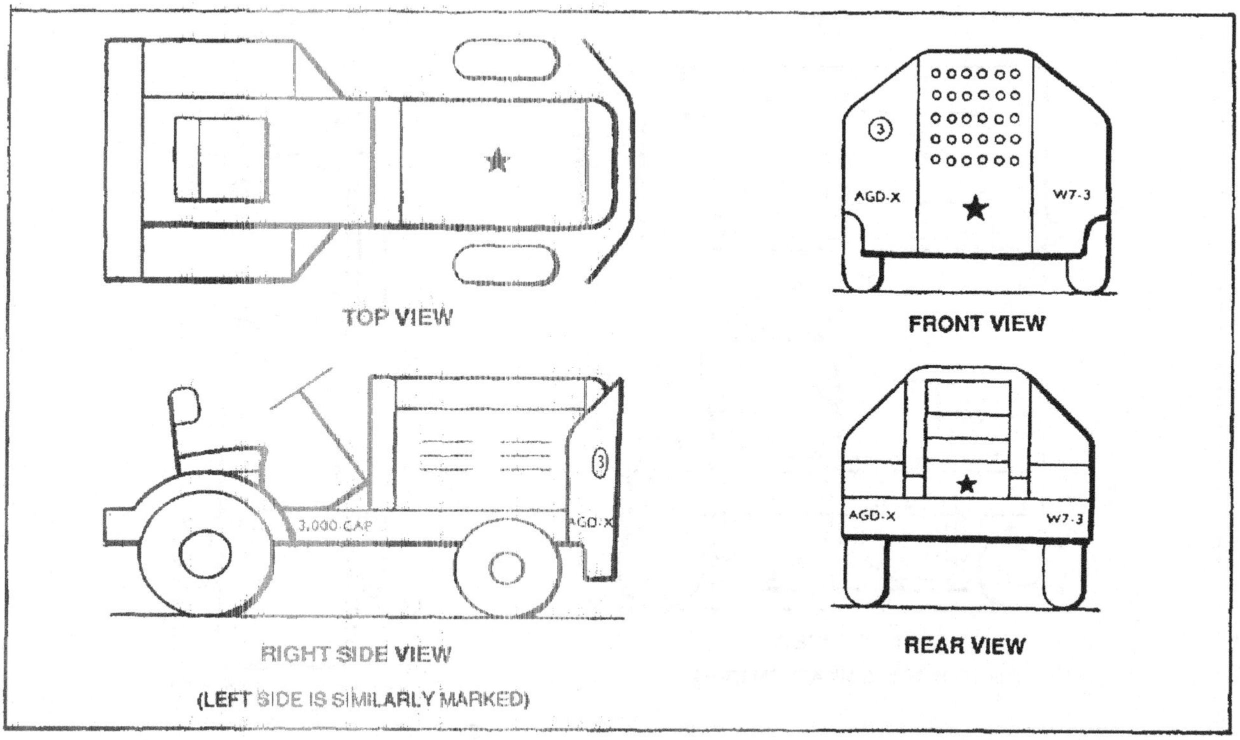

Figure 13. Marking locations -- materials handling equipment. (2 of 4)

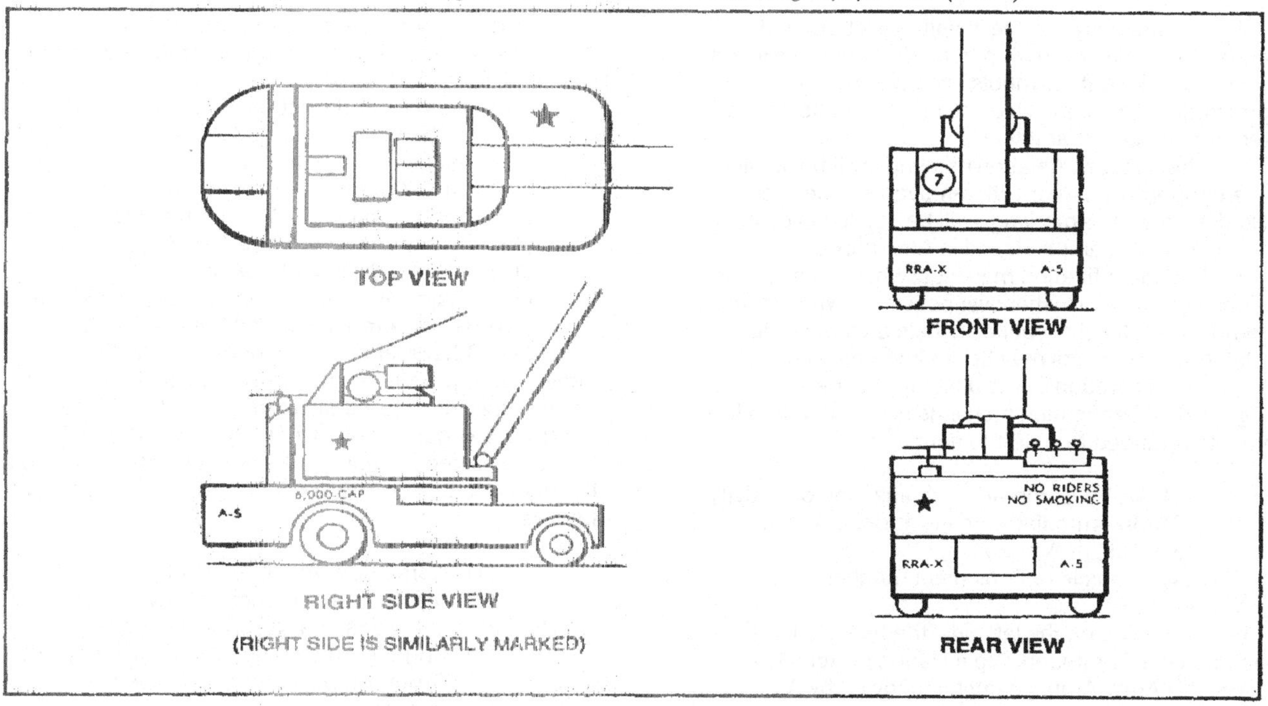

Figure 13. Marking locations -- materials handling equipment. (3 of 4)

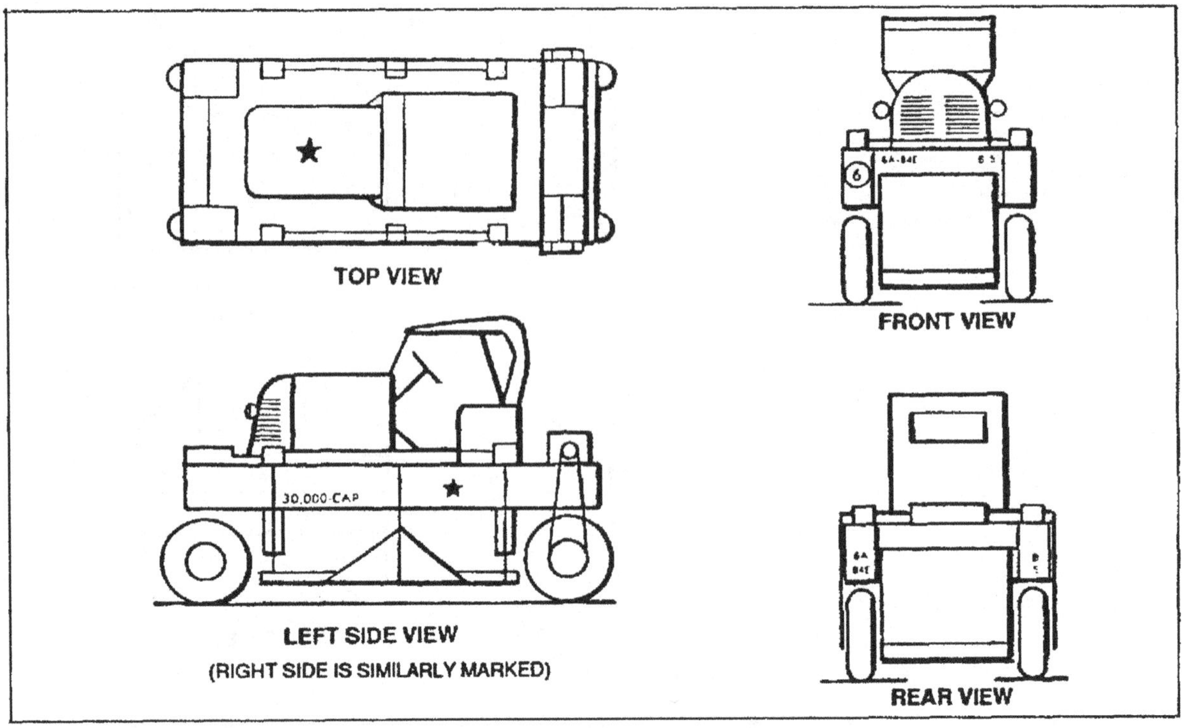

Figure 13. Marking locations -- materials handling equipment. (4 of 4)

d. The markings will be in uniform letters and numerals of the largest size practical, but not to exceed four inches in height. Symbols used will be approximately the same size and stroke widths as the letter and numeral characters.

e. The unit identification markings will be applied with a temporary paint or with removable adhesive-backed decals. The markings will be applied in colors as prescribed in paragraph 5e and in Appendix C.

f. Unit identification markings, when painted, will be applied by stencils whenever possible. When stencils or removable adhesive-backed decals are not readily available, the markings may be applied freehand with a brush at the discretion of the unit commander.

g. Unit identification markings are divided into four elements arranged from left to right:

(1) Major Command, Organization or Activity.
(2) Intermediate Organization or Activity.
(3) Unit or Activity
(4) Vehicle or Equipment number.

h. Beginning on the left side, the first marking denotes Major Command, Organization or Activity as follows: (1) Army-Arabic numeral followed by A.
(2) Corps-Roman numeral only
(3) Command
(4) Army Air Defense Command AADCOM
(5) Army Reserve Command-Arabic numeral as assigned, followed by C-R
(6) Training and Doctrine Command TRADOC
(7) Electronics Command ECOM
(8) Missile Command MICOM
(9) Tank-Automotive Command TACOM
(10) Test & Evaluation Command TECOM
(11) Armament Command AMMCOM
(12) Military District of Washington MDW
(13) Major Oversea Command. See AR 310-50.
(14) Group-Arabic numeral followed by GP
(15) Other Major Commands, Organizations, or Activities. Appropriate and non-conflicting letters and, where applicable, numerals to properly identify the Command, Organization or Activity. See AR 310-50.

i. The second unit marking on the left side denotes Intermediate Commands, Organizations or Activities as follows:

(1) Corps-Roman numeral only
(2)Division-Arabic numeral followed by appropriate service designation shown in I below.
(3) Brigade-Arabic numeral followed by BDE
(4) Group-Arabic numeral followed by GP
(5) Command

(6) Division Support Command DSC

(7) Army Posts, Camps and Stations. See AR 310-50.

(8) Army Depots and Arsenals. See AR 310-50.

(9) Reception Center RC

(10)Replacement Training Center RTC, preceded by appropriate service designation shown in 1 below.

(11) Battalion or Squadron organic to a larger force. Arabic numeral designating the battalion/squadron, followed by a slash (1), followed by the Arabic numeral designating the larger force, followed by the appropriate service designation shown in 1 below.

(12)Other Intermediate Command, Organization or Activity. Appropriate and nonconflicting letters, and where applicable, numerals to properly identify the Command, Organization or Activity. See AR 310-50.

j. The third markings, which will appear on the right side in the unit markings, denotes company, battery, or other small unit. Typical examples are shown below:

(1) Automatic Weapons.............................. AW
(2) Calibration... CAL
(3) Collection & Classification....................C&C
(4) Direct Support....................................... DS
(5) Explosive Ordnance Disposal..............EOD
(6) Floating Bridge....................................FLB
(7) General Supply...................................GSP
(8) Heavy Equip Maintenance.................. HEM
(9) Inventory Control Co. (center)...............ICC
(10) Light Truck......................................LTRK
(11) Missile...MSL
(12) Medium Boat......................................MB
(13) Panel Bridge..PB
(14) Petroleum... PET
(15) Reconnaissance ....R
(16) Repair Parts Supply..........................RPS
(17) Service...SV
(18) Support... SPT
(19) Supply & Service...............................S&S
(20) Transportation Motor Pool ....TMP
(21) Tire Repair...TR
(22) Other Units--Appropriate and non-conflicting letters, normally not to exceed three, which will easily identify the unit. AR 310-50 should be used as a guide when choosing a unit designation.

k. The last marking on the right denotes the vehicle number in the unit. In tactical type units, vehicles will be numbered in the sequence of their normal order of march. Vehicles in administrative use assigned to transportation motor pools or depot operations may be marked in any appropriate sequence. Trailers will be marked with the same number as the vehicle which normally tows them, preceded by the letter T.

l. Branch of Service. The following symbols will be used to denote the branch of service:

(1) Airborne...ABN
(2) Airmobile...AM
(3) Amphibious ....APH
(4) Antiaircraft...AA
(5) Armored .... Equilateral triangle with ....base down
(6) Cavalry..CAV
(7) Air Cavalry...ACAV
(8) Armored Cavalry........... Equilateral triangle with ....base down, followed by CAV
(9) Chemical..C
(10) Civil Affairs...CA
(11) Engineer...E
(12) Field Artillery..F
(13) Infantry............ I, preceded by a dash 1 in long
(14) Mechanized...MCH
(15) Medical...M
(16) Military Intelligence..MI
(17) Military Police...P
(18) Ordnance ....0, preceded by a dash 1 in. long
(19) Quartermaster..Q
(20) Signal...S
(21) Transportation..T

m. A list of Approved Abbreviations is found in AR 310-50.

n. Typical examples of the above markings are as follows:

(1) 1A-X-HQ-4: First United States Army, Hq & Hq Co, 4th vehicle.

(2) VII-M-X-9: VII Corps, Med. Det, 9th vehicle.

(3) 25-1/27-C-3: 25th Inf. Div, 1st Bn, 27th Inf., Co C, 3d vehicle.

(4) 4-2R8-A-6: 4th Div, 2d Recon Sq of 8th Cav, Trp A, 6th vehicle.

(5) 3-1/32-B-7: 3d Armd Div, 1st Tk Bn of 32d Armor, Co B, 7th vehicle.

(6) 10-2F7-C-2: 10th Inf. Div, 2d Bn of 7th FA, Btry C, 2d vehicle.

(7) 82ABN-820F-D-5: 82d Abn Div, 820th FA, Btry D, 5th vehicle.

(8) 2-1/50-A-8: 2d Armd Div, 1st AR Bn of 50th Inf., Co A, 8th vehicle 2-BDE.

(9) 2BDE-53APH-A-4: 2d Engr Amph Spt Brig, 53-2d Amph Spt Regt, Trans Boat Co A, 4th vehicle.

(10) 4GP-S-HQ-2: 4th Sig Gp, Hq & Hq Det, 2d vehicle.

(11) 1A-962-O-AM-5: First United States Army, 962 Ord Ammo Co, 5th vehicle

(12) 4A-483Q-REF-1: Fourth United States Army, 483d Q Refrigerating Co, 1st vehicle.

(13) 3A-1M-SUR-10: Third United States Army, lst Surg Hosp, 10th vehicle.

(14) 5A-531T-TRK-7: Fifth United States Army, 531st Trans Trk Co, 7th vehicle.

(15) 1AADC-52ADA-3HQ-2: 1st Rgn Army Air Def Comd, 52d Air Def Arty Rgt, Hc Btry of 3d Msl Bn, 2d vehicle.

(16) TC-FE-TMP-12: Trans Corps, Ft. Eustis, Trans Mtr Pool, 12th vehicle.

(17) 6A-FO-TMP-9: Sixth United States Army, Ft Ord, Trans Mtr Pool, 9th vehicle.

o. The unit identification markings will be applied on one line whenever possible, and will consist of each of the four basic categories described above, in the same order. Where space is limited, requiring the use of two lines, the Major Command marking will appear directly above the Intermediate Activity, and the Unit Identification will appear directly above the Vehicle Number. The first two categories will appear to the left of the last two categories. The following arrangements of markings shown in g above are permitted.

(1) (1)--(2)--(3)--(4)

(2) (1)--(3)
(2)--(4)

p. When directed on authority from the responsible commander for security purposes, the first and second categories listed in this paragraph will be completely removed from all vehicles and equipment being shipped with units outside the continental United States.

10. OPERATIONAL MARKINGS

a. Tactical markings are used to provide a more visible means for identifying vehicles of tactical units by the personnel of those units during tactical operations. Commanders of major organizations are responsible for determination of the type of tactical markings to be used by all units within their organizations and the location of the markings on the vehicles. Changes may be made only with the approval of the major commanders.

(1) These tactical markings may consist of stripes, geometrical figures, or other simple designs, including name, for individual vehicles.

...a kings will be of suitable size for ic... he ground at reasonable distances, and ... in any appropriate color using te... e movable adhesive-backed decal m...

... a markings should be designed to b... gible to enemy observation as pos... gs will not, in any way, represent or sug... al designation or distinctive insignia o... en record will be made or published des... n of markings used.

... is a sign to mark those vehicles whi... priority of movement over all other vehi... the responsible area commander. The... be authorized for any significant and ... and are valid only within the area und... of the responsible commander. They ... ed when no longer required for the pri... ey may not be displayed when the vehi... on a priority mission.

... ority sign will be made of sheet me... f an equilateral triangle with one poin... ce will be 12 inches long; the cor... ded to a 1/2inch radius, Figure 14.

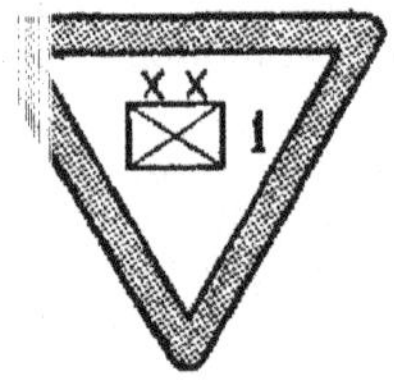

...ure 14. *Priority sign.*

... ckground color will be lusterless whit... nch border will be lusterless red 311... symbol of the organization, whose com... zed the priority, will be centered in the ... f the symbol will be lusterless red 311...

... bols used will conform to FM 21-30.

SECTION IV. SAFETY MARKINGS

11. SAFETY MARKINGS.

a. Vehicles used for the bulk transportation of gasoline, fuel oil or other flammable liquids will be marked on both sides and on the rear of the body with the word "FLAMMABLE" in 6-inch block letters, and directly below where space permits, or on the same line with the words "NO SMOKING WITHIN 50 FEET" in 3-inch block letters and numerals, green 383, lusterless black 37030. When the vehicle is commonly used on public highways in CONUS, Title 49, Sec 177.823, Code of Federal Regulations, requires 6-inch block letters in gloss red 11105 or 11136 for word "FLAMMABLE' and 3-inch block letters, "NO SMOKING WITHIN 50 FEET" also in gloss red 11105 or 11136. Gloss white 17875 will be used as a background extending one inch all around the lettering. If available, reflective markings should be used. Markings may be permanently installed by painting or by use of adhesive-backed decals. Vehicles used infrequently on public highways must employ removable or reversible signs which are securely fastened while in use. These safety markings will be applied upon receipt by the using service.

b. Vehicles carrying explosives or other dangerous material will be marked as follows:

(1) When used for the transportation of such dangerous material over public highways, vehicles will be marked or equipped with appropriate placards in accordance with AR 55-355.

(2) When used for the transportation of such dangerousmaterial to training areas, vehicles will be marked or equipped with placards on both sides and ends. The word "Explosives" will be marked in white letters not less than 3 inches high on a red background in accordance with AR 385-63. Reflectorized placard should be used. These are available as follows:

ADHESIVE-BACKED

"Explosives A"-NSN 7540-01-054-5090
"Explosives B"-NSN 7540-01-054-7237
"Explosives C"-NSN 7540-01-054-7238

PLACARD

"Explosives A"-NSN 7540-01-028-5124
"Explosives B"-NSN 7540-01-029-5669

c. When authorized by the responsible commander, stripes of a contrasting color may be applied to certain vehicles, Figure 15, and equipment used in a nontactical area, that because of size, construction or function, presents a possible hazard. These stripes are applied to permit vehicles to be more readily seen by drivers of other vehicles. They are normally applied to the rear of such vehicles; they may, however, be also applied to the front or certain side surfaces if required.

(1) The stripes will be applied across the face of the bumper, or across the width of the vehicle on a flat surface above the bumper in a band not to exceed two feet in height. Other appropriate surfaces may be utilized when necessary.

(2) Width of the stripes will be 4 inches with a 4 inch space between stripes. They will be inclined at a 45' angle in opposite directions from the center line of the vehicle to form an inverted V pattern.

(3) Reflective red and white adhesive-backed stripes conforming to Federal Specification LS-300 will be utilized.

(4) Gloss black 17038 will be used for stripes on vehicles painted yellow or other light color.

d. All motorized materials handling equipment will be marked with the legends "No Smoking" and "No Riders."

(1) These legends will be placed on a surface facing the operator.

(2) The markings will be applied in 1-inch letters in lusterless white 37875 for vehicles in tactical use, and in gloss red 11105 or 11130 for nontactical vehicles.

(3) The markings may be applied horizontally to read left to right or vertically to read top to bottom.

e. Fuel and oil dispensing vehicles and equipment used for servicing aircraft will be marked with the grade of fuel or oil on each side of the tank. Marking will be lusterless white 37875 in the following sizes:

(1) Semitrailers.................................6 inches high
(2) Trucks...4 inches high
(3) 2-Wheel trailers ....3 inches high

NOTE

For hard top vehicles, Figure 16, with Explosive Ordnance Disposal marking (EOD) roof must be painted in lusterless white so as to reduce the internal temperature. This will also reduce bleed off of the liquid nitrogen.

f. Examples of such markings follow:
(1) Gasoline (number) grade
(2) Oil (number) grade
(3) Water alcohol
(4) Explosive Ordnance Disposal (EOD)

Where available and certified, the capacity of tank compartments in gallons will be marked below and to the right of the cupola covers: "( ) Gals Cap" in 2-inch letters and numerals.

12. TECHNICAL MARKINGS.

a. The prescribed vehicle tire pressure will be marked on the dash of all motor vehicles and pneumatic-tired equipment. This marking will also be applied to the fenders above each wheel or, if fenders are not used, to the body or frame immediately above each wheel. On towed vehicles or equipment, this marking will be applied to the body or frame immediately above each wheel. On tandem axle vehicles, the marking will be centered between the two wheel sets.

(1) The markings will consist of the numerals representing the proper pressures, preceded by the letters TP.

(2) Letters and numerals will be 1 inch high and in the same color as the unit markings.

b. When maximum speed is specified by a manufacturer, or in an applicable technical manual, and when appropriate operation plates are not available, the maximum speed of the vehicle in its highest range and gear ratio will be marked on the dash.

(1) The marking will consist of the numerals representing maximum speed, preceded by the abbreviation "Max Speed."

(2) Letters and numerals will be 1 inch high and in the same color as the unit markings.

c. All vehicles and equipment processed with special arctic lubricants, and/or other materials for operation in cold climates, will be marked with a temporary marking as follows:

(1) The words "Arctic Lubricated" will be marked in the following order of preference:
(a) Adjacent to the name or identification plate.
(b) On other conspicuous location in order that the markings may be readily seen prior to operation or use. Specific locations may be prescribed in applicable technical publications for the vehicle or equipment.

(2) Letters of the marking will be 1 inch high and in the same color as the unit markings.

Figure 15. Safety marking calibration liftgate.

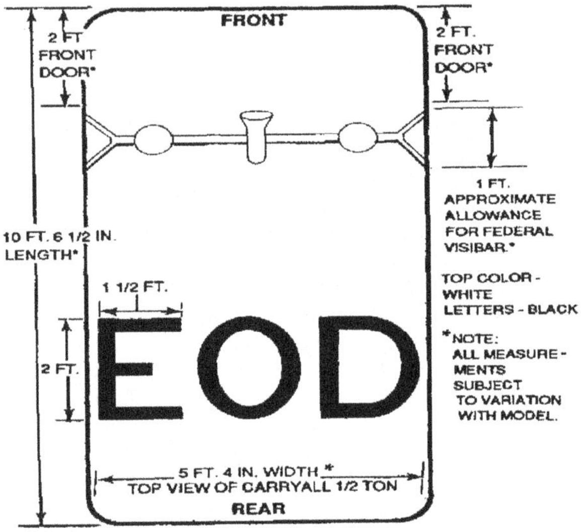

Figure 16. Hardtop vehicle with explosive ordnance disposal marking.

(3) The marking will be kept in a legible condition until the vehicle or equipment is deprocessed, and will then be thoroughly removed.

d. All tactical use wheeled vehicles which are equipped with an engine hood safety hook will be marked with the words "Warning: Secure Hood in Raised Position with Safety Hook Before Servicing Engine."

(1) The marking will be applied at the rear area on the underside of the hood panel.

(2) Letters of the marking will be 1 inch high and white.

e. Boom extension markings are required to assure operation within boom radii limits specified on the crane safe load data plates.

(1) Boom markings will be applied in a permanent method of marking. Color will be the same as the unit markings.

(2) Procedure for truck wreckers M60 and M108.

(a) Retract boom and retain in horizontal position.

(b) Measure, scribe, and mark a 3/4-inch vertical stripe along the outer edge of the shipper, extending the stripe onto the boom, Figure 17.

(c) To the right of and parallel with the stripe, mark "8 Ft" in 2-inch letters.

(d) Mark a horizontal arrow on the shipper pointing to the stripe.

(e) Extend the boom to its maximum length.

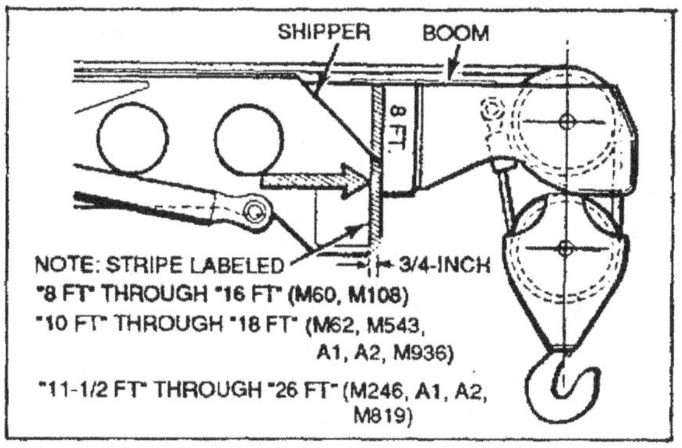

Figure 17. Boom markings (boom retracted).

marked, and labeled "12 Ft." Each succeeding stripe will be marked consecutively through "26 Ft."

f. This paragraph applies to tactical transport vehicles, and is intended to reduce the possibility of fuel spillage due to fuel tank overfilling and subsequent fire hazard, regardless of fuel type.

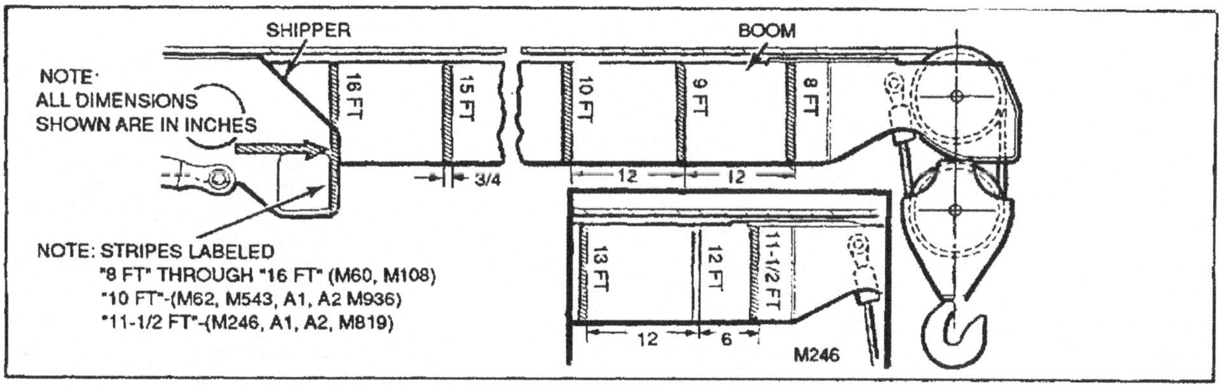

Figure 18. Boom markings (boom extended).

(f) Measure from the "8 Ft" stripe. Scribe and mark 3/4inch vertical stripes at intervals of 12 inches along the boom (Figure 18).

(g) Mark the second stripe "9 Ft," and each succeeding stripe consecutively through "16 Ft," in each case to the right of and parallel to each stripe.

(3) The procedure for Truck wreckers M62, M936, and M543 is the same, except that the stripe will be marked "10 Ft" and each succeeding stripe will be marked consecutively "11 Ft" through "18 Ft." (4) The procedure for Truck wrecker M246, Al, A2, and M819 is the same, except that the first stripe will be marked "11 1/2 Ft." The second stripe is to be measured, scribed,

(1) If the fuel tank filler cap is a relief valve type, but does not bear a warning decal reading "Pressurized: Open Slowly," the warning note "Pressure: Open Slowly" will be marked in 1-inch lusterless black letters on the fuel tank near the filler pipe.

(2) On fuel tanks, where a mark would be readily visible, measure 2 inches down from the top of the tank and mark a 1/4 x 8-inch line, with the following note in 1-inch lusterless black letters: "Caution: Do Not Fill Above Line to Allow For Expansion," Figure 20.

(3) On vehicles where a mark on the fuel tank would not be readily visible, mark the following note in 1-inch lusterless black letters as near as possible to the filler pipes: "Caution: Do not Overfill--Allow for Expansion."

Marking Instructions for M984E1 C

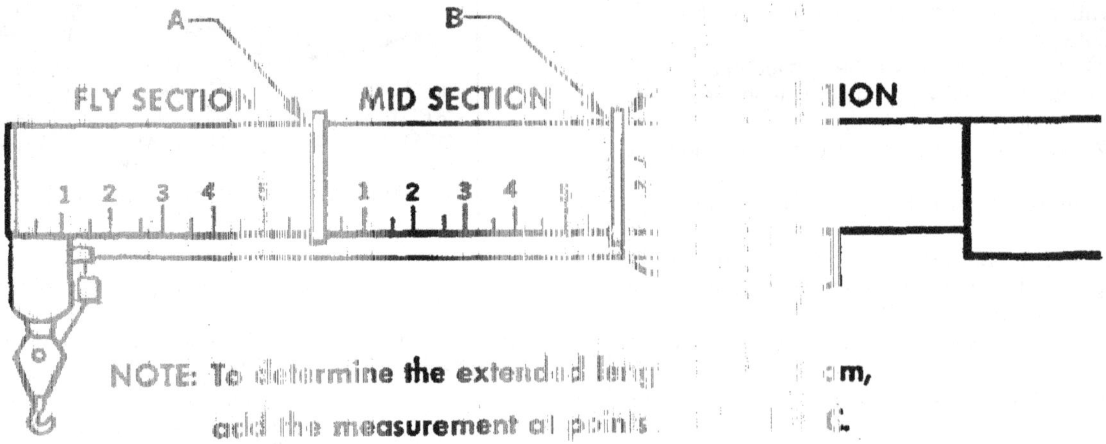

NOTE: To determine the extended leng… …m, add the measurement at points … C.

FORMULA: A+B+C= EXTENDED L…

NOTE: The mid section and fly section are … …ne.

1. Place the crane in operation condition (TM 9-2320-279-10-2).
2. Raise boom to horizontal position with extensions fully extended.
3. Measure and mark a 3/4-inch vertical strip along the outer edge of the three boom sections.
4. On the base section boom to the right of and parallel to the 3/4-inch strip mark 7' 2" in two-inch letters.
5. … …he outer edge of the boom … at 6-inch intervals a vertical strip … Mark the first strip 2 inches high … section and the second strip 4 … …ating the 2-inch and the 4-inch … …plete length of the section.
6. … …ch strip, in two inch letters mark … …e feet measurement (1, 2, 3, etc).

Figure 19. Boom Marking Instructi…

Figure 20. Fuel tank markings.

g. All calibration vehicles will be marked with a calibration insignia plate as provided in kit ORD 7911573. These plates will be applied on each side of the M820, M934 and M935 Expansible Vans, between the second and sixth rails, Figure 21. On other types of calibration vehicles (such as vans and panel trucks), the plates will be applied to the most appropriate surface, depending on surface size and window location.

h. All motorized materials handling equipment will have the capacity marked on each side. The marking will consist of the numerals showing the applicable pounds followed by the letters CAP. The actual weight of the equipment will also be marked on each side, in numerals, to show the applicable pounds, followed by the letters MTWT. Letters and numerals will be one inch high in the same color as the registration marking.

i. Other necessary technical instructions may be marked on appropriate surfaces as required or authorized by the responsible command.

(1) Such markings will be applied in letters and numerals normally one inch in height, but not less than 1/2 inch.

(2) Color of the markings will be lusterless green 383 or black, whichever affords the greater contrast with the surface color. If practicable, the color chosen will be the same as the registration marking.

(3) Temporary instructions will be applied with removable vinyl markings or gasoline soluble paint.

13. WEIGHT CLASSIFICATION.

a. All single vehicles with a gross weight of over 3 tons, and all trailers with a rated payload exceeding 1-1/2 tons, are classified for the purpose of bridge crossing per FM 5-36. Each single vehicle and vehicle combination has a classification number for empty, cross-country loading. The classification numbers are used to determine whether a bridge, correspondingly classified, can be safely crossed.

b. All applicable self-propelled vehicles subject to camouflaging, not towing other vehicles, will carry a circular sign of 9-inch diameter with the vehicle classification number in black on a green background. This will normally be located on the front of the vehicle toward the right side and below the driver's line of vision.

c. All applicable self-propelled vehicles subject to camouflaging, towing other vehicles, will carry a sign similar in size, design, and location except that the classification number of the vehicle combination will be shown with the letter C in black above the number. A 6-inch green circle with tone basic vehicle classification number in black will be marked on the right side of the vehicle in a conspicuous location. The towed vehicle will be similarly marked with its basic classification number.

d. When the classification number of the vehicle is not subject to change, the marking may be applied directly on a suitable surface. Use adhesive-backed markings or paint with stencils. Side markings may be applied by painting or by using adhesive-backed decals. On vehicles not subject to camouflaging, painted in gloss colors, the markings will be gloss yellow 13538; gloss black 17038; and gloss red 11136. On vehicles not subject to camouflaging, painted in semigloss or lusterless colors, the markings will be lusterless yellow 33538; gloss black 17038; and gloss red 11136. When the marking is applied directly to vehicles painted in the same or similar color, the circumference of the circle will be outlined with a black circle 3/4 inch wide.

e. In usages where the classification number may change, a sign kit will be utilized which consists of a 9-inch metal disk, a retaining plate, and two sets of five different numeral plates. By installing the retaining plate, the numeral plates will be locked in place. The sign has black numerals on a green background and a black C on the top of the retaining plate, Figure 22.

f. The following weight classification sign kit is being supplied initially with mounting brackets and hardware.

NAME	PART NO.	NSN	QTY
KIT: vehicle class sign, w/brackets and hardware Consisting of:	6-1-2248	9905-00-565-6267	1
6 ea. BOLT, MACHINE: hex-hd, S, cd or zn-pltd 5/16-24 UNF-2A x 3/4 in.	MS90726-32	5306-00-225-9087	
2 ea. BOLT, MACHINE: hex-hd, low-carb-S cd- or zn-pltd, 5/16-24UNF-2A x 1-3/4 in.	MS90726-39	5306-00-225-9094	
1 ea. BRACKET: unit, lg*			
1 ea. BRACKET: unit, short*			
1 ea. DISK: vehicle class sign, 9-in.			
6 ea. NUT, PLAIN, HEXAGON: S, cd- or zn-chromatic-fin, 5/16-24UNF-2B-1/2 w, 17/24 thick	MS51968-5	5310-00-880-7746	
2 ea. PLATE: numeral, 0 one side, 1 other side			
2 ea. PLATE: numeral 2 One side, 3 other side			
2 ea. PLATE: numeral, 4 One side, 5 other side			
2 ea. PLATE: numeral, 6 one side, 7 other side			
2 ea. PLATE: numeral, 8 one side, 9 other side			
1 ea. PLATE: retaining, C on one side			
6 ea. WASHER, LOCK: int-teeth, S, cd-or zn-pltd, 5/16 in. screw size	MS35333-41	5310-00-167-0721	
Hardware items not required for the specific installation procedure will be returned to local stock. *Brackets not required for the specific installation procedure will be scrapped.			

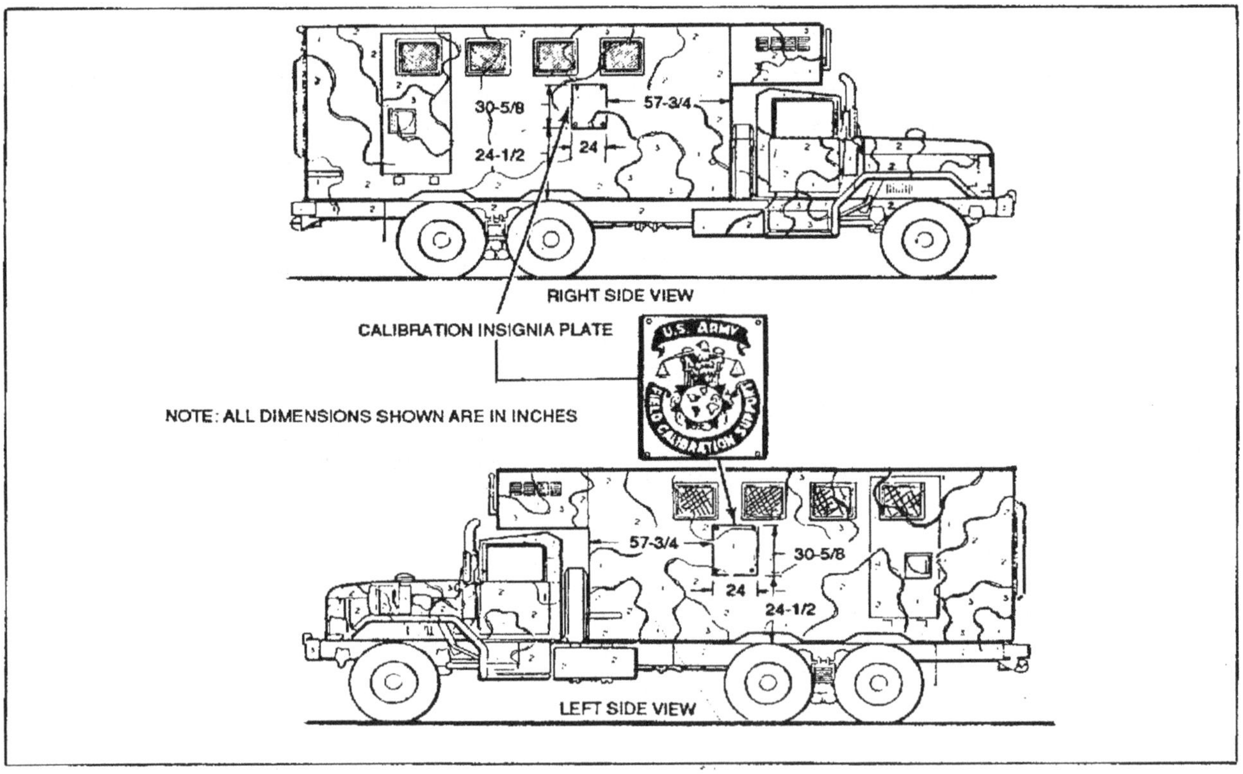

Figure 21. M929, M920, M921, and M934 Expandable Van (Calibration).

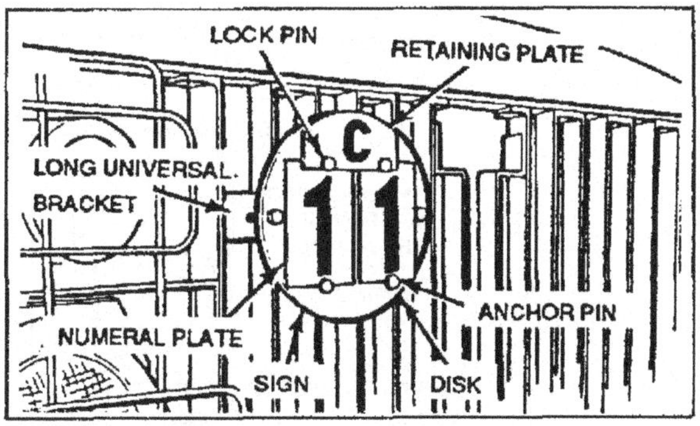

Figure 22. Vehicle weight classification sign installed.

g. In addition, a replacement kit itemized below, is authorized for the replacement of lost or damaged numeral plates and retaining plates:

Name	National Stock No.
KIT: Retaining and Numeral Plate Replacement, for Vehicle Class Signs	9905-00-565-6268

Consisting of:

1 each PLATE: numeral, 0 one side, 1 other side
1 each PLATE: numeral, 2 one side, 3 other side
1 each PLATE: numeral, 4 one side, 5 other side
1 each PLATE: numeral, 6 one side, 7 other side
1 each PLATE: numeral, 8 one side, 9 other side
1 each PLATE: retaining, C on one side

h. A vehicle weight classification sign kit, as described herein, will be installed on the front of each vehicle except:

(1) All trailers.

(2) Vehicles having a gross weight of 3 tons or less.

(3) Vehicles having standard classification numbers higher than 99.

(4) Vehicles having rated payloads of 3/4 to 1-1/2 tons. This exception does not apply to vehicles authorized for towing loads.

(5) Vehicles with fixed loads (such as tanks, self-propelled guns, etc.).

i. Front classification signs for vehicles included in h(3), (4), and (5) above, will be painted directly on the vehicle. Side classification signs, where applicable, will also be painted directly on the vehicle.

j. These vehicle weight classification sign kits can be installed on vehicles in different ways. For example, in the case of Cargo Truck, the sign can be installed on the brush guard, on the splash shield, or in an upright position on the fender. The choice of method is left to the using organization.

(1) Location.

(a) The classification sign will be installed on the vehicle in one of the following locations:

1. Right front fender.
2. Right side of brush guard.
3. Right front section of hull.
4. Right side of center front bumper splash shield.

(b) The sign will be placed below the driver's line of vision and will be positioned so that it will not block headlamps, marker lights, blackout lights, air intakes, etc.

(2) Installation.

(a) On brush guard.

1. Paint the long universal bracket (channel) with rust-inhibiting enamel.

2. After the enamel has dried, position the bracket so that the two holes in the bracket align with two bolt holes (9 o'clock positions) in the sign. Assure that the flat bearing surface of the bracket faces toward the sign. Install one bolt (1-3/4 inches long) P/N 123798, NSN 5306-00-012-3798, one washer P/N 138538, NSN 5310-00-272-4119, and one nut P/N 120368, NSN 5310-00-012: 0368. Turn nut two or three turns.

3. With the bracket in a vertical position, insert it between two bars of the brush guard at the desired location. With one hand, reach behind the brush guard and adjust the bracket to a horizontal position. With the other hand, install another bolt (1-3/4 inches long), washer, and nut. Tighten both bolts, but apply only enough torque to secure the sign to brush guard. Excessive torque will cause the sign disk to bend. Figure 22 shows the sign installed on brush guard.

CAUTION

If it is found that the vehicle classification sign seriously obstructs airflow to the radiator, as indicated by engine overheating, remove sign from brush guard and install sign elsewhere.

(b) On the sloping fender, install the sign as indicated in 1 and 2 below only when the slope of the fender is less than 30° from vertical. When the slope of the fender is more than 30° from vertical, install the sign in a vertical position, as outlined in (c) below, to insure readability of the sign.

1. Position the sign in the desired location on the fender and, using the sign as a template, drill two 1 1/32-inch holes in fender to align with bolt holes in the sign at the 9 o'clock and 3 o'clock positions.

2. Install the sign to the fender with two bolts (3/4 inch long), P/N 120741, NSN 5306-00-012-0741 , two washers, P/N 138538, NSN 5310-00-272-4128, and two nuts, P/N 120368, NSN 5310-00-012-0368. Tighten bolts securely, Figure 23.

(c) In a vertical position on fender or on splash shield, alter the long universal bracket (channel), Figure 24.

1. Paint both the altered bracket and the short universal bracket (angle) with rust-inhibiting enamel.

2. After the enamel has dried, position the long leg of the short bracket on the back of the sign and align two holes in bracket with the two lower bolt holes in the sign.

NOTE

When installing the sign on a horizontal surface, position the short leg of the bracket so that it faces rearward, Figure 25. When installing the sign on a sloping surface, position the short leg of bracket so that it faces forward, Figure 26.

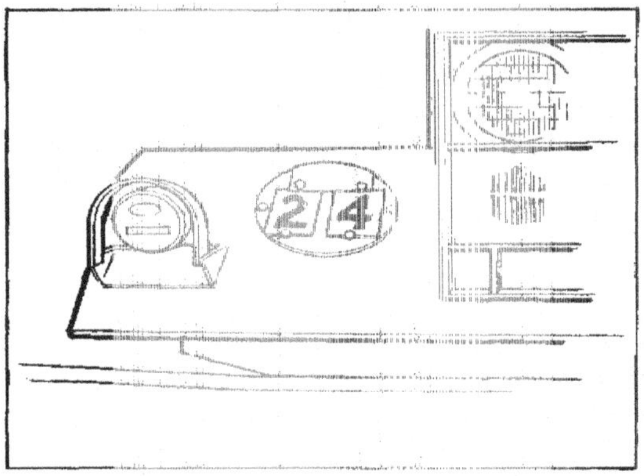

Figure 23. Sign installed on sloping fender (slope angle less than 30° from vertical).

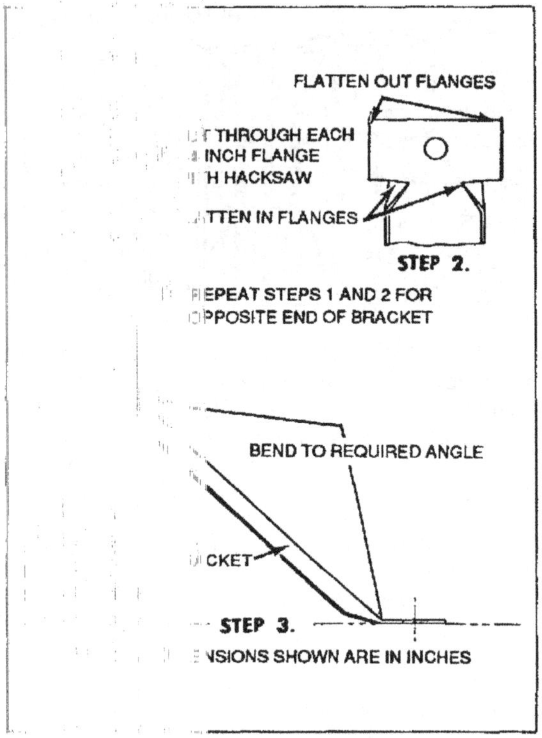

Figure 24. Alteration of bracket.

Figure 25. Sign installed on splash shield.

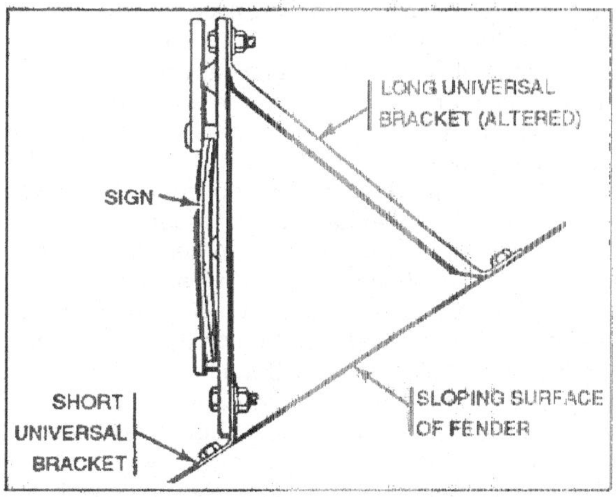

Figure 26. Sign installed on sloping fender (slope angle exceeds 30°from vertical).

3. Install bracket to sign with two bolts (3/4 inch long), P/N 120741, NSN 5306-00-012-0741 two washers P/N 138538, NSN 5310-00-272-4119, and two nuts P/N 120368, NSN 5310-00-012-0368.

4. Position the sign in the desired location on the fender or splash shield. Bend short leg of bracket, as required, to conform to the contour or slope of fender.

5. Using the short leg of the bracket as a template, drill two 11/32-inch holes in fender or splash shield.

6 Attach the sign to the fender or splash shield with two bolts (3/4 inch long), two washers and two nuts.

7. Secure the altered long universal bracket to the sign with one bolt (3/4 inch long), one washer, and one nut. (Insert bolt in hole located between two stacks of numeral plates on sign).

8. Drill an 11/32-inch hole in fender or splash shield to correspond with the hole in the altered bracket and secure the bracket with a bolt (3/4 inch long), washer, and nut. Figure 25 shows sign installed on splash shield and Figure 26 shows sign installed on sloping fender.

k. Instructions for unlocking and locking the retaining plate are as follows:

(1) When unlocking retaining plate, use the thumb of one hand and press down firmly on one stack of numeral plates, Figure 27. With the other hand, press gently on the retaining plate and pivot the plate away from the lock pin. To remove retaining plate completely, press down firmly on the remaining stack of numeral plates and free retaining plate from the other lock pin.

(2) To lock retaining plate, make sure that five numeral plates are tacked on each anchor pin with the two top plates showing the desired classification number. If the vehicle is to be used in combination, position the retaining plate so that the C on one side of the plate will be visible. If the vehicle is to be used singly, position the retaining plate so that the C will not be visible. Press down firmly on the top of one stack of numeral plates and slide the retaining plate onto the lock pin until the plate snaps in place. Repeat the procedure for the remaining stack of plates and lock the retaining plate to the remaining lock pin. Refer to FM 5-36 for vehicle weight classification data.

Figure 27. Unlocking the retaining plate.

14. OPERATION AND DESIGNATION PLATES

a. These plates, as provided on vehicles and equipment, will be kept clear and free from paint at all times.

b. Operation plates contain data or instructions on vehicle operation and will be installed within the normal view of the operator when in an operating position. The plates will include:

(1) Road speed plates. To prescribe the maximum permissible speed for operation in each gear ratio and range.

(2) Shift diagram plates. To show the location of each position of the transmission, transfer, and/or power takeoff.

c. Operation plates containing precautionary instructions to avoid personnel injury or equipment damage will have a red background. This applies to new plates of photosensitive anodized aluminum, conforming to type H, MIL-SPEC 15024.
Existing plates will be replaced only when no longer serviceable.

d. Designation plates contain data pertinent to the vehicle or equipment, but are not required for actual operation. They will be located in convenient position without creating confusion with operation plates. The plates will include:

(1) Name or identification plates. To show nomenclature, manufacturer, model, serial number, stock number, contract number, US registration number, and date of delivery to the Government.

(2) Data plates. To show empty weight, maximum payload and/or gross weight, weight on each axle and/or maximum towed load for highway and cross-country, center of gravity empty and loaded, shipping cubage, overall and reducible dimensions, and other pertinent linear and angular dimensions.

(3) Servicing data plates. To show fuel capacity and minimum octane, cooling system capacity, draining instructions, tire pressures for terrain conditions, lubricant data for different temperature ranges, and other servicing data for the vehicle, components, or equipment installed.

(4) Responsible agency plates. To show the Commodity Command or Agency responsible for the procurement and/or maintenance of the vehicle or its chassis, body, and/or equipment.

(5) Publication plates. To show the manuals (operator's maintenance, and supply) and lubrication order applicable to the vehicles.

e. Designation plates may be combined plates when space conditions warrant.

15.AMBULANCE MARKINGS--TACTICAL USE VEHICLE.

a. Truck, Ambulance, 1-1/4-Ton, 4x4 M1010, Figure 28, will have on the left and right side outer panel of the ambulance body, a red cross on a white field decal, P/N 23-0196, Cage 25022, NSN 7690-01-165-0541. Also on each of these panels will be a decal, P/N 230193, Cage 25022, NSN 7690-01-165-0542.

(1) The Ambulance Body roof panel will have a red cross on a white field decal, 36 x 36 inch, P/N 230198, Cage 25022, NSN 7690-01-165-0544.

(2) The Ambulance Body left and right rear doors, will have a red cross on a white field decal, P/N 230200, Cage 25022, NSN 7690-01-165-6901.

b. The Truck Ambulance M996 and M997, Figure 29, (HMMWV) has attached reversible markings, See TM 9-2320-280 series manuals for details.

c. Truck, Ambulance, 1/4-Ton, 4 x 4, M718 (series), will have an 18-inch red cross on a 20-inch-square white field marked on each canvas side curtain directly to the rear of the canvas side doors, Figure 28.

(1) An 18-inch red cross on a 20-inch-square white field will be marked on the rear of the vehicle. The upper half of this marking will be applied on the canvas rear doors and the lower half on the tailgate, centered with the centerline of the vehicle, in order that the full marking is shown when the rear doors are closed.

(2) A 45-inch red cross on a 48-inch-square white field will be marked centered on the canvas roof.

d. Truck Ambulance, M1010 will have the word "Ambulance" marked in 3-inch-high-white letters, centered three-quarters of an inch above the windshield.

(1) A 3-inch red cross on a 4-inch-square white field will be marked to each side of, and in line with, the word "Ambulance" above the windshield.

(2) A 33-inch red cross on a 36-inch-square white field will be marked on each side of the body below the roof joint and centered above the rear wheels.

(3) A 33-inch red cross on a 36-inch-square white field will be marked on the rear door panels directly below the windows and centered on the vertical centerline of the vehicle.

(4) A 60-inch red cross on a 64-inch-square white field will be marked on the roof 14 inches from the rear of the vehicle, Figure 28.

e. Vehicles other than the ones enumerated, that are intended for such usage and that are affected by international agreement concerning such vehicles, will be marked in a similar manner.

f. The National Symbol will not be applied to Medical Service vehicles. The sizes of the red crosses specified are measured across opposite ends. Each limb of the cross will be a square having sides equal to one-third the size of the cross.

g. The following colors will be used for ambulance markings, as applicable:

(1) Gloss red 11136 and gloss white 17875 on vehicles painted gloss or semigloss olive drab.

(2) Lusterless red 31136 and lusterless white 37875 on vehicles painted lusterless forest green.

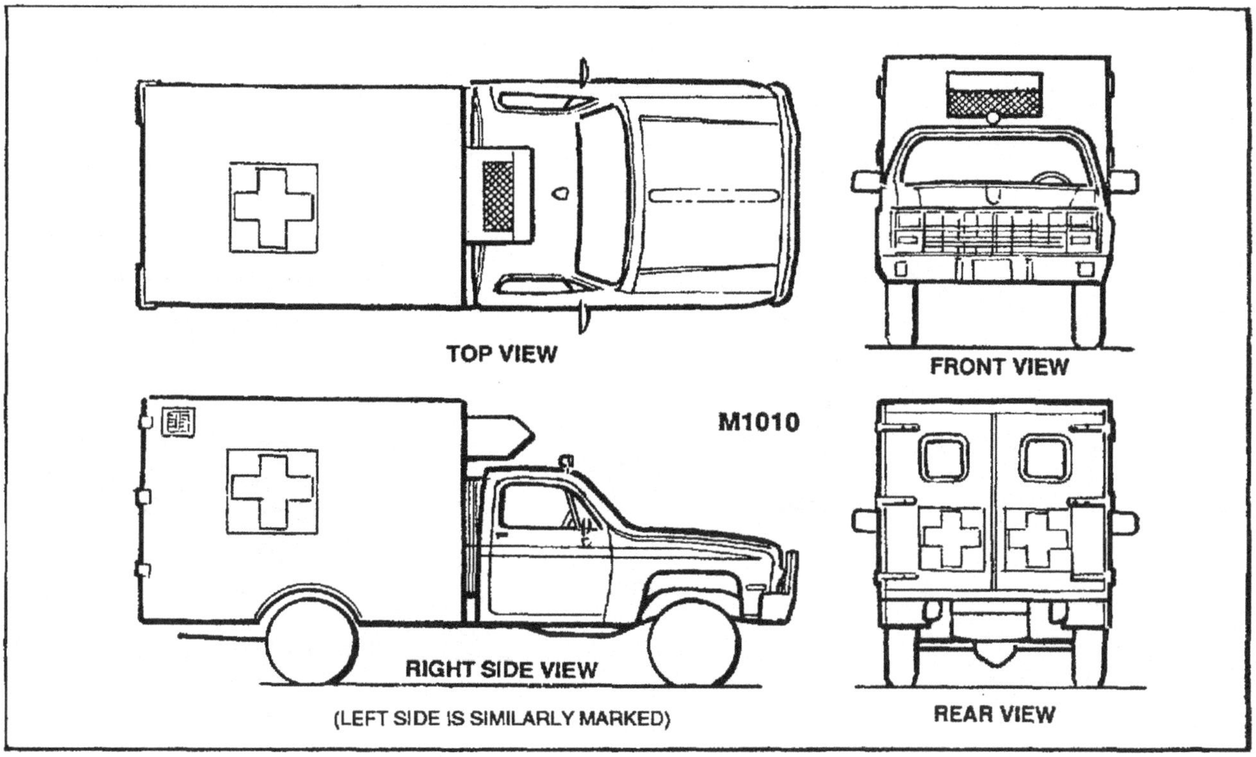

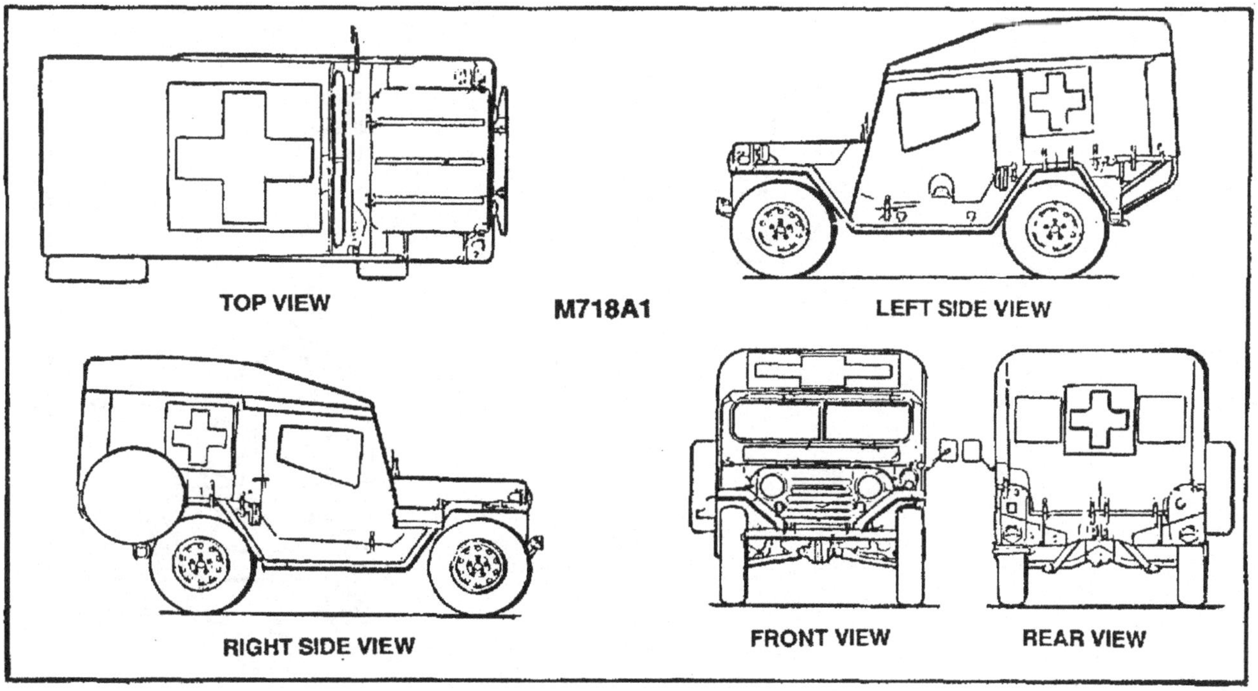

Figure 28. Ambulance marking military design vehicle.

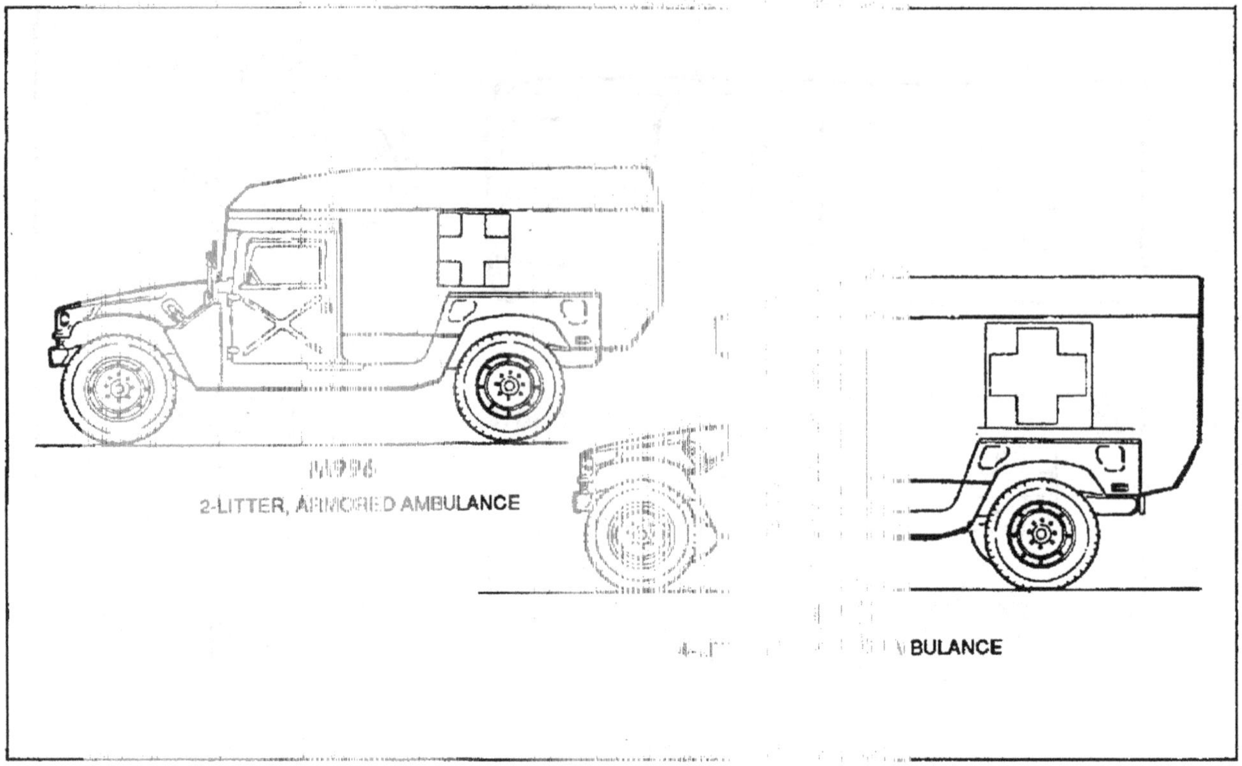

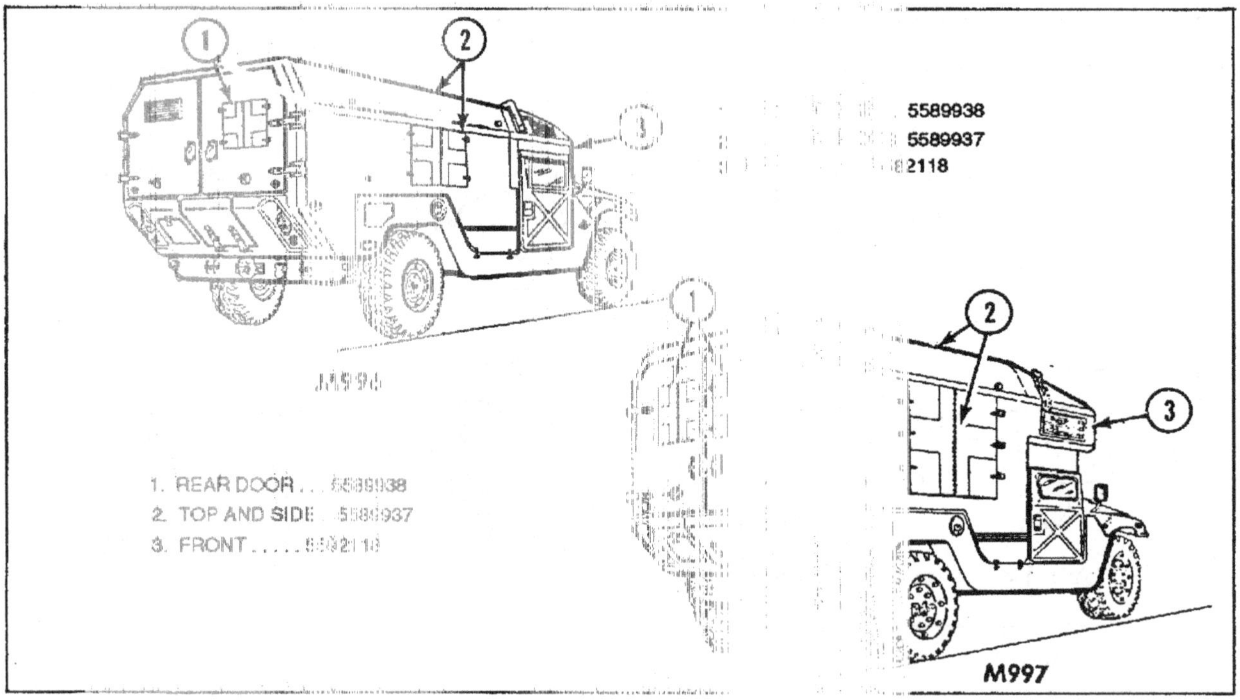

Figure 29. M996 2-Litter and M997 4-Litter Armored Ambulance.

16.MILITARY POLICE MARKINGS--TACTICAL USE VEHICLES.

a The HMMWV, M998 series and CUCV, M1008 when used as Military Police Tactical Vehicles, should be treated as such for identification. The words Military Police will be 3inch block letters and used in reference to the markings of the HMMWV, M998 series and the CUCV, M1008 Vehicles.

(1) CUCV: The tops of the "Military Police" will be located 9-1/2 inches from the top crease of the left and right passenger doors and centered horizontally on the doors, Figure 30. The tops of the words "Military Police" will be 3-1/8 inches below the cargo door handle in the rear and centered horizontally, Figure 30. The words "Military Police" will be located immediately above the leading edge of the hood and centered horizontally, Figure 30.

(2) HMMWV: The bottom of the words "Military Police" will be located 10-1/2 inches above the rear wheel well and centered using 2-inch block letters, Figure 30. The bottom of the words "Military Police" will be located 1-1/2 inches above the bottom edge of the cargo shell door and evenly spaced on either side of the door latch using 3-inch block letters, Figure 30. The words "Military Police" will be centered on the front hood using 2-inch block letters, Figure 30.

b. Truck, Utility, 1/4-Ton 4 X 4, M151 Series: a white disk, 16 inches in diameter with the words "Military Police" in 3inch black letters located around the circumference of the disk, 1/2 inch from the disk edge, will be attached to the center of the spare wheel on the rear of the vehicle. The sign will be locally fabricated from 16-gage sheet metal or thicker. The disk will be attached by means of a bracket made from strap-iron, 1 inch wide and 1/8 inch thick.

(1) M151 series truck will have a detachable white sign, 5 inches high and 36 inches long, with the words "Military Police" in 3-inch-high black letters, mounted on the windshield panel below the glass. The sign will be fabricated from 16-gage sheet metal or thicker and will be attached to the windshield with sheet metal screws.

(2) Motorcycle, solo, will be marked with a white band, 3 inches high, with the words "Military Police" in 2-inch-high black letters, across the front of the windshield directly above the horizontal support. When not obstructed by lights or sirens, the band may be placed below the support.

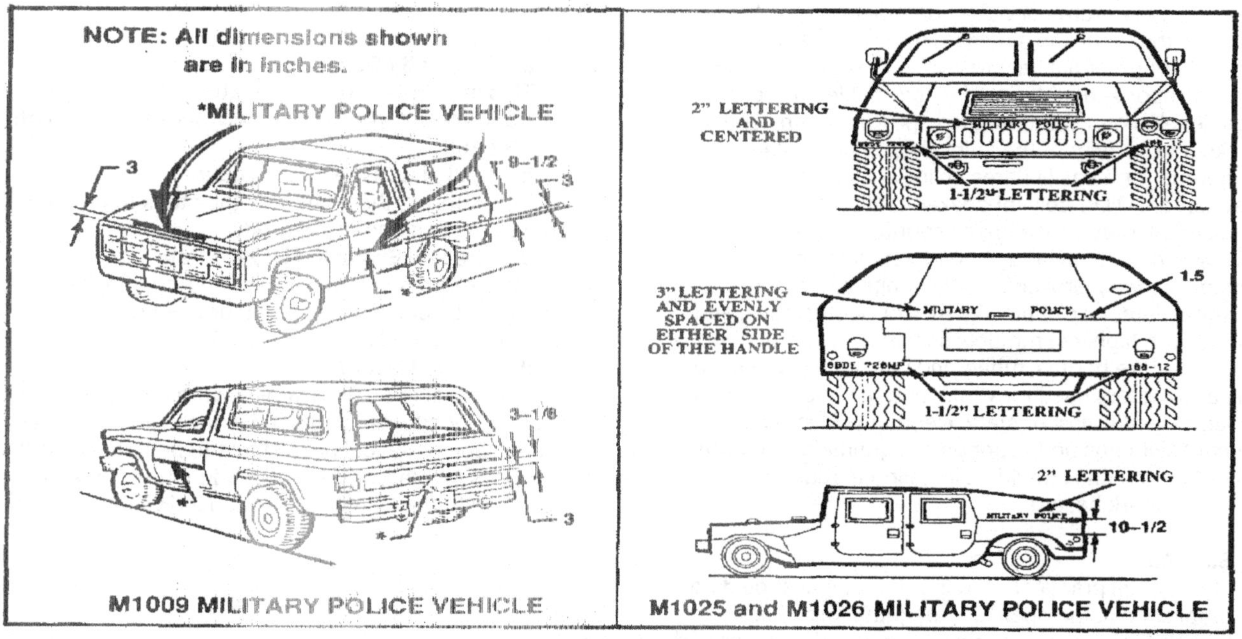

Figure 30. Military Police Markings.

SECTION V. SAFETY SUMMARY

17. GENERAL

a. Fumes.

Thinners used with paints and enamels may have harmful effects. Continued breathing of fumes during and after painting operations should be avoided. Toxic fumes will persist, in some cases, for many days indoors after painting operations. Avoid inhaling solvent vapors.

b. Contact with Paint Materials .

Avoid skin contact with paint materials, particularly if there are cuts or open wounds on the hands. Unwashed hands may convey poisonous matter to food. Many painting materials contain lead sulfide, lead carbonate, chromium compounds, and other poisonous materials. Lead chromate is particularly dangerous, and like other lead paints, may enter the system during eating, or handling. Many of the thinners are also poisonous and can enter the human system through the skin, through inhalation, or through carelessness in handling food. Wear personal protection equipment (PPE) as recommended in paragraph 19a.

c. Fire .

The mist that comes from a spray gun is highly flammable. A spark will cause it to flash. Smoking is prohibited in paint shops. Open cans containing paint removers, thinners, paints, and paint materials are a fire hazard. Empty drums or other containers in which solvents, thinners, and similar materials have been shipped are potential hazards since they often contain enough vaporized material of a flammable nature to cause explosions. Accumulated spray in booths and in cracks and comers of the paint shop is particularly dangerous for it easily flares up in spontaneous combustion. Oil or solvent-soaked cloths, if not promptly disposed of, may cause fire by spontaneous combustion. Fires which occur in spray booths result from six principle causes: broken electric lamps and other electrical defects; cleaning interior of booths, fans, and motors with highly flammable solvents; accumulations of deposits in the booths, tubes, and vent pipes; defective fans and motors used for ventilating the booths; poorly designed vent tubes; or static electricity. Electrical mercury switches and vapor proof lighting fixtures with protective screens should be installed to reduce electrical spark hazard.

d. Safe Air .

If it is at all practical, painting of material should be accomplished in the open air. Adequate forced draft ventilation for indoor work should be provided to carry off fumes. Respirators should be worn during all spray-painting operations. Refer to paragraph 19b for specific respiratory protection required.

e. Safe Practices.

Preparations containing benzene should not be used for spraying. Only vapor proof lamps should be used where spraying is being done. Paints should be stored in a steel cabinet in a small building away from work buildings. Once opened, cans containing paint removers, thinners, paints, and paint materials should be covered tightly before being stored or put away overnight. Do not apply heat or flame to drums, cans, or other containers that have contained solvents, thinners, or other paint materials. Observe safe operating procedures at all times, particularly when handling cleaning materials. For disposal of unserviceable paints, refer to paragraph 21.

18. MATERIAL SAFETY DATA SHEETS (MSDS)

MSDS are prepared by the manufacturer and should accompany each single shipment or batch of paint, primer or thinner. It is mandatory that personnel working with these substances read this information. Because of the variations involved, MSDS must be reviewed for each shipment procured on a single purchase order. MSDS must be filed in a location readily accessible to workers exposed to the substances. MSDS also assist management by directing attention to need for specific control engineering, work practices and protective measures to ensure safe handling and use of the material. Along with the products ingredients and specific protection information, the MSDS contain the following data:

a. Reactivity Data.

This informs the paint user about the stability, hazardous decomposition, or polymerization properties of the coating.

b. Spill al d Disposal Procedures.

This informs the paint user the steps to be taken for proper spill or disposal methods.

c. Fire and Explosion Hazard Data .

This informs the paint user about the flash point of the product, special fire fighting procedures, and the extinguishing media.

d. Health Hazards.

Personnel should be familiar with emergency and first aid procedures -s outlined in the product's MSDS. This includes medical procedures to be followed if the product is inhaled, or if the procedure has come in contact with the skin or eyes of an individual.

19. CONTROL MEASURES

a. Personal Protection .

(1) Personal Protection Equipment (PPE) used in conjunction with respiratory protection equipment (paragraph 19b) includes cloth coveralls, eye protection, and head coverings. Cloth gloves are suitable unless cellosolve acetate (2-ethoxyethyl acetate) is present in the paint, solvent, or primer. When this solvent is present, silicon rubber gloves are recommended. Spot painters applying paint by brush or roller must wear clothing and gloves affording full skin coverage. Persons who clean mixing and painting accessories should wear eye protection and rubber gloves to preclude solvent absorption and defatting of the hands caused by the thinner.

(2) The use of barrier creams is acceptable; however, their usefulness in preventing the absorption of solvent through the skin is not documented. If a solvent with a skin notation is being used, then silicon rubber gloves must be used. Barrier creams are useful in preventing paint from adhering to the skin and in combating the "dryness" associated with the defatting action of most solvents. Solvents must never be used to remove paint/coating from the skin.

(3) Work clothing should be provided. After completing painting or sanding operations, hygienic showers should be taken prior to changing into street clothing.

b. Respiratory Protection .

NOTE

For the purpose of determining respiratory protection required for a given operation, a confined space is any of the following sites:

- **Any area where dilution ventilation cannot take place, or where airflow is obstructed**
- **Under or in vehicles/equipment, whether indoors or outdoors**
- **Indoor areas (not in spray paint booth) of less than 10,000 cubic feet**
- **Indoor areas where ceiling height is less than 16 feet a**
- **Touchup paint areas with partitions, balconies, or any structural barriers obstructing cross-ventilation**
- **Any area where outside air is not mechanically distributed at a minimum rate of 3.5 cfm per square foot of bay/room/area where touchup painting takes place**

(1)Spray Painting Indoors. An approved airline respirator is the standard respirator to be worn when paint spraying indoors; however, alternatives are permitted when spray operations are not conducted in a confined space and valid sampling results document the personal exposure levels.

(a) Large Vehicular or Walk-in Booths. If the diisocyanate concentration is below the standard, a full facepiece chemical cartridge respirator with a paint prefilter is adequate. If the solvent concentration is less than 10 times the standard, a paint-spray respirator is adequate. If the pigment containing lead or chromate concentration is less than 10 times the standard, an organic vapor respirator with a HEPA filter is adequate.

(b) Spray Cabinet or Conveyor-type Booth. If the contaminant (solvent, pigment, or diisocyanate) concentration is below the standard, no respiratory protection is required. If either the solvent or diisocyanate concentration exceeds the standard, an airline respirator is required. If the pigment containing lead or chromate concentration exceeds the standard, but is less than 10 times the standard, an organic vapor respirator with a HEPA filter is required.

(c) Spray Painting Outdoors. If in a confined space, an airline respirator is required. If not in a confined space, a paint-spray respirator is required.

(2) Brush or Roller Painting Indoors or Outdoors. If not in confined space, an organic vapor respirator, unless deemed unnecessary by preventive medicine/industrial hygiene personnel. In a confined space, an approved airline respirator is required. Application of a water-base paint does not normally require respiratory protection; however, local preventive medicine/industrial hygiene personnel will determine the requirements.

(3) All Painting. If respiratory protection other than an approved airline respirator is used, then exposure must be documented by preventive medicine/industrial hygiene personnel.

(4) Sampling. Sampling is done during the painting of different vehicles and equipment, when the booths' filters are clean and dirty, when changes or modifications occur in or to the booths, and for each painter. The sampling will be performed according to the US Army Environmental Hygiene Agency's Technical Guide No. 141 and in sufficient numbers to be statistically valid. Sampling methods other than those listed in the Tech Guide, i.e., NIOSH methods, may be used upon approval by the laboratory providing services. The documentation will be revalidated every 2 years if the initial sampling results are below one-half of the standard and every 6 months if greater than one-half of the standard. Additional monitoring is required whenever there has been a production, process, control or personnel change or whenever there is any other change that may result in new or additional exposures.

c. Ventilation.
The use of respiratory protection equipment does NOT waive the requirement for engineering control measures. The ventilation design specifications for spray paint booths are in TB MED 514.

d. Preferred Paints.
Whenever possible, Type II or Type III aliphatic polyurethane paints (PUP) should be substituted for Type I since they contain no lead or chromate. Non-lead-containing primers should be used when available.

e. Warning Labels.
Warning labels are required on products containing lead or chromium IV.

20. MEDICAL SURVEILLANCE General. Medical surveillance to detect adverse health effects will be determined by the installation medical authority (IMA) based on the specific constituents of the coating. In general, medical surveillance is required for anyone who works more than 30 days per year in either a paint spraying operation or in a brush or roller application when respiratory protection is required. Personnel involved in painting at direct support, general support, and depot levels of maintenance will normally require surveillance. Vehicle/equipment operators and unit maintenance section personnel usually do not perform enough brush touchup painting to warrant medical surveillance.

21. DISPOSAL

a. General.
Unserviceable paints including aliphatic polyurethane paint (PUP)/epoxy polyamide enamel (EPE) blend must be discarded as a hazardous waste according to AR 420-47. Note that because they contain ignitable solvents, these unusable blends of PUP are defined as hazardous wastes by the Resource Conservation and Recovery Act (RCRA) of 1976 as amended by the Hazardous and Solid Waste Amendments of 1984. Unserviceable component B is a hazardous waste per RCRA due to ignitable solvent content; not because it contains HDI.

b. Method of Disposal.
The method used to dispose of this waste stream depends on the types of paint used. When the paints contain no hazardous heavy metals, the liquid portion of the waste stream may be able to be discharged into the sanitary sewer and the sludge disposed of in a sanitary landfill. The environmental coordinator should be contacted to ensure state and local ordinances are not violated. If heavy metal-based paints have been used, samples of the liquid and sludge should be analyzed for the characteristic of extraction procedure (EP) toxicity (heavy metals) and disposed of accordingly.

[illegible] sewer serving a paint processing ope[illegible] s to a government-operated sewage treat[illegible]), notify the STP operator of the app[illegible] al loadings of total organic carbon, bio[illegible] demand, and total processing ope[illegible] charge.

[illegible] sewer serving a paint processing ope[illegible] s to privately owned treatment works, pret[illegible] ewaters may be required by local regu[illegible] The installation environmental coo[illegible] determine such cases with public en[illegible] ors having primacy over gove[illegible] ions. Pretreatment regulations have bee[illegible] the USAEHE Water Quality Infor[illegible]. 13.

22. [illegible] RDS

[illegible]

CAUTION

• Co[illegible] ot be applied at temperatures belo[illegible] they will not cure properly.

[illegible] azards.

CAUTION

• M[illegible] haust parts, breathers, etc., [illegible] nt dust, solution, water, or [illegible] from entering the engines of [illegible]

• D[illegible] s on electronic equipment as [illegible] changes in performance.

• D[illegible] eum based products on natural [illegible] e destructive agents.

• D[illegible] quipment containing any [illegible] nents to apply coatings [illegible] 1,1,1 trichloroethane (such as [illegible] e m).

23. [illegible] TIONS

[illegible] information is contained in the follo[illegible]:

[illegible] ventive Medicine.

[illegible] HA Safety and Health Standards.

[illegible]-R-1, Hazardous Materials Han[illegible] Criteria.

[illegible], Occupational and Environmental Heal[illegible] otection Program.

e. TB MED 514, Occupational and Environmental Health Requirements for Painting Operations.

SECTION VI. CAMOUFLAGING PROCEDURES

24. CAMOUFLAGE PATTERN PAINTING

a. This section covers methods of applying Camouflage Paint Patterns (CPP) to Army materiel. It also covers procedures for inspecting applied CPP. Equipment consists mainly of brushes and guns; paints consist of Chemical Agent Resistant Coatings (CARC), such as those listed in Appendix B. The CARC topcoats are suited for camouflage painting and protect military material against visual and infrared detection and chemical agent contamination. The CARC camouflage topcoats are MIL-C-46168 and MIL-C-53039. The CARC epoxy primers are MIL-P-52192, MIL-P-23377, MIL-P-53022, and MIL-P-53030.

b. Information on specific finish systems and kinds of paint to be used for various applications is contained in Appendix B. Surface preparation is covered in paragraph 27. Application techniques are covered in paragraphs 30-32. For additional information on camouflage, refer to the following publications:

AR 750-1	Army Materiel Maintenance Management
FM 5-20	Camouflage
TM 5-200	Camouflage Materials

c. Major items to be camouflage-painted are weapons systems, vehicles, construction equipment, and materials handling equipment. The painting procedures described in this chapter do not apply to aircraft. Three-color camouflage patterns are created for all combat, combat support, and combat service support equipment having an area greater than nine square feet on one or more sides. New items of tactical equipment normally will be CARC coated in a three-color camouflage pattern at the factory or depot. Some items, however, may be painted with lusterless CARC green 383 when the pattern has not been developed, Items painted in CARC green 383 or alkyd enamel will only be three-color camouflage CARC coated when a complete repainting is required for maintenance purposes.

d. Only DS/GS Support and Depot level personnel are authorized complete painting and repainting of components with any topcoat or primer. DS/GS Direct Support personnel are authorized complete painting/repainting only if they use an OSHA approved spray-paint booth. Unit level personnel are permitted to use topcoats and primers for touchup efforts only. This includes all CARC coatings.

25. REASON FOR PATTERN PAINTING

a. All military vehicles and equipment have characteristic shapes and shadows. These shapes and shadows contrast with the material surroundings and make the object stand out. Pattern painting using wavy, irregular patches of camouflage colors does much to break up the characteristic shapes of the equipment by reducing contrasts with sod and vegetation, pattern shape, and placement. Patterns have been designed for each type of vehicle to cut off sharp corners, avoid straight, vertical, and horizontal lines, and extend shadows in shapes similar to natural features and vegetation; however, the accuracy with which the CPP is applied completely determines how well the CPP camouflages the equipment.

b. Pattern painting is not a magic, cure-all camouflage technique, but it makes the item much harder to see and recognize as a military object. It also provides an excellent base for applying further natural camouflage such as tree limbs, shrubs and grass.

c. The theory behind pattern design is to provide a camouflage system for equipment that is universal and can he adapted to various geographical, seasonal, and terrain changes by the changing of one or possibly two colors. For instance, temperate zone colors of green, brown and black may be changed to desert colors by changing green to tan.

NOTE

CARC colors should not be mixed with one another as this will alter their individual effectiveness when applied to the end item. Blending is not permitted.

d. These resigns also lend themselves to touchup painting with better results than are now obtainable from touchup of the current olive-drab vehicles. Slight mismatches in color are expected at times and will not be noticeable except upon close inspection. Likewise, minor abrasions and scaling of surfaces will be equally inconspicuous. Marring and surface lightening due to handling is characteristic of camouflage coatings and does not impede camouflage or infrared properties. This is typical of low gloss and sheen coatings in dark colors and is considered satisfactory.

26. PATTERN DESIGNS

See Figure 31, for example of the CPP for the M113A1, Personnel Carrier. Within each area is a number that stands for the color to be painted. There are five views of the vehicle; front, back, left side, top, and right side. Most designs will show these five views. The base, or "primary" color is Color #2 and is usually applied first over all surfaces. Color patches #1 and #3 a then applied over the primary coating. The approximate distribution, in percent, of the three colors is: green--44 percent, black--41 percent, and brown--15 percent.

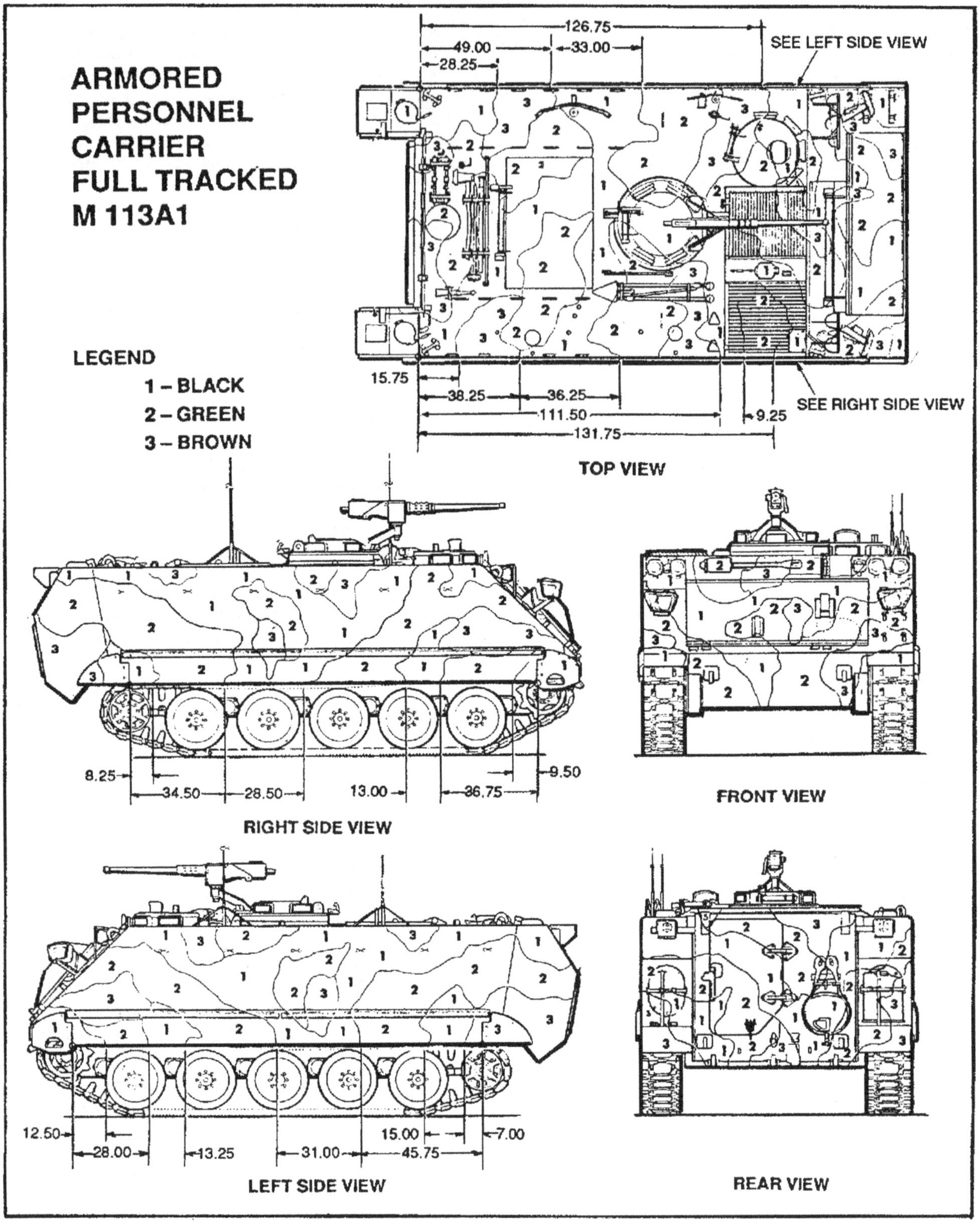

Figure 31. Pattern Painting Design for the M113AI Personnel Carrier.

27. PREPARATION FOR PATTERN PAINTING

a. The first step in pattern painting an item is to thoroughly clean the surfaces to be painted. Applying paint to anything less than a totally clean surface will result in rapid peeling, cracking, and scaling of the new coating.

NOTE

Check for cleanliness by the red litmus or water break test before applying CARC materials.

b. Clean with detergent or solvent and thoroughly rinse or steam clean to insure a totally clean surface suitable for a durable coating. Grease, oil, dirt, markings, and all loose and scaling paint must be removed. Reference TM 43-0139.

c. Sanding of rough areas to remove the oxidized surface from the old paint will also make the new paint job look better and last longer. If bare metal is exposed following sanding and cleaning, apply an epoxy primer to the bare metal. The primer should be allowed to thoroughly dry before CARC is applied.

d. All glass, grease fittings, and items which can be damaged by paint must be masked with masking tape and/or paper.

NOTE

Grease should not be used for masking because it can easily be smeared onto other parts of the equipment and prevent the paint from sticking.

e. The following is a list of items needed for pattern painting:

(1) Vehicle or other item of equipment.
(2) Pattern-painting design for the specific vehicle or other item of equipment.
(3) Spray guns, if applicable.
(4) Paint.
(5) Thinners as required by TM 43-0139.
(6) Mixing buckets.
(7) Paint brushes, 1/2-inch, 1-inch, and 3-inch.
(8) Paddles or sticks to mix paint.
(9) Painters' masks.
(10) Masking tape for masking small items.
(11)Brown (kraft) paper for masking windshields, etc.
(12) Detergent to clean vehicles.
(13) Chalk for marking paint areas.
(14) Steel wool or sandpaper.
(15) Rags.
(16) Compressed air source, if applicable..
(17) Water separators and fittings with pressure gage.
(18) Wrenches for spray guns, if applicable.

28.OUTLINING THE CAMOUFLAGE PATTERN ON THE VEHICLE

a. The black bands, located at the visual center of each side in a pattern, are the key to the three-color camouflage systems.

These bands must be in the correct place and must be the correct width.

b. To ensure correct placement and width, each pattern has between 10 and 25 critical reference points. Critical reference points are based on fixed features on the item and are reflected on the vehicle patterns in Section VIII.

c. All dimensions are listed in inches

d. All dimensional tolerances are limited to +1.00 inch.

e. The right side view is the primary view of the CPP drawings.

f. In the top view of the paint pattern drawings, the horizontal is established as being from front-to-back along the longest dimension of the item.

g. The edge of the band must be within one inch of a critical-location reference point and the pattern must be no more than one inch wider or narrower than stated at a critical-width reference point The rest of the item will mirror the camouflage pattern as much as possible.

h. Inspections will only be measured at reference points and will evaluate the overall effect at 50 feet. In addition, overspray which can be discerned at 50 feet will be considered sloppy painting and will be corrected.

i. Outlining the camouflage pattern can be done in a variety of ways. Successful methods include freehand, template, wire-form and viewgraph, Figure 32.

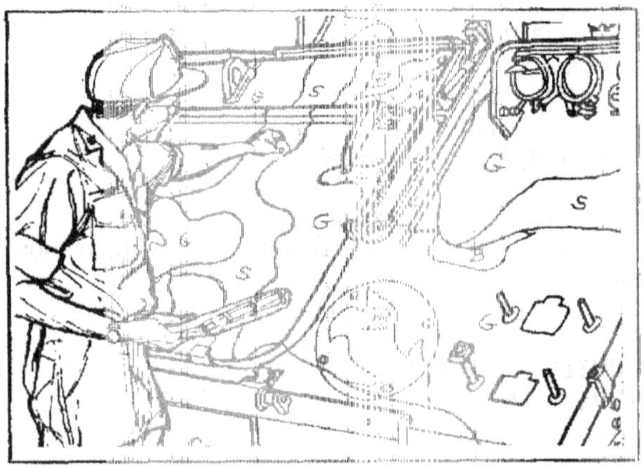

Figure 32. Applying Pattern with Chalk.

29. CAMOUFLAGE PAINTING

WARNING

The local safety office and preventive medicine support activity must be consulted prior to initial CARC painting.

The paints and thinners to be used in pattern painting are to be obtained through normal supply channels. Pattern painting will begin upon receipt of the necessary pattern designs and paints. Use the equipment and procedures described in TM 43-0139 to accomplish the painting.

30. TECHNIQUES OF MIXING AND THINNING

a. Method.

The best, quickest, and easiest method of painting is by spraying. Paint rollers are used on large surfaces when spraying is impractical. Paints are brushed on when other methods are impractical or special equipment is not available.

In general, the use of brushes is confined to touchup jobs.

b. Readiness.

In most cases, paints are issued ready mixed. Hence color blending is not required. CARC paints MIL-C-46168 and MIL-C-22750, however, are issued in a two-component form and require accurate mixing techniques.

c. Preservation.

(1) Stirring. Stir paints well before use. If the liquid portion (vehicle) has separated from the pigment, pour off

mos ... ion into a clean container. Stir the thic ... (pigment) in the bottom until all chur ... and dissolved. Restore the poured off p ... a time, stirring constantly with a lifting and ... 'Box ... ghly by pouring it from one con ... several times, stirring the paint for a few ... n each transfer.

NOTE

... lacquer, as this will ... of the liquid portion by ...

(... n paint stands over a period of time ... n over the surface and the pigment may ... s to the extent that stirring will not mix ... nts properly. In such cases, strain the ... ainer into a clean container, disca ... left in the strainer. Do not strain CAR ... ver. CARC coatings which cannot be pr ... ll be resealed and disposed of as haza ...

(... is necessary to thin paint, use a spra ... prescribed thinner. Because of its vola ... evaporate from the paint film, leavi ... e same ratio of vehicle to pigment per ... rface as the paint would have prov ... ng. The warmer and drier the wea ... nner is needed because heat tends to th ... is required in cold weather to has ... hardening of the film. Thinner sho ... care, as the less used, the more dura ... oat will be. ... thane coatings may be thinned up to 20 p ... e with thinner MIL-T-81772, Type I, or w ... rer's recommended thinner. Epoxy prim ... nded (admixed) four to one by volu ... P-52192 and MIL-P-53022, may be thin ... ent by volume with epoxy thinner MIL ... Epo ... rimer which are blended one to one by v ... IL-C-22750 and MIL-P-23377, usua ... tory spray viscosities, but may be thin ... ounts of MIL-T-81772, Type II, if nece ... sed to thin MIL-P-53030. (... contain a slow-drying vehicle may requ ... nner.

31. TIPS ON PAINTING

Certain basic precautions are applicable to paint, varnish, enamel, and lacquer. The following should be observed at all times:

a. Do not paint over an unclean surface. Be sure all dirt, rust, scale, etc., are removed.

b. Do not fail to stir paint thoroughly.

c. Do not mix one paint with another unless instructed to do so.

d. Do not fail to follow instructions that appear on containers, particularly those concerning safety, the addition of thinner, and the application instructions.

e. Do not apply paint or varnish unless the drying conditions are satisfactory.

f. Do not paint in wet or extremely cold weather, below 50' F (10' C).

g. Do not apply abnormally heavy coats.

h. Do not add too much thinner.

i. Do not use paint buckets, cans, paint rollers, spray guns, or brushes which are not clean.

j. Do not apply cold paints on varnishes.

k. Do not leave old paint and oil-soaked cloths laying around in the paint shop; they are a fire hazard.

l. Do not fail to clean brushes, paint rollers, and spray guns immediately after using.

m. Do not smoke when painting. Do not smoke near paint storage areas or paint booths.

n. Do not release the tops of pressure-feed material containers before releasing the air pressure.

o. Do not use electrical connections that show any inclination to becoming loose.

p. Do not pour paint out of a container in a manner that obscures the label.

q. Do not fail to strain paint before using, if required.
CARC paints cannot be strained, however.

r. Do not fail to remove all traces of wax from surfaces where paint or varnish is to be used.

s. Do not paint without proper respiratory equipment and ventilation.

t. Do not waste paint by spraying beyond the item being coated.

u. Do not paint over a moist or wet surface.

v. Do not paint between the ground strap and hull of tanks.

w. Do not paint on operator-instruction plates.

32. TIPS ON PAINTING WITH CARC

The following precautions should be observed, in addition to those listed in paragraph 31 above, when applying Chemical Agent Resistant Coatings (CARC):

a. Spray 'nes for epoxy applications should not be used with polyurethane coatings without complete flushing or cleaning wit] solvents.

b. Test for cleanliness when applying CARC with the red litmus or war-r break test.

c. Remem3er to notify the local safety office and preventive medicine support activity prior to initial CARC painting. This also applies to all spray painting operations, regardless of the material used.

d. Do not use CARC for items like manifolds and mufflers that exceed 400' F. Do not use CARC on rubber, lacquer coatings, or vinyl.

e. Use silicon rubber (impervious), not cloth, gloves when applying CARC.

f. Only DS/GS General Support and Depot level personnel are authorized complete painting and repainting of components with topcoats and primers. Unit and DS/GS Direct Support level personnel are only pemitted to use topcoats and primers for touchup efforts.

g. When using CARC, mix only the amount needed to do the job (i.e. don't open a large container for a small job) because unused CARC must be disposed of and cannot be stored.

h. Epoxy-polyamide coatings build up thickness quickly.
Thick films are detrimental for good adhesion. Do not apply CARC beyond its thickness tolerances.

33. MARKINGS TO BE APPLIED TO EQUIPMENT

After the equipment has been pattern painted, only the following markings are to be applied:

a. **Unit identification.**

b. National Symbol.
All stars will be removed and replaced by a 3-inch star in contrasting lusterless black 37030 or green 383, front and rear.
Placement should be centered on the equipment, on line with
unit ID markings. On the rear of wheeled vehicles, the star may
be placed on the tailgate.

c.Agency Identification and Registration Number.
All U.S. Army identification and registration numbers will be removed from the item's exterior surfaces. The identification and registration number shall be placed on any appropriate interior area which is visible from outside a locked or secured item.
Combat Vehicles and Trailer Tankers: "registration number" will be applied to both sides of the vehicle. The numbers will be painted to contrast with color patch painted over, Figure 32A.

NOTE
The identification legend "US ARMY" is no longer required.

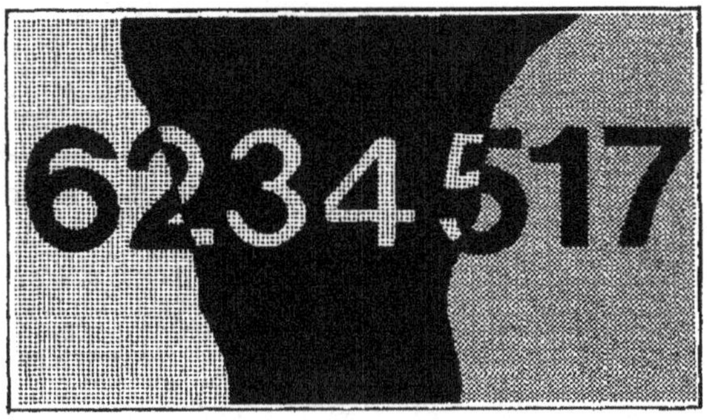

Figure 32A. Markings on Camouflaged Combat Equipment

d. Safety and Instructional Markings.
Markings such as tire pressure, fuel type, and fill level may be retained in contrasting lusterless black 37030 or green 383 (34094), letters no larger than one inch. Markings directly related to troop safety, such as wrecker boom capacity and danger zones, must be evaluated by safety personnel.

e. CARC and Alkyd Markings.
Equipment with data plates shall have the word "CARC," as applicable, painted in black or green contrasting color in a conspicuous area as near the data plate as possible. Painting shall be in block letters, as large as possible, not to exceed one inch.

All major items having a log book shall have the Equipment Control Record, DA Form 2408-9, annotated in the lower left corner of Block 21, Remarks, to reflect the word "CARC", as applicable, and the date applied.

SECTION VII.
VEHICLE CAMOUFLAGE PATTERNS

INSPECTION NOTES FOR ALL CAMOUFLAGE PATTERNS.

- **ALL DIMENSIONS ARE IN INCHES.**
- **O DENOTES A LOCATION ON THE ITEM THAT IS USED AS A REFERENCE POINT. THE DISTANCE FROM THIS REFERENCE POINT TO THE EDGE OF THE BLACK BAND IS THE LOCATING DIMENSION. TOLERANCE ON THIS DIMENSION IS ± 1.00 INCH.**
- **THE HORIZONTAL DISTANCE ACROSS A BLACK BAND IS THE BAND WIDTH. IF THIS DIMENSION IS SHOWN THE TOLERANCE IS + 1.00 INCH.**
- **Δ LOCATES A PHYSICAL FEATURE THAT THE PATTERN CONTOUR INTERSECTS WITHIN + 1.00 INCH. NO LOCATING DIMENSION IS REQUIRED.**
- **INSPECTION OF THE CAMOUFLAGE PATTERN WILL CONSIST OF CHECKING ALL LOCATING DIMENSIONS AND BAND WIDTHS SHOWN ON INSPECTION SHEET AS A MINIMUM. ALTERNATE OR ADDITIONAL REFERENCE POINTS, LOCATING DIMENSIONS AND BAND WIDTHS MAY BE INCLUDED IN THE INSPECTION. DIMENSIONS WILL BE DEVELOPED BY SCALING FROM PATTERN DRAWINGS. DO NOT USE INSPECTION SHEET TO DEVELOP DIMENSIONS. THE TOLERANCE FOR ALL DEVELOPED DIMENSIONS WILL BE + 1.00 INCH.**

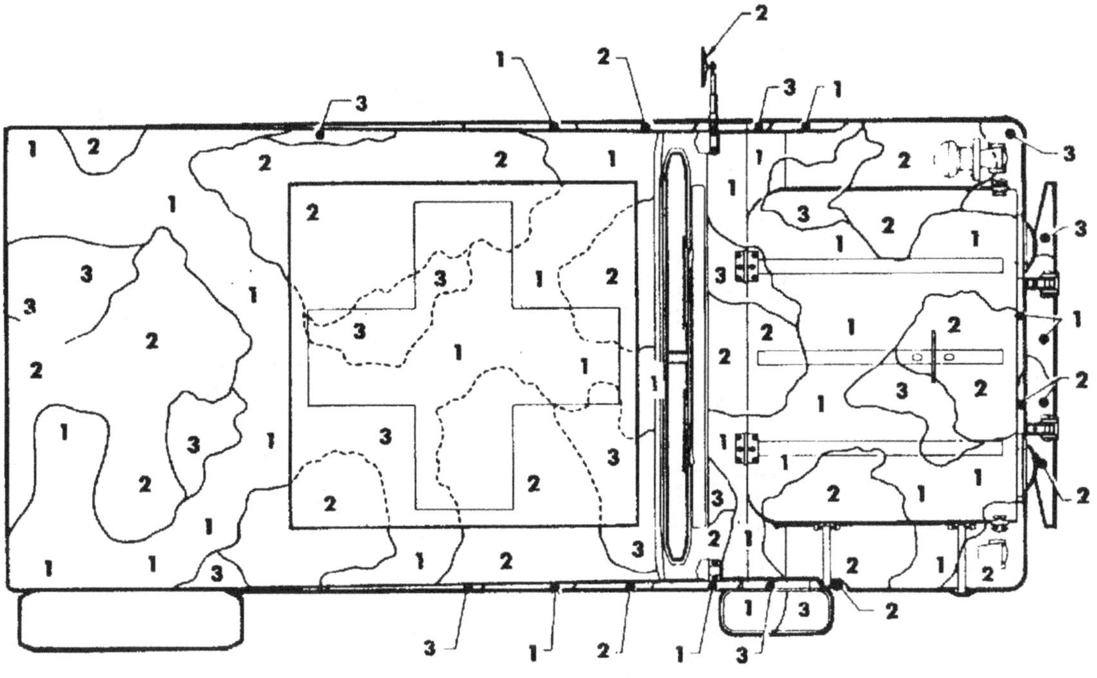

TOP VIEW

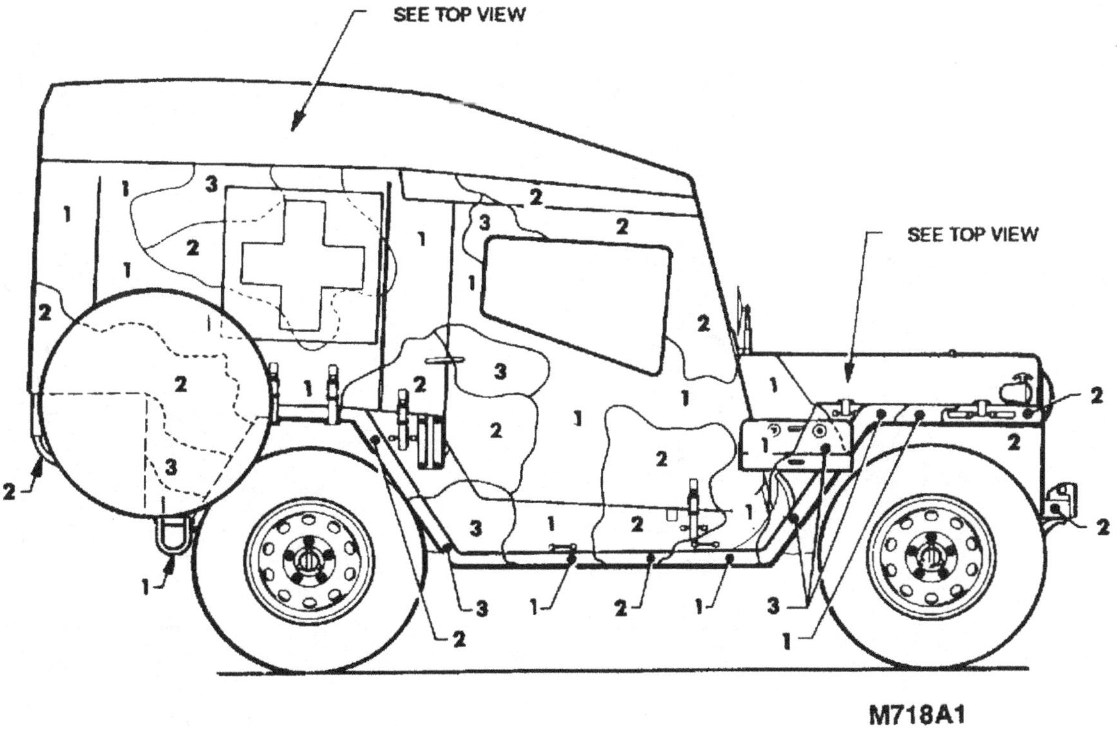

RIGHT SIDE VIEW

Figure 33. Truck, ambulance: 1l4-ton M718A1. (1 of 2)

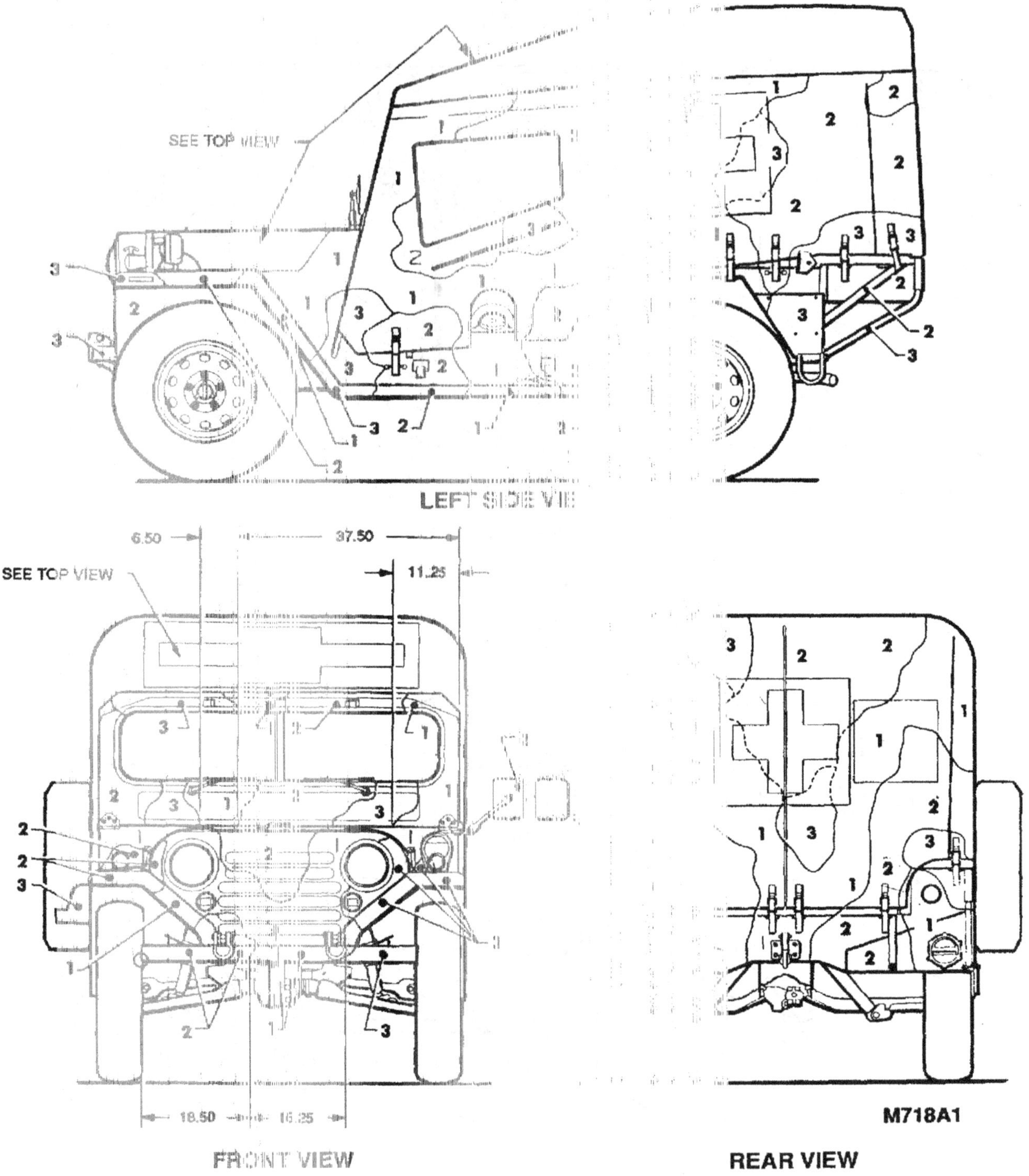

Figure 33. Truck, ambulance: 1/4-ton ... 2)

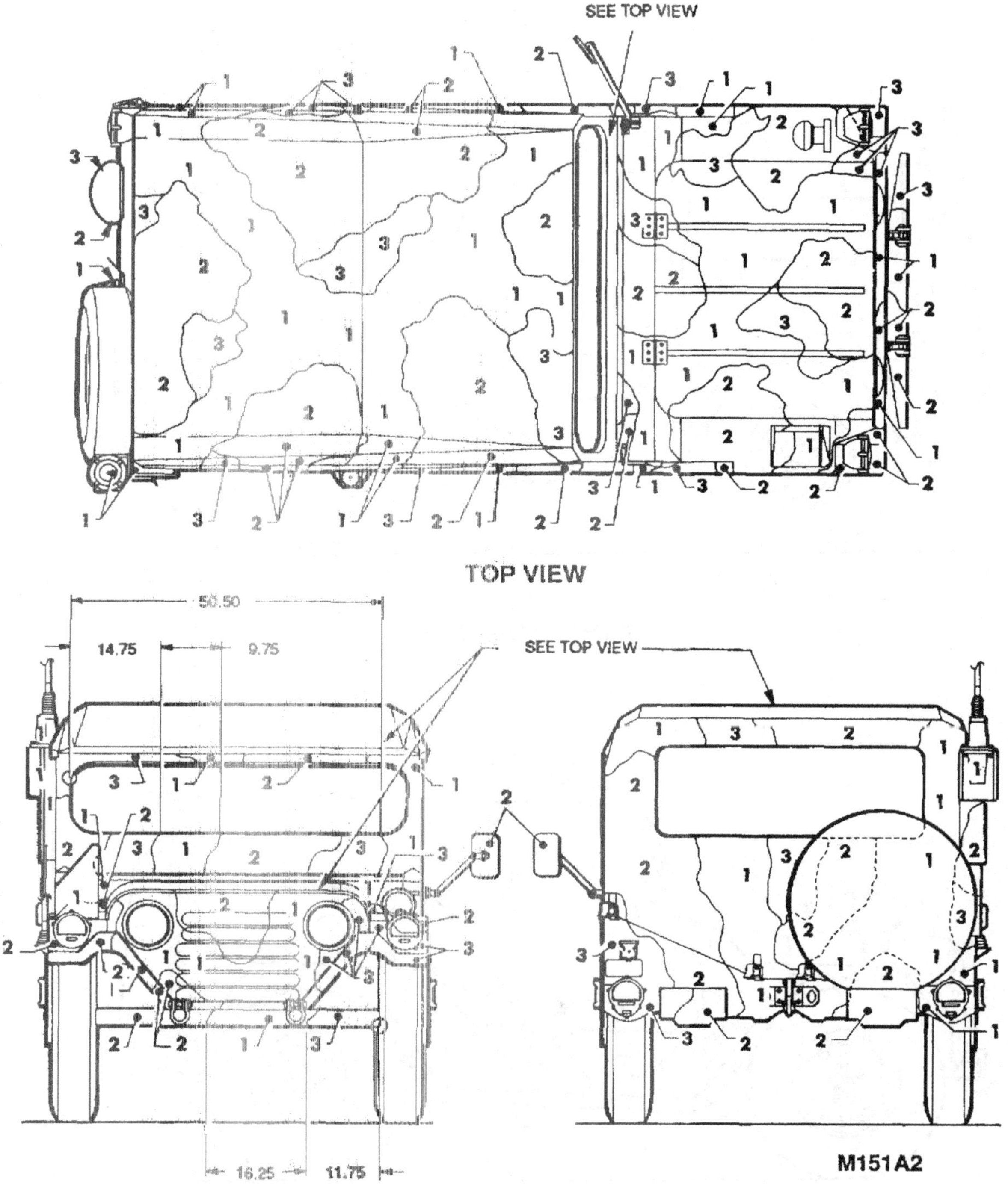

Figure 34. Truck, utility: 1-1/4-ton M151A2. (1 of 2)

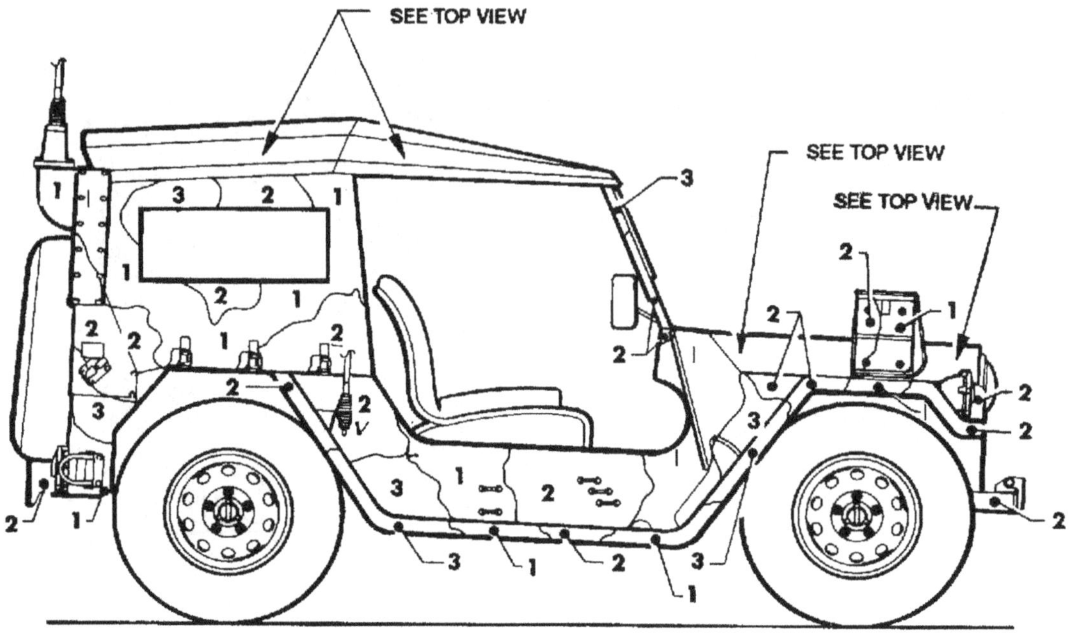

RIGHT SIDE VIEW

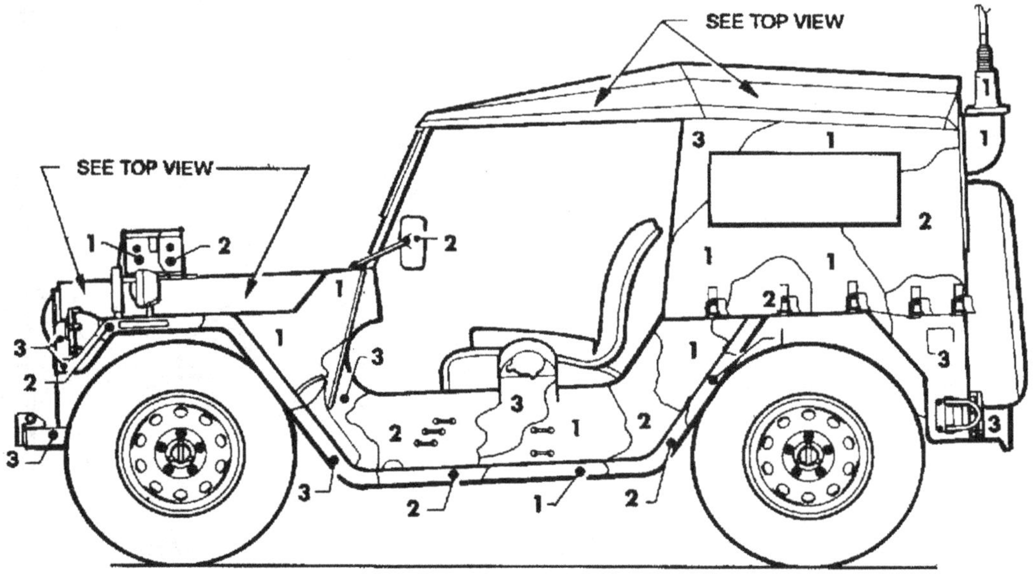

M151A2

LEFT SIDE VIEW

Figure 34. Truck, utility: 1-1/4-ton M151A2. (2 of 2)

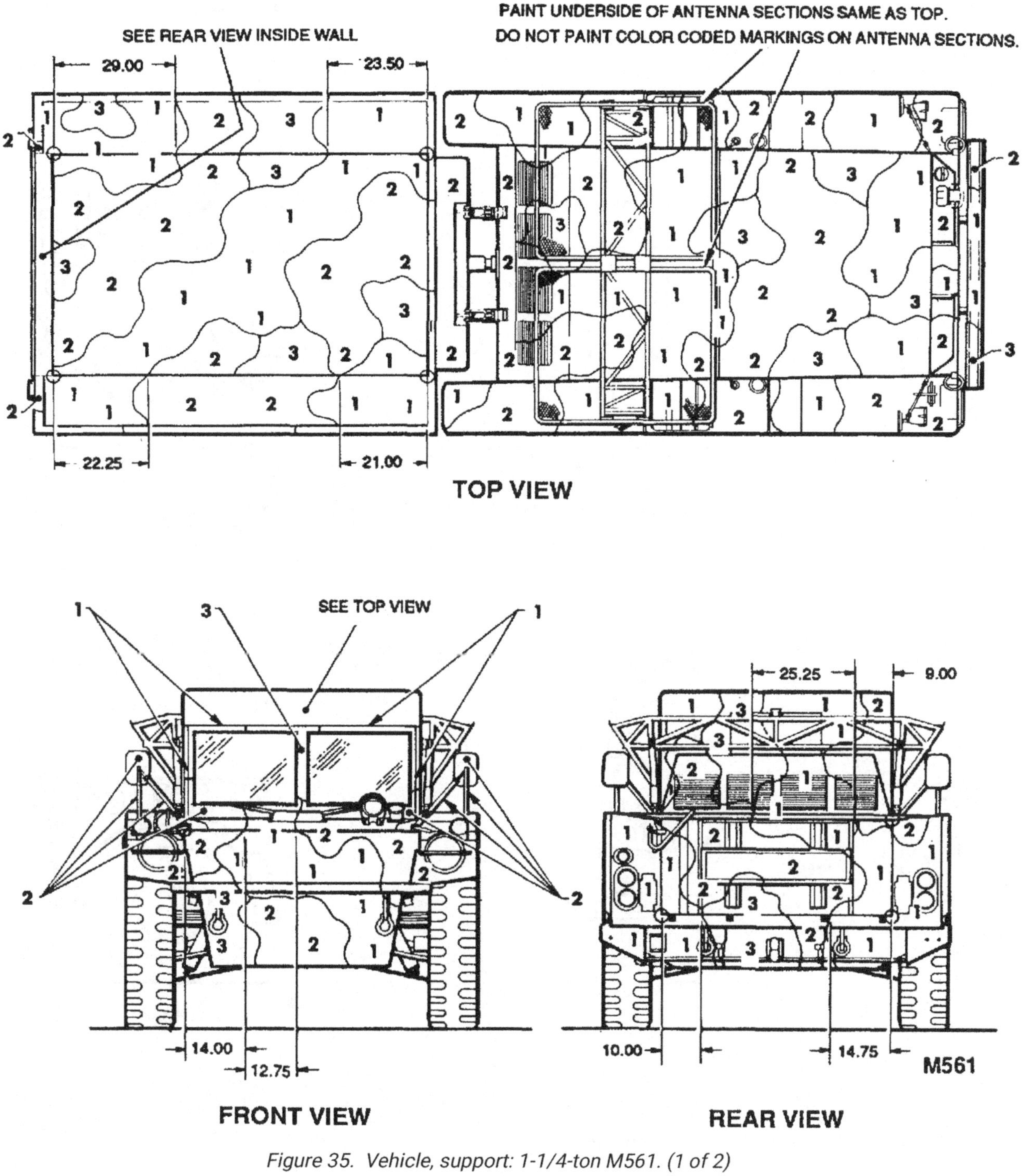

Figure 35. Vehicle, support: 1-1/4-ton M561. (1 of 2)

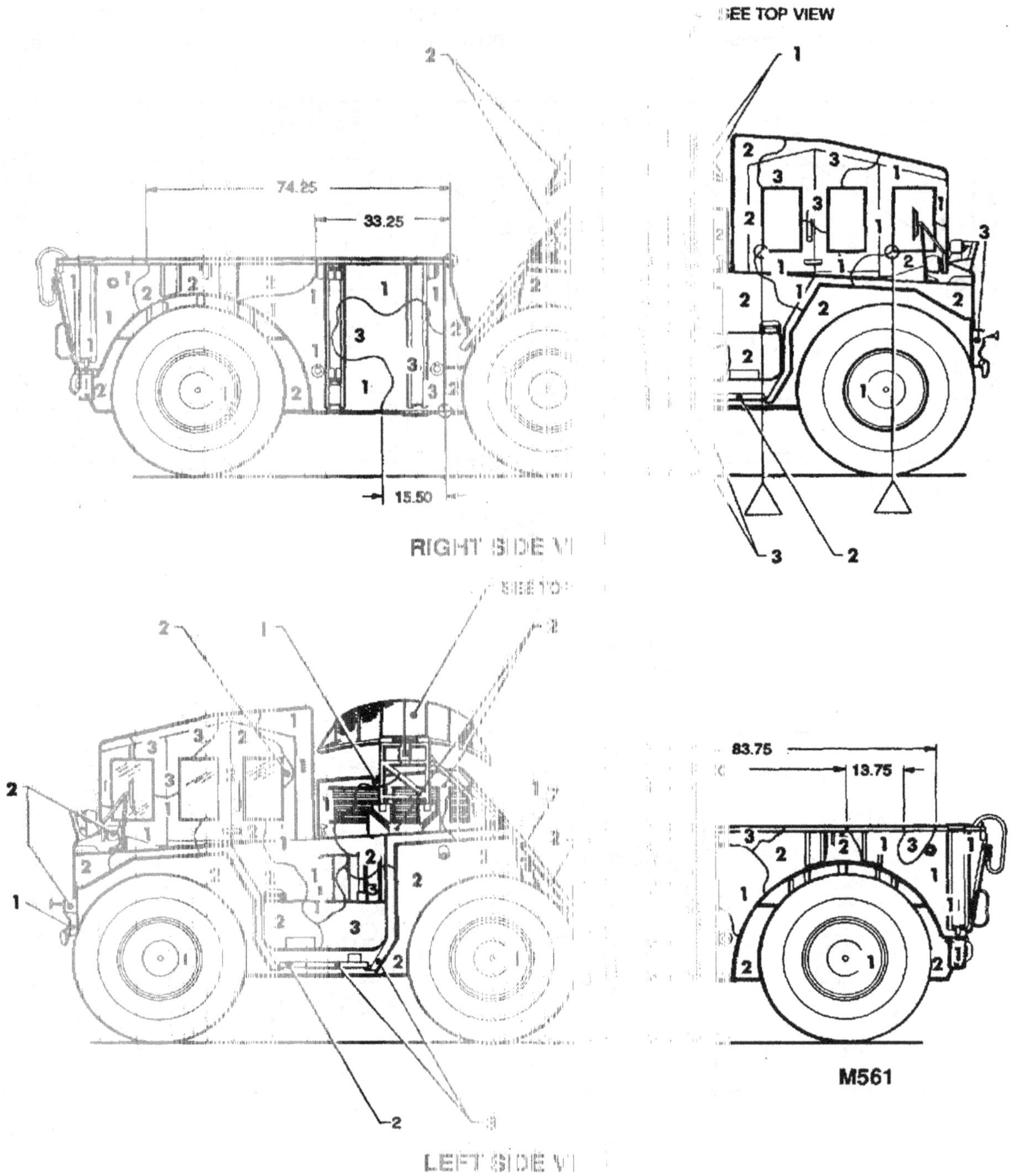

Figure 35. Vehicle, support: 1-1/4-t

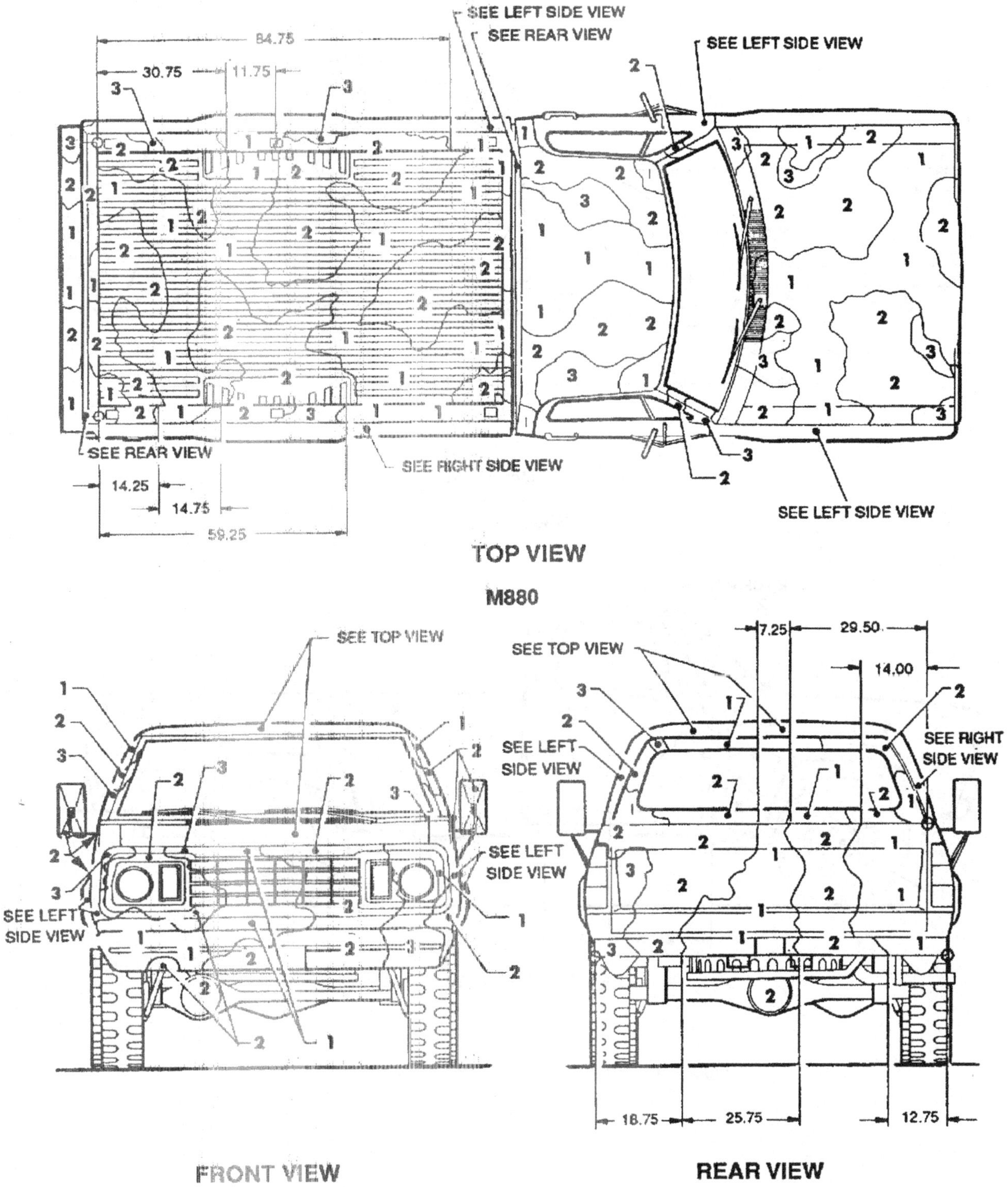

Figure 36. Truck, cargo: 1-114-ton M880. (1 of 2)

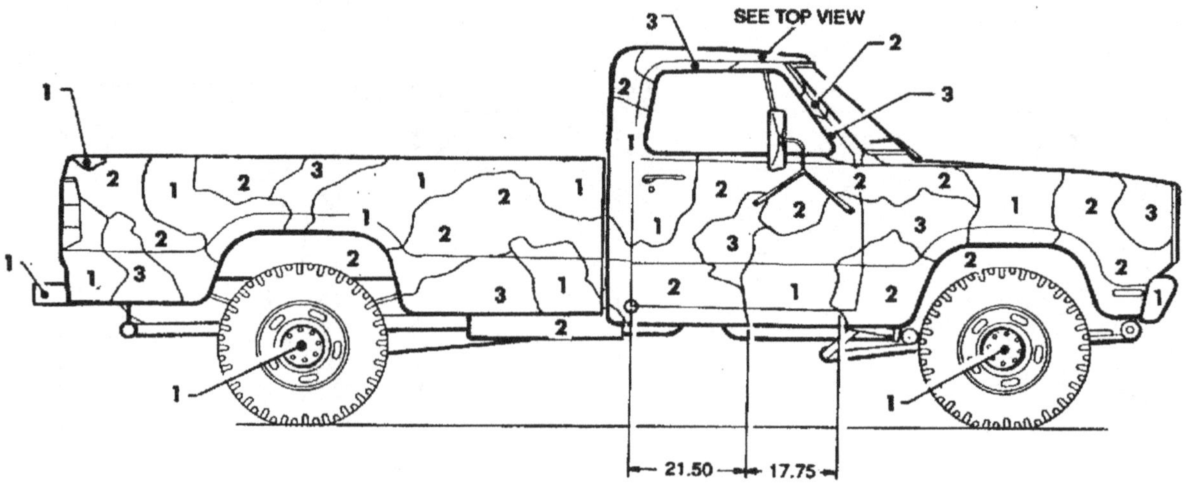

RIGHT SIDE VIEW

M880

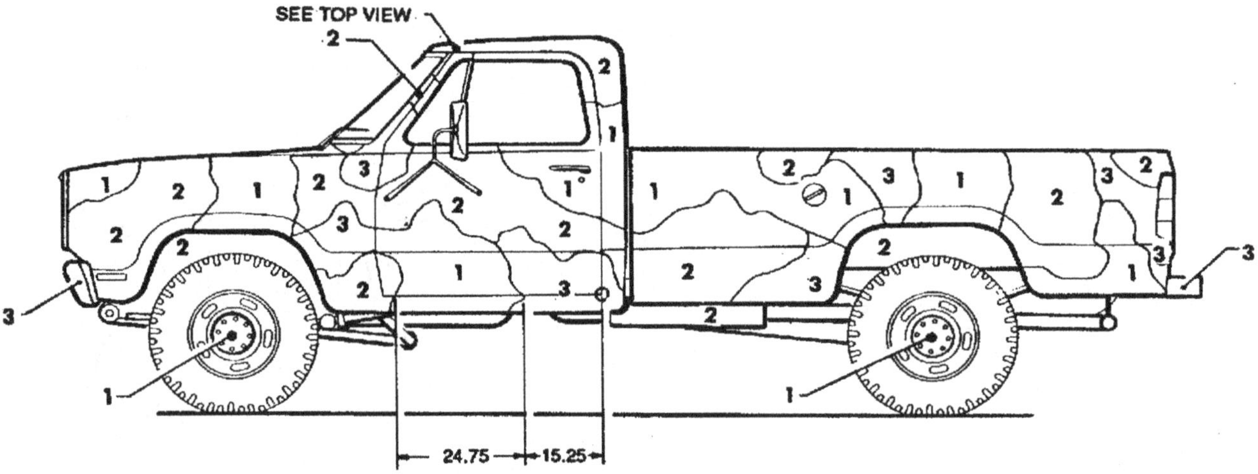

LEFT SIDE VIEW

Figure 36. Truck, cargo: 1-1/4-ton M880. (2 of 2)

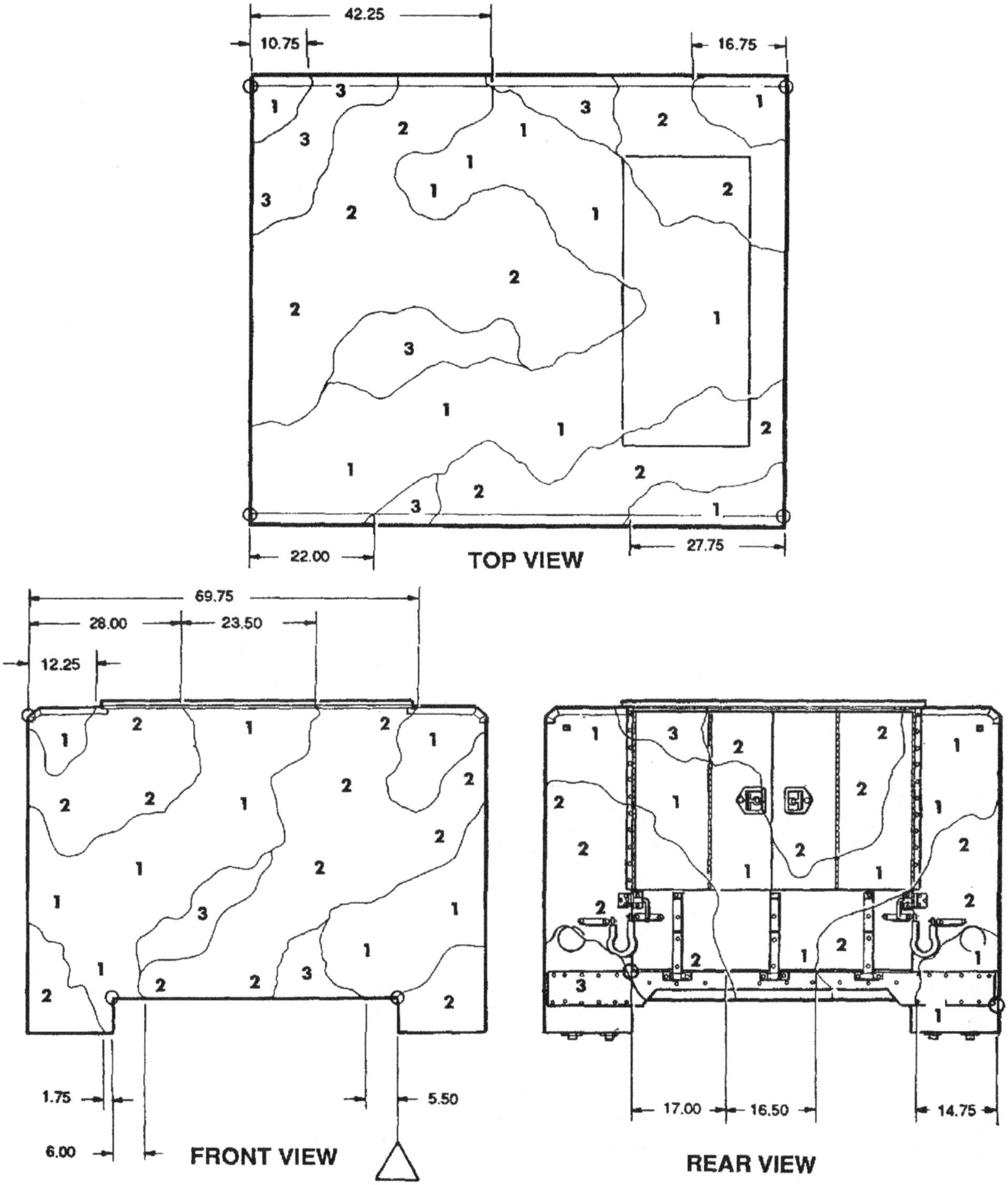

Figure 37. Shop equipment, contact maintenance truck mounted. (1 of 2)

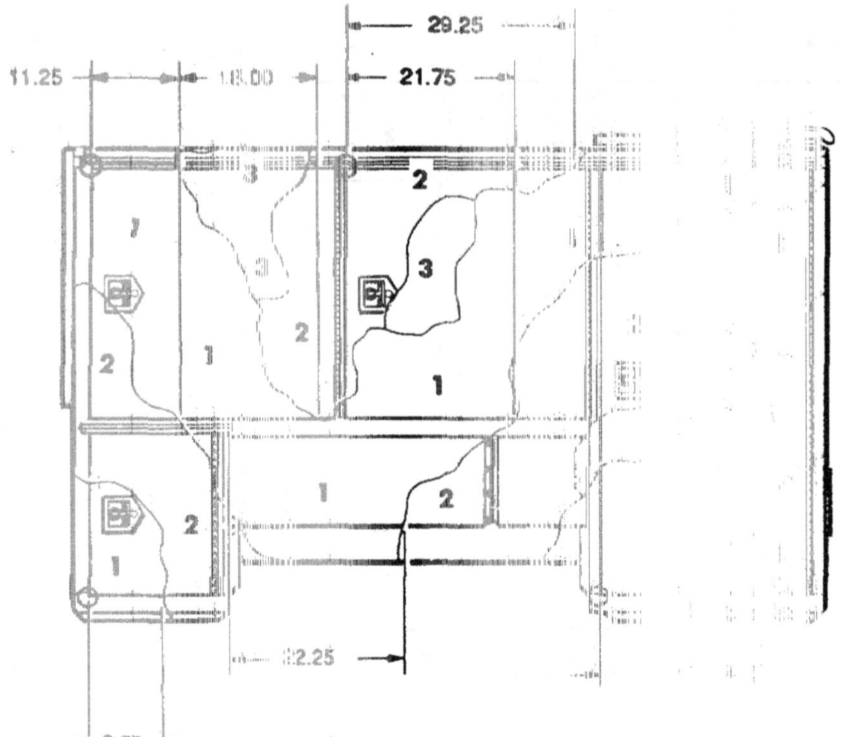

RIGHT SIDE VIEW

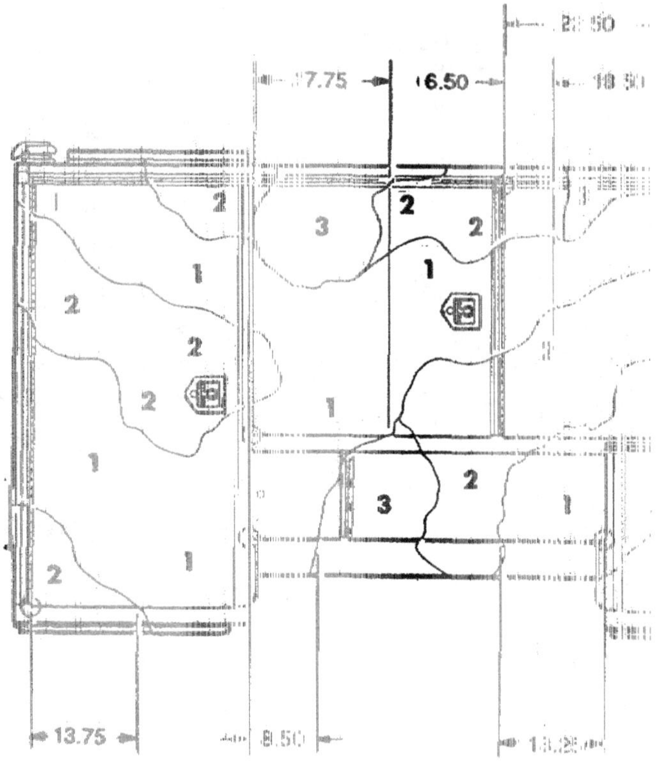

LEFT SIDE VIEW

Figure 37. Shop equipment, contact maintenance [illegible] 2 of 2)

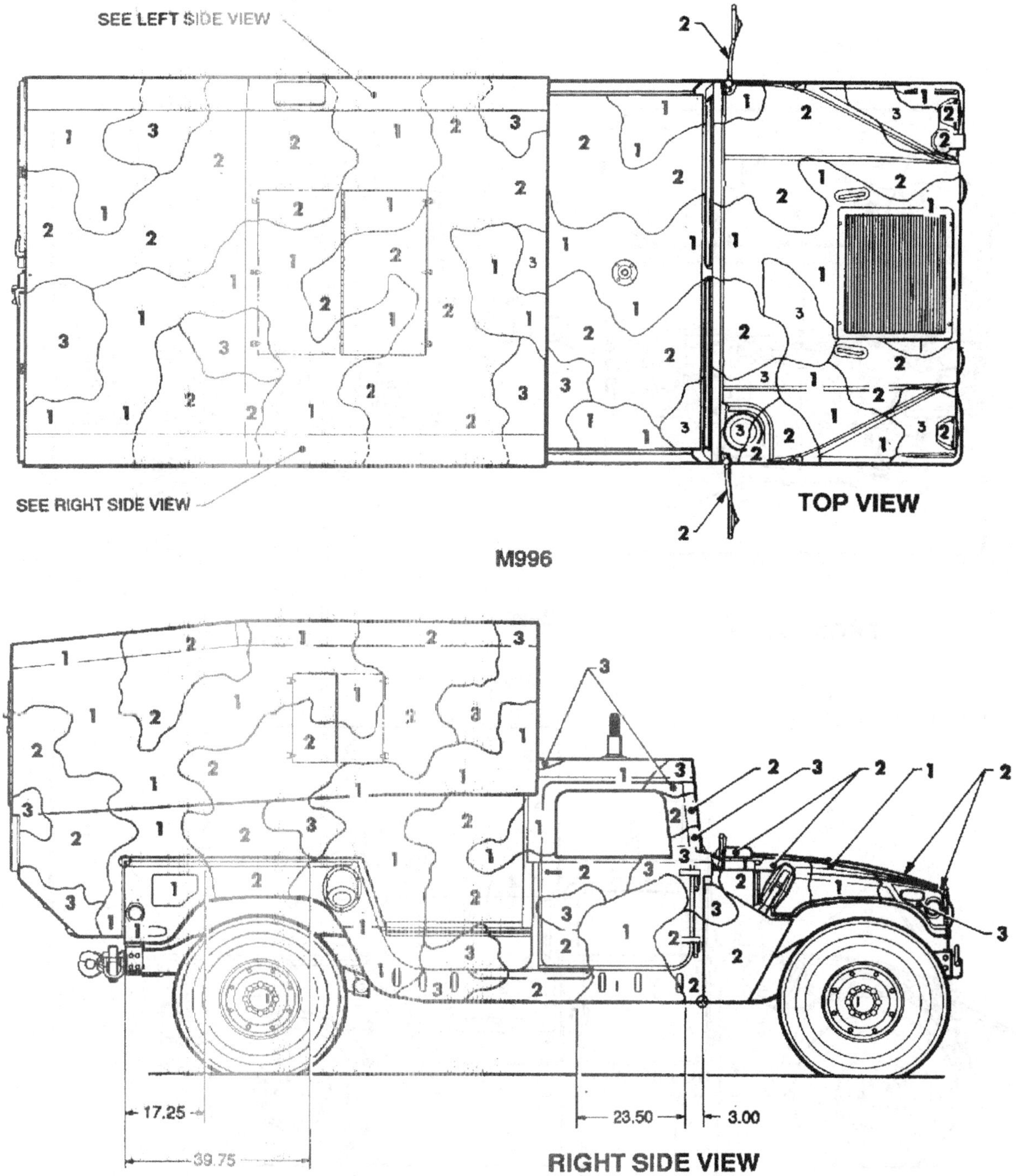

Figure 38. Ambulance, 2-litter, hard top, 4 x 4, M996. (1 of 2)

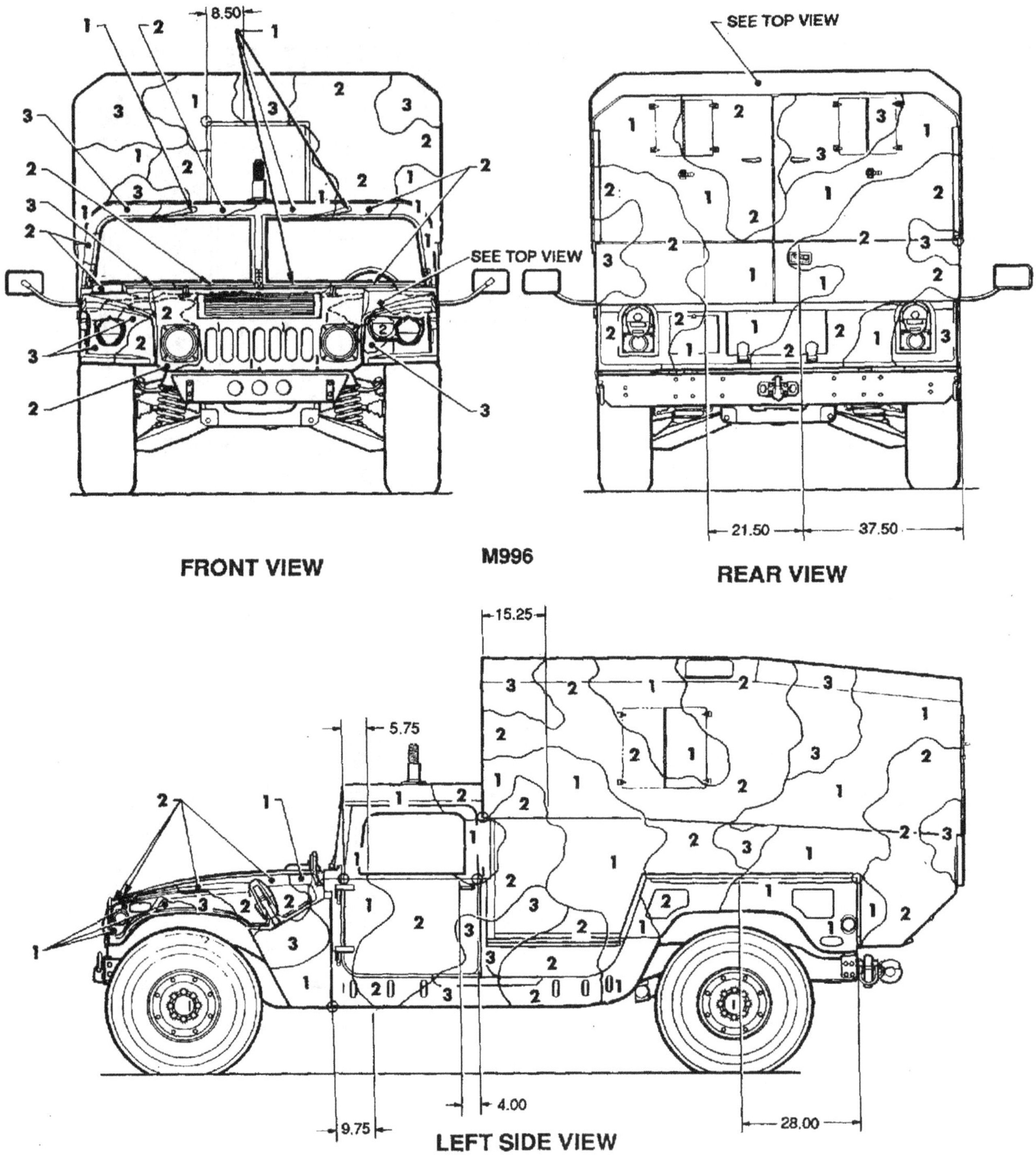

Figure 38. Ambulance, 2-liter, hard top, 4 x 4, M996. (2 of2)

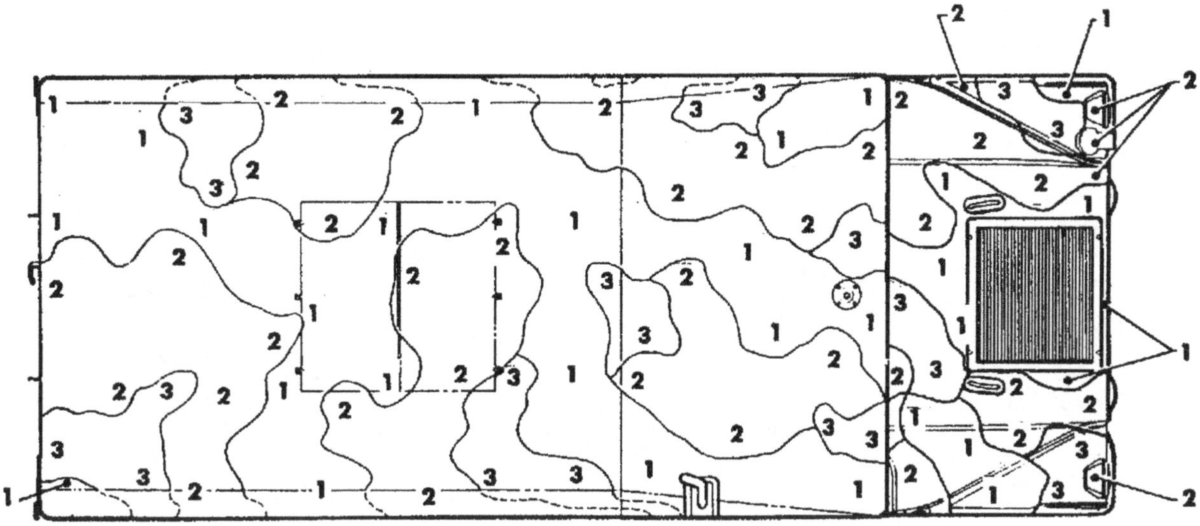

TOP VIEW

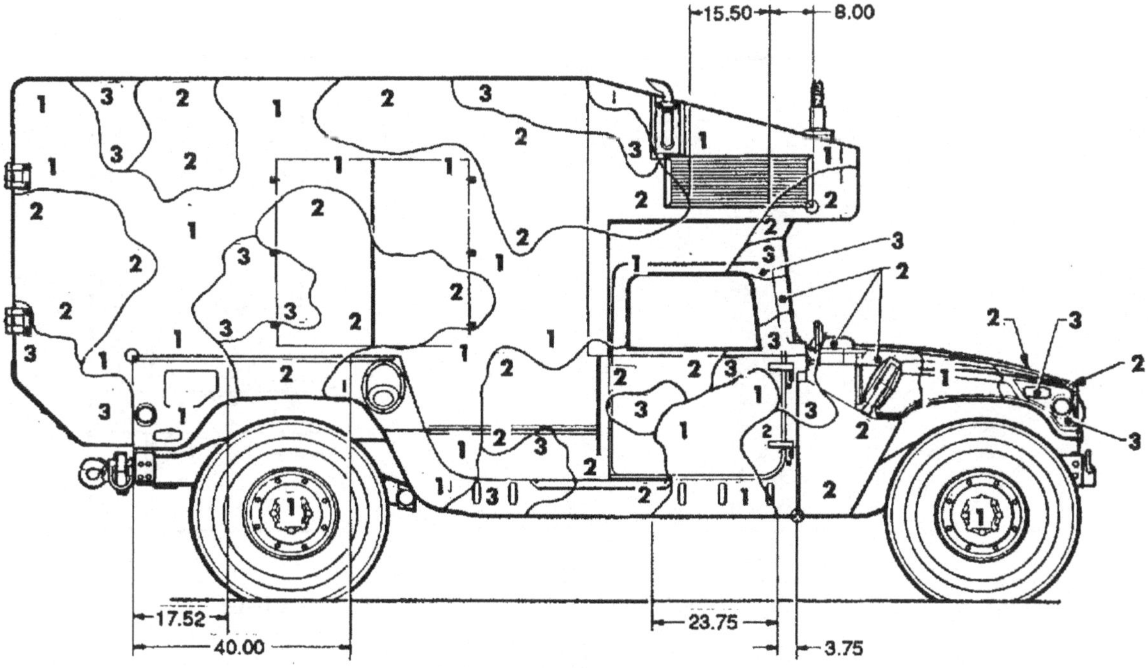

RIGHT SIDE VIEW

M997

Figure 39. Ambulance, 4-litter, hard top, 4 x 4, M997. (1 of 2)

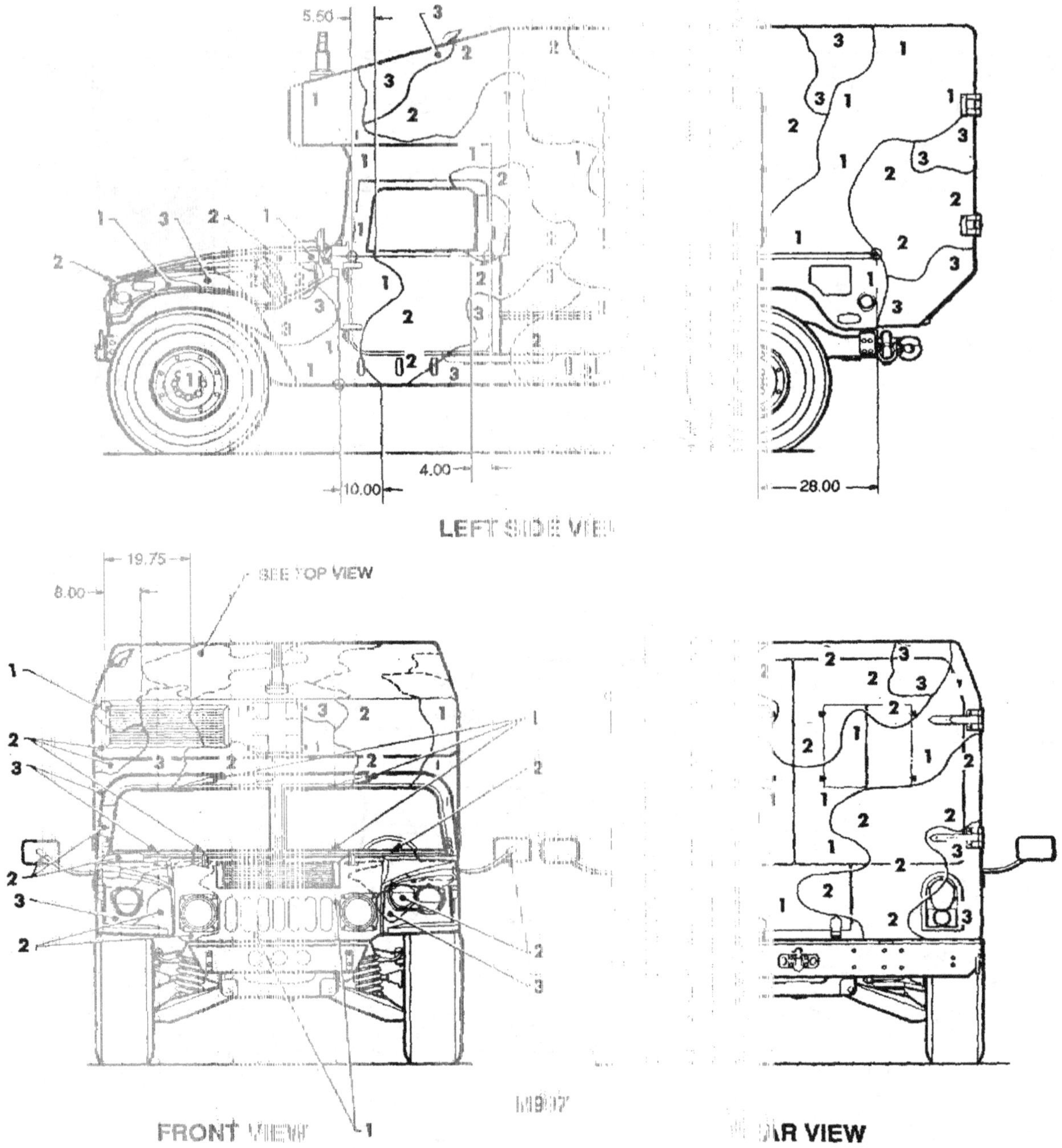

Figure 39. Ambulance, 4-litter, hard top, [illegible] of 2)

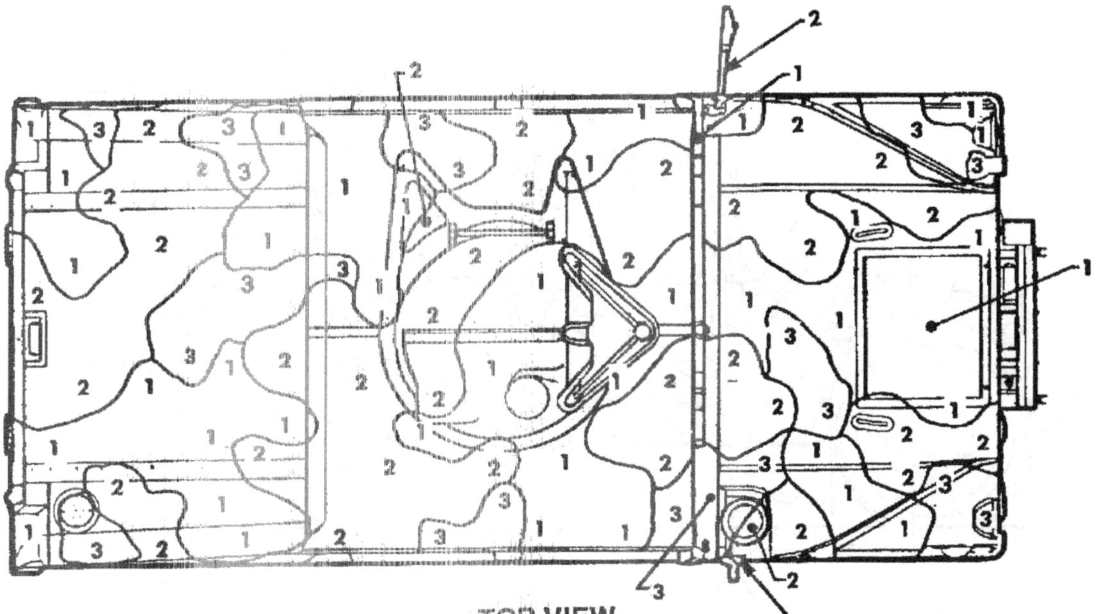

TOP VIEW

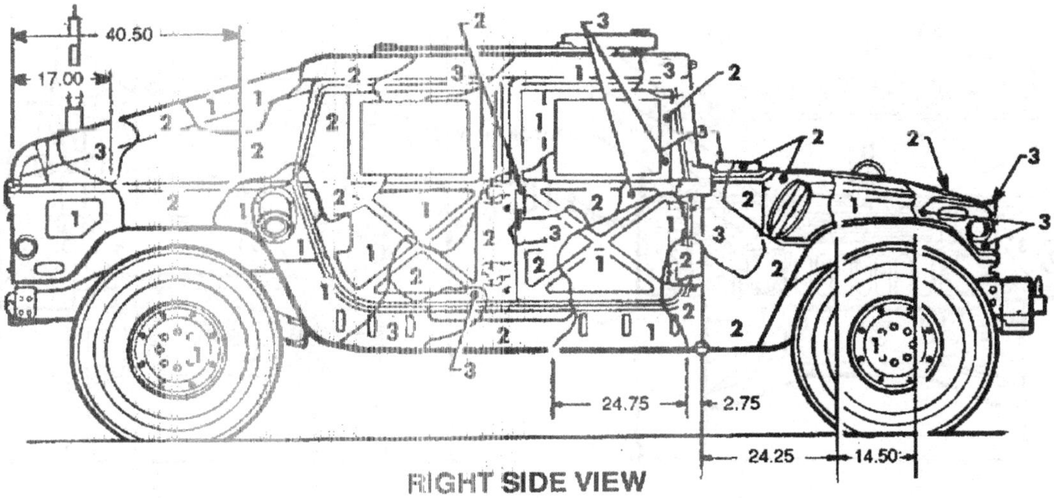

RIGHT SIDE VIEW

M1025 **M1026** **M1044** **M1045** **M1046**

Figure 40. Truck, utility: armament carrier, armored 1-1/4-ton, 4 x 4, M996, M1025, M1026 wlwinch, M1044, M1045, and M1046. (1 of 2)

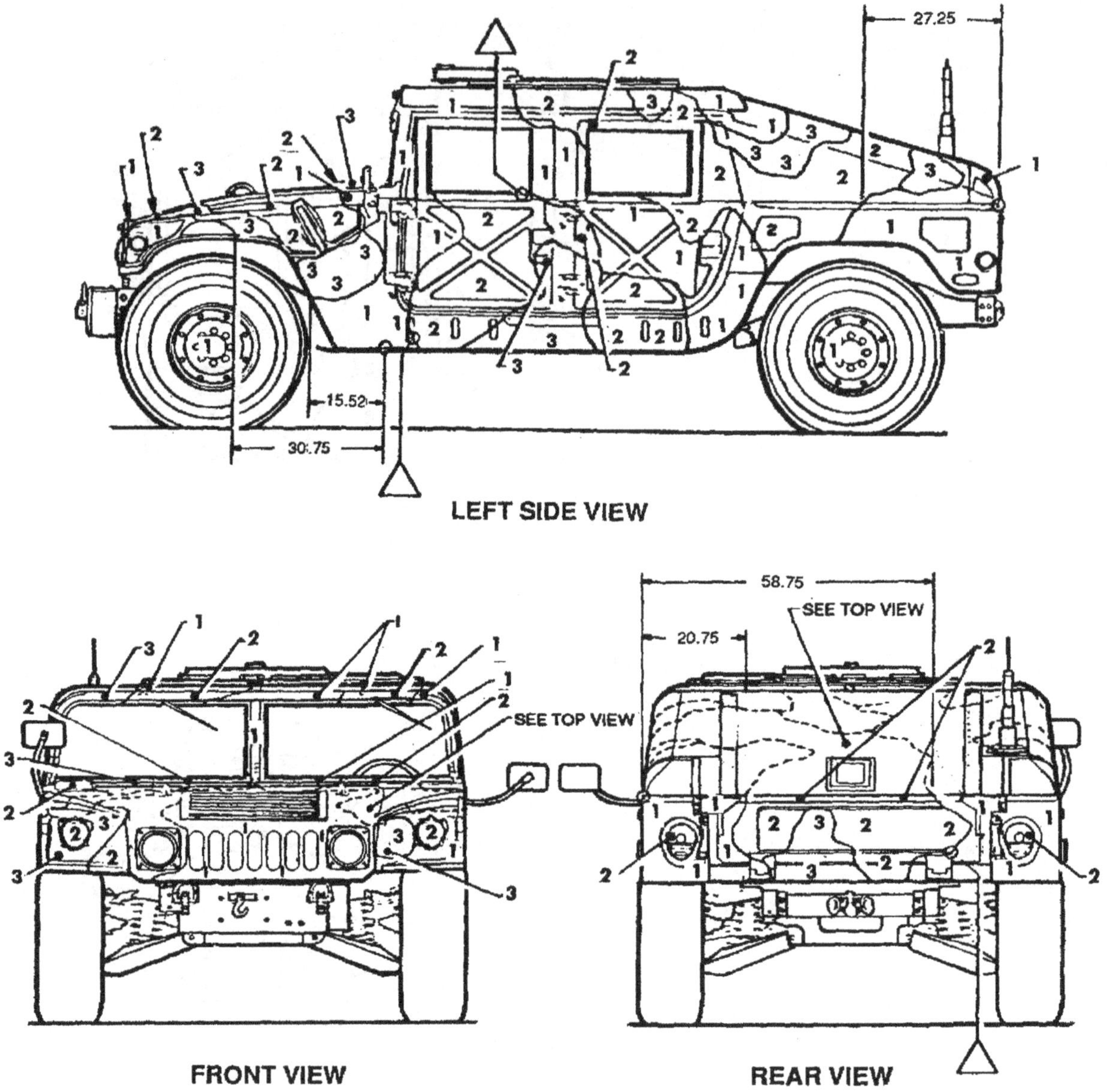

M1025 **M1026** **M1044** **M1045** **M1046**

Figure 40. Truck, utility: armament carrier, armored 1-1/4-ton, 4 x 4, M996, M1025, M1026 w/winch, M1044, M1045, and M1046. (2 of 2)

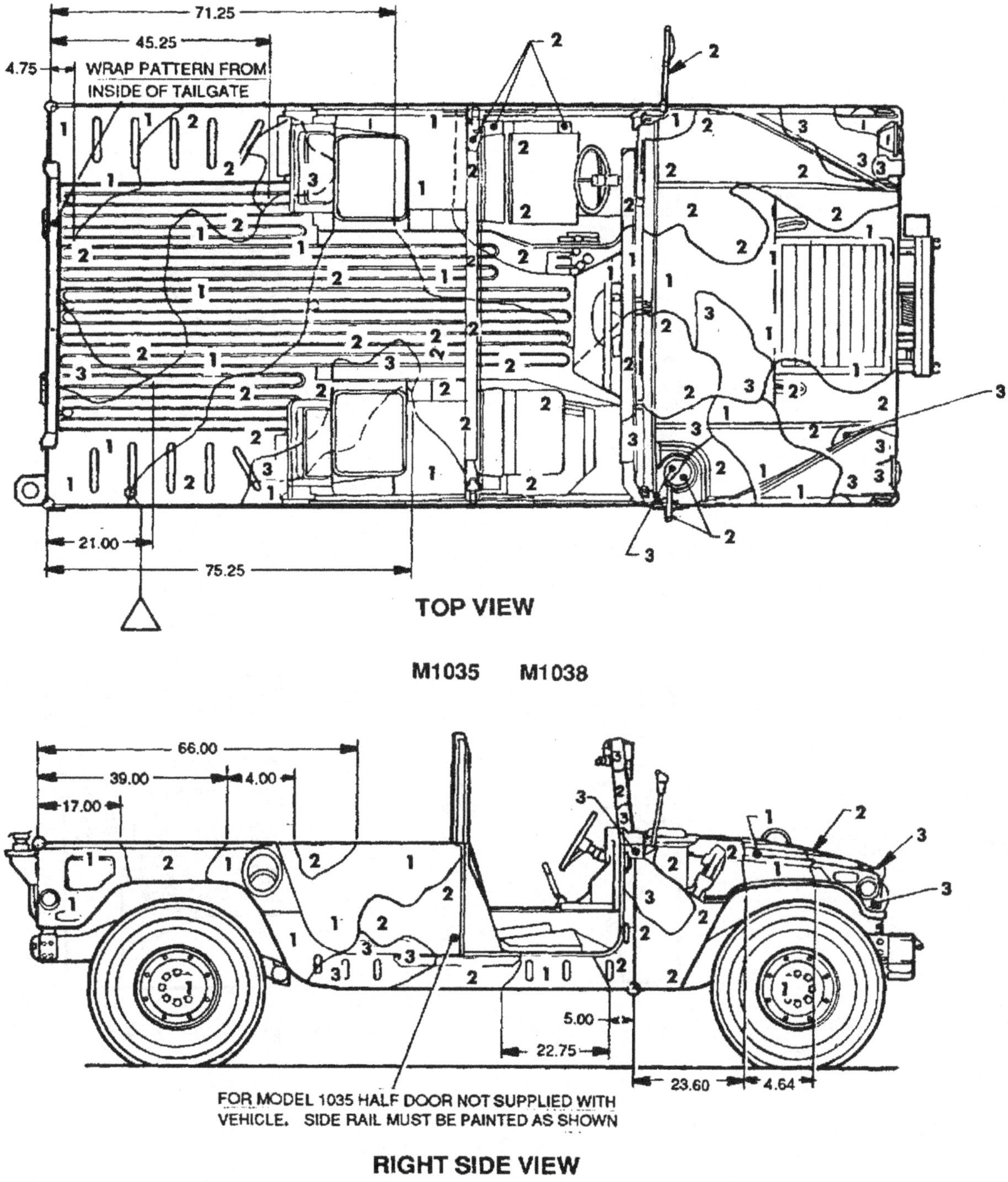

Figure 41. Truck, utility: cargo/troop carrier, 1-1/4-ton 4 x 4, M1035, M1038. (1 of 2)

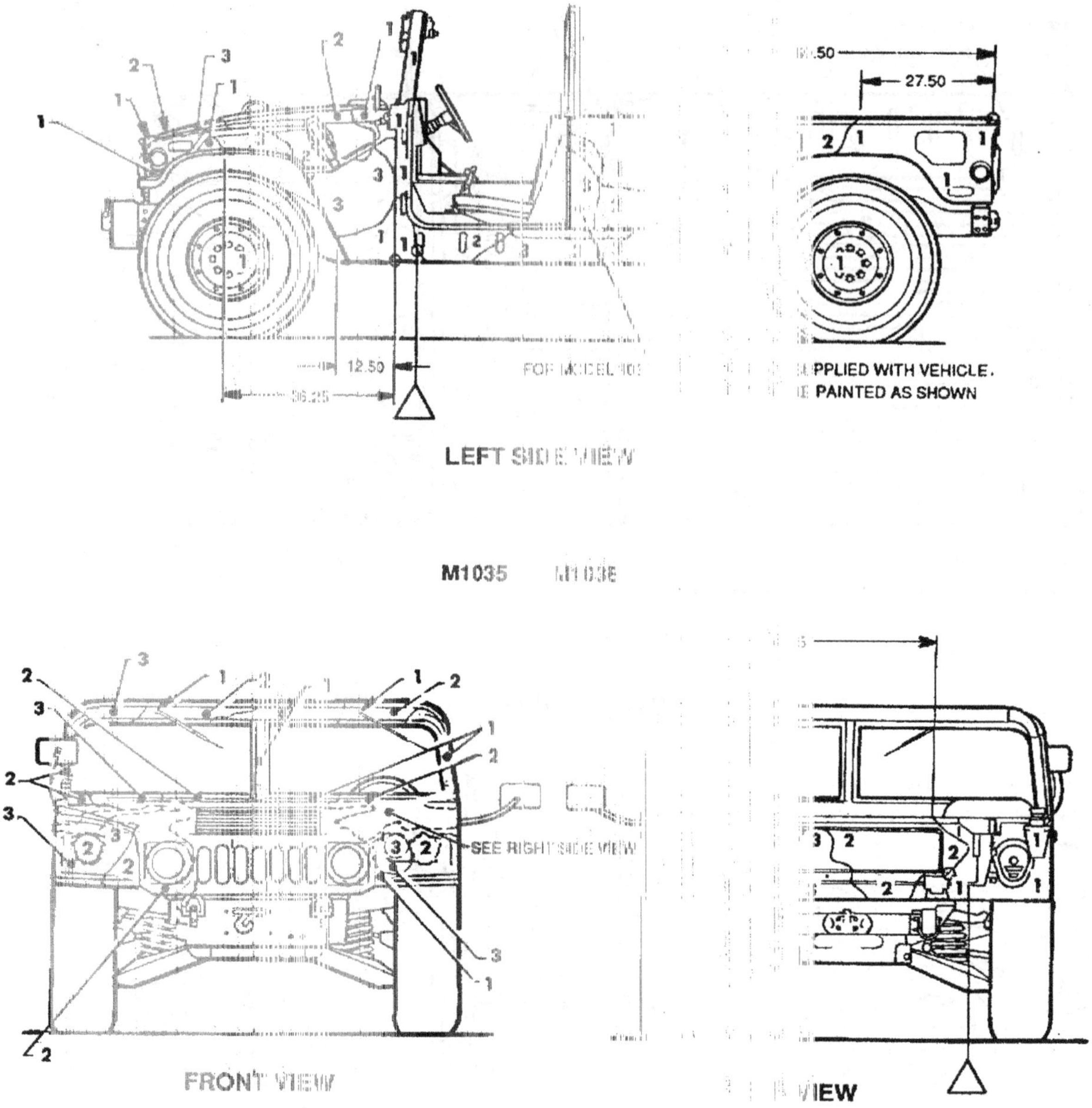

Figure 41. Truck, utility: cargo/troop carrier, 1 1/4-t... M1038. (2 of 2)

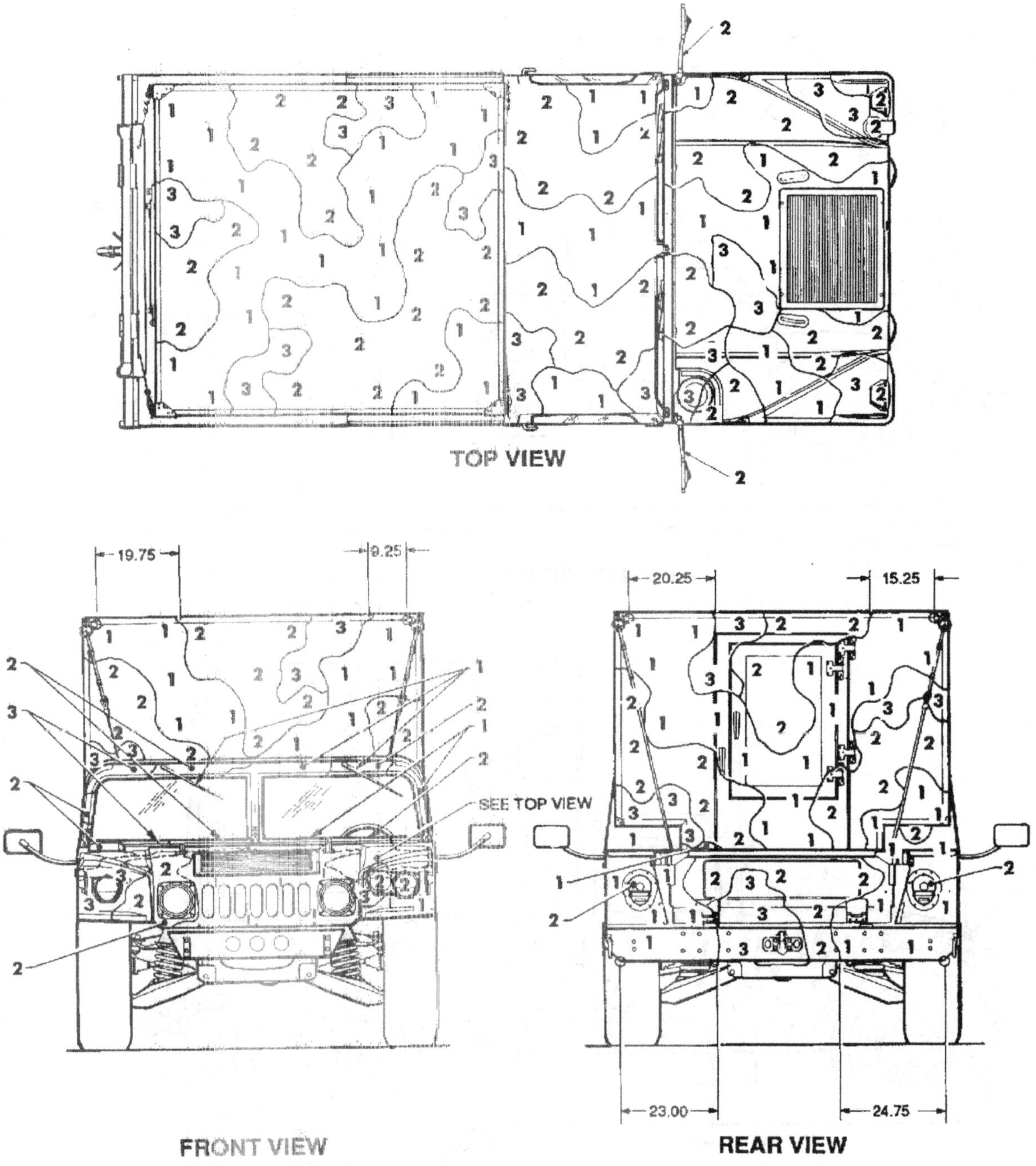

M1037 M1042

Figure 42. Truck, utility: 5250 shelter carrier, 4 x 4, M1037, M1042. (1 of 2)

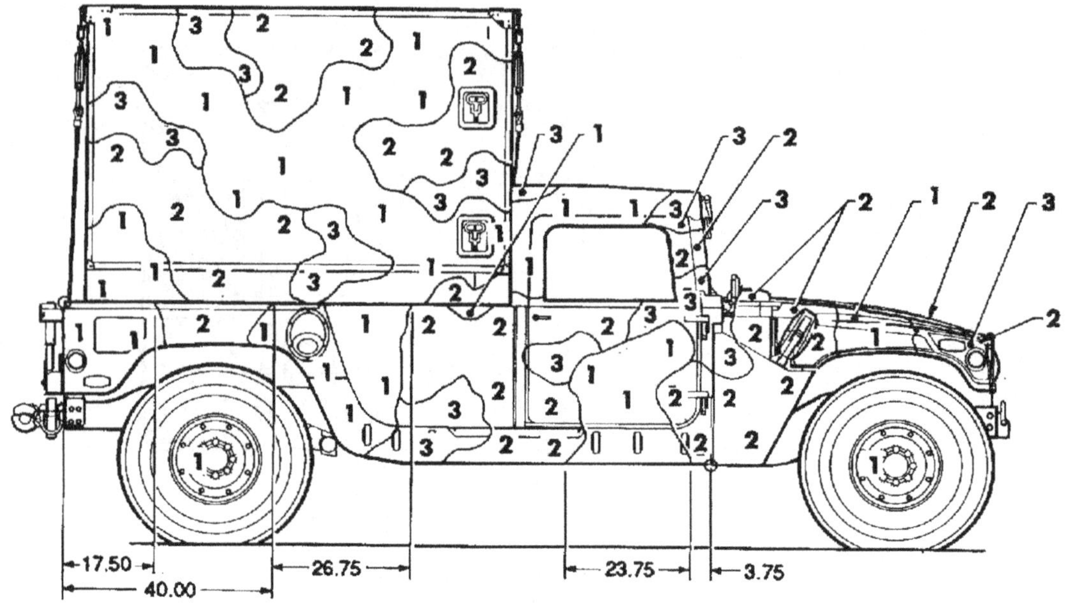

RIGHT SIDE VIEW

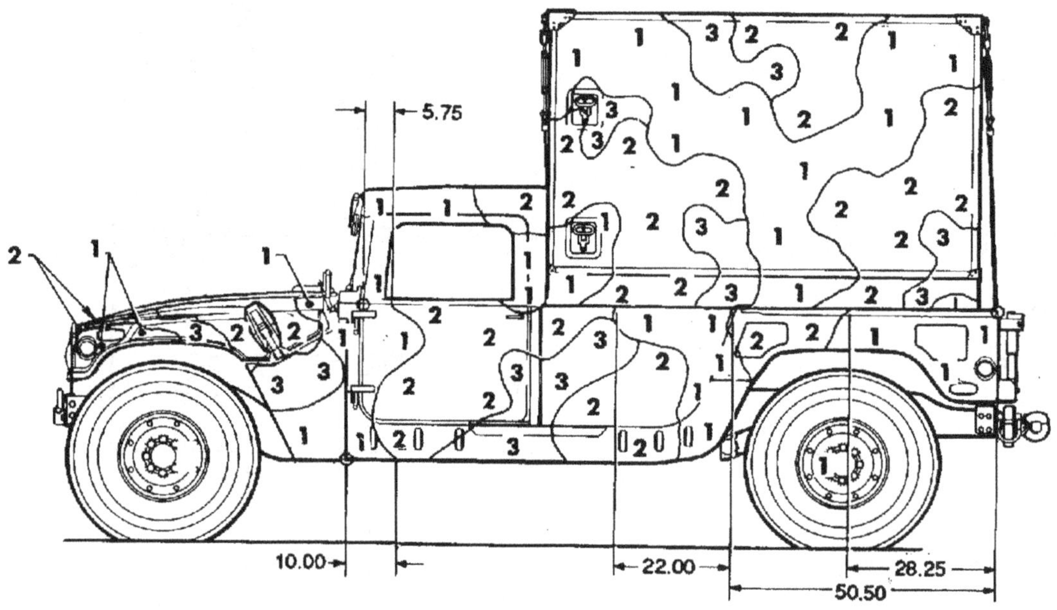

LEFT SIDE VIEW

M1037 M1042

Figure 42. Truck, utility: 5250 shelter carrier, 4 x 4, M1037, M1042. (2 of 2)

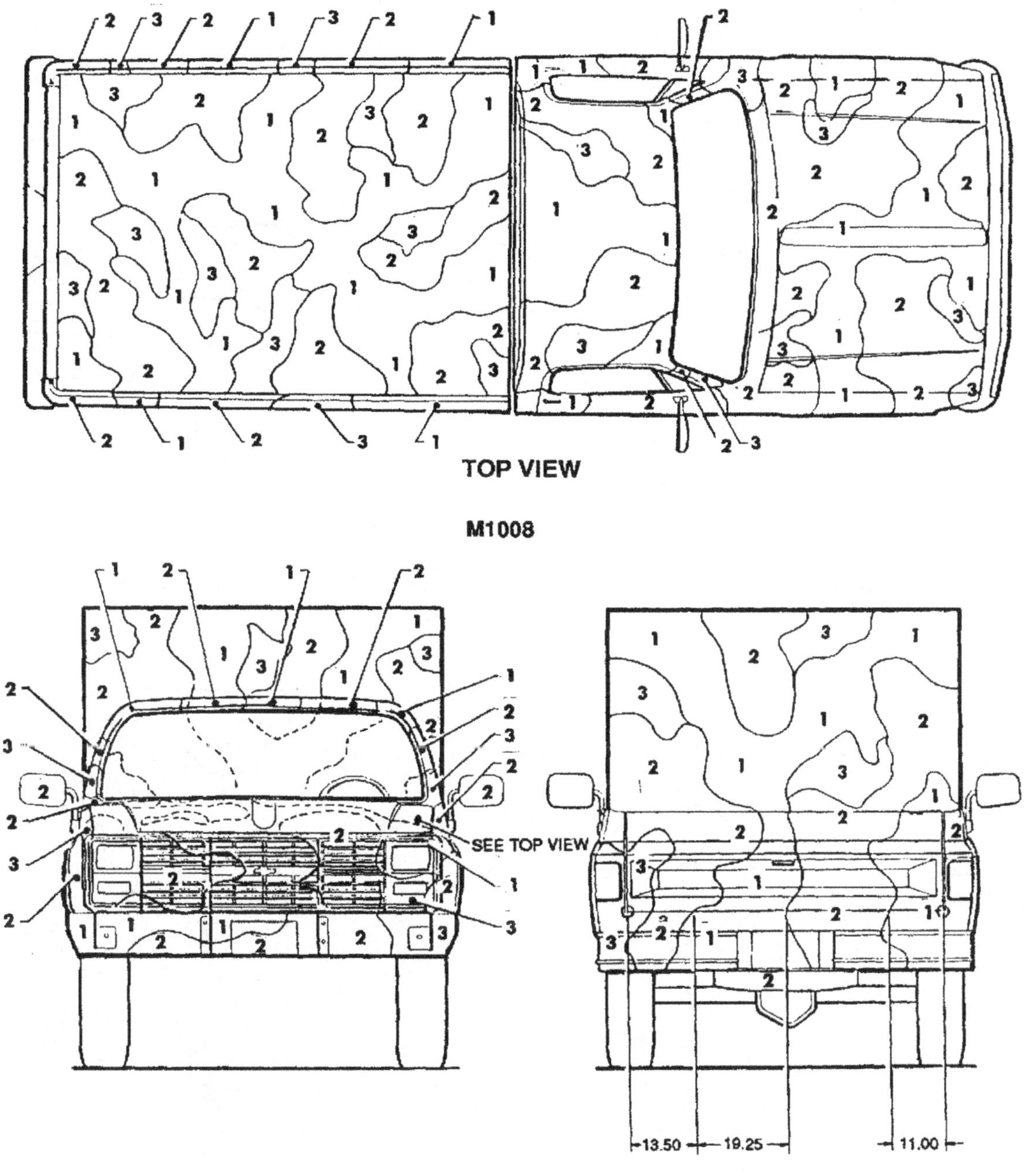

Figure 43. Truck, cargo: canvas shelter, 1-1/4-ton, 4 x 4, M1008. (1 of 2)

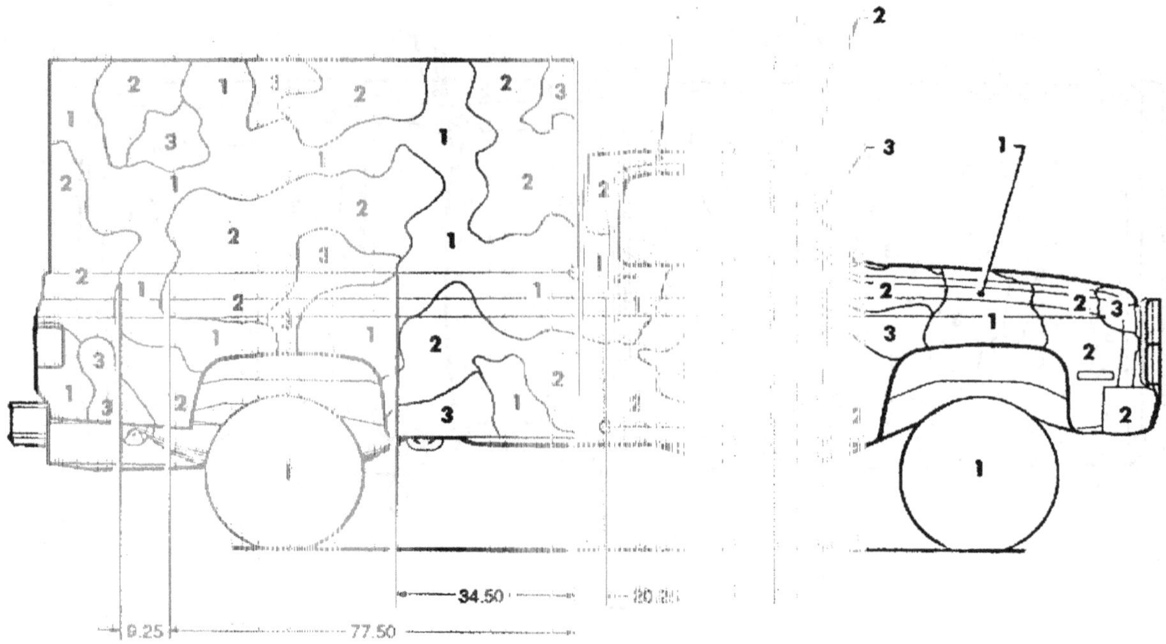

RIGHT SIDE VIEW

M1008

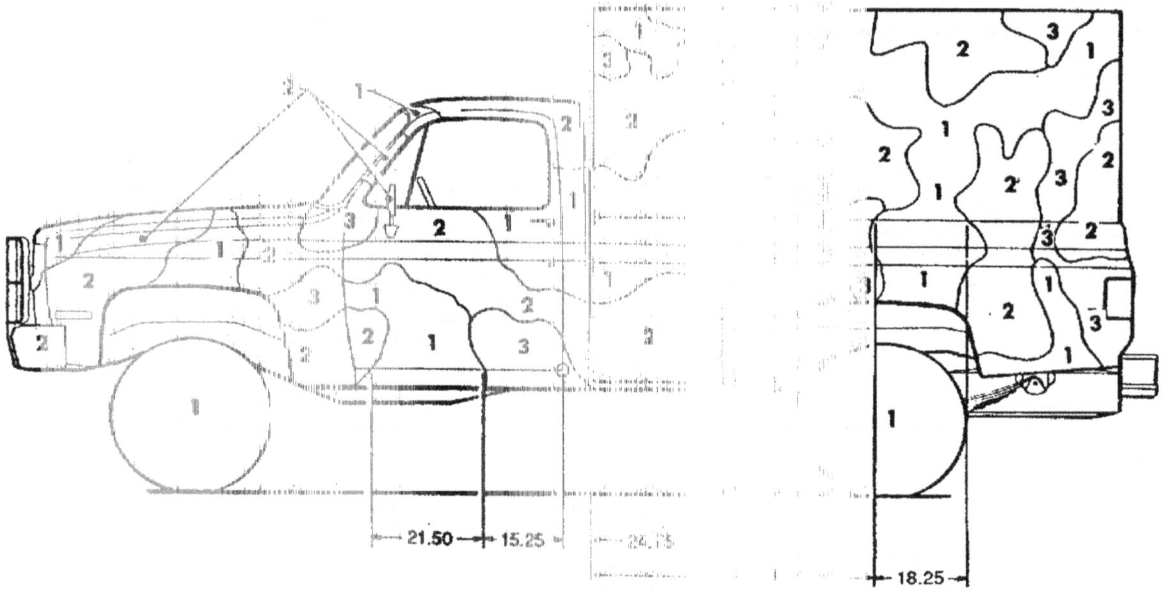

LEFT SIDE VIEW

Figure 43. Truck, cargo: canvas shelter, 1-1/4-ton, M1008. (2 of 2)

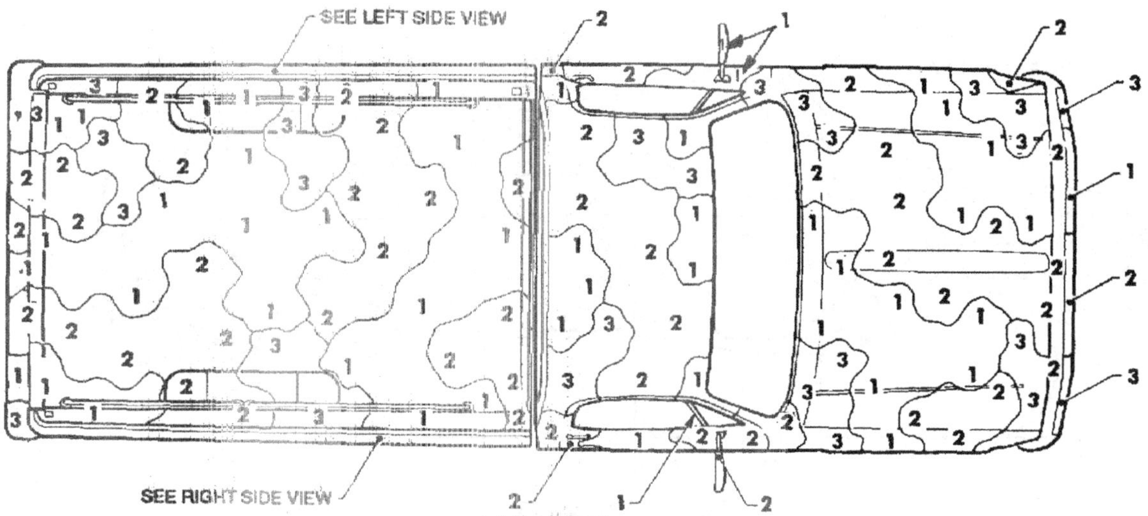

TOP VIEW

M1008

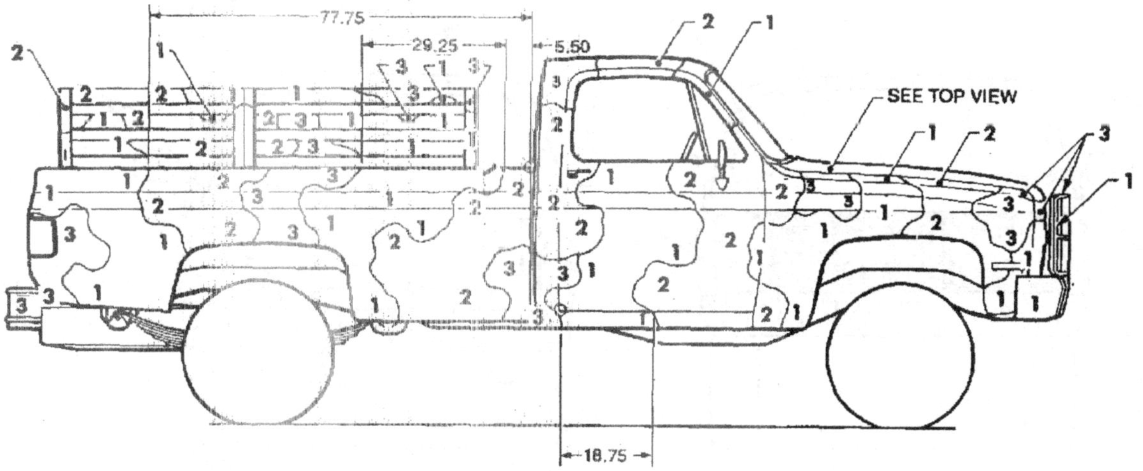

RIGHT SIDE VIEW

Figure 43a. Truck, cargo: 1-1/4-ton, 4 x4, M1008. (1 of 2)

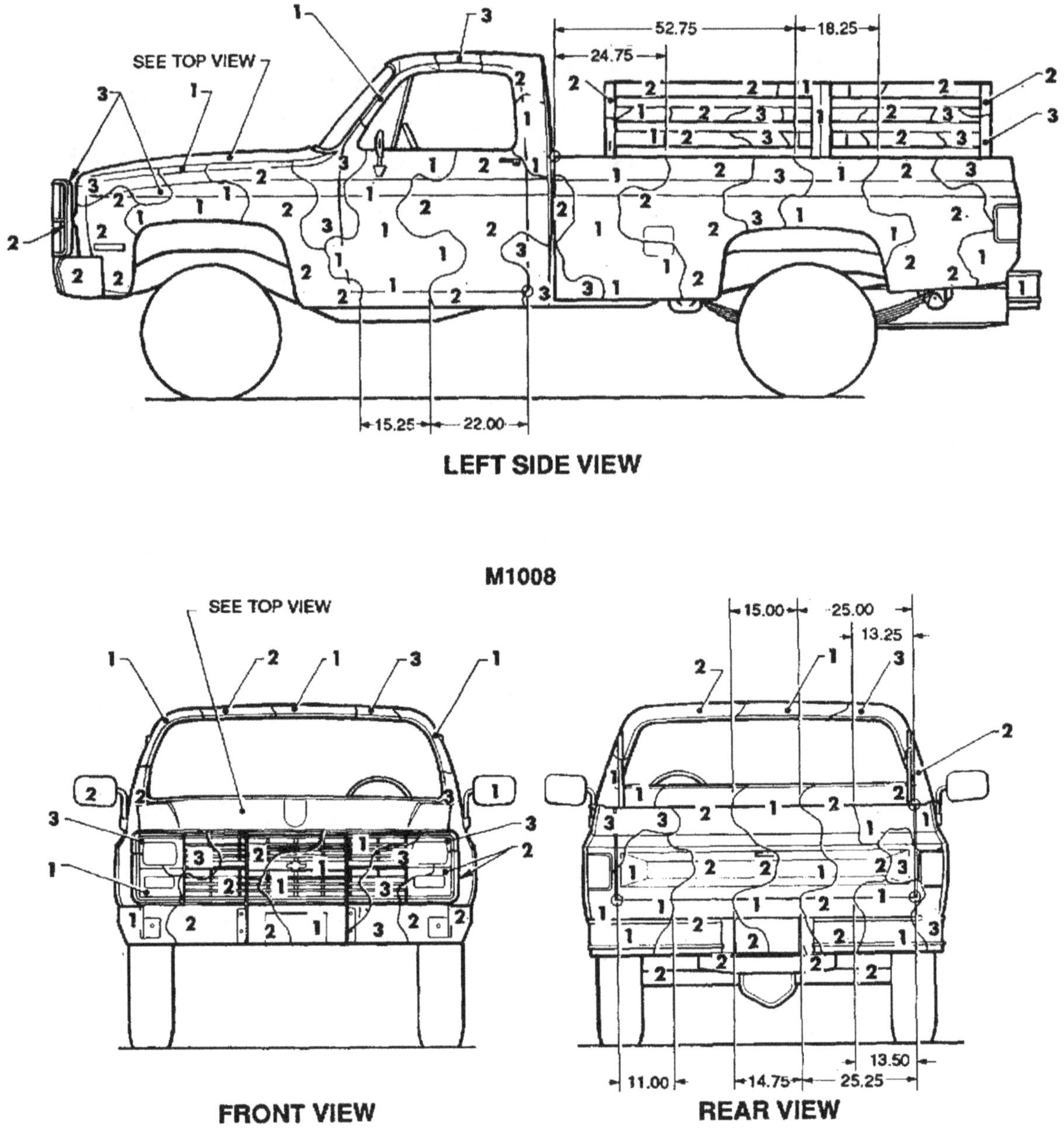

Figure 43a. Truck, cargo: 1-1/4-ton, 4 x 4, M1008. (2 of 2)

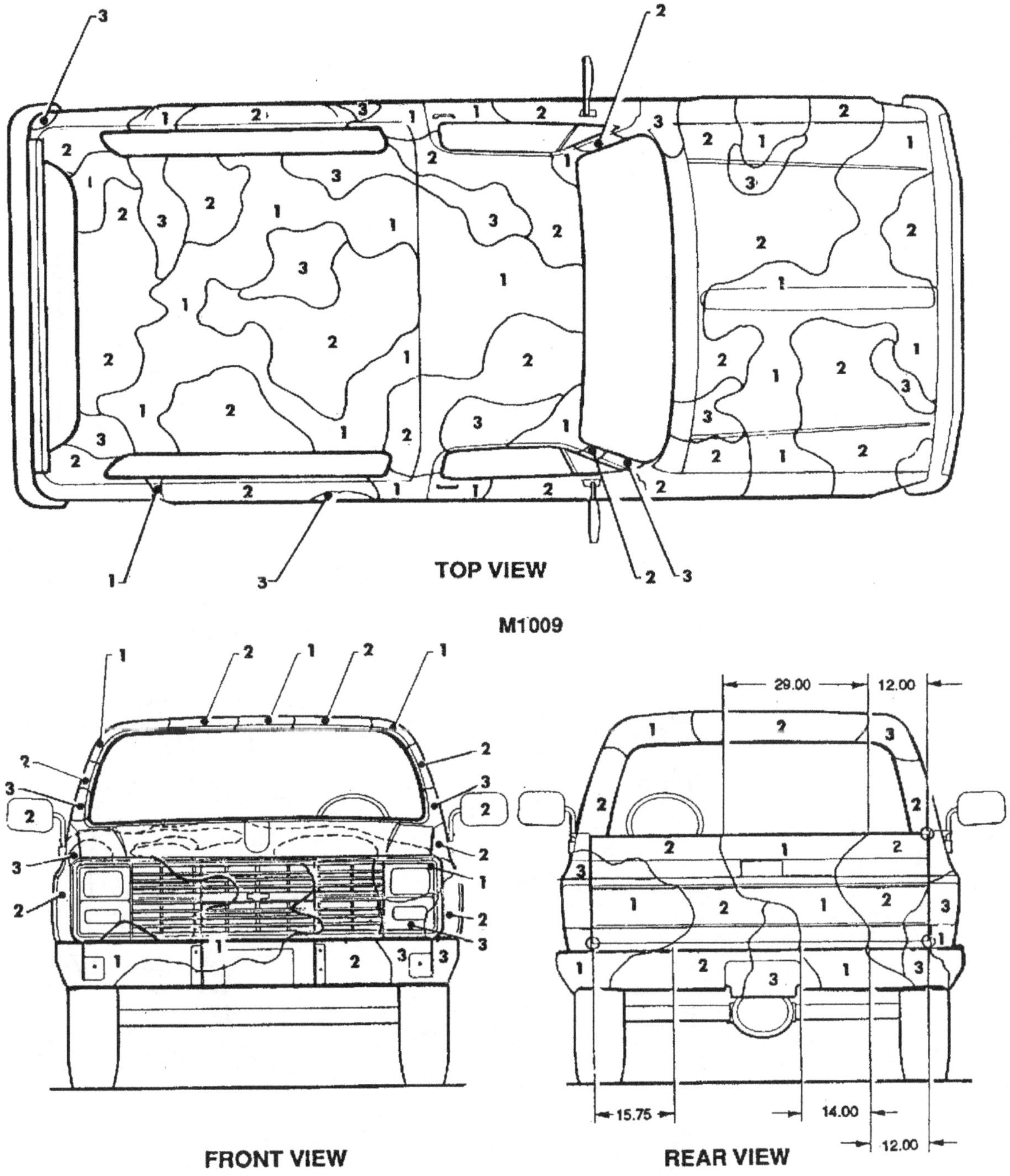

Figure 44. Truck, utility: 3/4-ton, 4 x 4, M1009. (1 of 2)

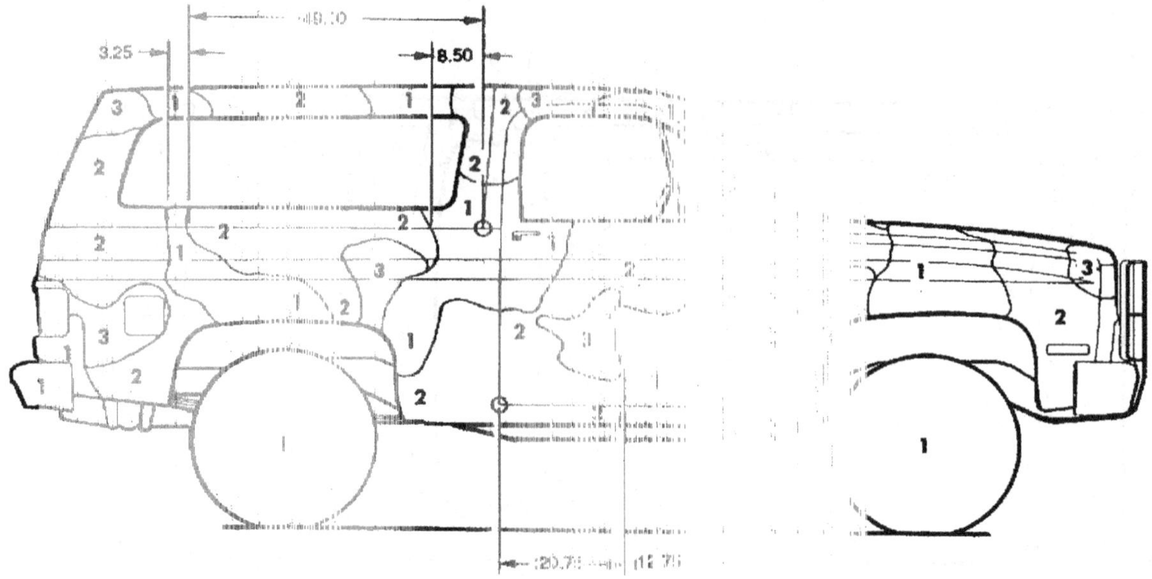

RIGHT SIDE VIEW

M1009

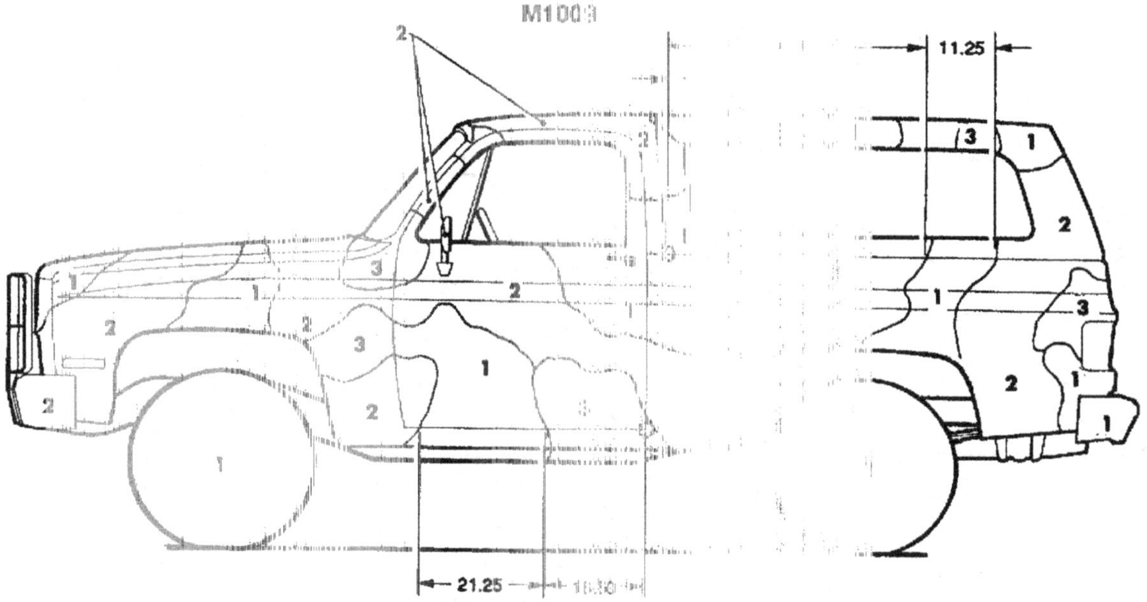

LEFT SIDE VIEW

Figure 44. Truck, utility: 3/4-ton, 4 x 4, [illegible]

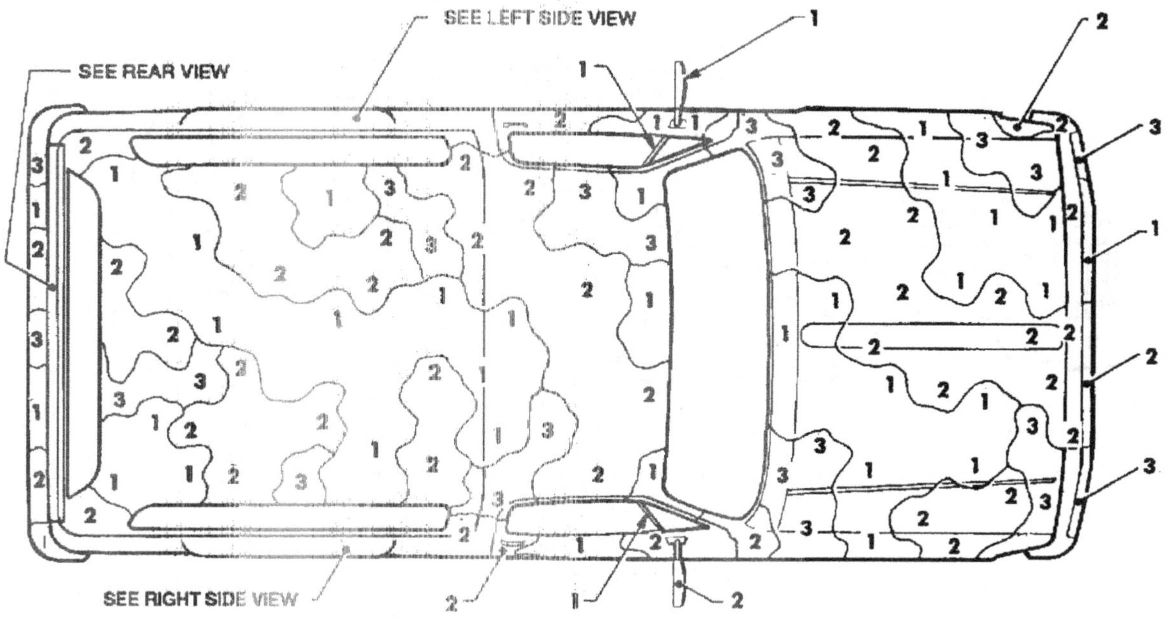

TOP VIEW

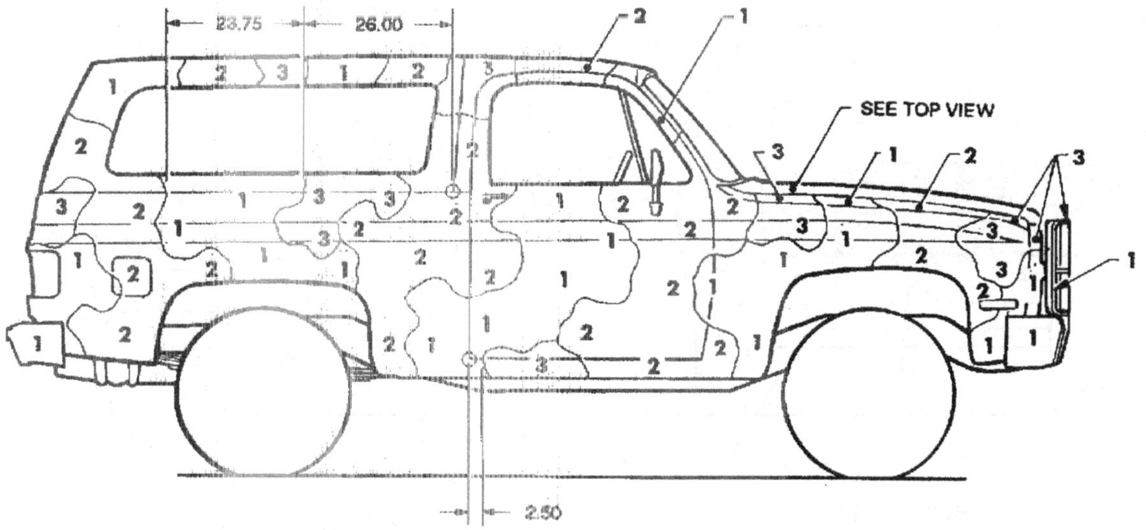

RIGHT SIDE VIEW

Figure 44a. Truck, utility: 3/44-ton, 4 x 4, M1009. (1 of 2) (complete repaint)

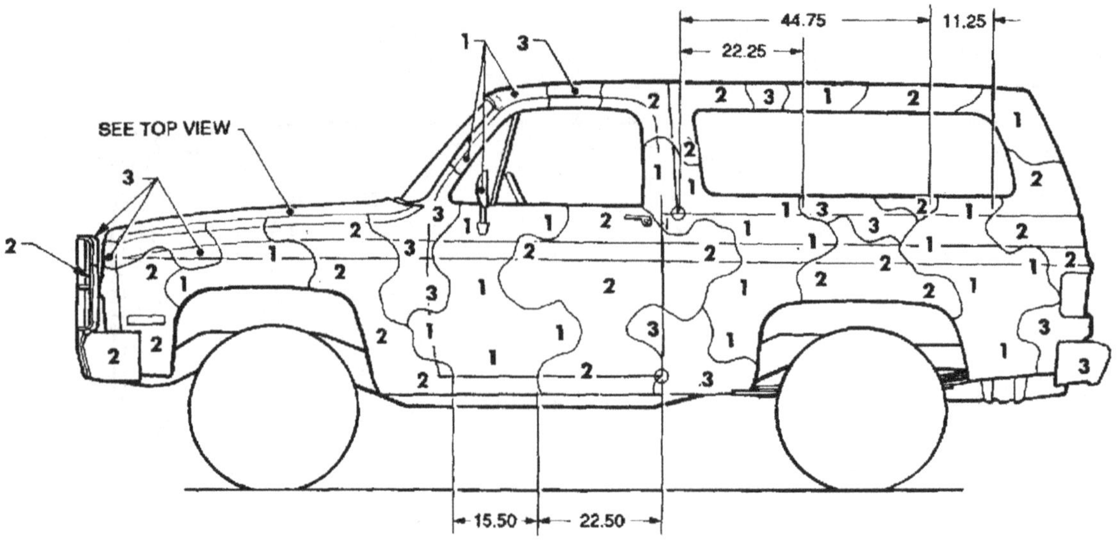

LEFT SIDE VIEW

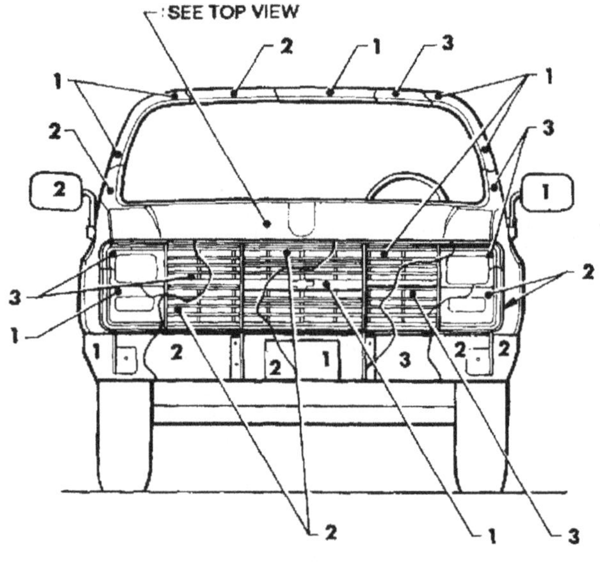

FRONT VIEW

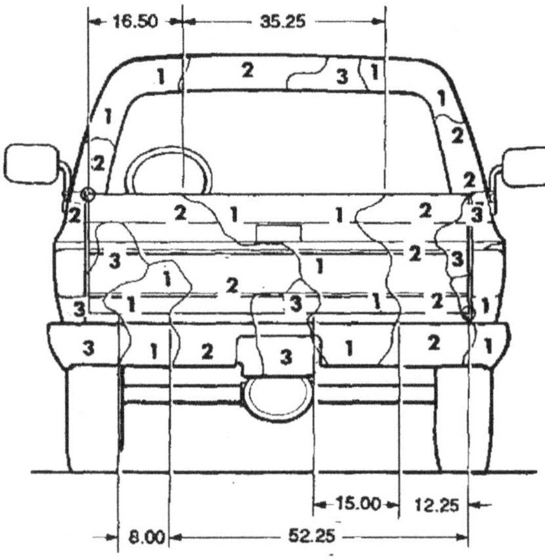

REAR VIEW

Figure 44a. Truck, utility: 3/4-ton, 4 x 4, M1009. (2 of 2) (complete repaint)

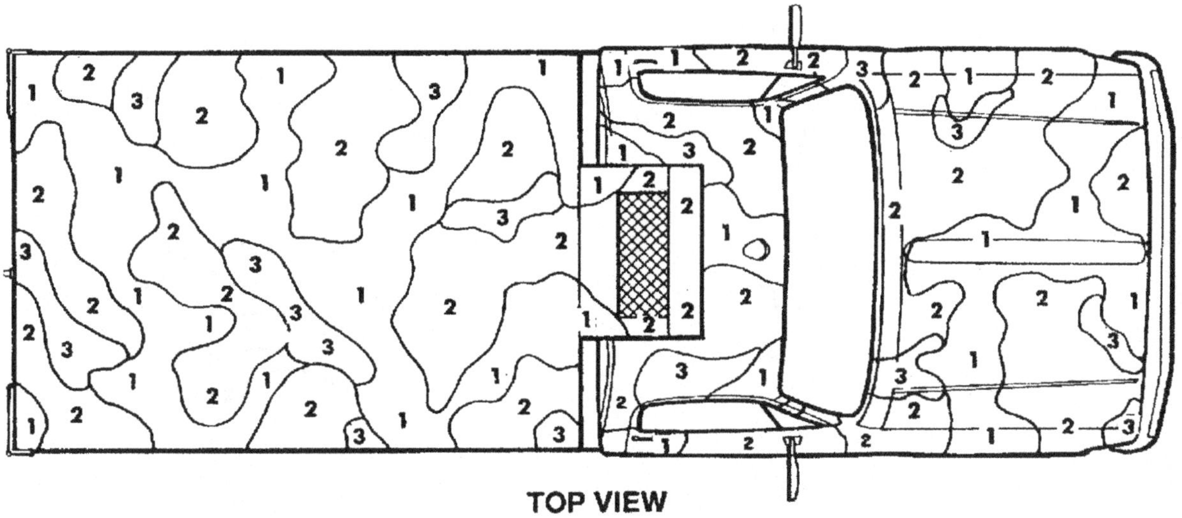

TOP VIEW

M1010

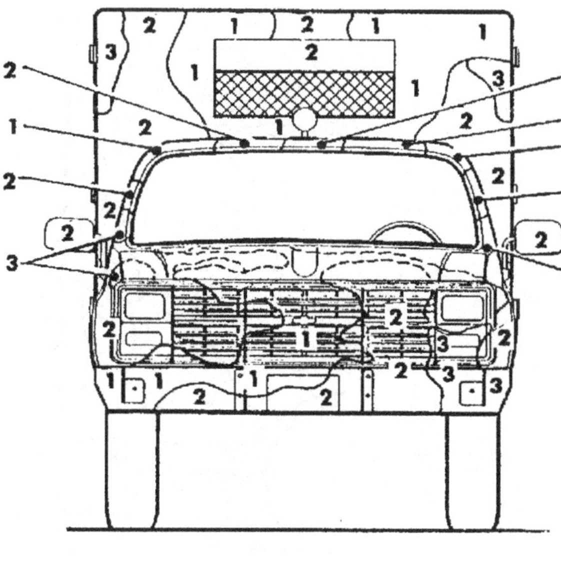

FRONT VIEW

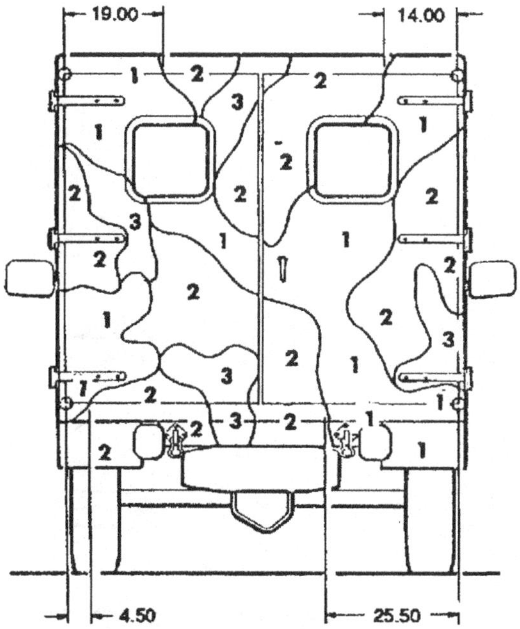

REAR VIEW

Figure 45. Truck, ambulance: 1-1/4-ton, 4 x 4, M1010. (1 of 2)

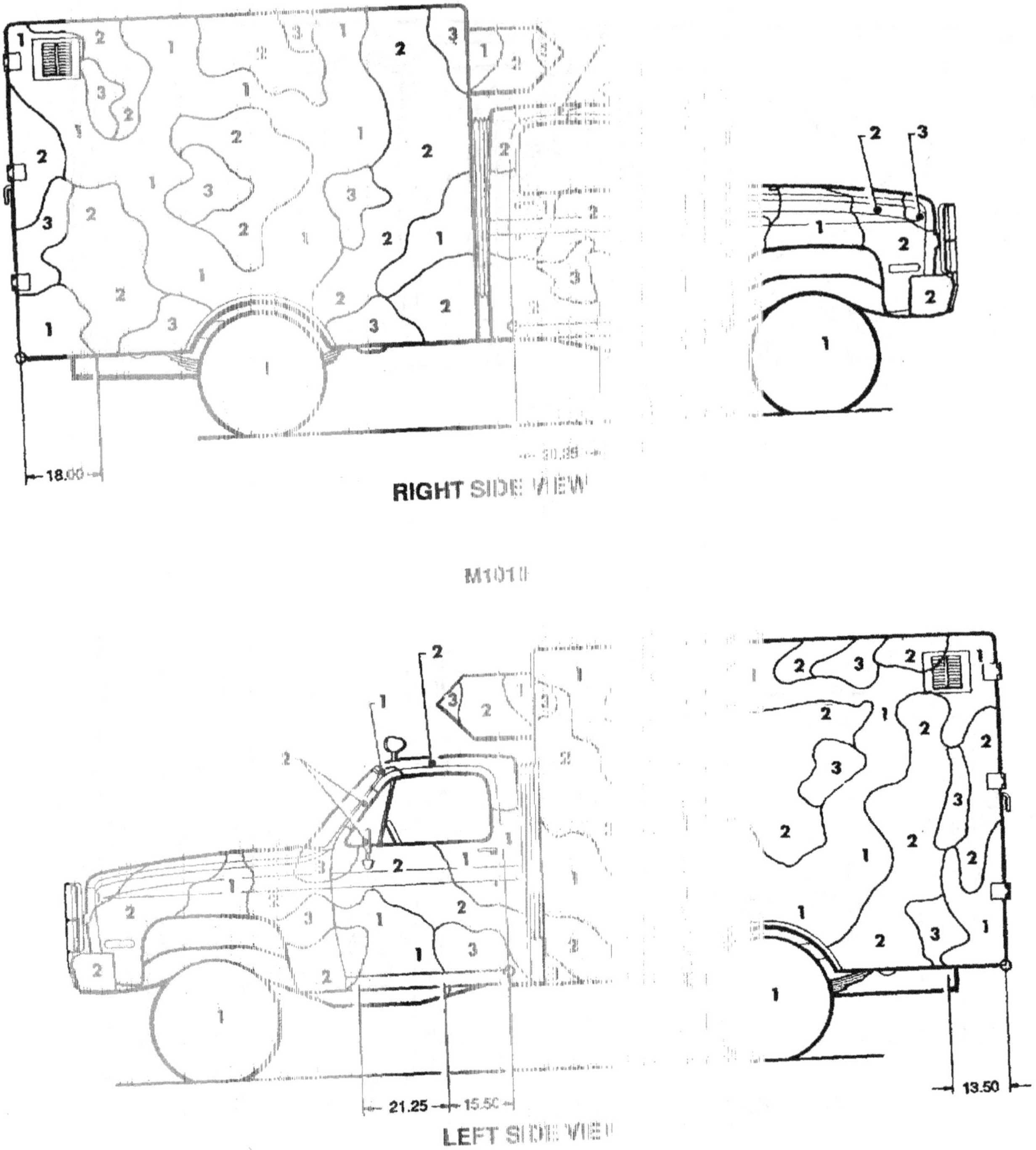

Figure 45. Truck, ambulance: 1-1/4-ton, 4 [illegible] of 2)

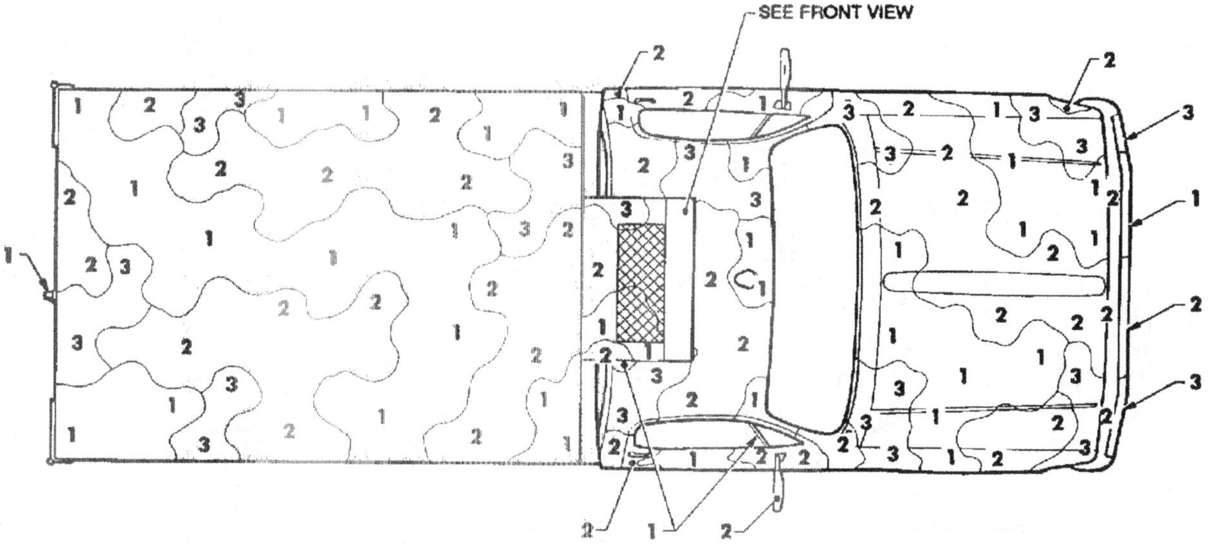

TOP VIEW

M1010

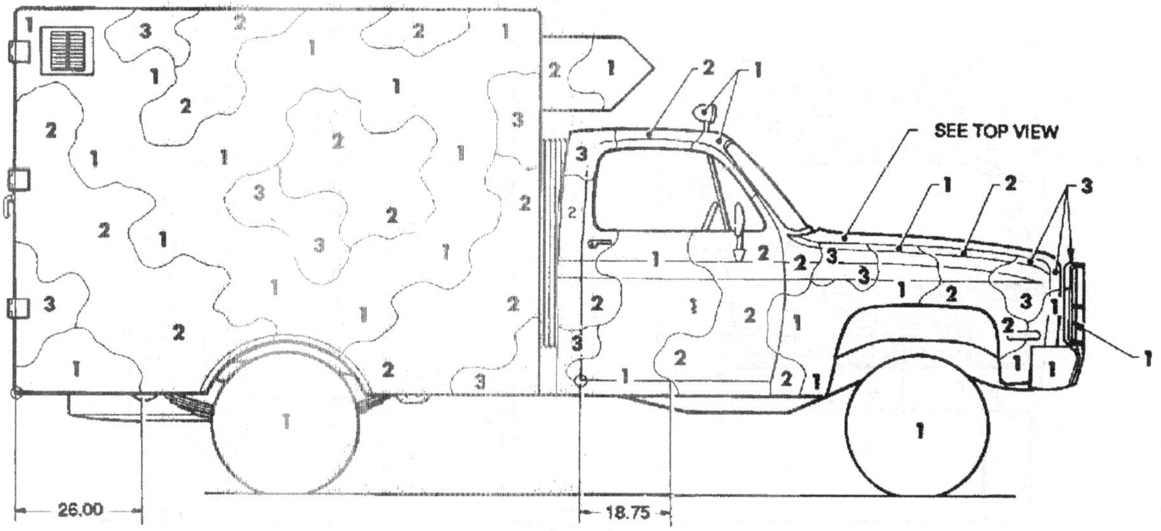

RIGHT SIDE VIEW

Figure 45a. Truck, ambulance: 1-1/4-ton, 4 x 4, M1010. (1 of 2) (complete repaint)

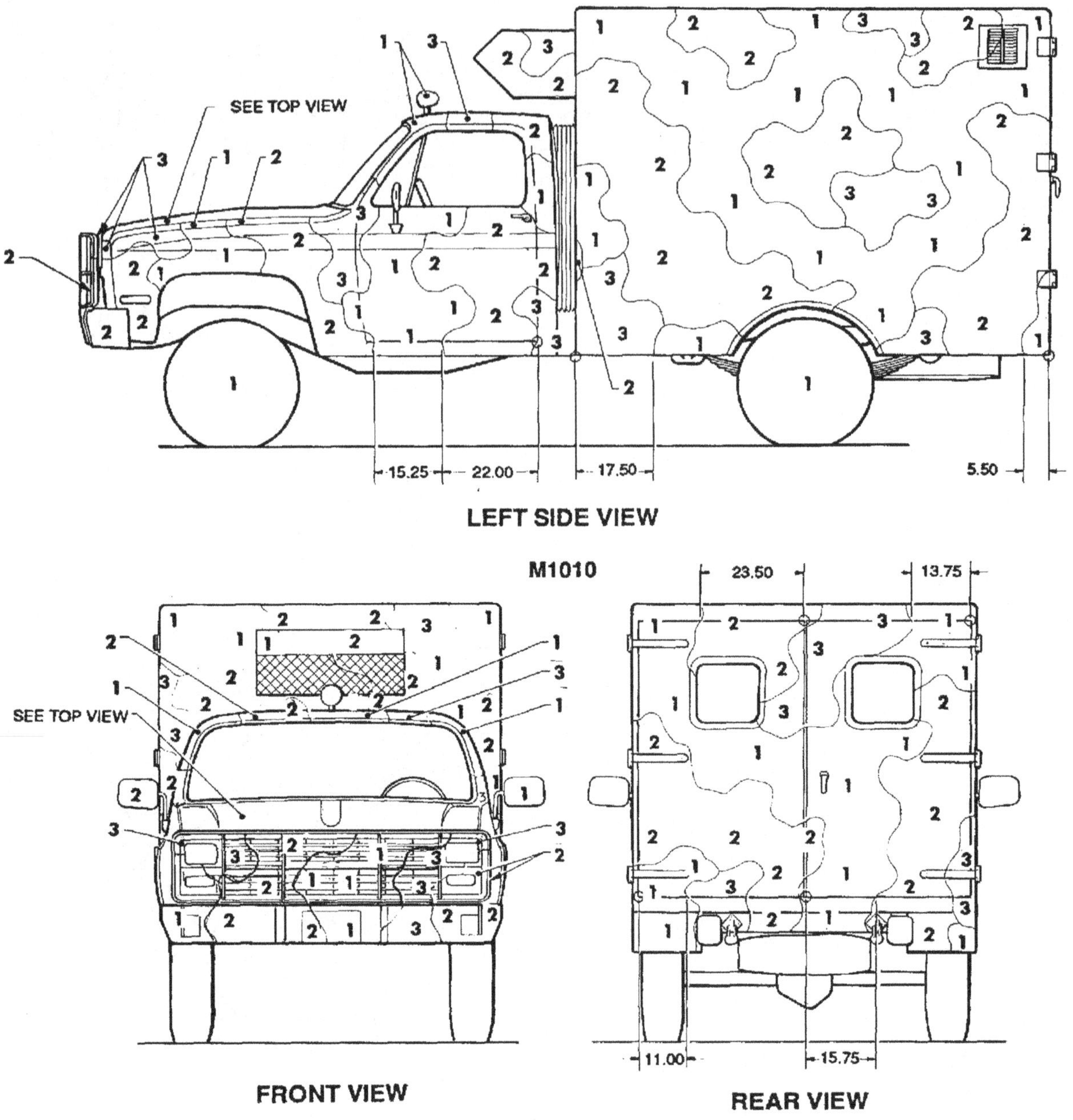

Figure 45a. Truck, ambulance: 1-1/4-ton, 4 x 4, M1010. (2 of 2) (complete repaint)

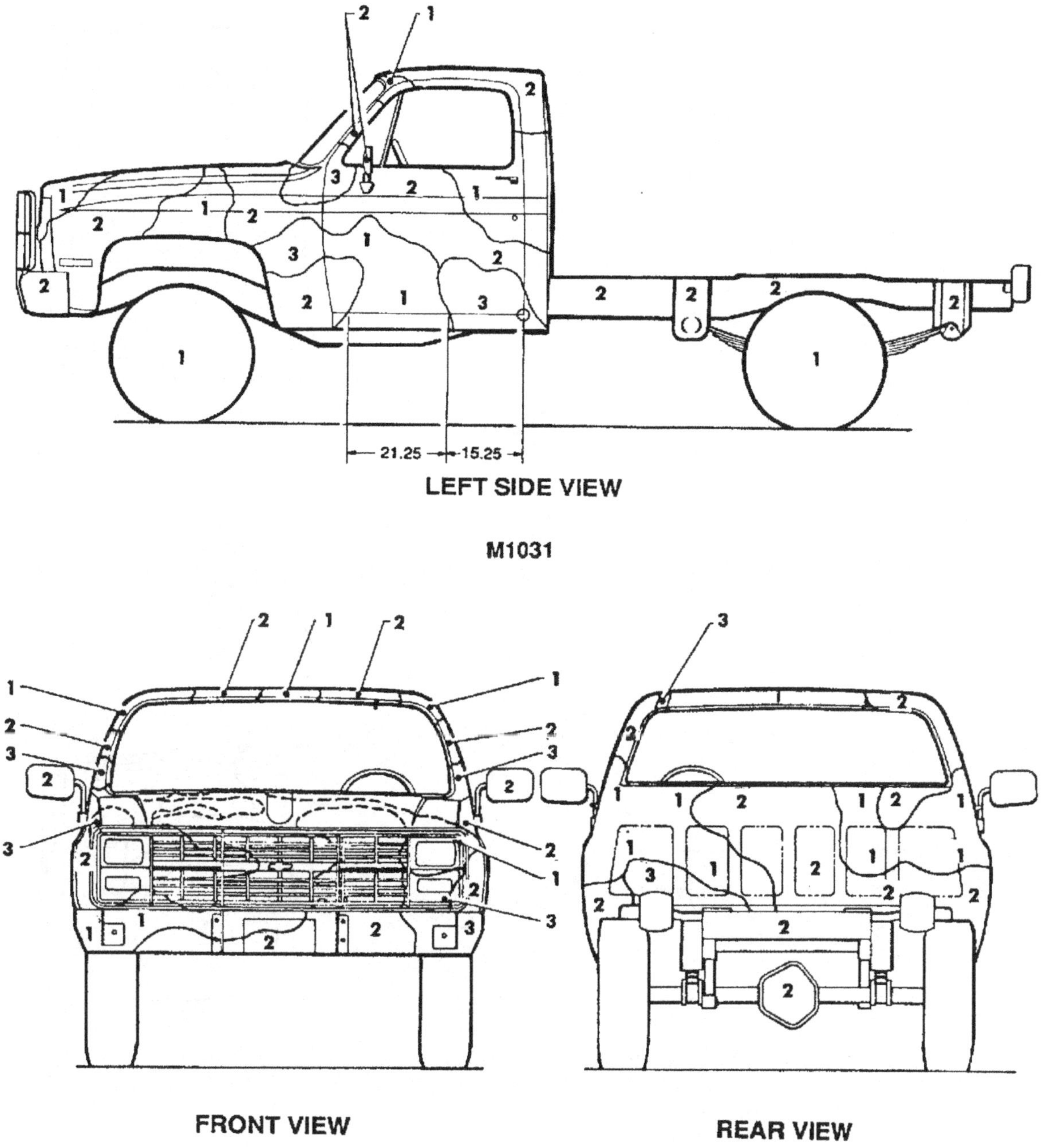

Figure 46. Truck, chassis: type D, 1-1/4-ton, M1031. (1 of 2)

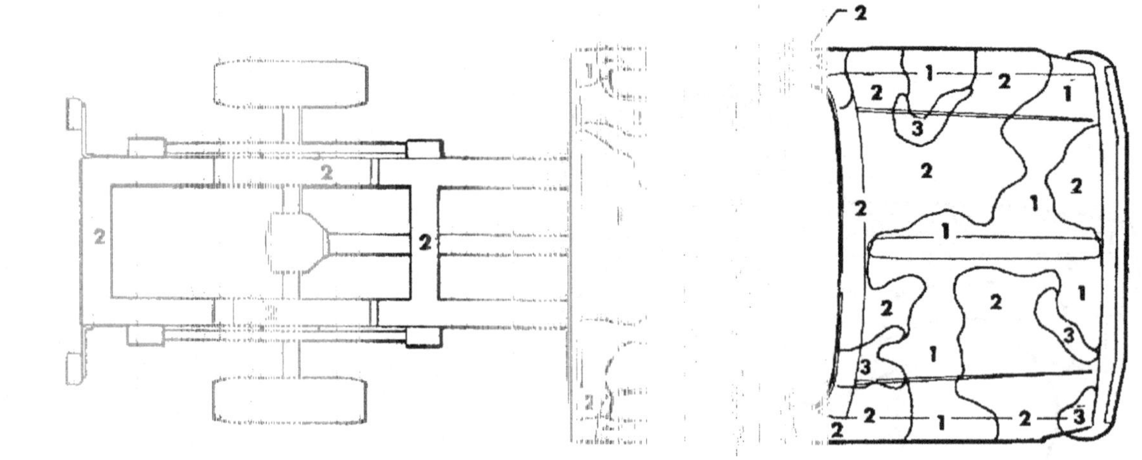

TOP VIE[illegible]

M1031

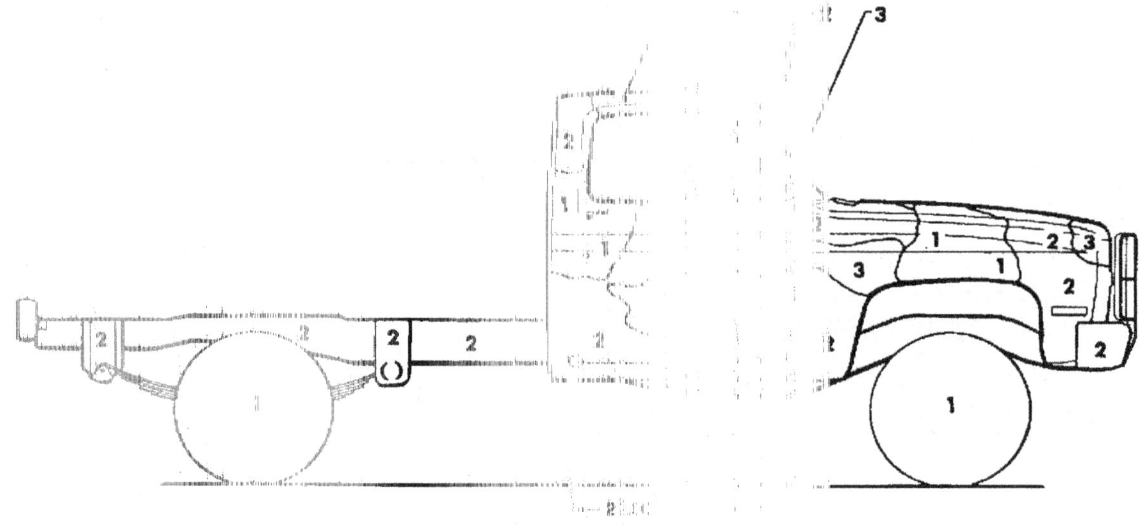

RIGHT SIDE V[illegible]

Figure 45. Truck, chassis: type D, 1-1/4 [illegible] of 2)

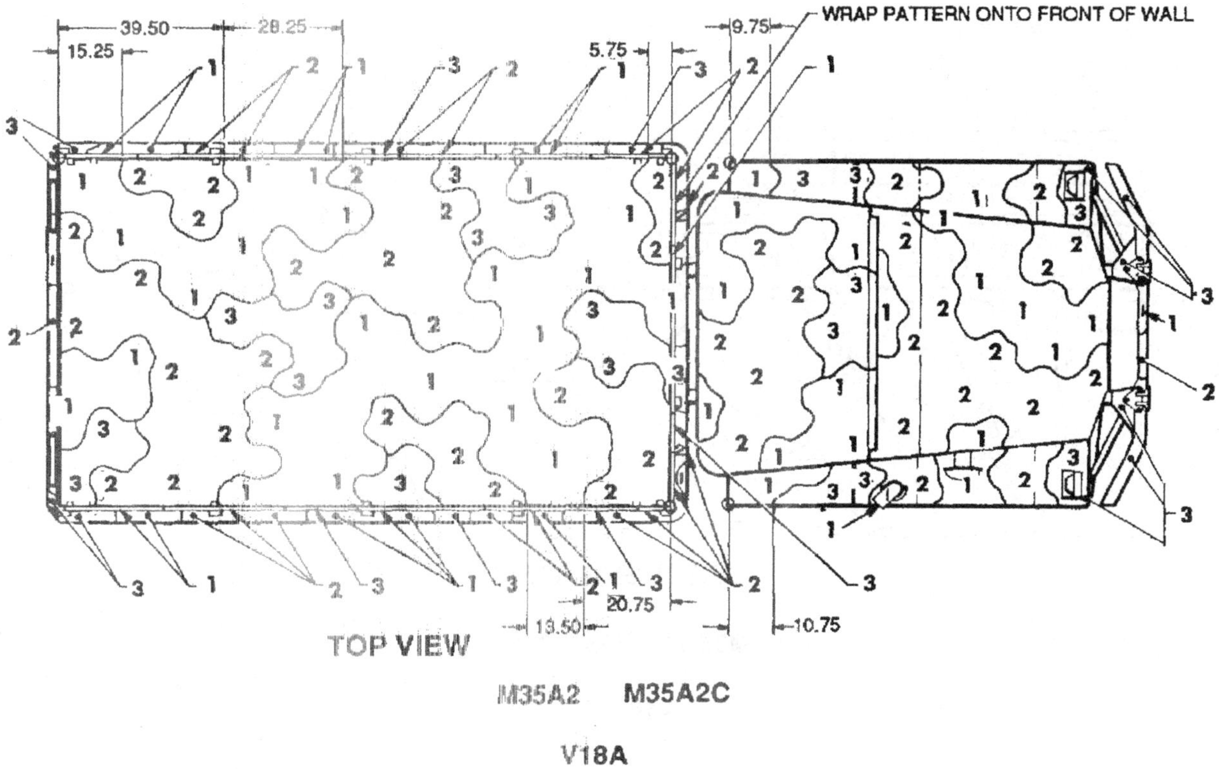

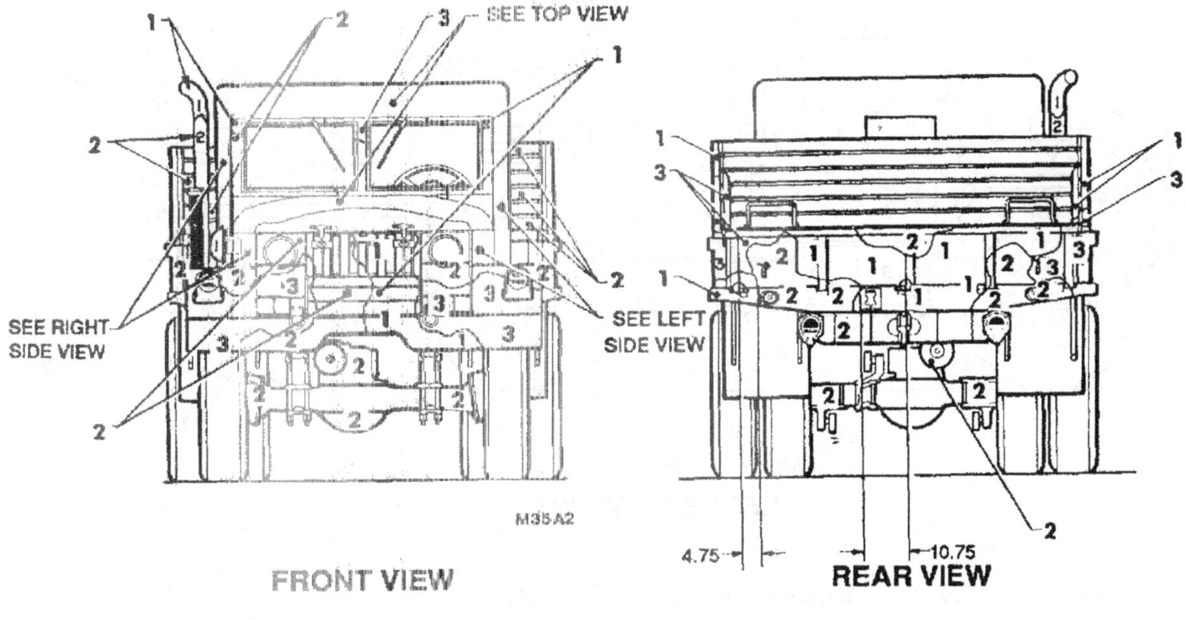

Figure 47. Truck, cargo/personnel: 6 x 6, 2-1/2-ton, M35A2, M35A2C, and V18A. (1 of 2)

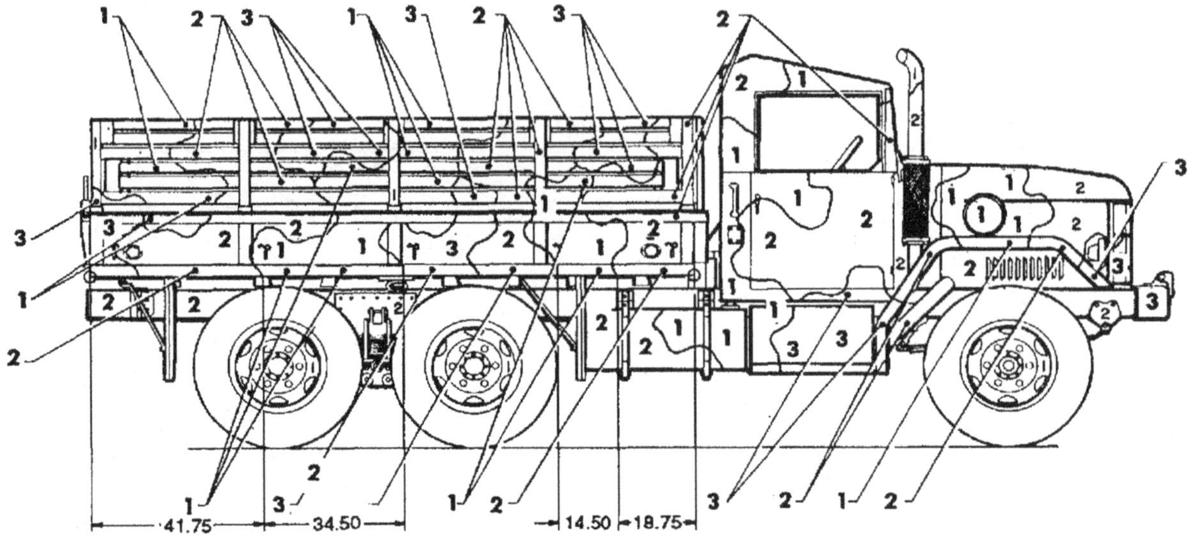

RIGHT SIDE VIEW

M35A2 M35A2C

V18A

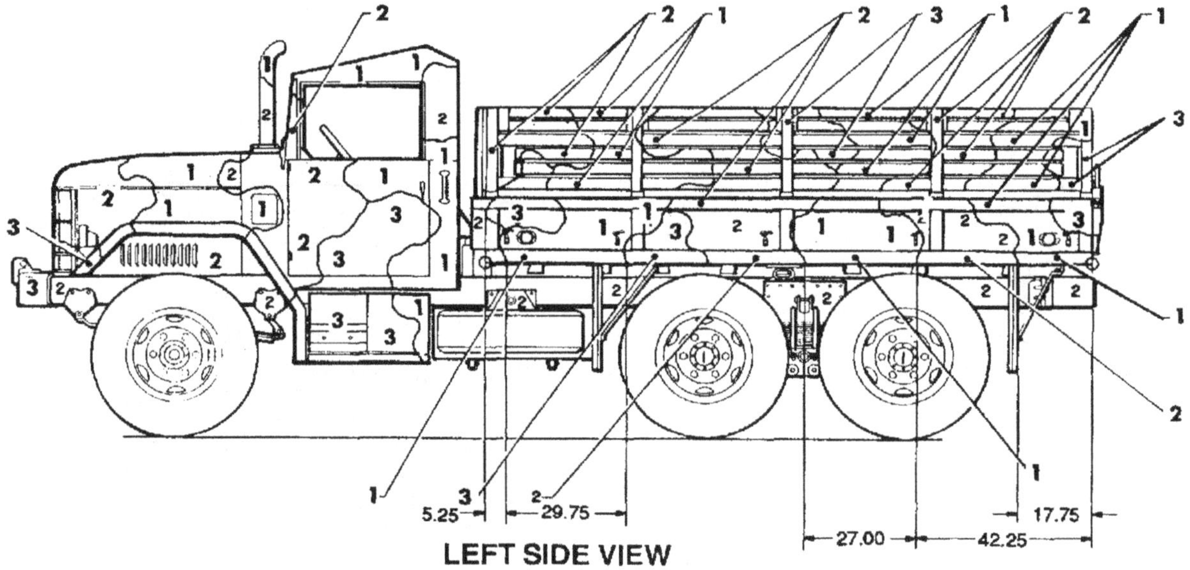

LEFT SIDE VIEW

Figure 47. Truck, cargo/personnel: 6 x 6, 2-1/2-ton, M35A2, M35A2C, and V18A. (2 of 2)

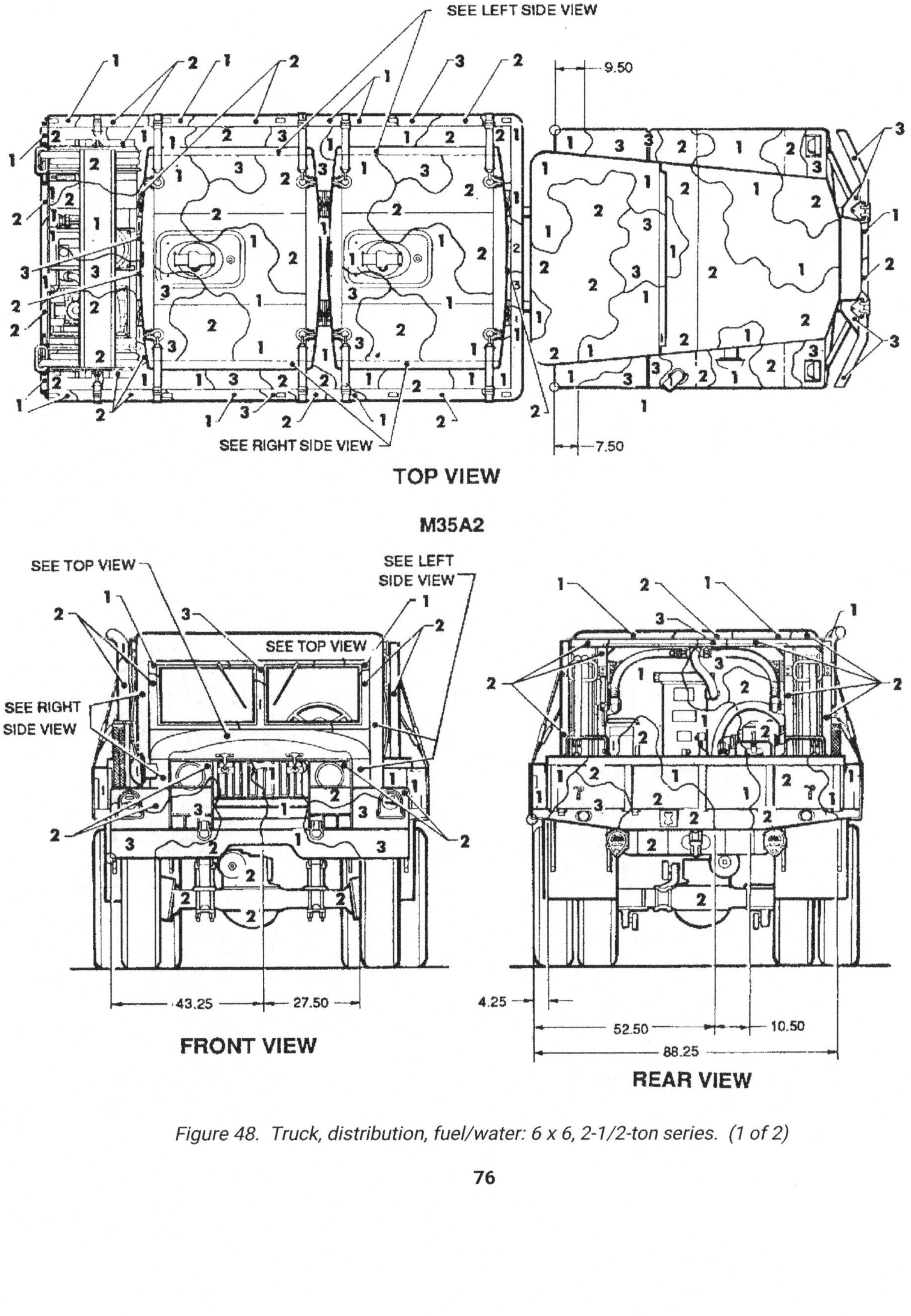

Figure 48. Truck, distribution, fuel/water: 6 x 6, 2-1/2-ton series. (1 of 2)

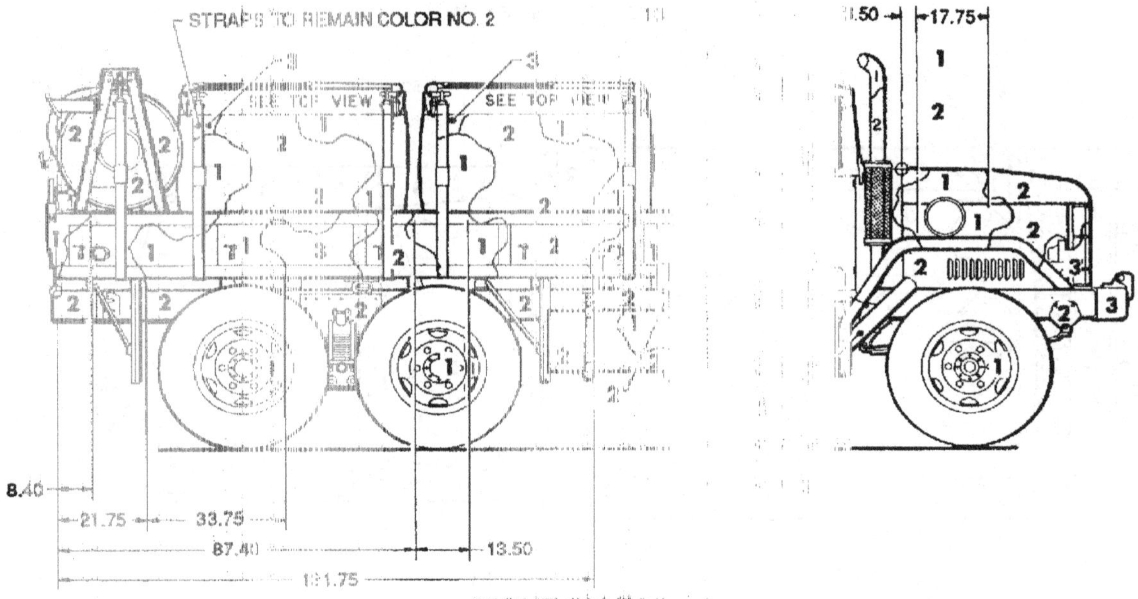

M35A2

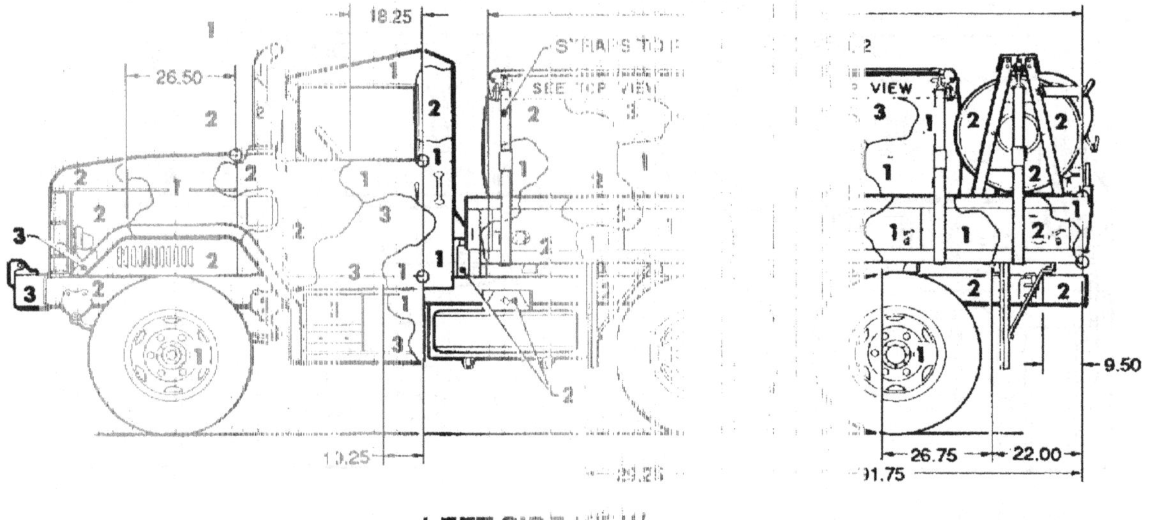

Figure 48. Truck, distribution, fuel/water 6 x 6, ... (2 of 2)

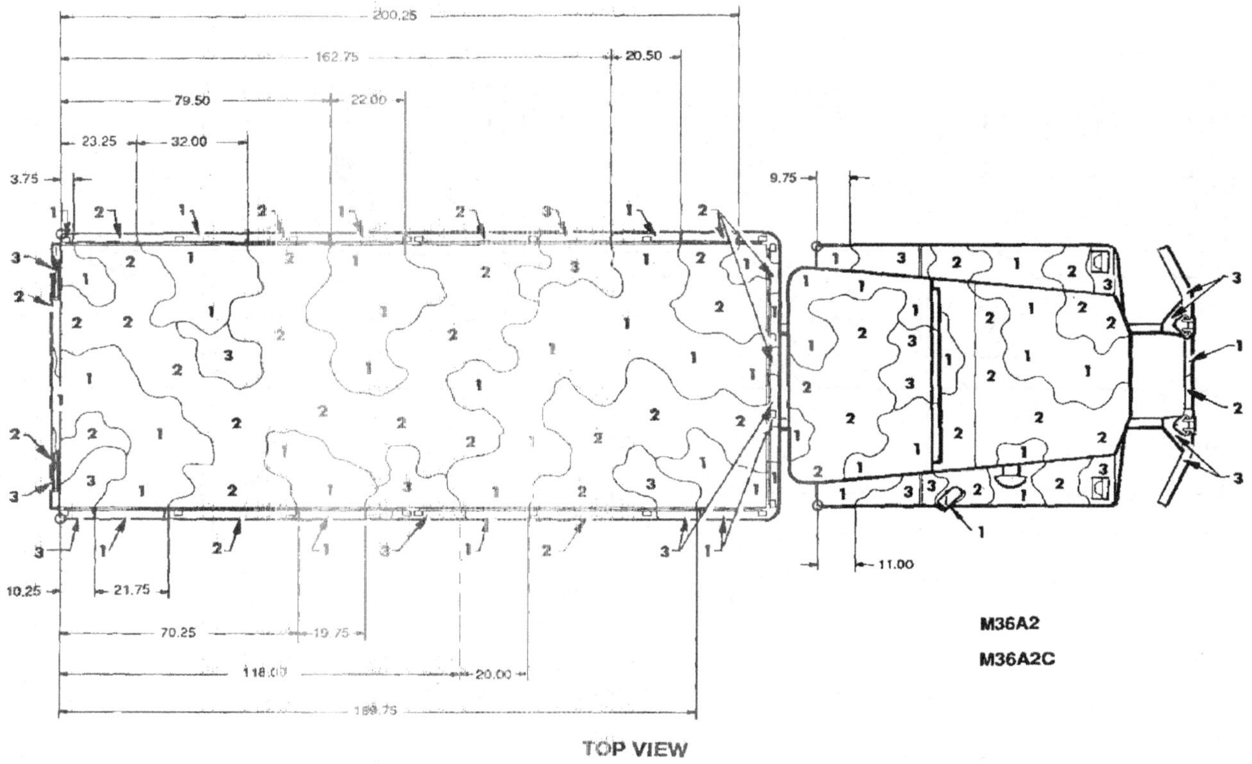

Figure 49. Truck, cargo: 2-1/2-ton SLWB, M36A2 and M36C. (1 of 4)

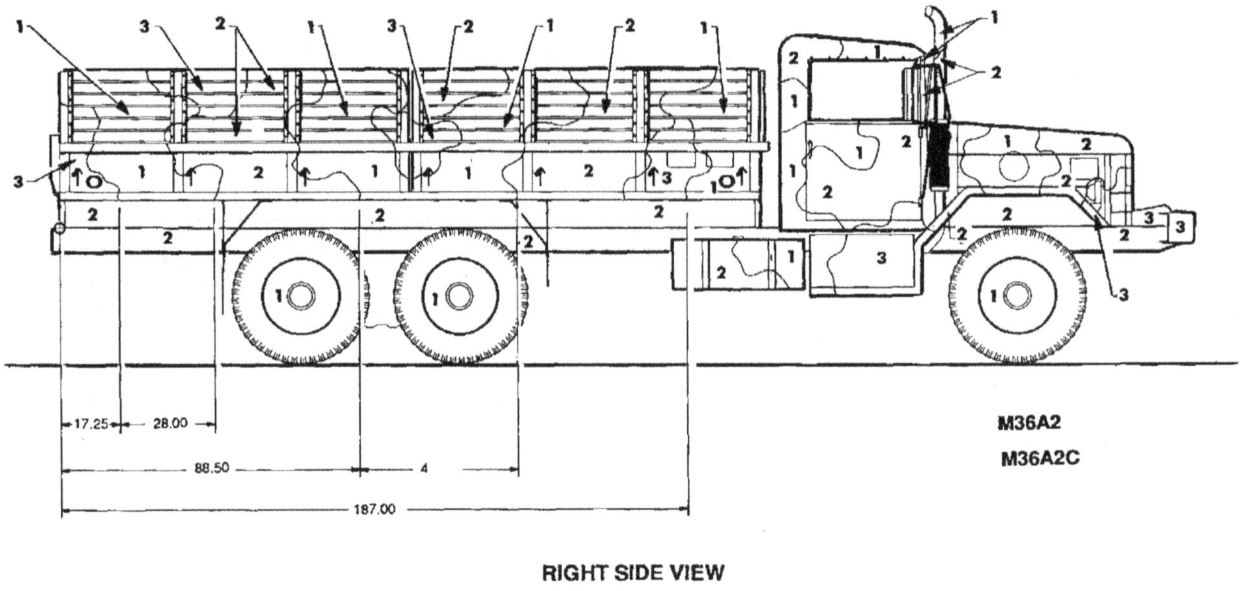

Figure 49. Truck, cargo: 2-1/2-ton SLWB, M36A2 and M36C. (2 of 4)

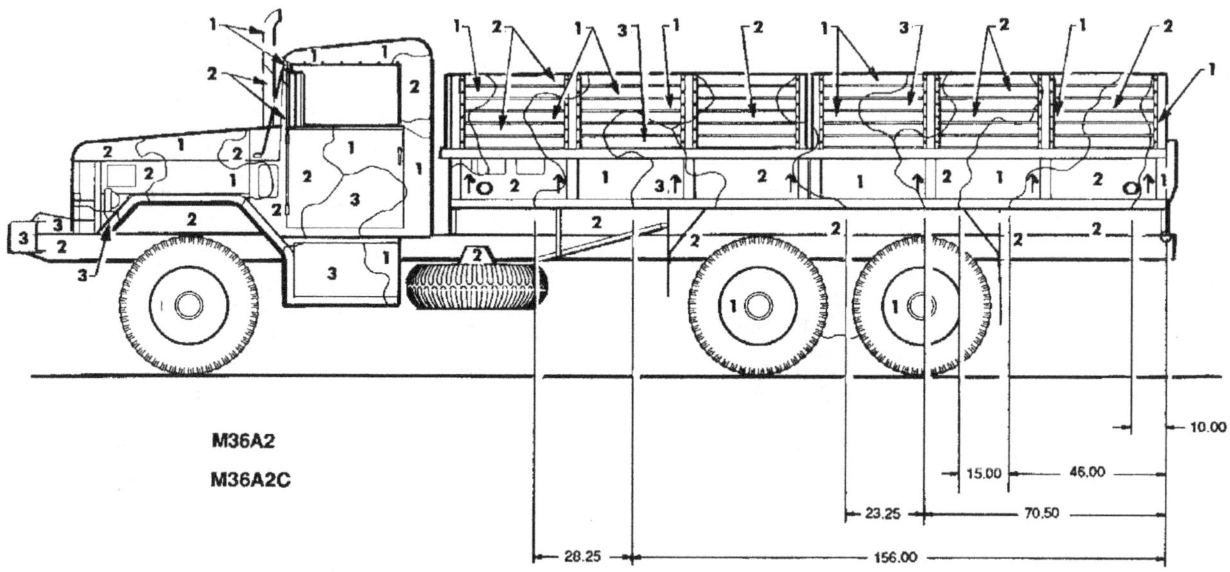

LEFT SIDE VIEW

Figure 49. Truck, cargo: 2-1/2-ton SLWB, M36A2 and M36C. (3 of 4)

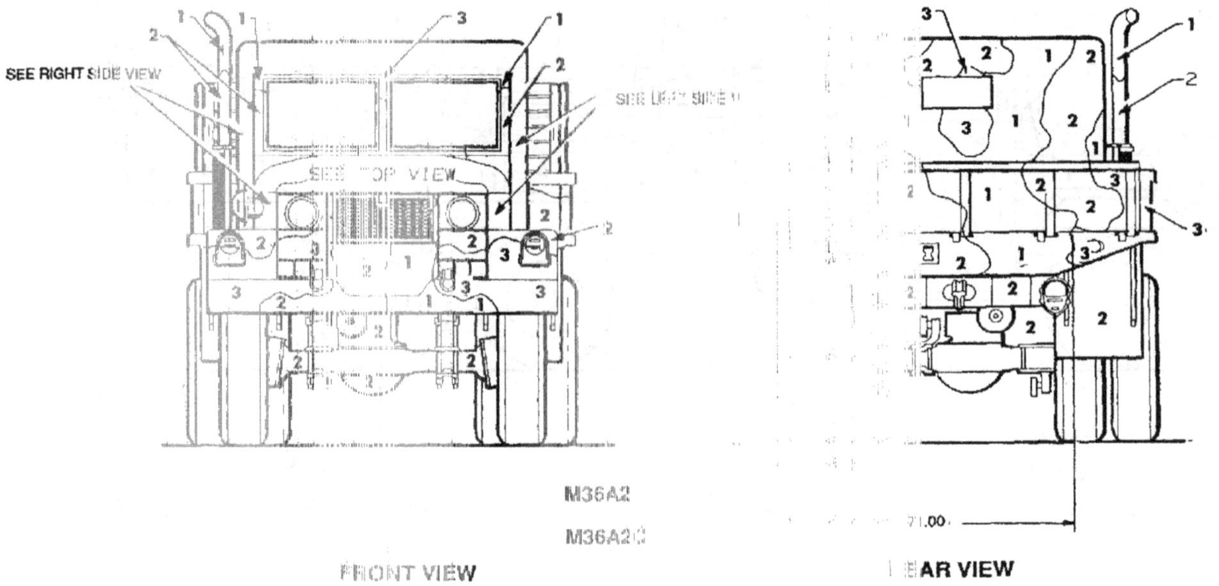

Figure 49. Truck, cargo: 2-1/2-ton SLWB, M35A2

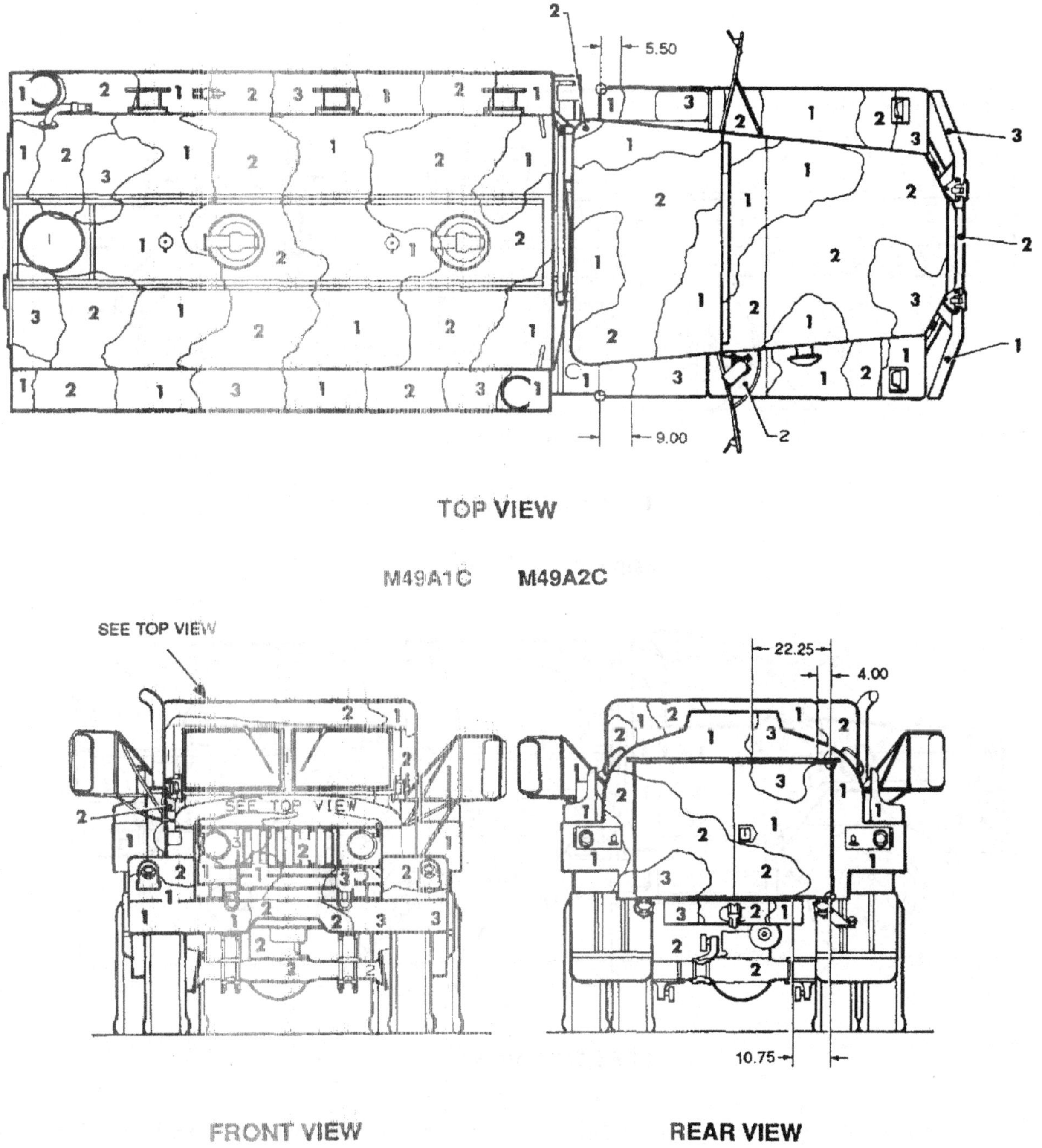

Figure 50. Truck, tanker, fuel: 2-1/2-ton, wo/winch M49A1C and M49A2C. (1 of 2)

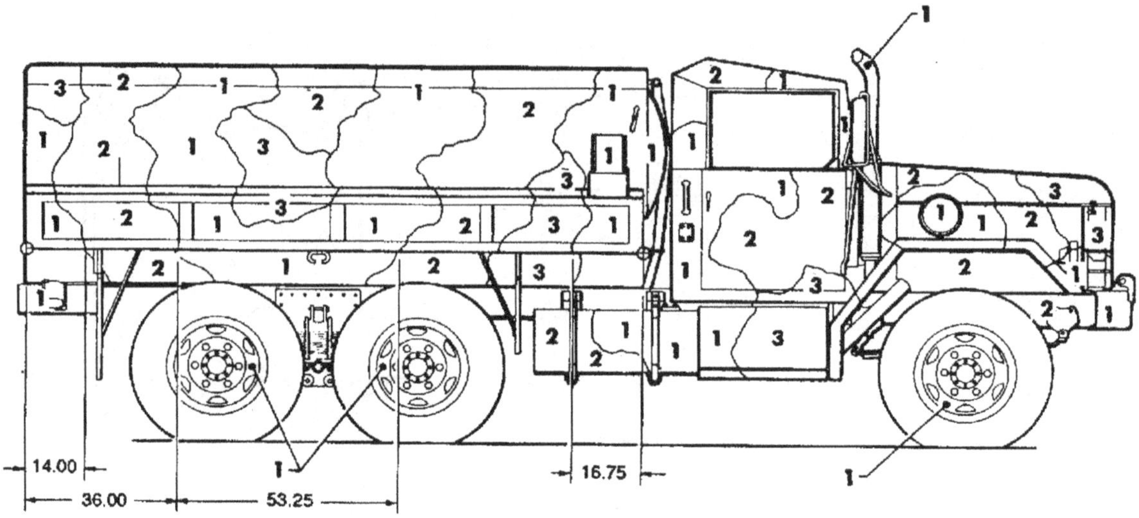

RIGHT SIDE VIEW

M49A1C M49A2C

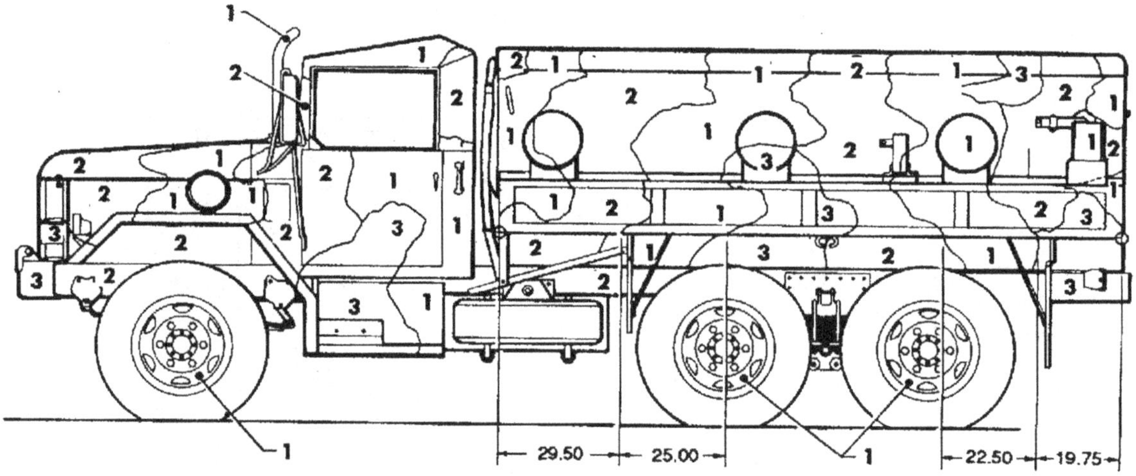

LEFT SIDE VIEW

Figure 50. Truck, tanker, fuel: 2-1/2-ton, wo/winch M49A1C and M49A2C. (2 of 2)

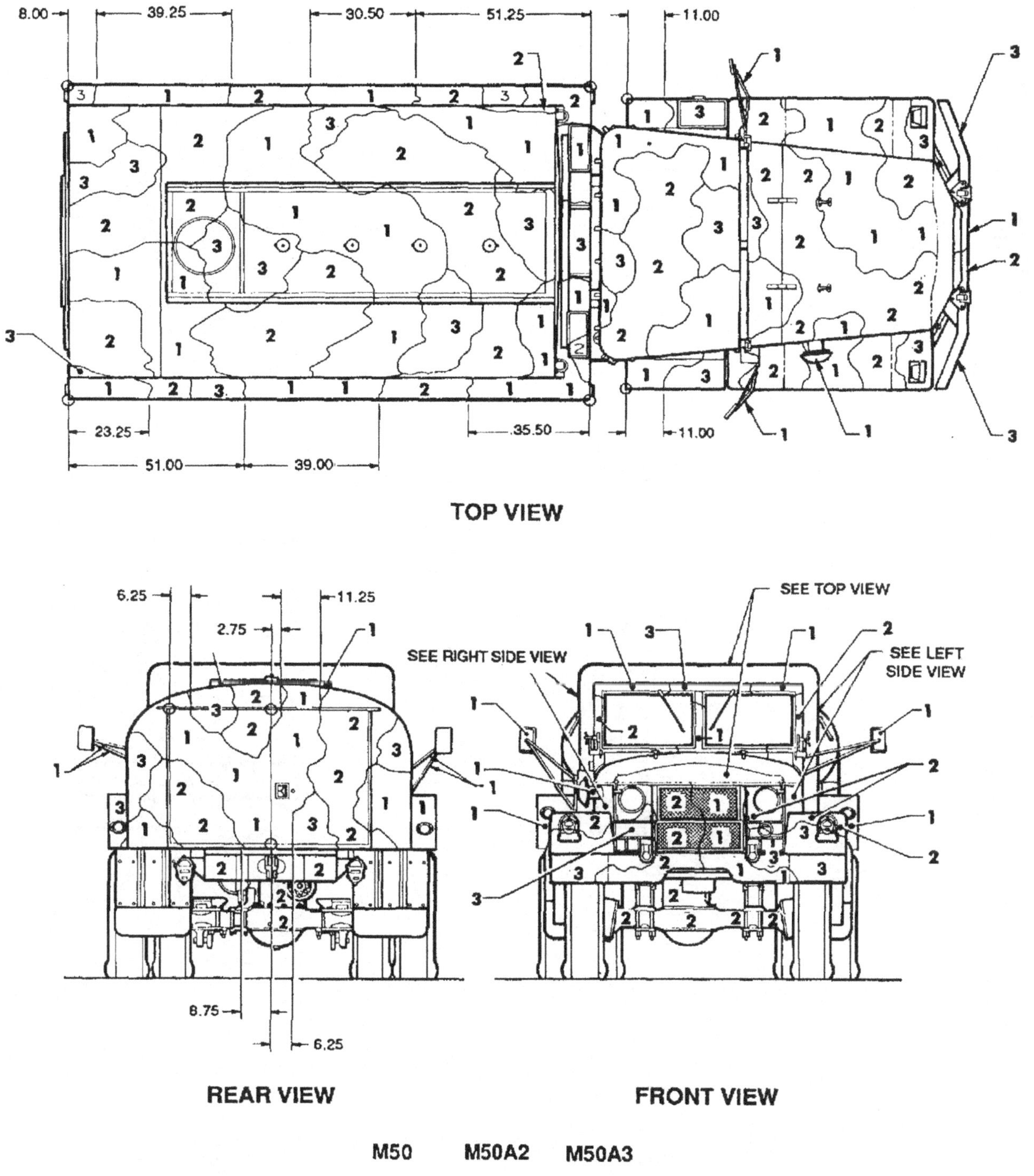

Figure 51. Truck, water distribution: 2-1/2-ton, M50, M50A2 and M50A3. (1 of 2)

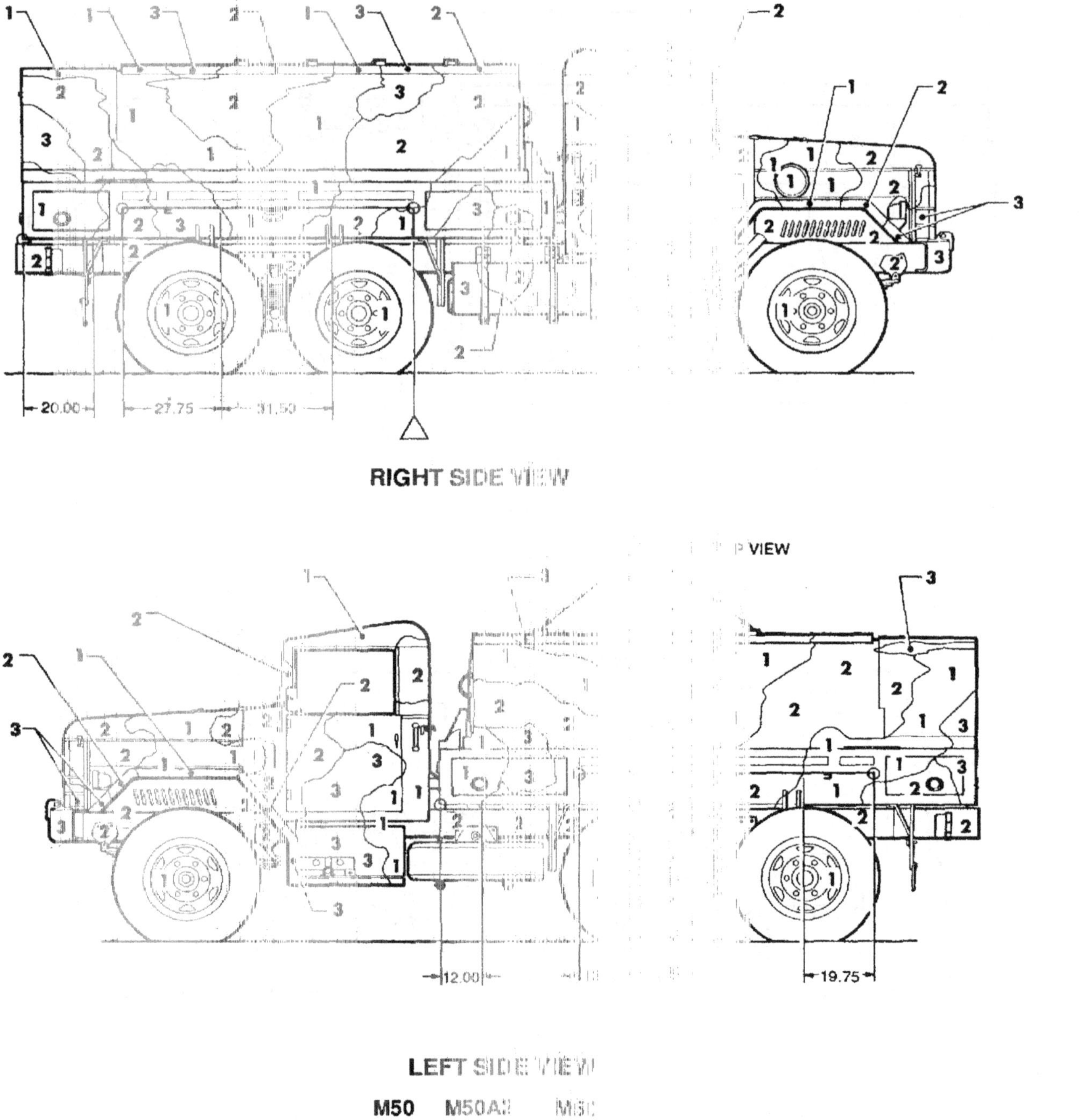

Figure 51. Truck, water distribution: 2-1/2-ton, M5[illegible] [illegible]50A3. (2 of 2)

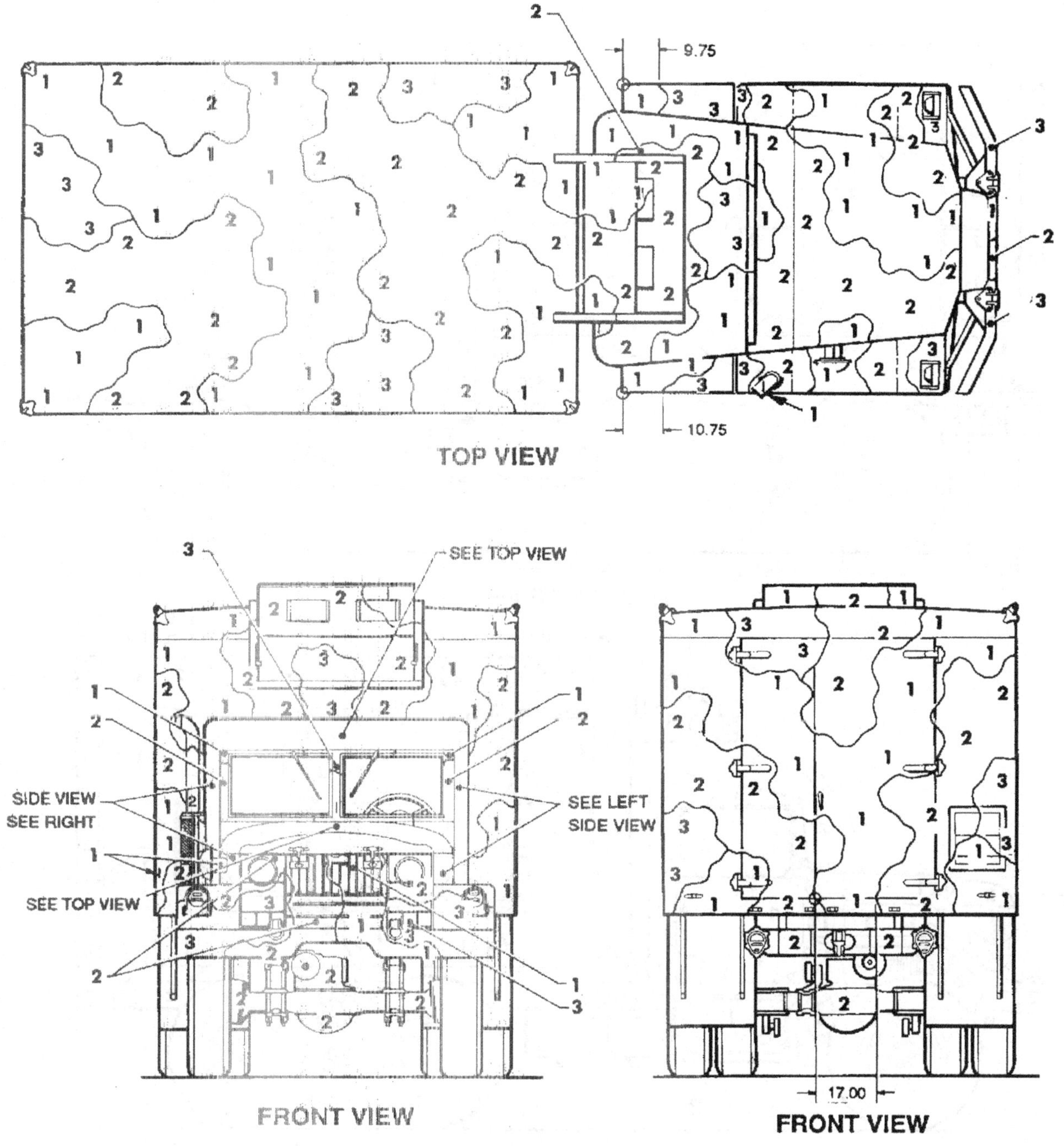

M109A3 M185A3

Figure 52. Truck, shop van: 2-1/2-ton, 6 x 6, M109A3 and M185A3. (1 of 2)

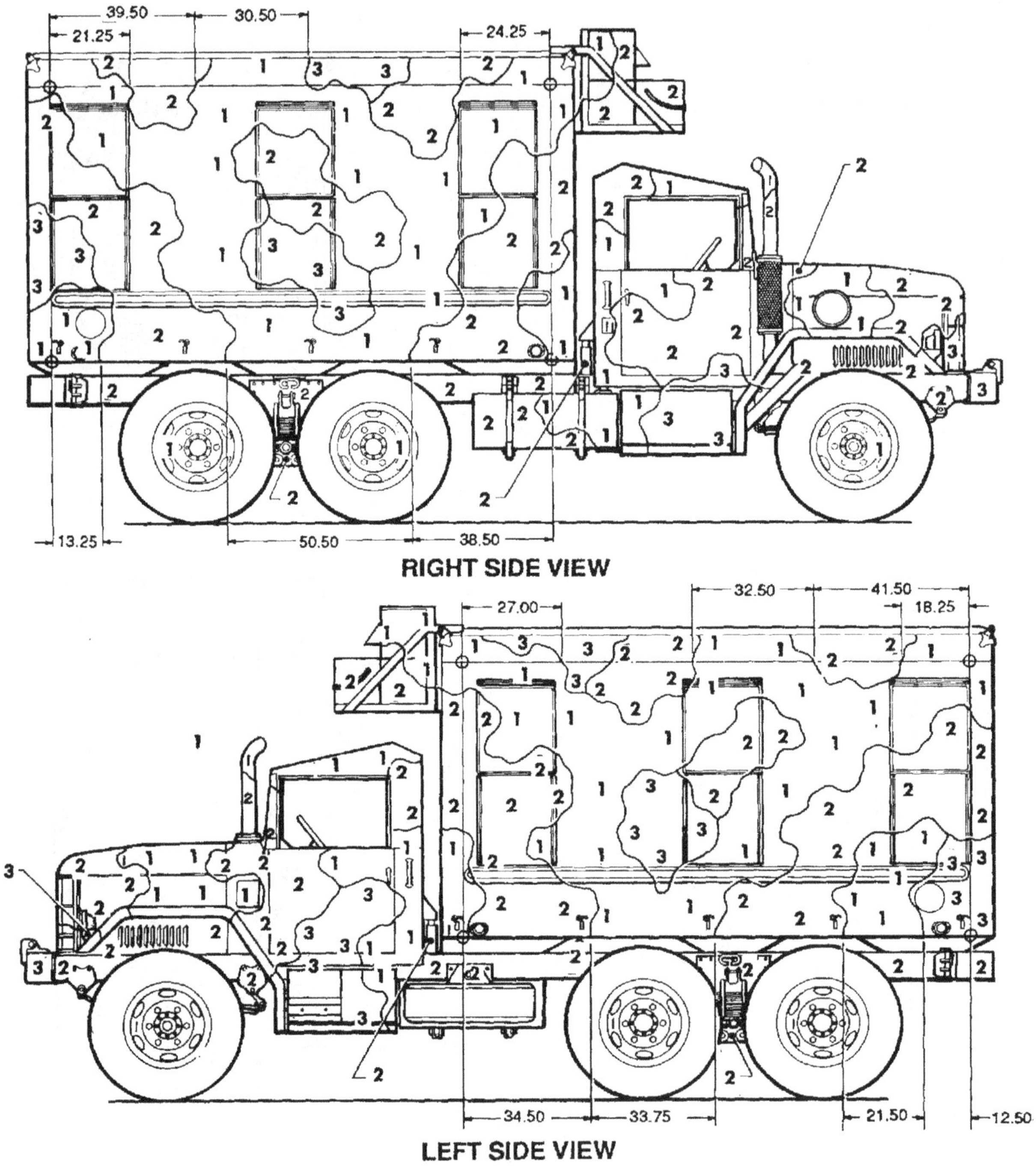

M109A3 M185A3

Figure 52. Truck, shop van: 2-1/2-ton, 6 x 6, M109A3 and M185A3. (2 of 2)

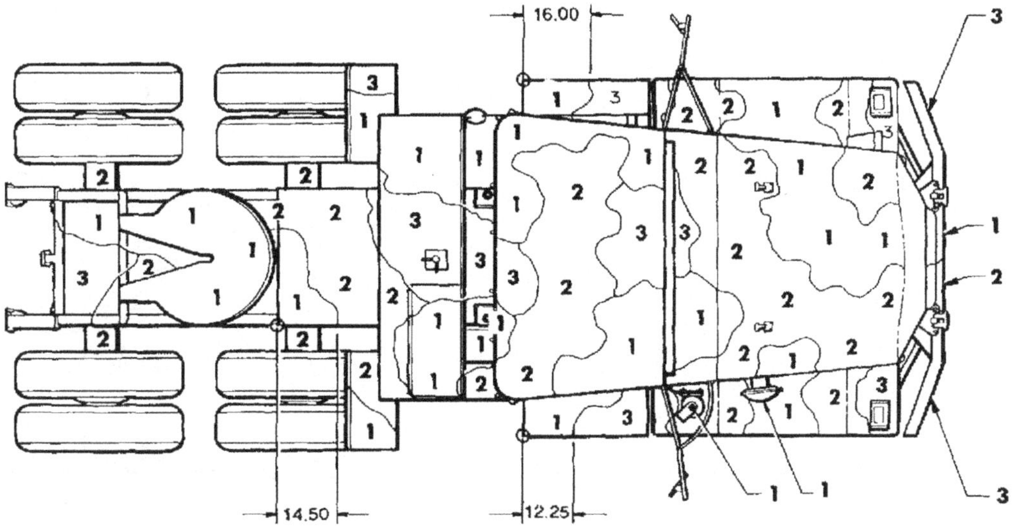

TOP VIEW

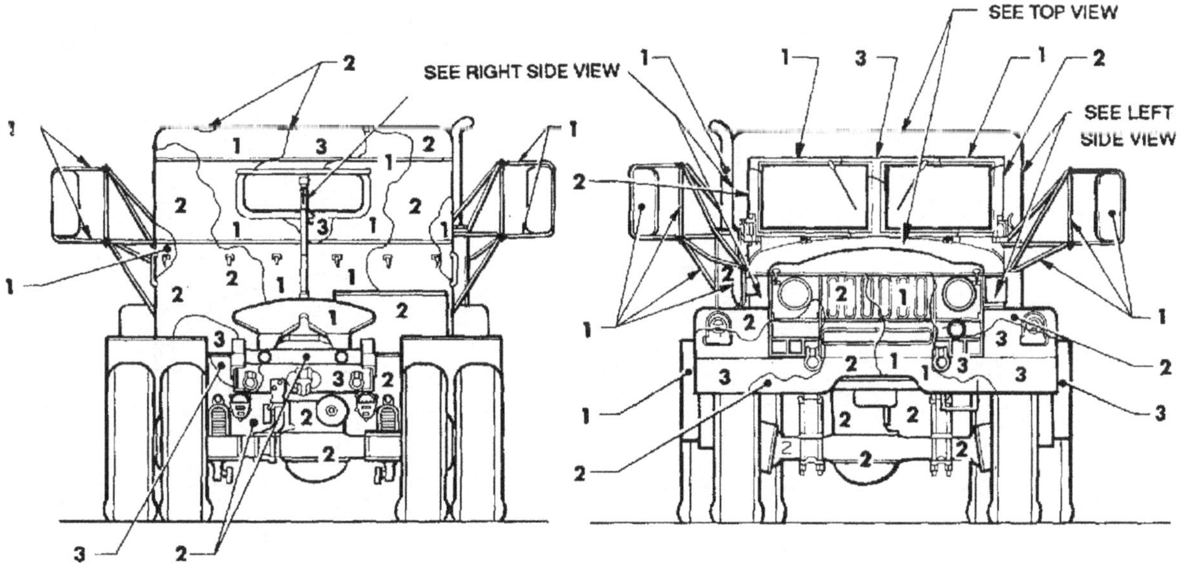

M44 M275A2

REAR VIEW

FRONT VIEW

Figure 53. Truck, tractor: 2-1/2-ton, M44, M275A2. (1 of 2)

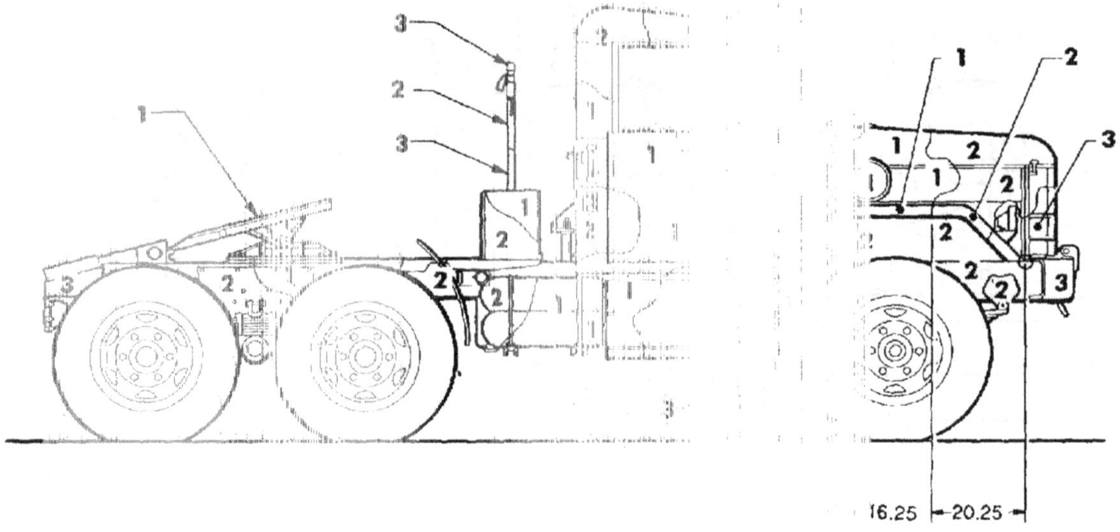

RIGHT SIDE VIEW

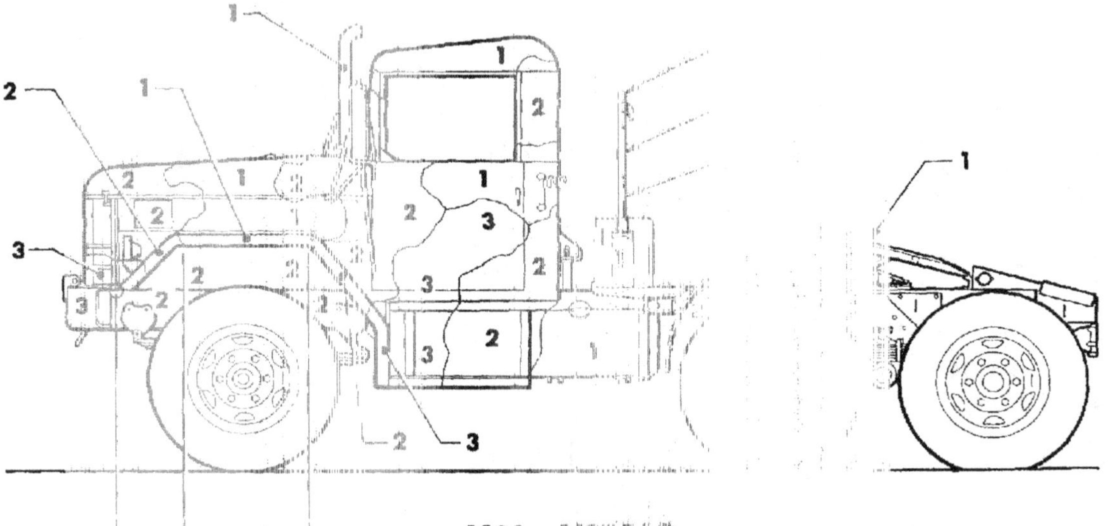

M44 M275A2

LEFT SIDE VIEW

Figure 53. Truck, tractor: 2-1/2-ton, M44 [illegible] of 2)

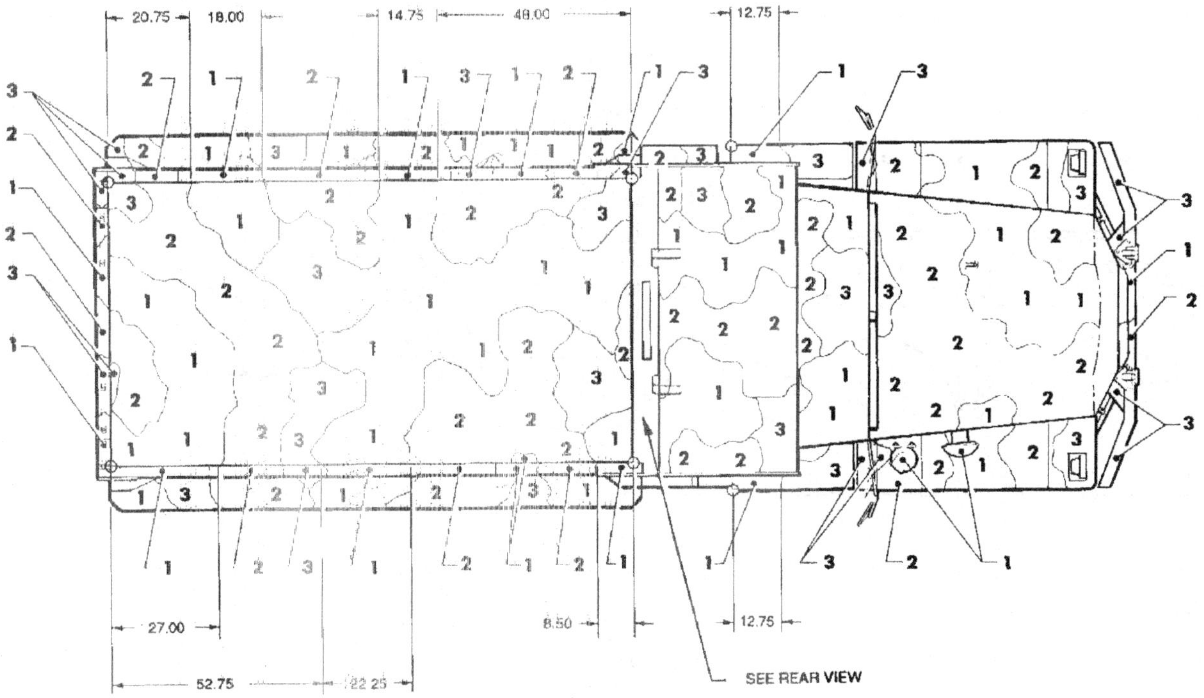

TOP VIEW

M342A2

Figure 54. Truck, dump: 2-1/2-ton, M342A2. (1 of 4)

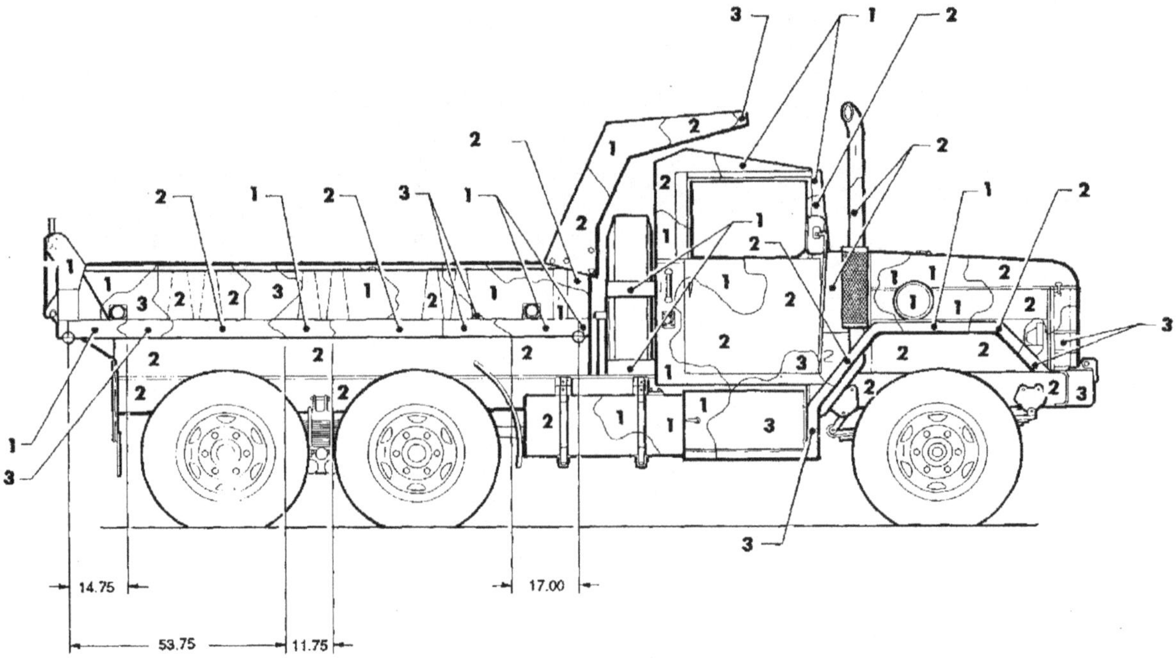

RIGHT SIDE VIEW

M342A2

Figure 54. Truck, dump: 2-1/2-ton, M342A2. (2 of 4)

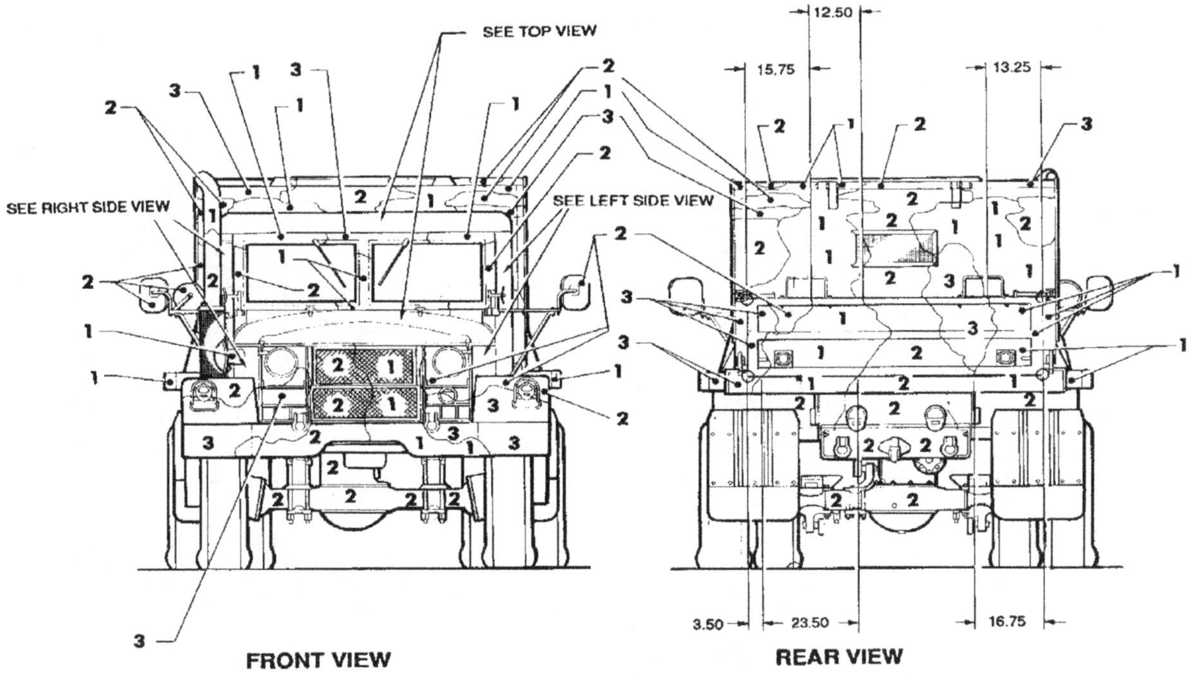

M342A2

Figure 54. Truck, dump: 2-1/2-ton, M342A2. (3 of 4)

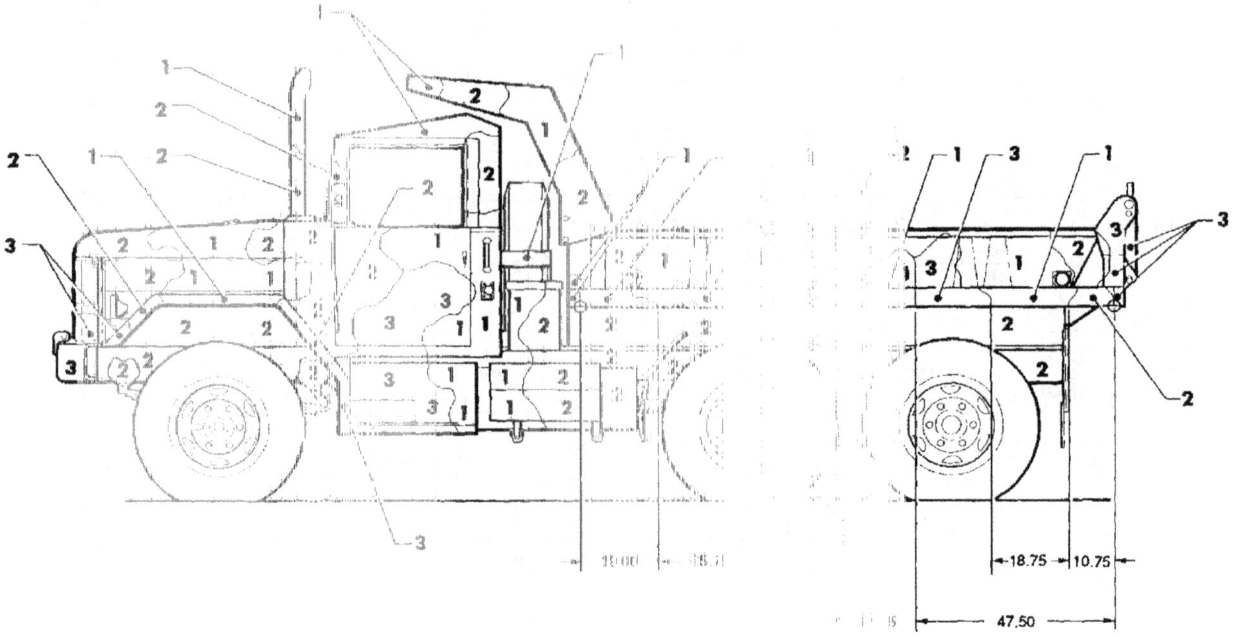

LEFT SIDE VIEW
M342A2

Figure 54. Truck, dump: 2-1/2 ton, M342

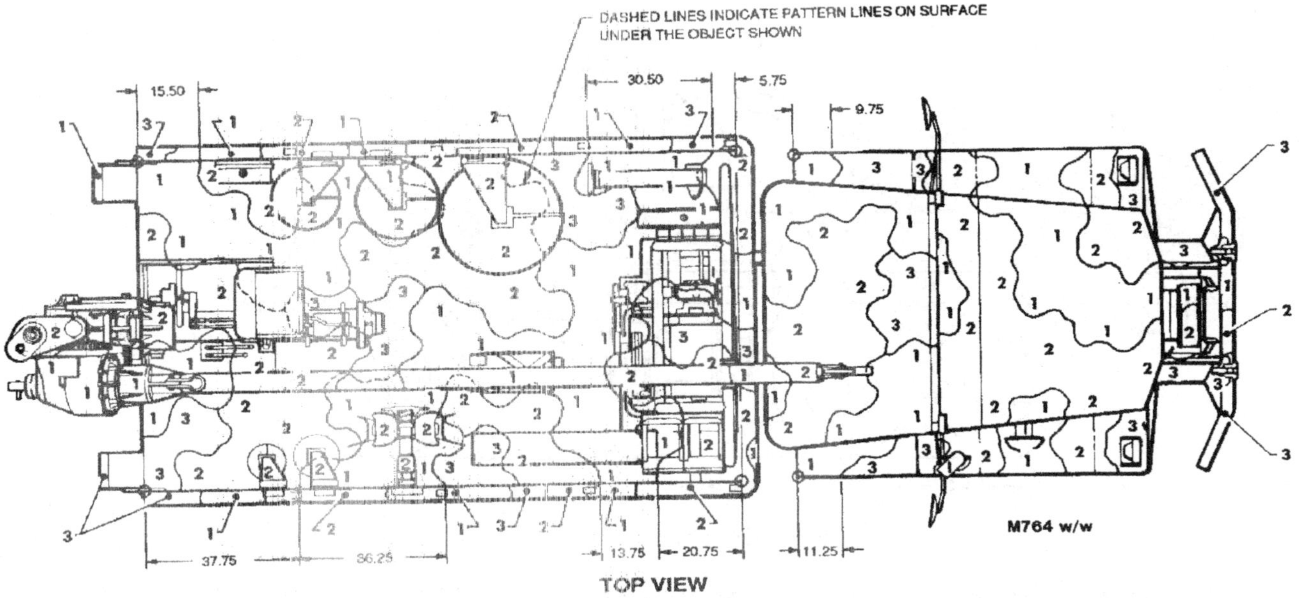

Figure 55. Truck, maintenance, earth boring machine and pole setter: 2-1/2-ton, 6 x 6, M764 w/winch. (I of 4)

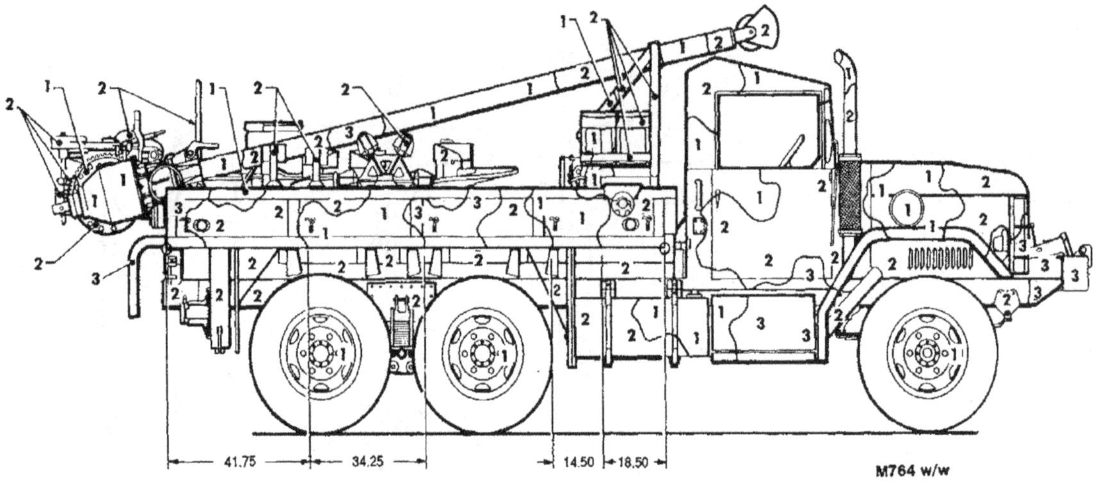

RIGHT SIDE VIEW

Figure 55. Truck, maintenance, earth boring machine and pole setter: 2-1/2-ton, 6 x 6, M764 wlwinch. (2 of 4)

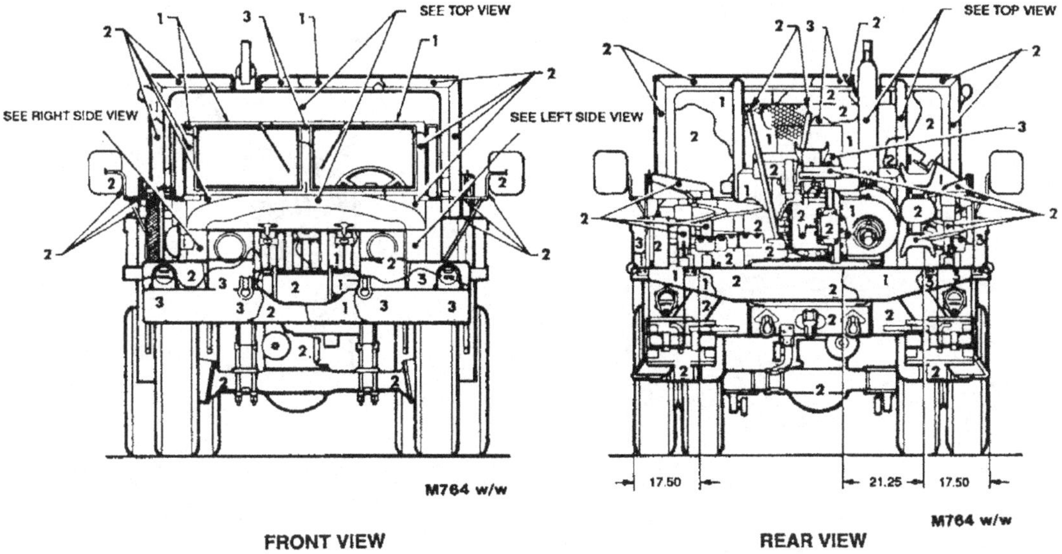

Figure 55. Truck, maintenance, earth boring machine and pole setter: 2-1/2-ton, 6 x 6, M764 wlwinch. (3 of 4)

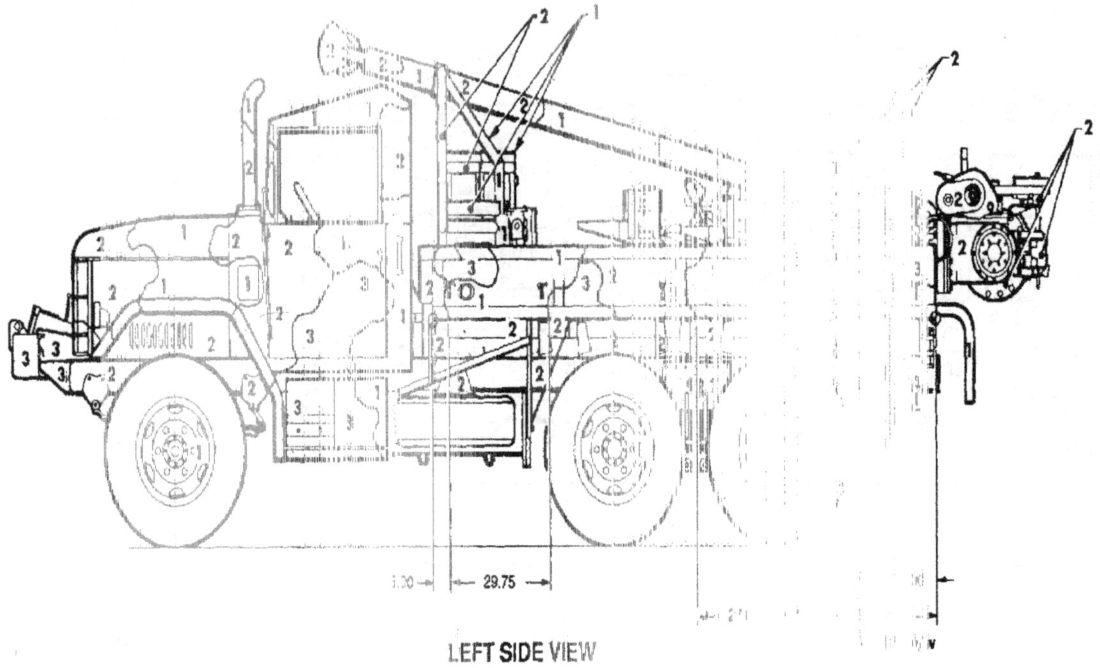

Figure 55. Truck, maintenance, earth boring machine and pole setter ... winch. (4 of 4)

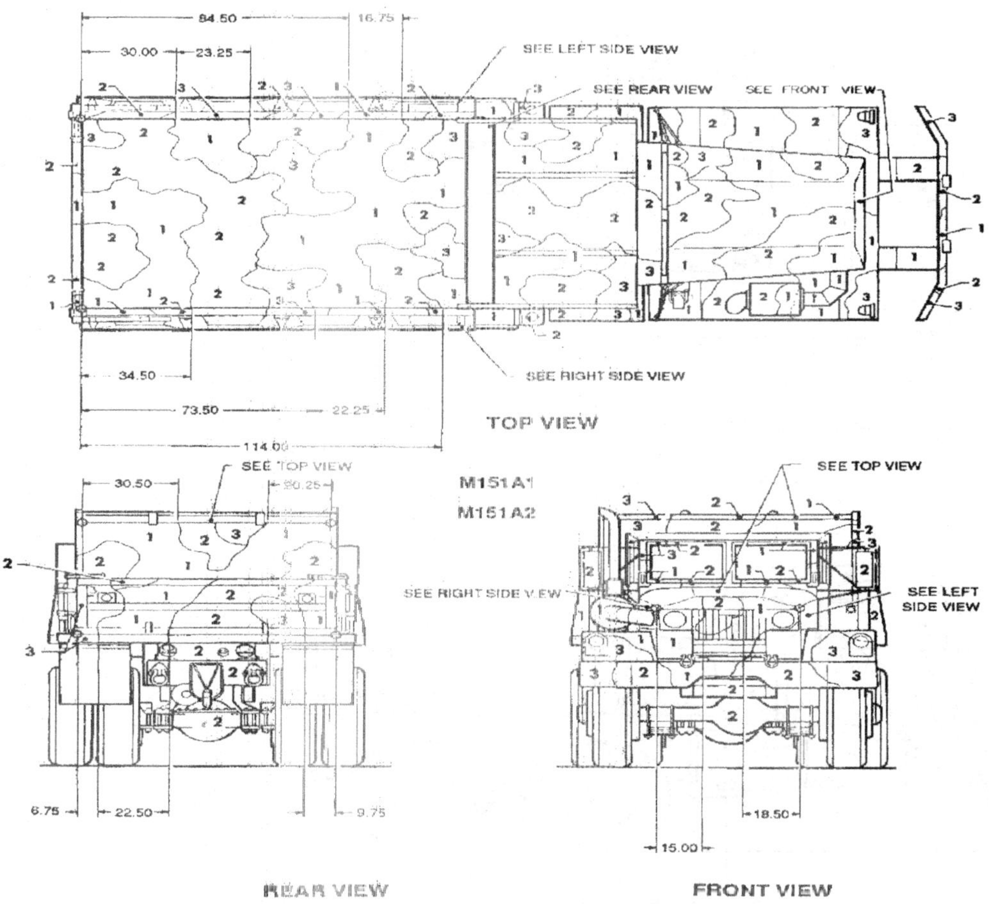

Figure 56. Truck, dump: 5-ton, w/winch, M51AI and M51A2. (I of2)

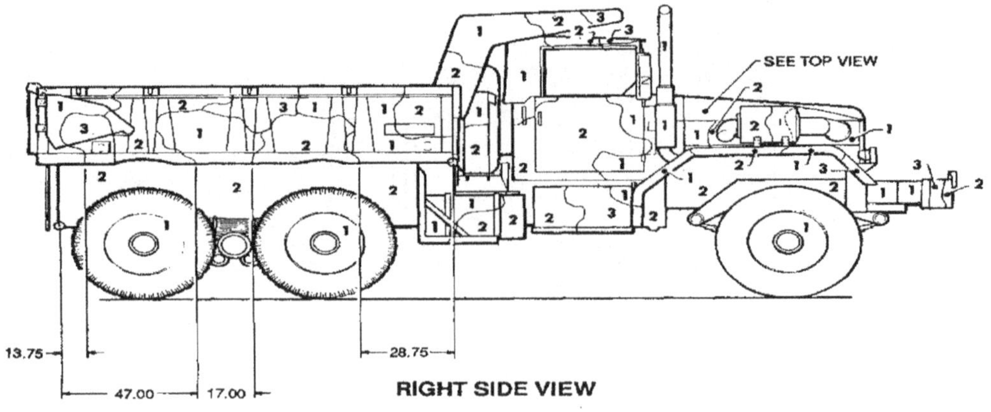

RIGHT SIDE VIEW

M151A1
M151A2

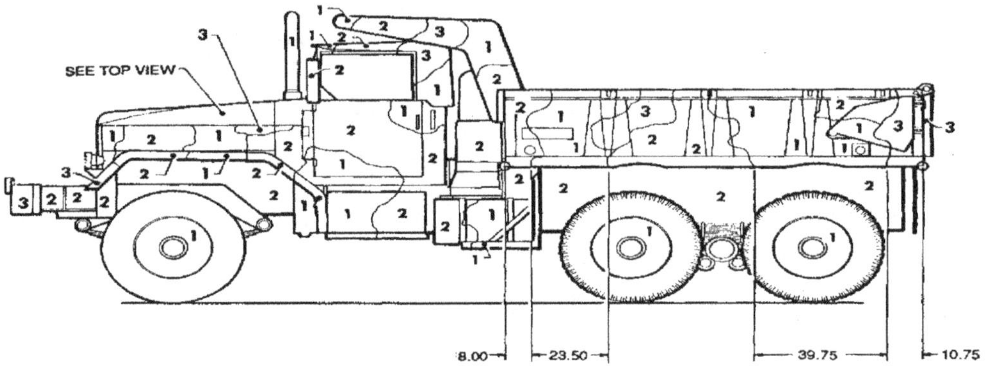

LEFT SIDE VIEW

Figure 56. Truck, dump: 5-ton, w/winch, M51AI and M51A2. (2 of 2)

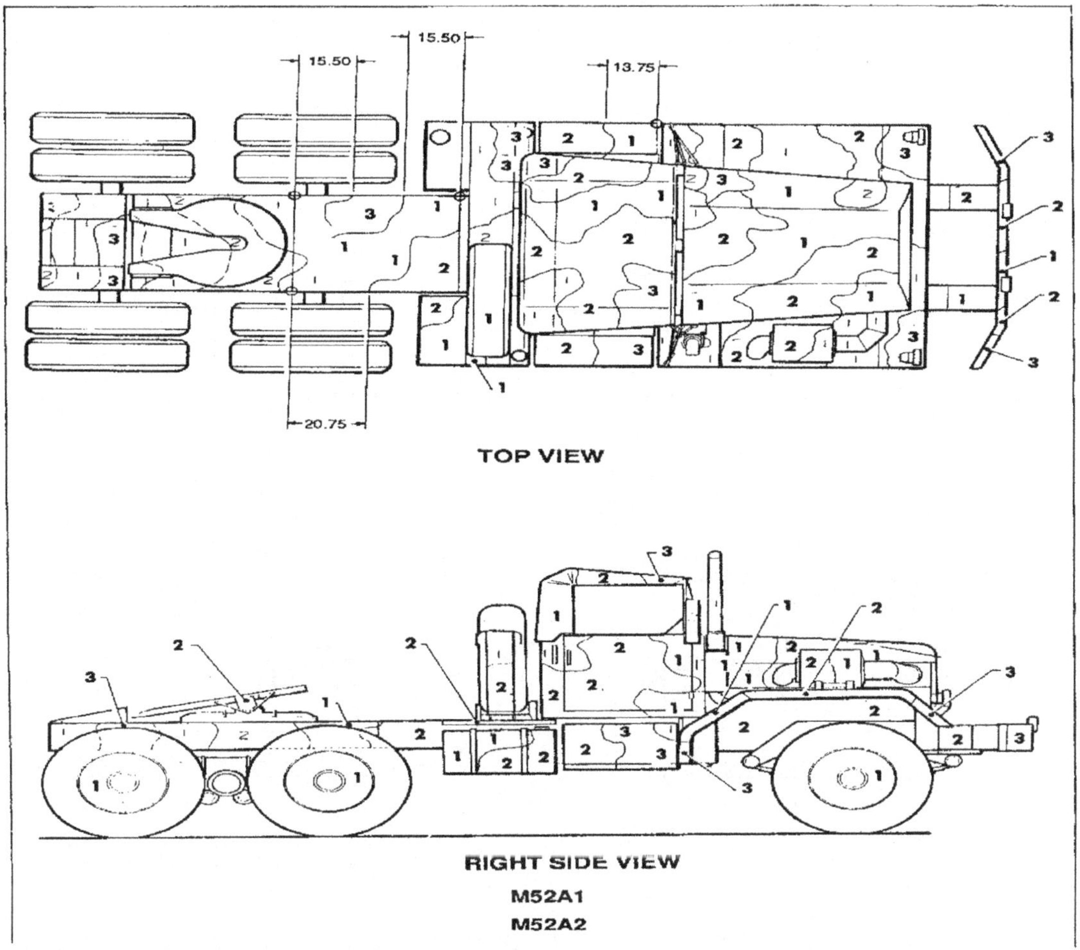

Figure 57. Truck, tractor: 5-ton, wo/lwinch, M52Al and M52A2 (I of2)

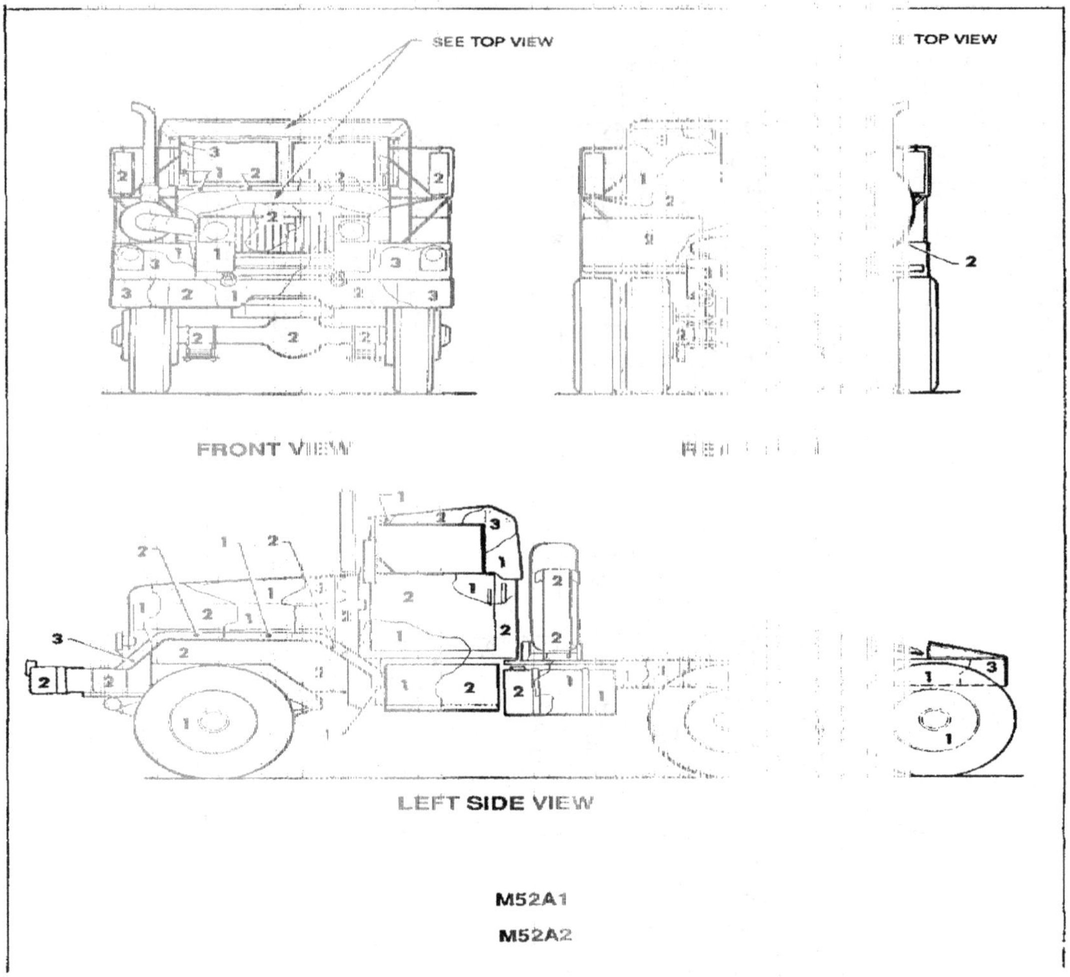

Figure 57. Truck, tractor: 5-ton, wo/lwinch, M52A1 and M52A2

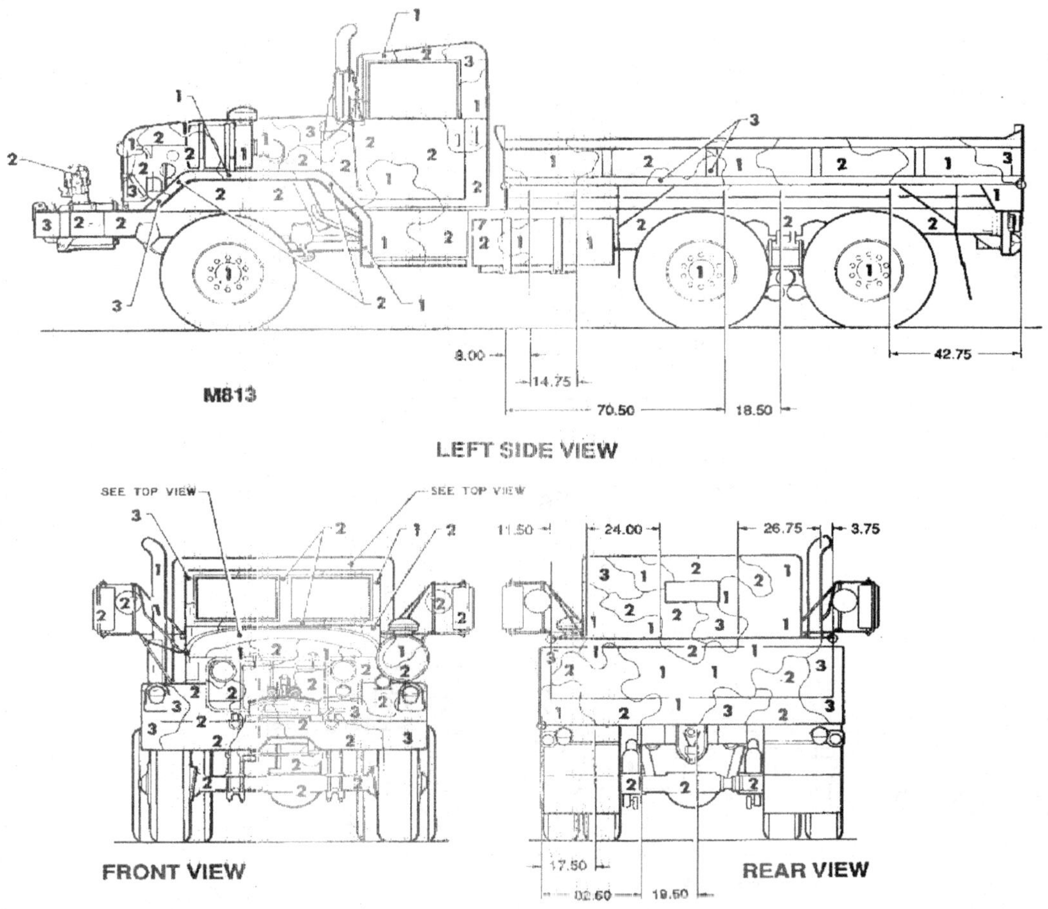

Figure 58. Truck, cargo: 5-ton, w/winch, M813 and M131Al. (I of 2)

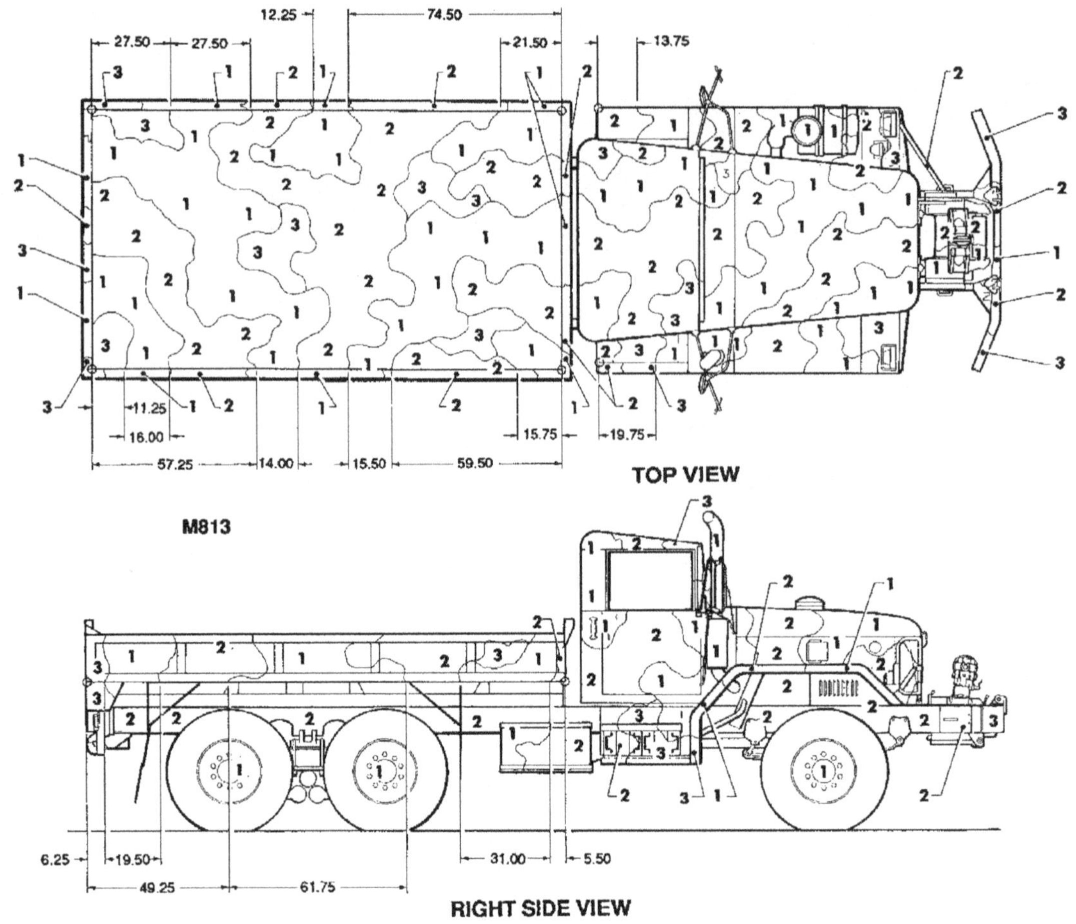

Figure 58. Truck, cargo: 5-ton, wlwinch, M813 and M131AI. (2 of2)

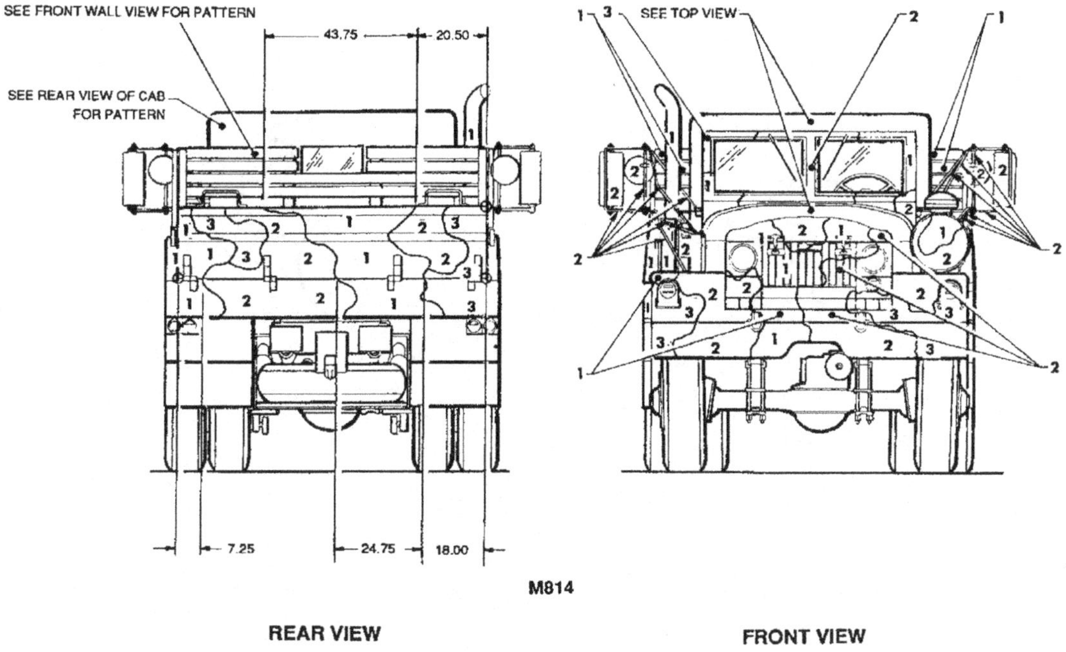

Figure 59. Truck, cargo: 5-ton, 6 x 6, M814 and M55A2. (I of 4)

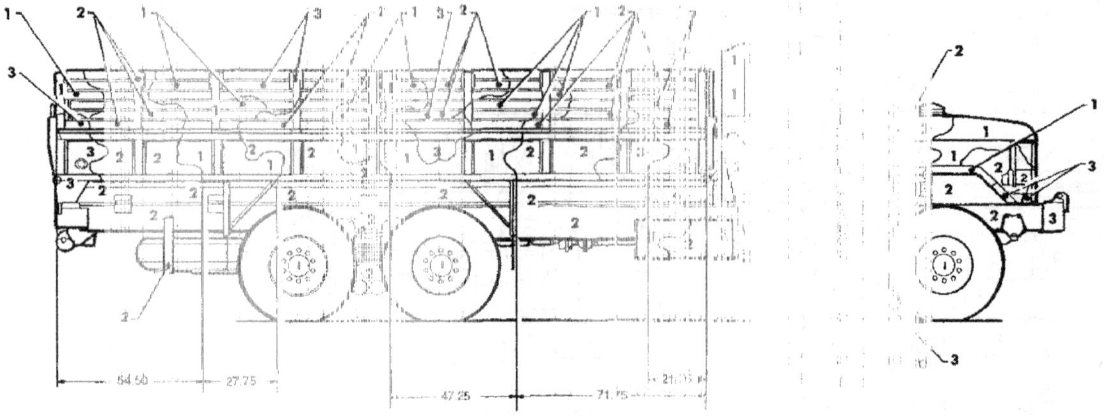

RIGHT SIDE VIEW

M814

Figure 59. Truck, cargo: 5-ton, 6 x 6, M814 and M55A.

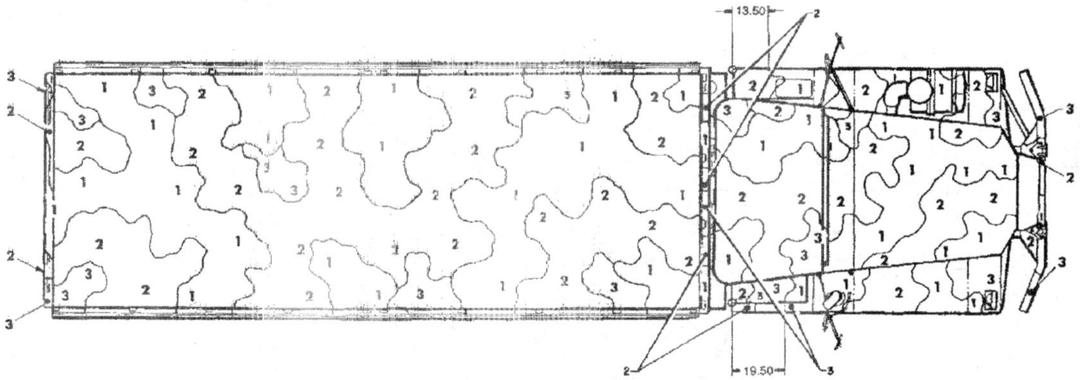

M814

Figure 59. Truck, cargo: 5-ton, 6 x 6. M814 and M55A2. (3 of 4)

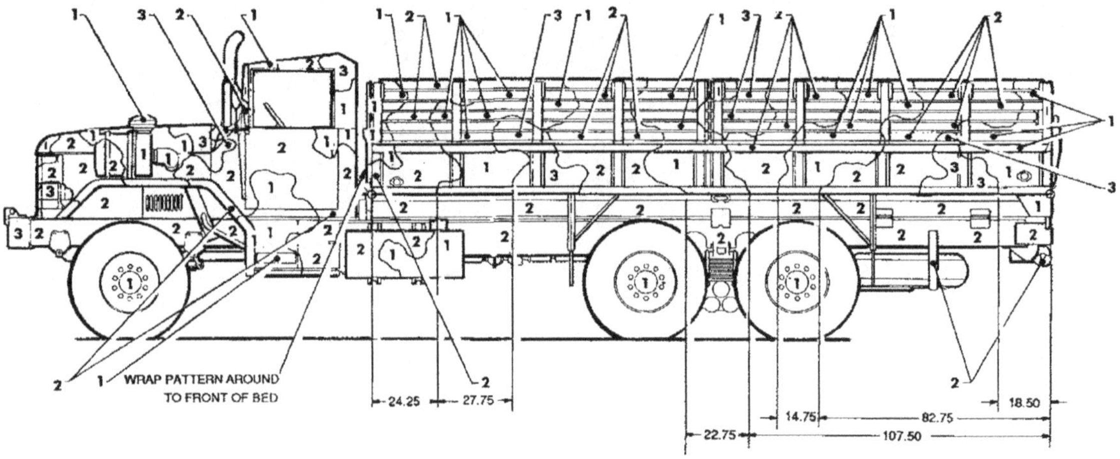

LEFT SIDE VIEW

M814

Figure 59. Truck, cargo: 5-ton, 6 x 6, M814 and M55A2. (4 of 4)

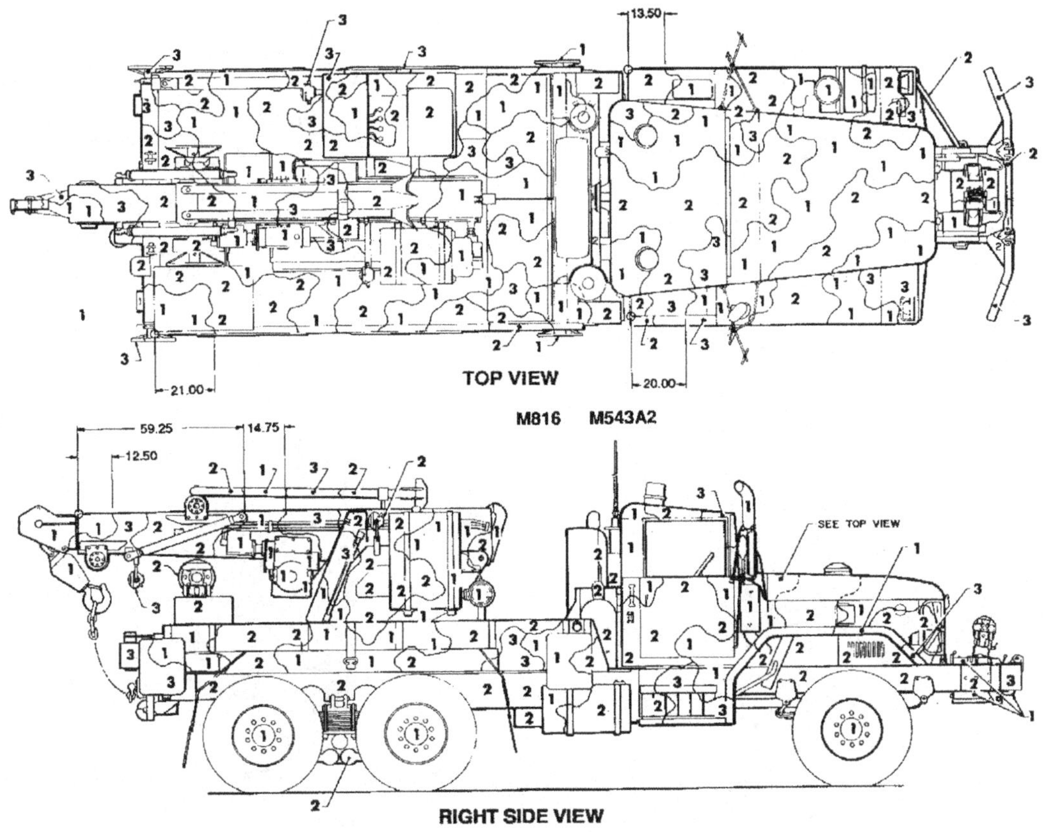

Figure 60. Truck, wrecker: 5-16n, M816 and M543A2. (1 of2)

TB 43-0209

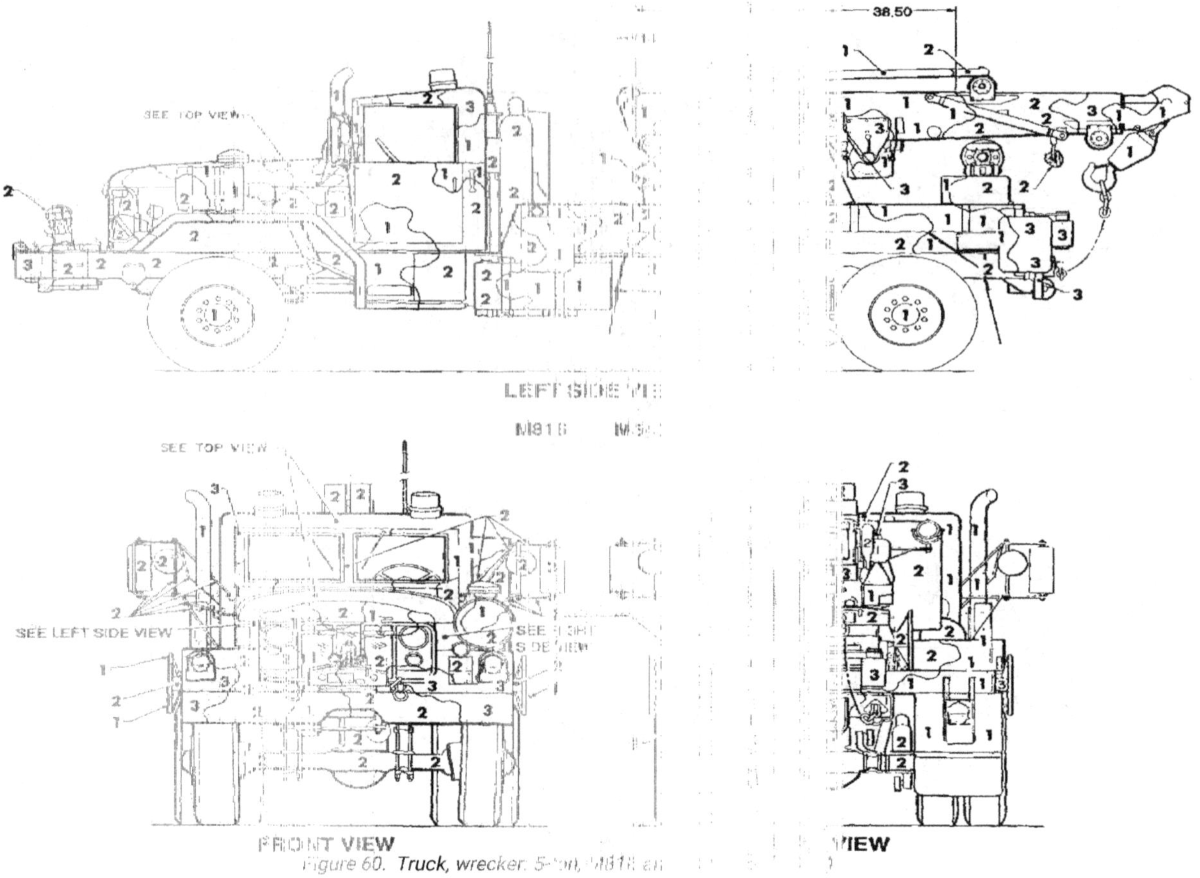

Figure 60. Truck, wrecker: 5-ton, M816 an

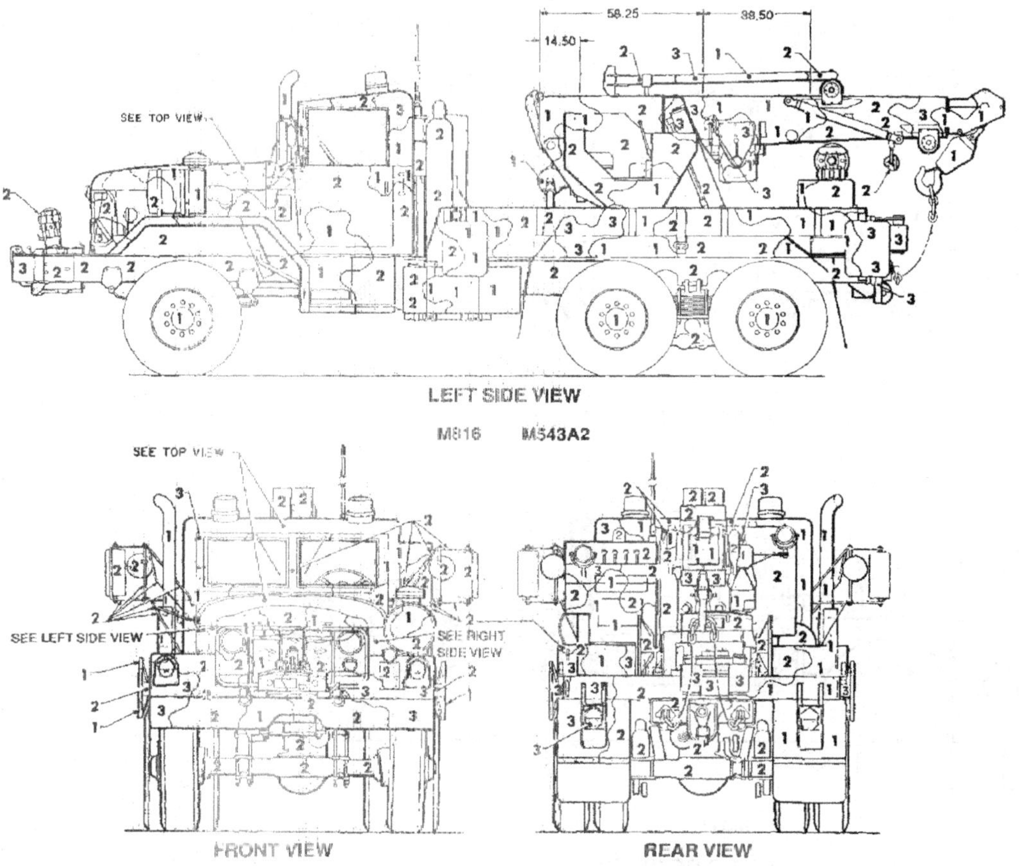

Figure 61. Truck, dump: 5-ton, M817, wlwinch. (l of2)

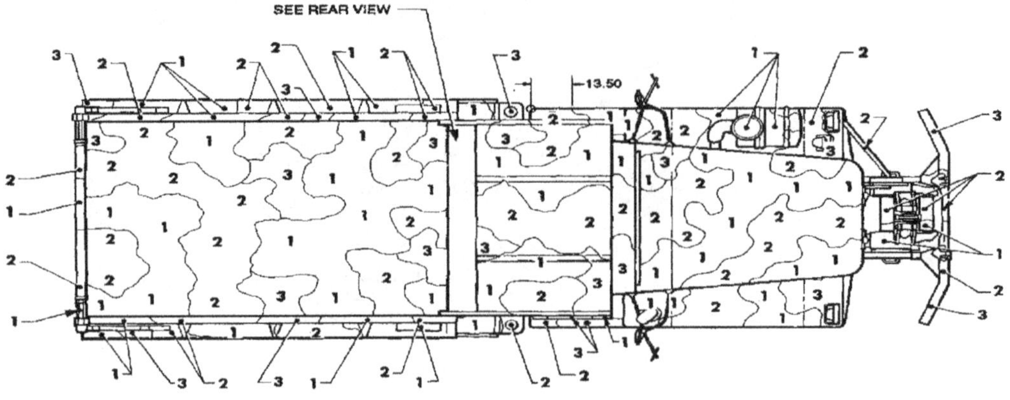

TOP VIEW

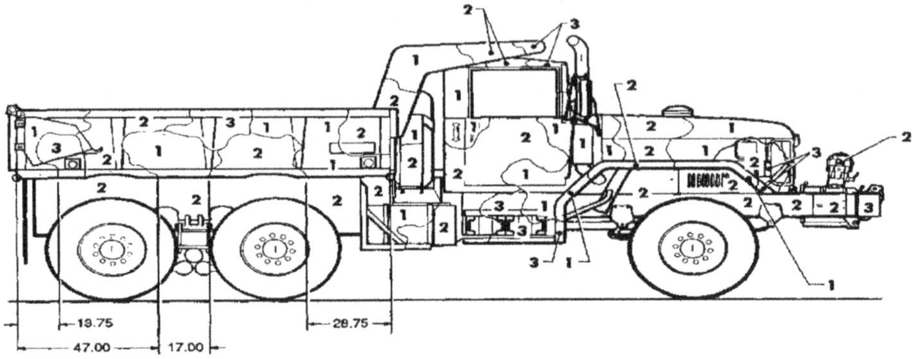

M817

RIGHT SIDE VIEW

Figure 61. Truck, dump: 5-ton, M817, wlwinch. (2 of 2)

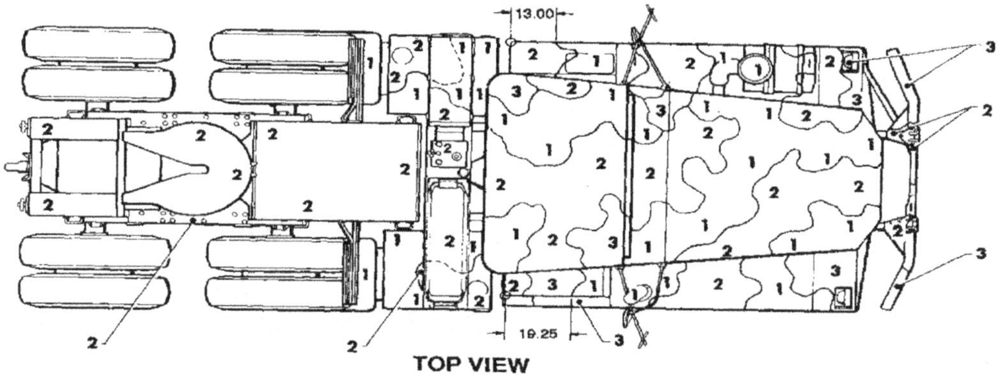

M818

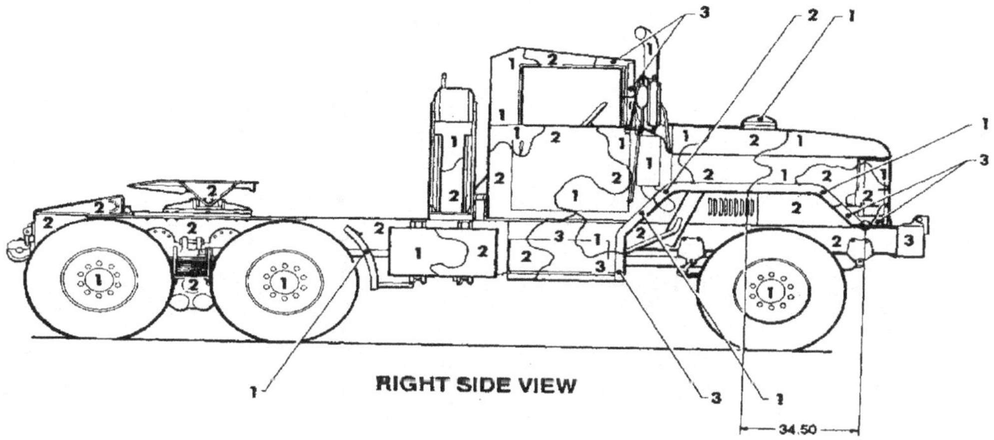

Figure 62. Truck, tractor: 5-ton, 6 x 6, M818. (I of 2)

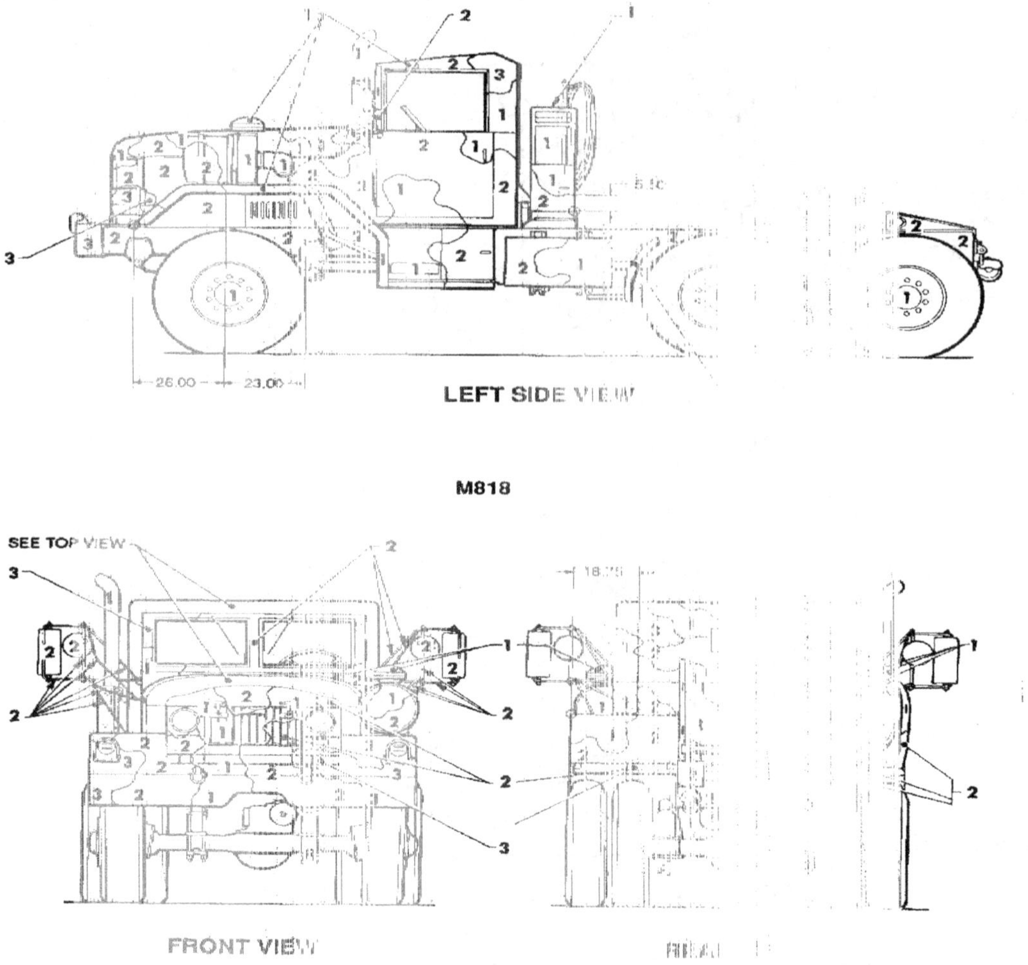

Figure 82. Truck, tractor: 5-ton, 6 x 6, M818.

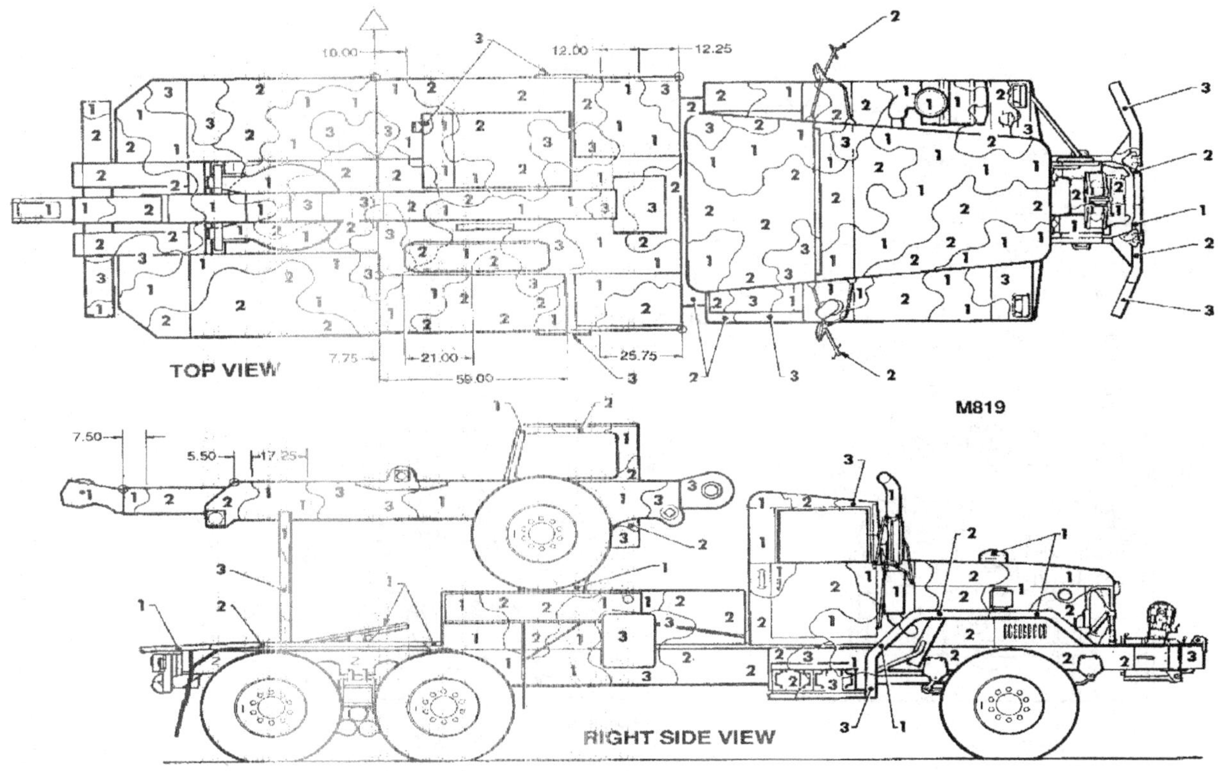

Figure 63. Truck, tractor wrecker: 5-ton, w/winch, M819. (1 of 2)

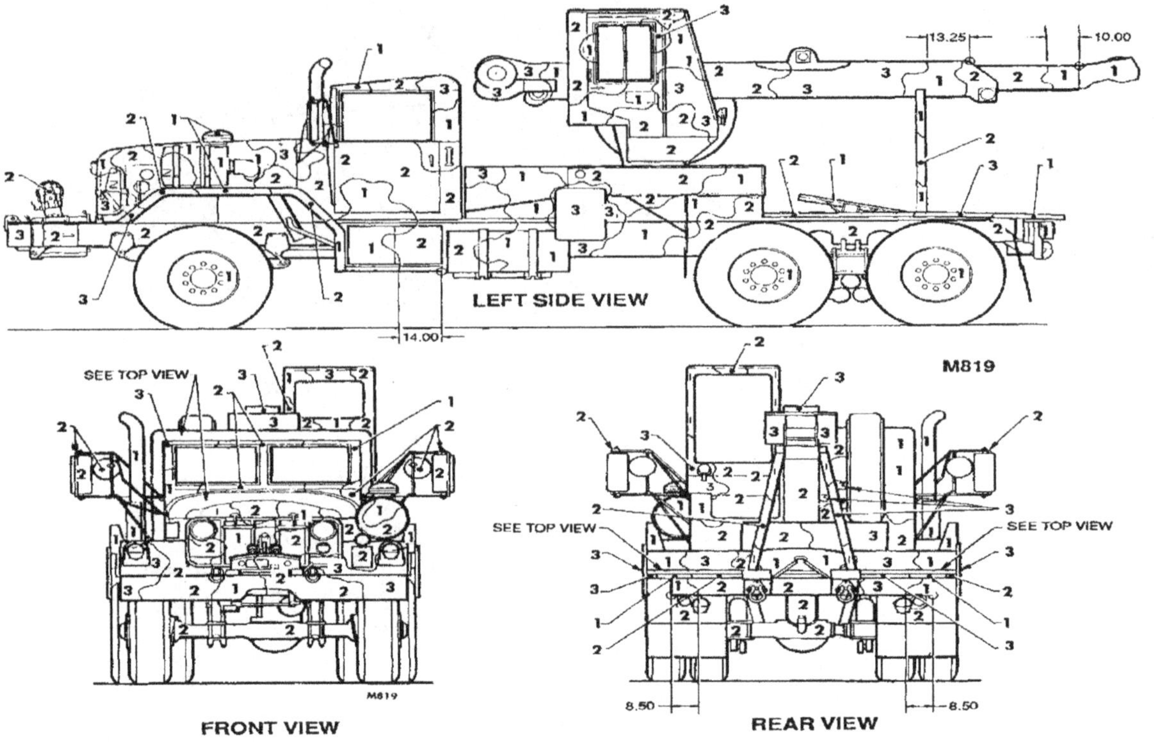

Figure 63. Truck, tractor wrecker: 5-ton, wlwinch, M819. (2 of 2)

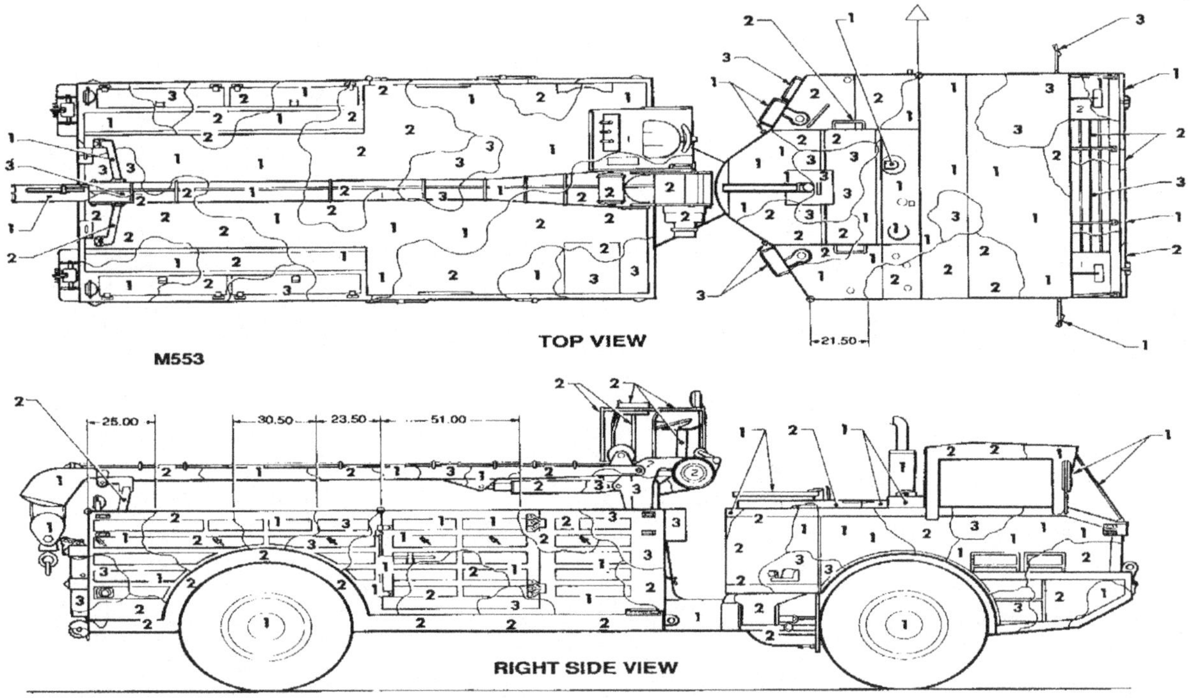

Figure 64. Truck, wrecker: 10-ton, 4 x 4, M553. (1 of 2)

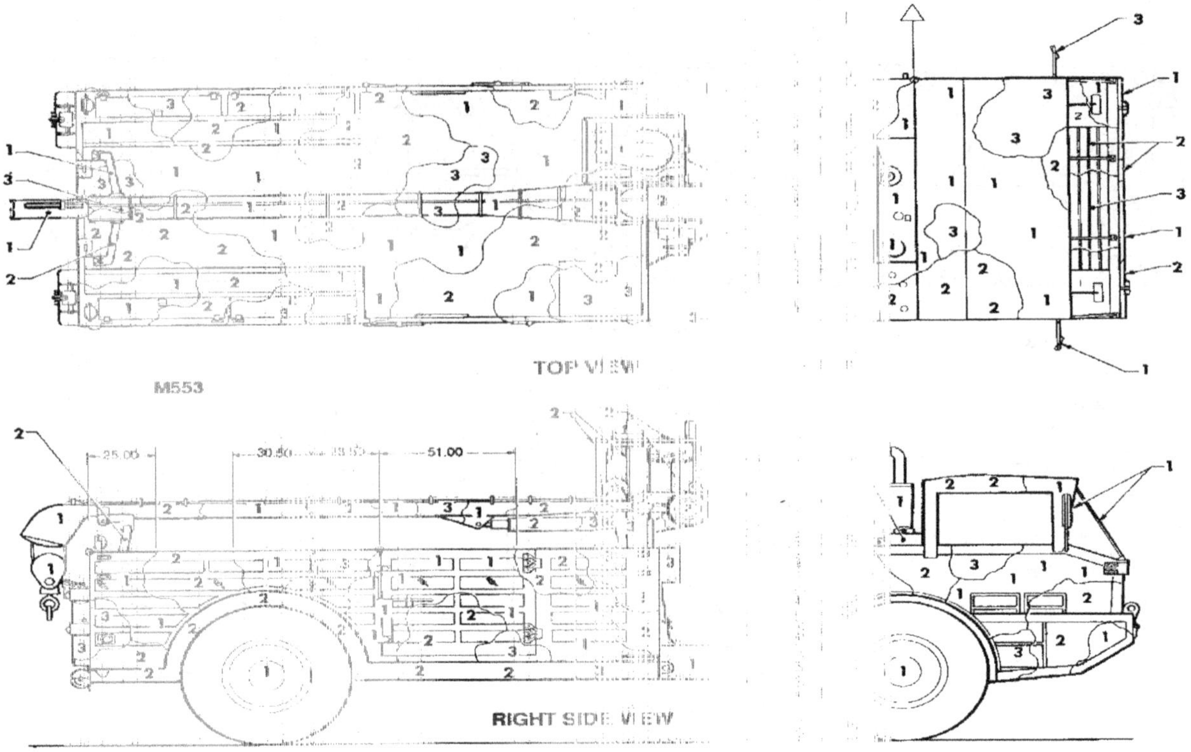

Figure 64. Truck, wrecker 10-ton, 4x4, (

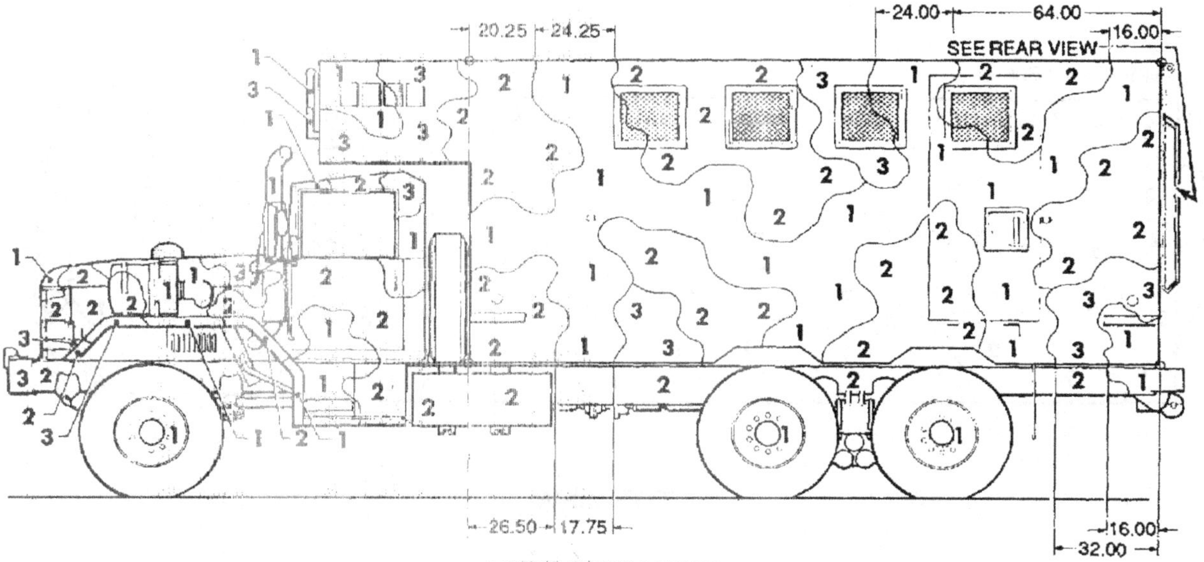

LEFT SIDE VIEW

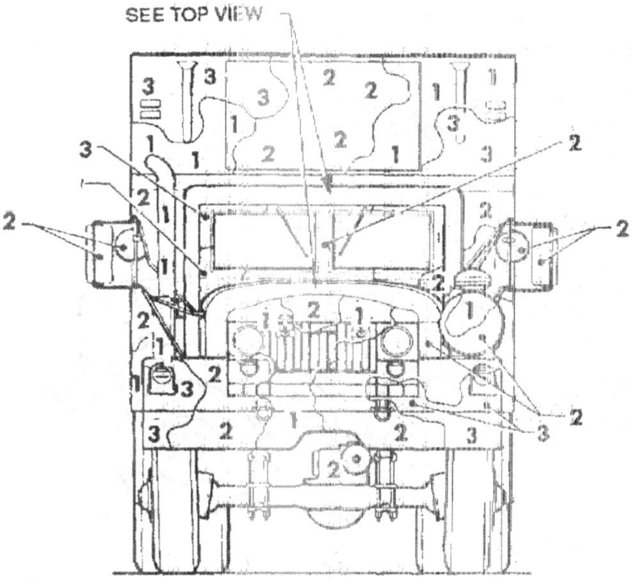

FRONT VIEW

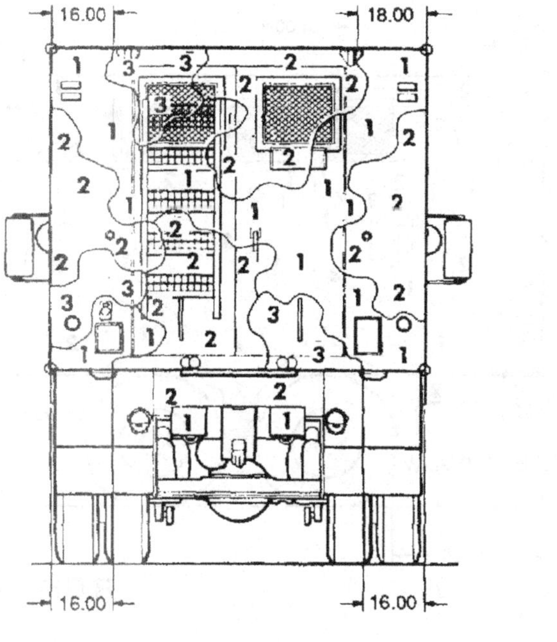

REAR VIEW

Figure 65. Truck, van, expansible: 5-ton, M820A2. (l of 2)

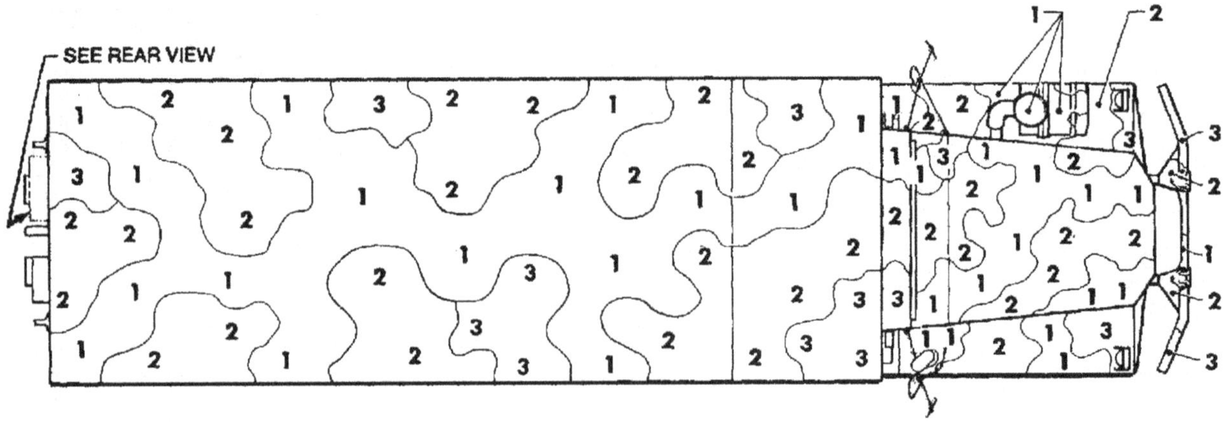

TOP VIEW

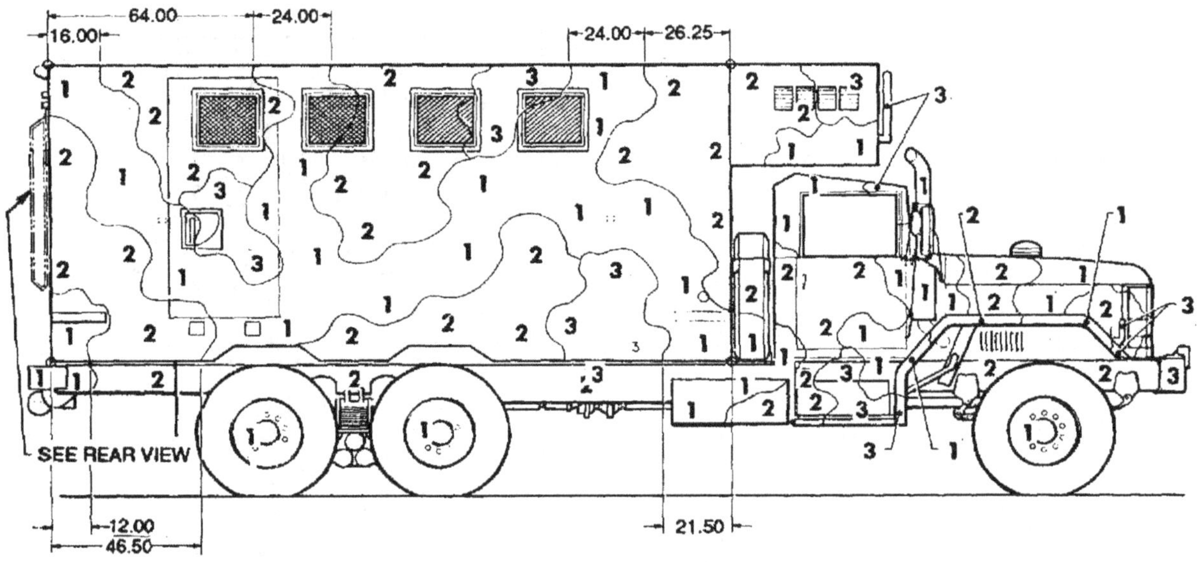

RIGHT SIDE VIEW

Figure 65. Truck, van, expansible: 5-ton, M820A2. (2 of2)

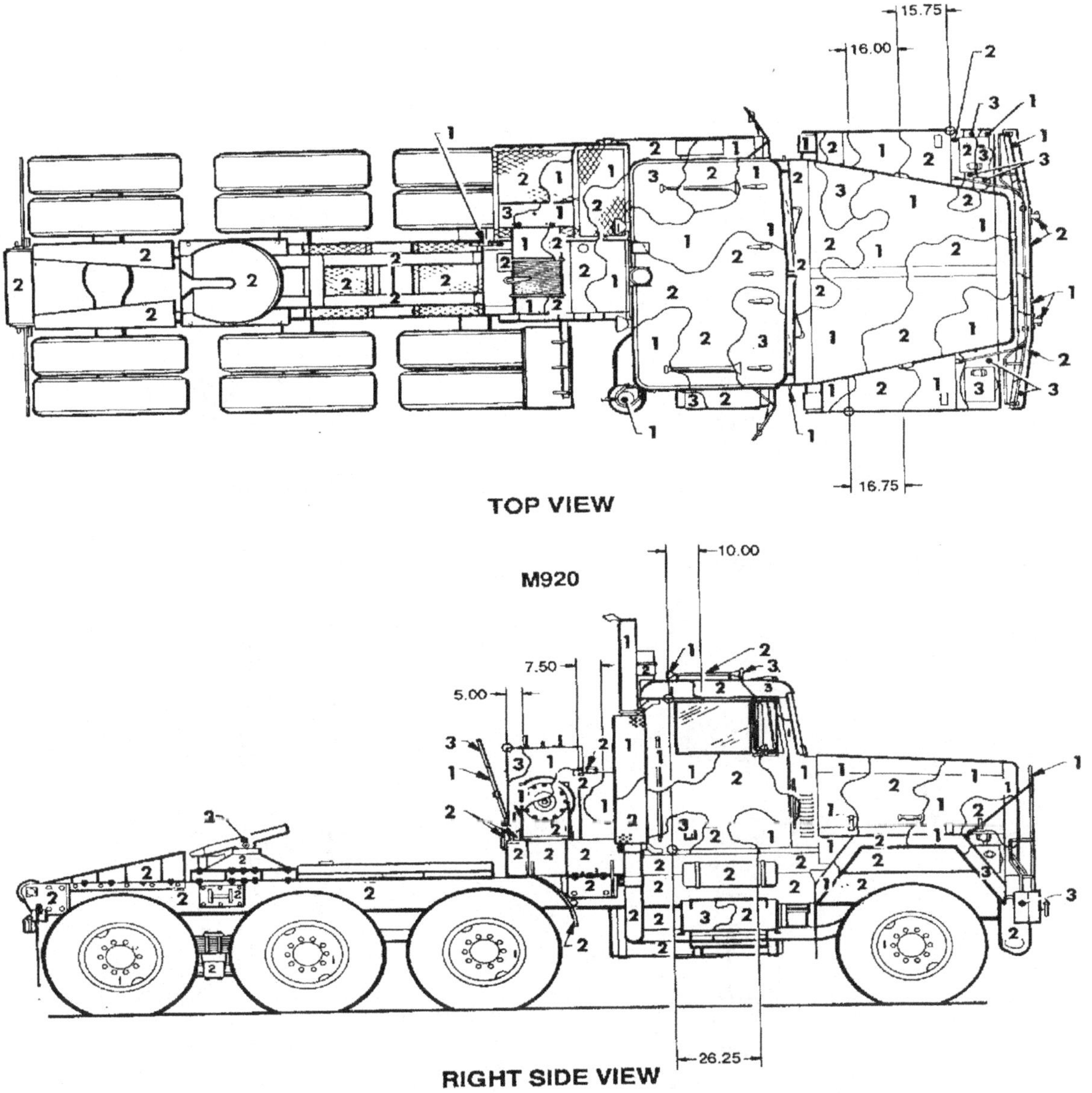

Figure 66. Truck, tractor: 5-ton, M920. (I of2)

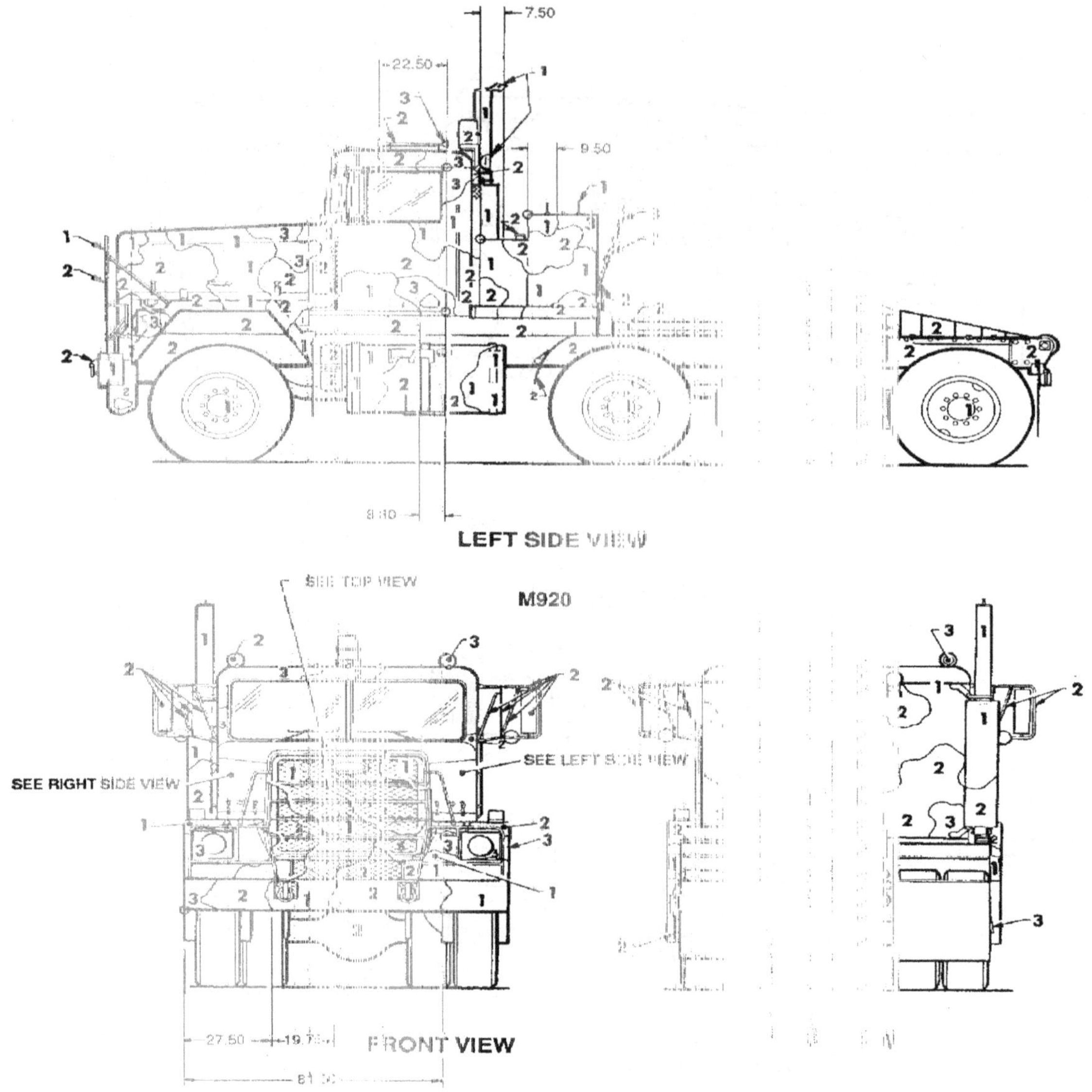

Figure 66.

Truck, tractor: 5-ton, M920.

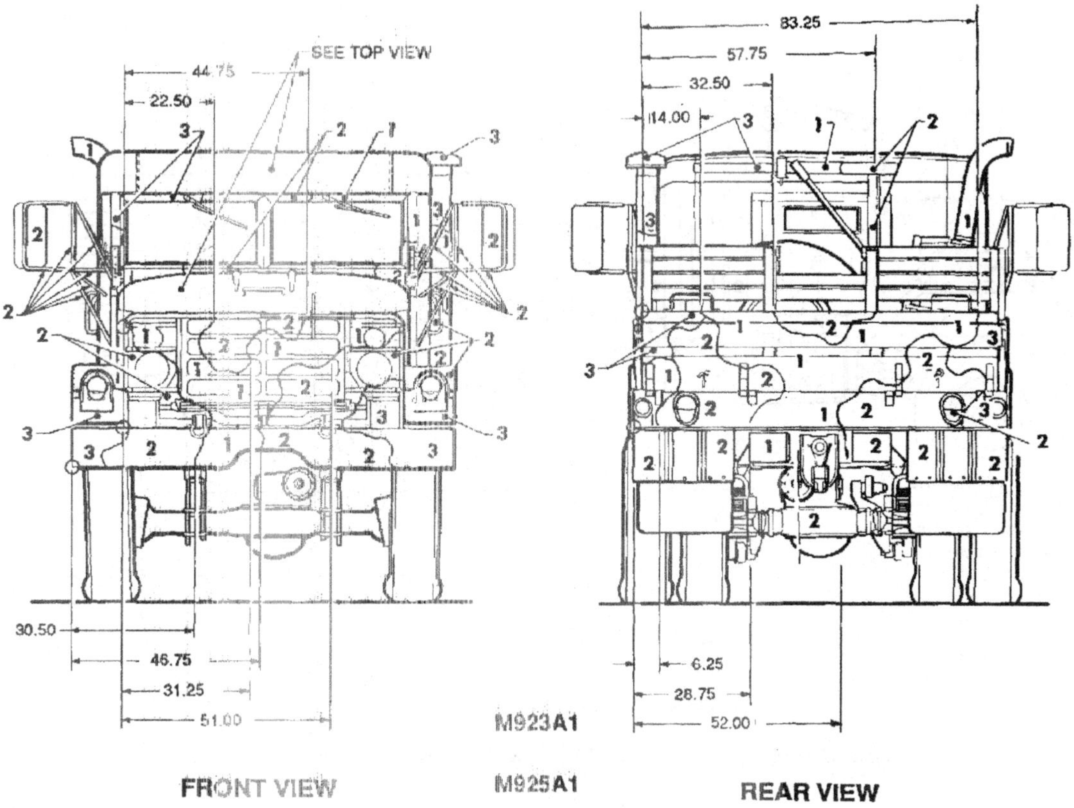

Figure 67. Truck, cargo: 5-ton, M923A1 and M925A1. (1 of 4)

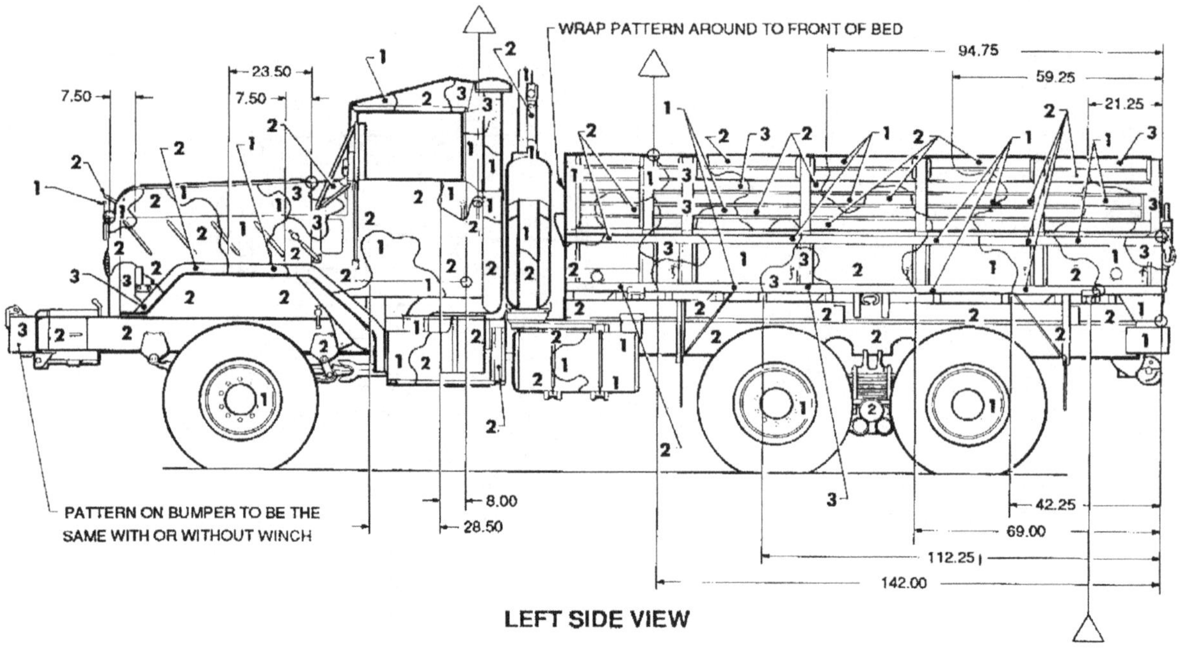

LEFT SIDE VIEW

M923A1 M925A1

Figure 67. Truck, cargo: 5-ton, M923A1 and M925A1. (2 of 4)

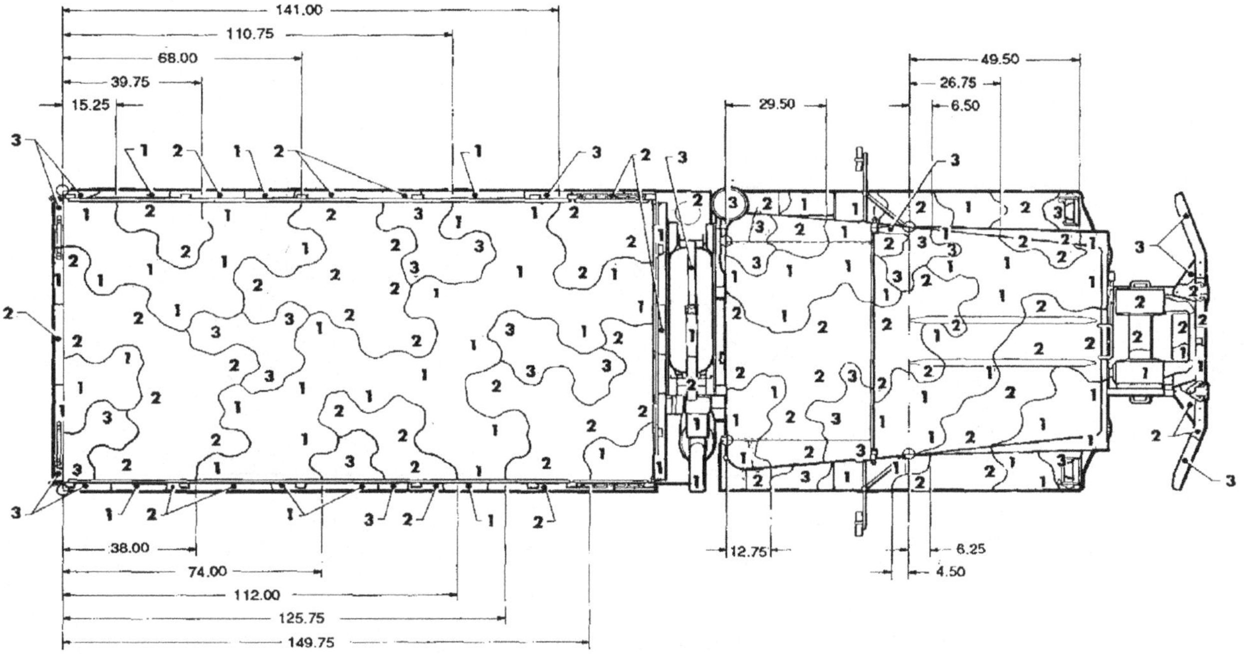

TOP VIEW

M923A1 M925A1

Figure 67. Truck, cargo: 5-ton, M923A1 and M925AI. (3 of4)

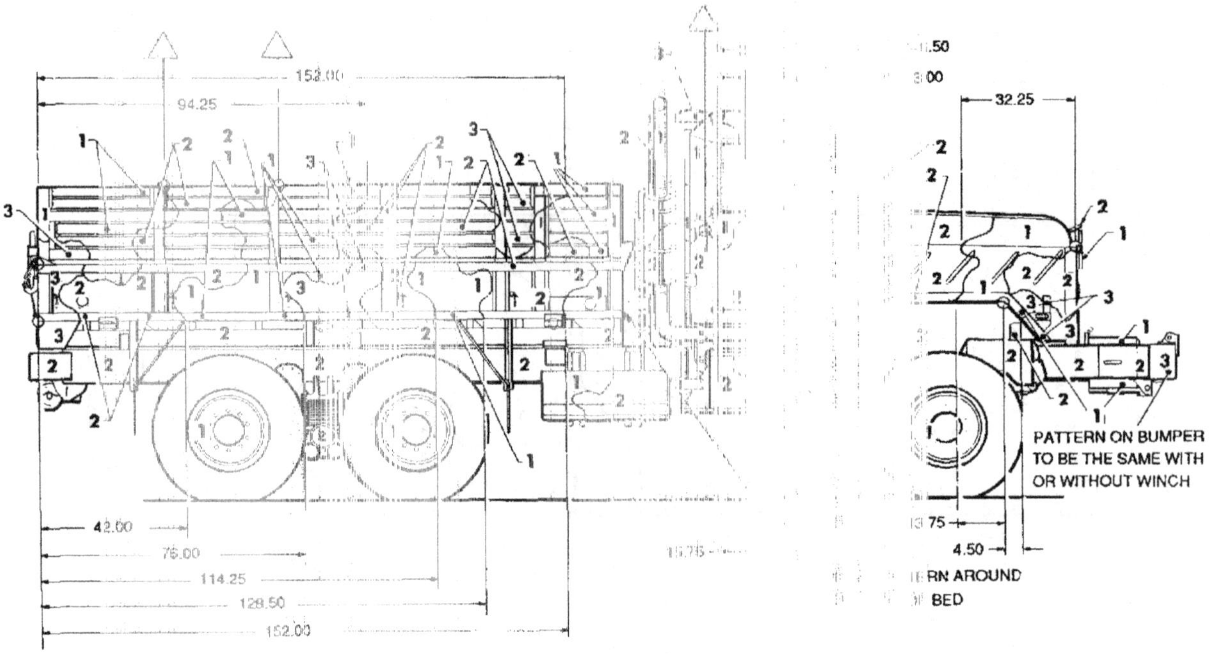

RIGHT SIDE VIEW

M923A1

Figure 67. Truck, cargo: 5-ton, M923A1 and

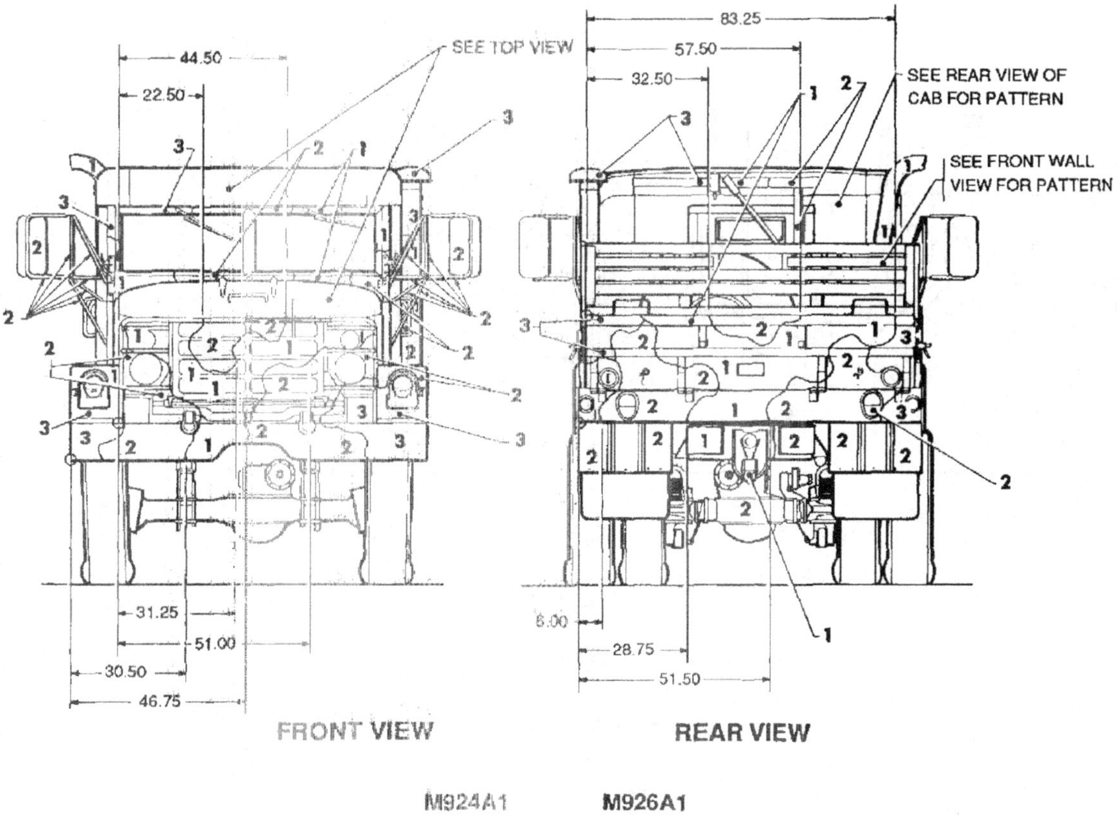

Figure 68. Truck, cargo: 5-ton, M924A1 and M926A1. (1 of 4)

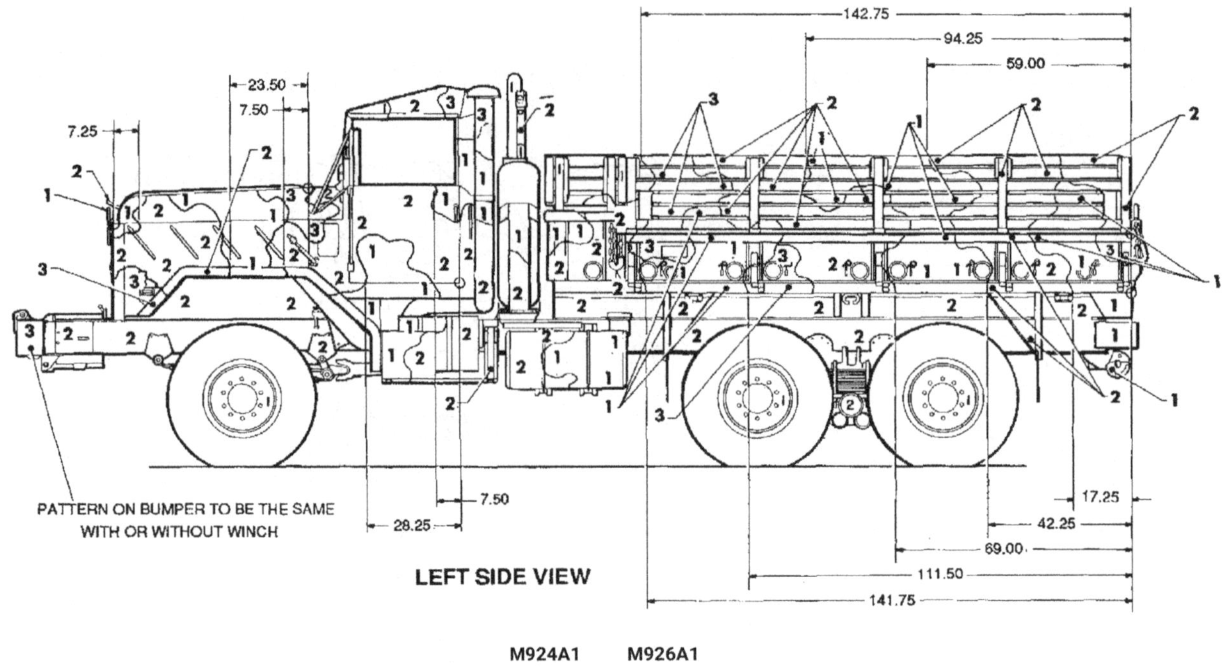

Figure 68. Truck, cargo: 5-ton, M924AI and M926AI. (2 of 4)

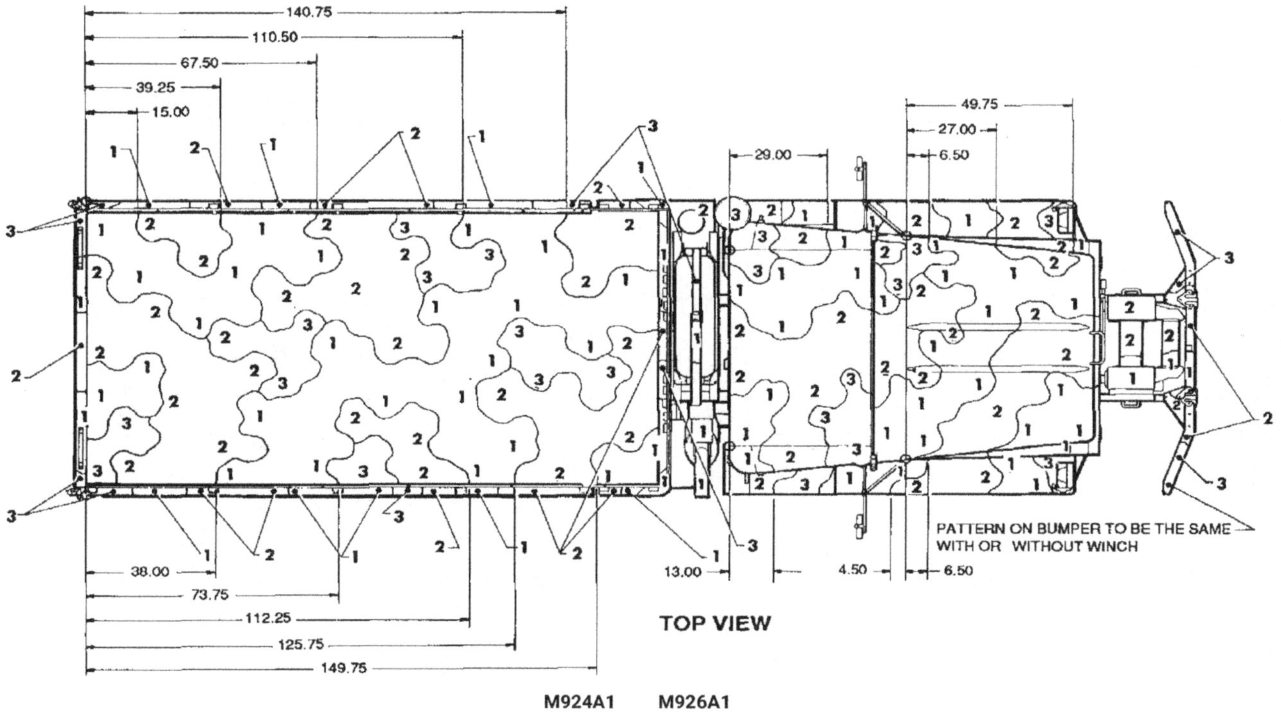

Figure 68. Truck, cargo: 5-ton, M924AI and M926A1. (3 of 4)

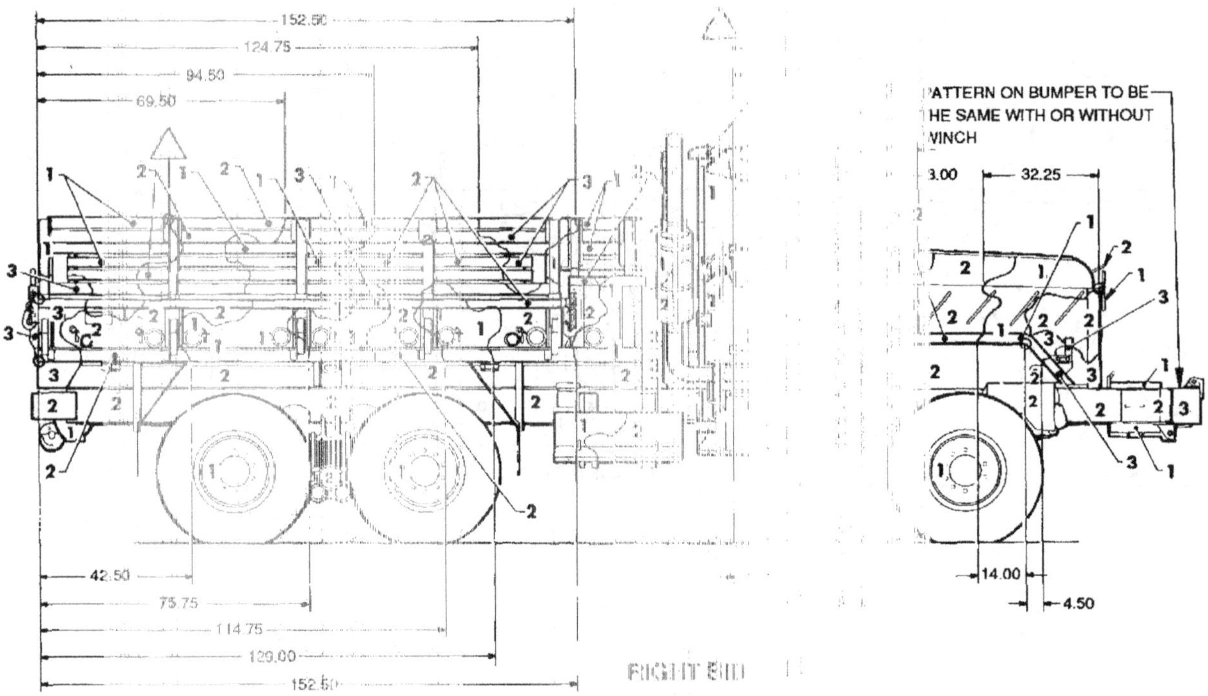

M924A1 M925A1

Figure 68. Truck, cargo: 5-ton, M924A1 and M925A1

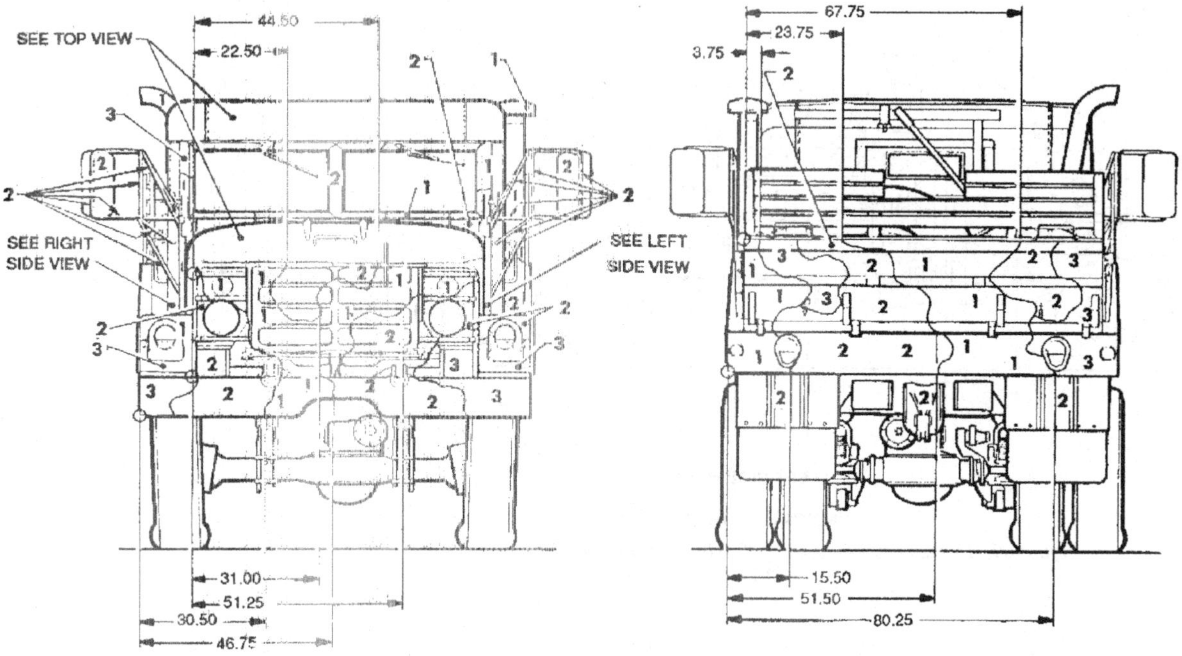

FRONT VIEWM928/A1 REAR VIEW M928/AI

Figure 69. Truck, cargo: 5-ton, XLWB, M927A1 and M928A1. (I of 4)

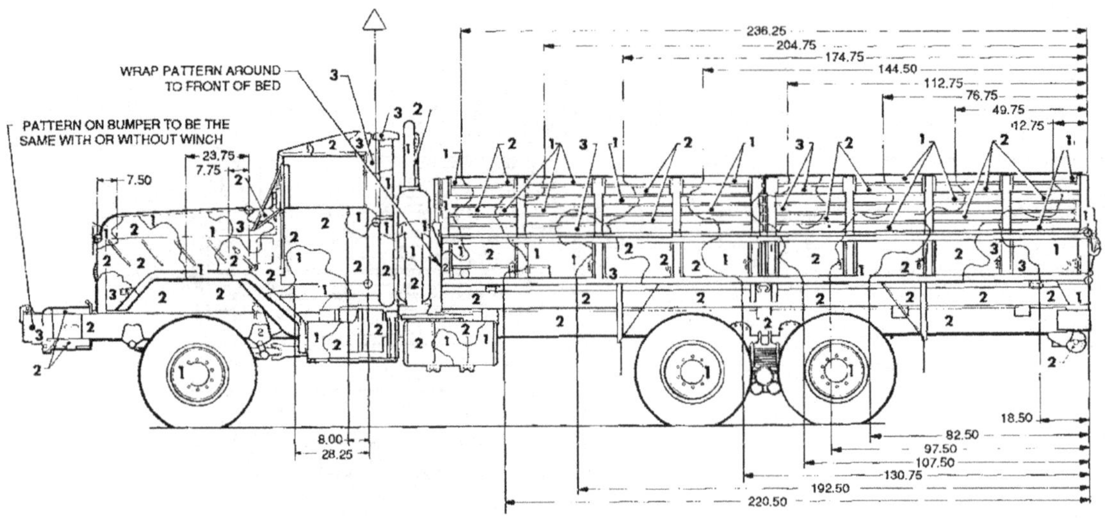

LEFT SIDE VIEW

M927A1 M928A1

Figure 69. Truck, cargo: 5-ton, XLWB, M927A1 and M928AI. (2 of 4)

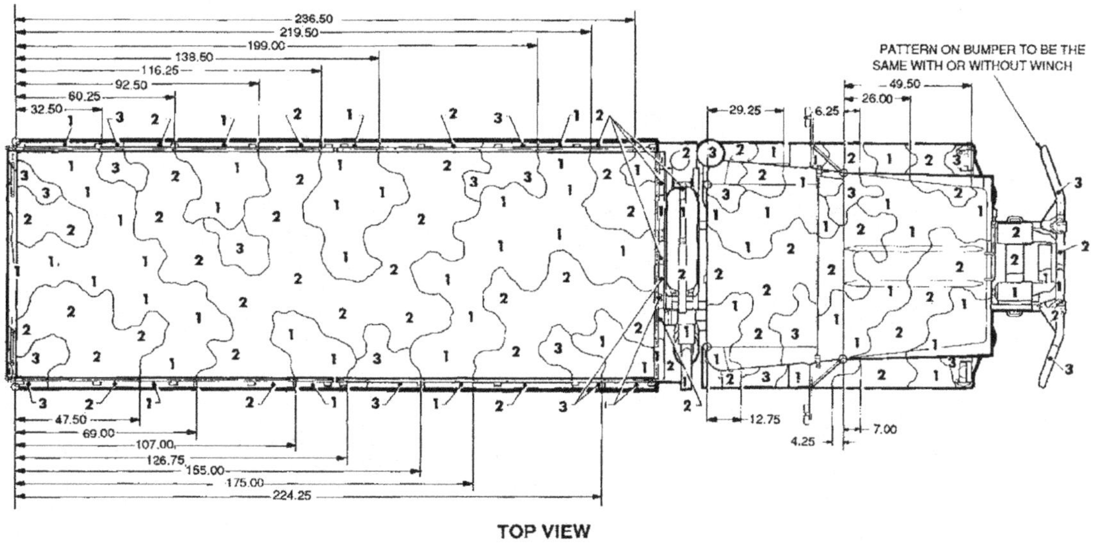

M927A1 M928A1

Figure 69. Truck, cargo: 5-ton, XLWB, M927A1 and M928A1. (3 of 4)

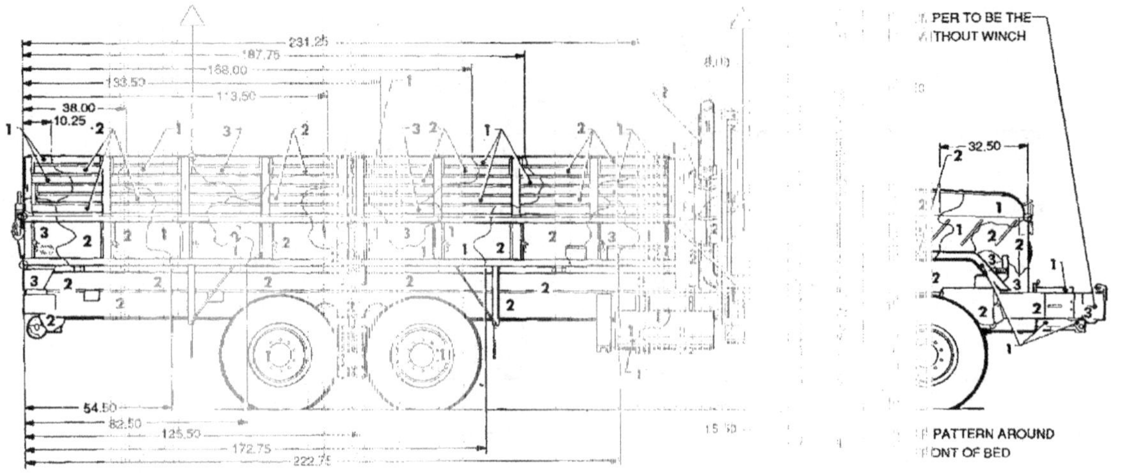

RIGHT SIDE VIEW

M927A1 M928A1

Figure 69. Truck, cargo: 5-ton, XLWB, M927A1 and M928A1

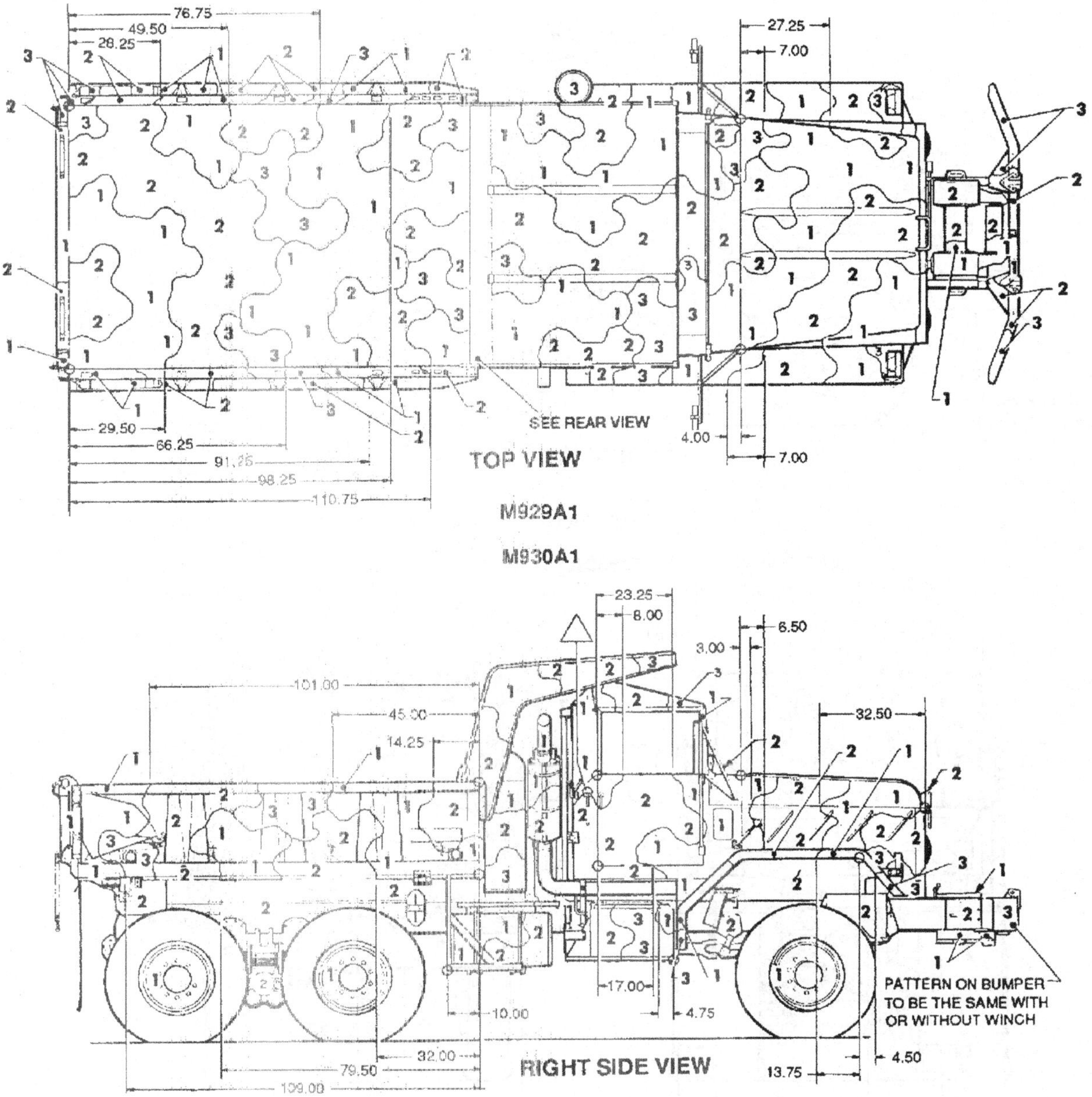

Figure 70. Truck, dump: 5-ton, M929AI and M930AI. (1 of 2)

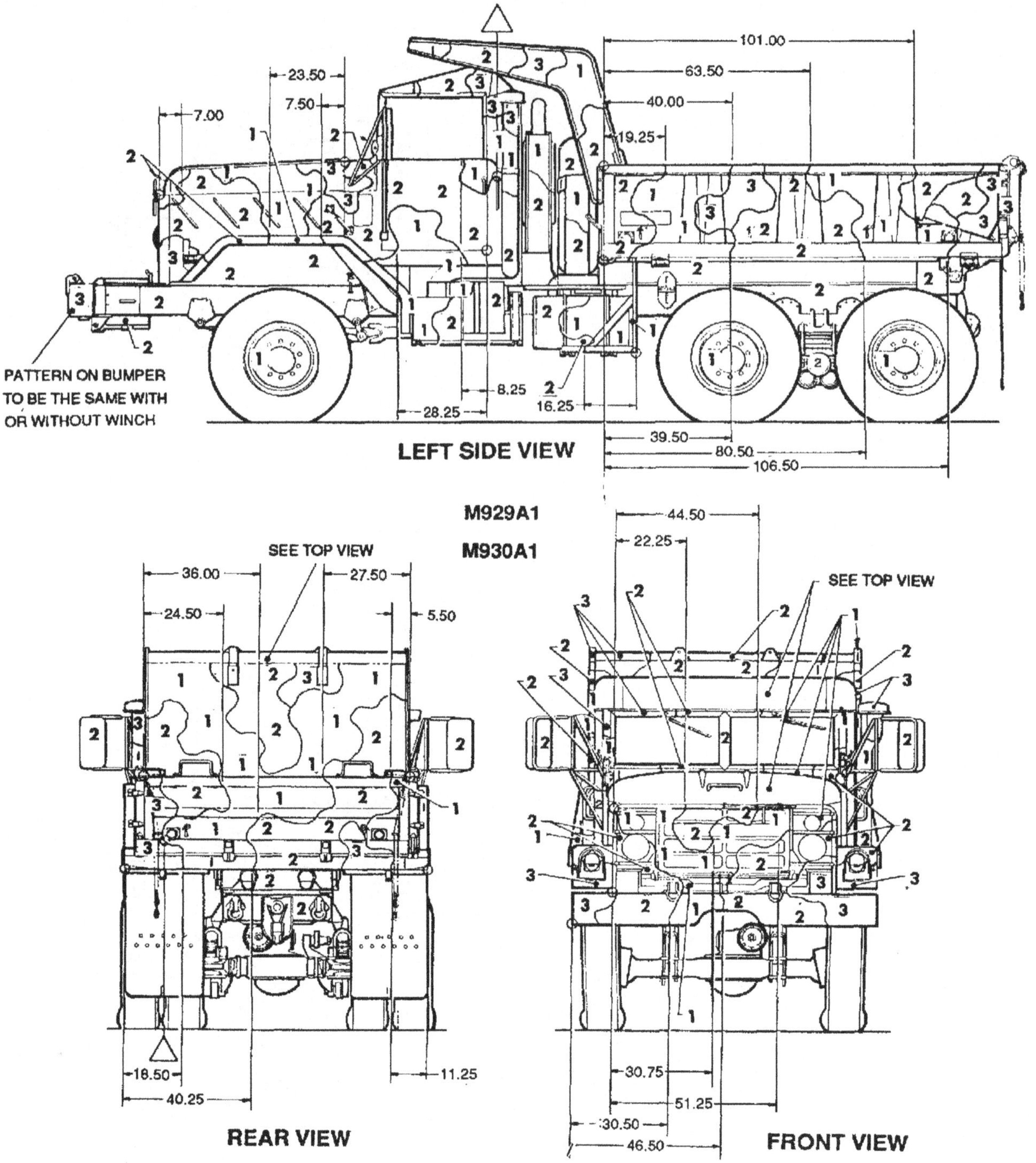

Figure 70. Truck, dump: 5-ton, M929AI and M930AI. (2 of2)

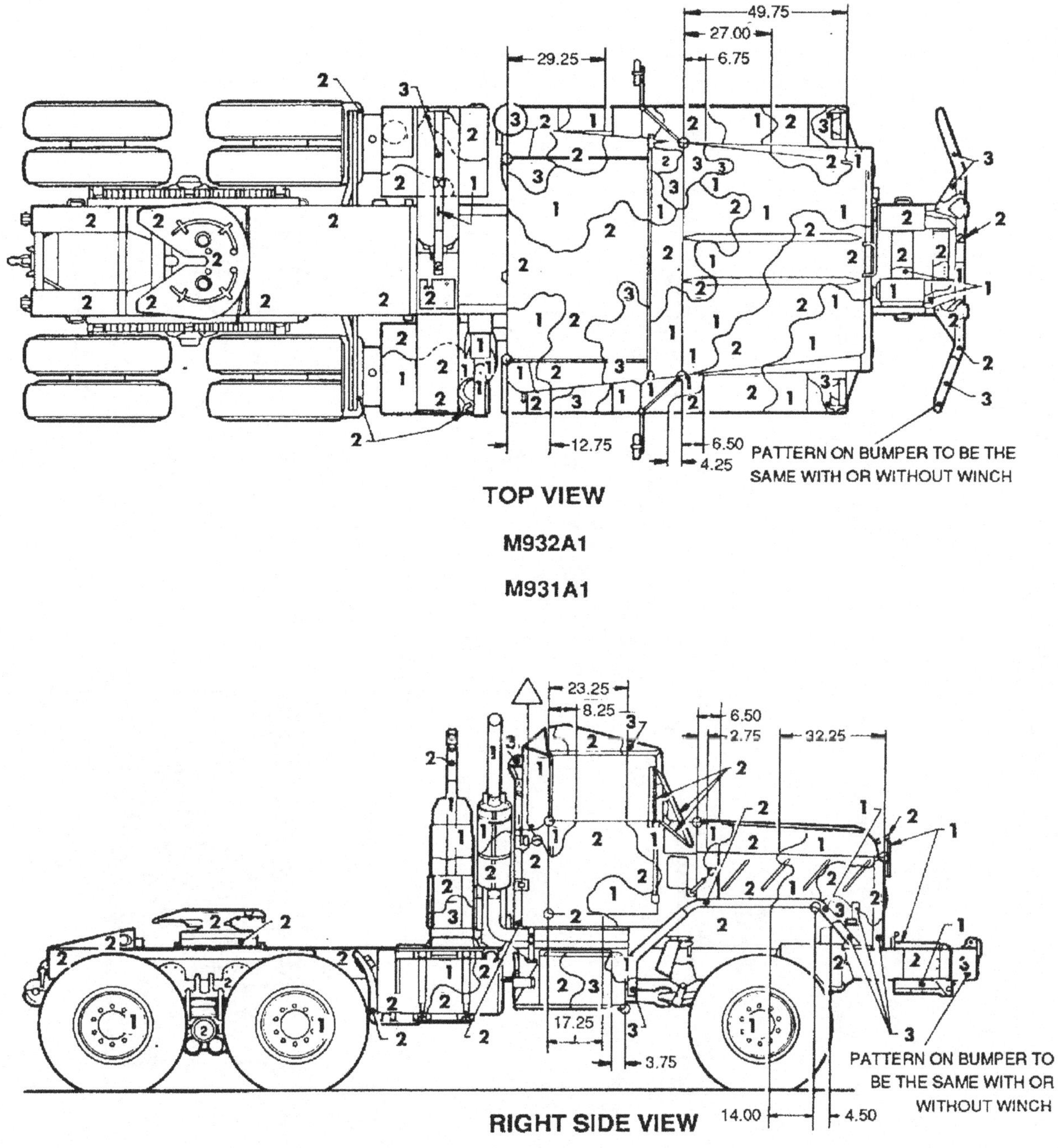

Figure 71. Truck, tractor: 5-ton, M931A1 and M932 Al. (1 of 2)

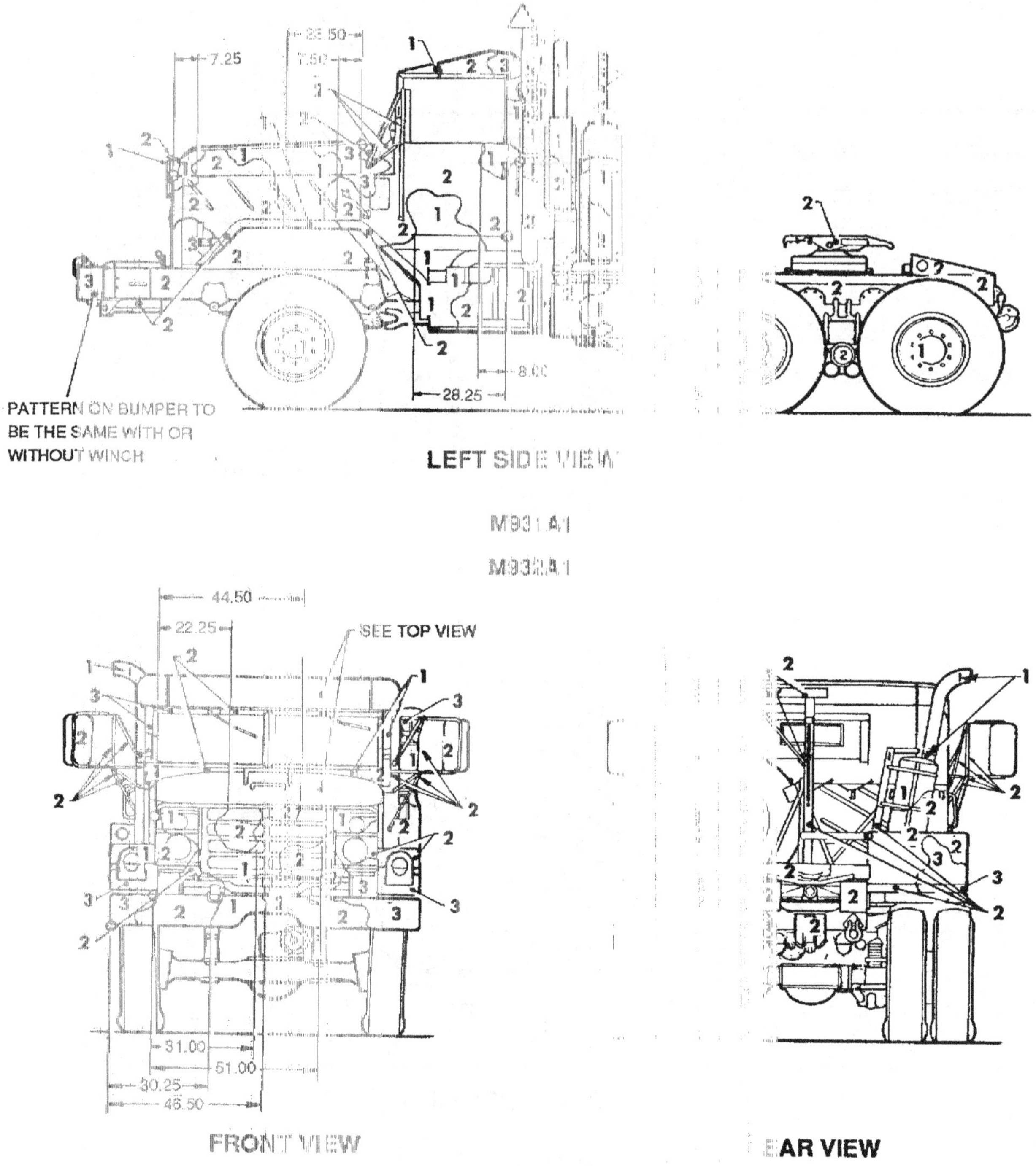

Figure 7[illegible]. Truck, tractor: 5-ton, M931A1 [illegible] 2 of 2)

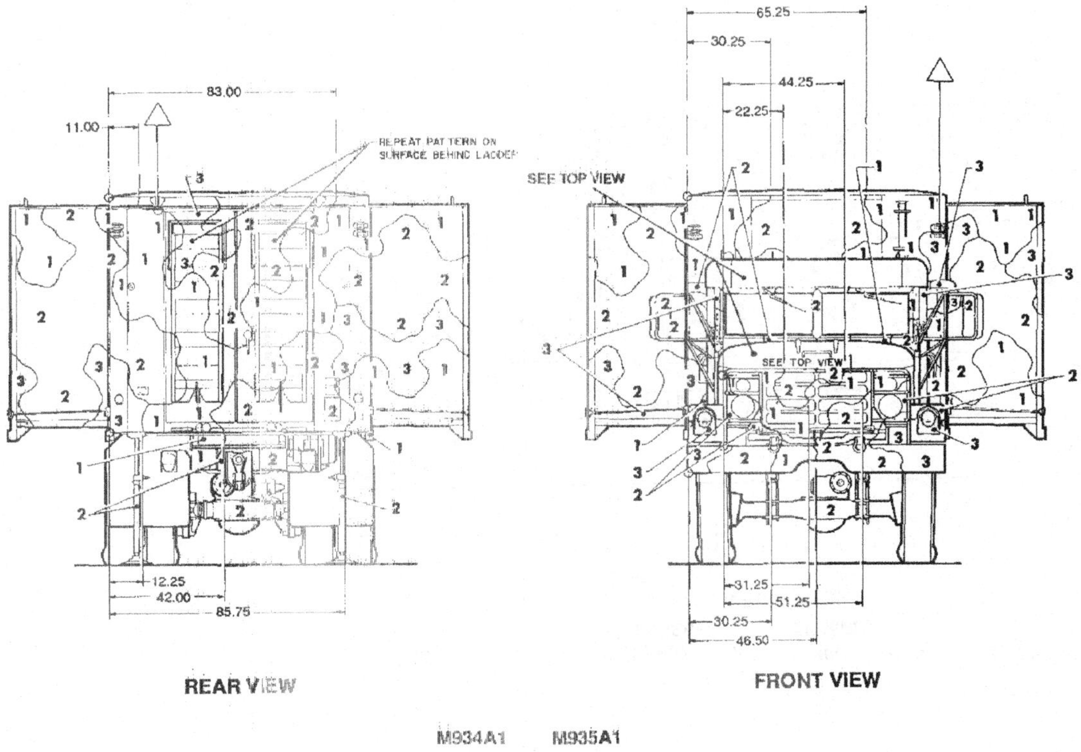

Figure 72. Truck, van, expansible: 5-ton, M934A1 and M935AI. (1 of 4)

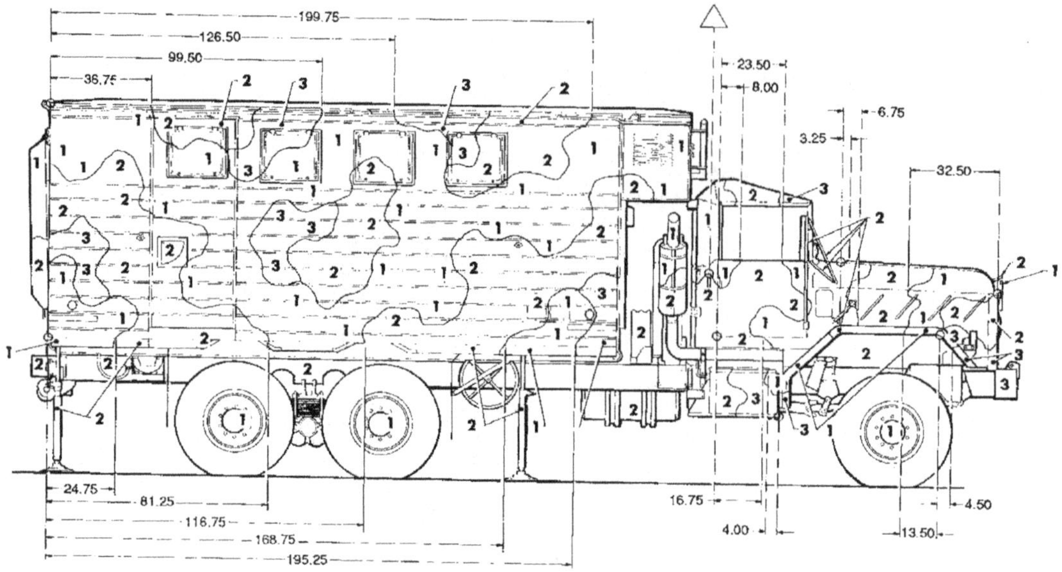

RIGHT SIDE VIEW

M934A1 M935A1

Figure 72. Truck, van, expansible: 5-ton, M934AI and M935A1. (2 of 4)

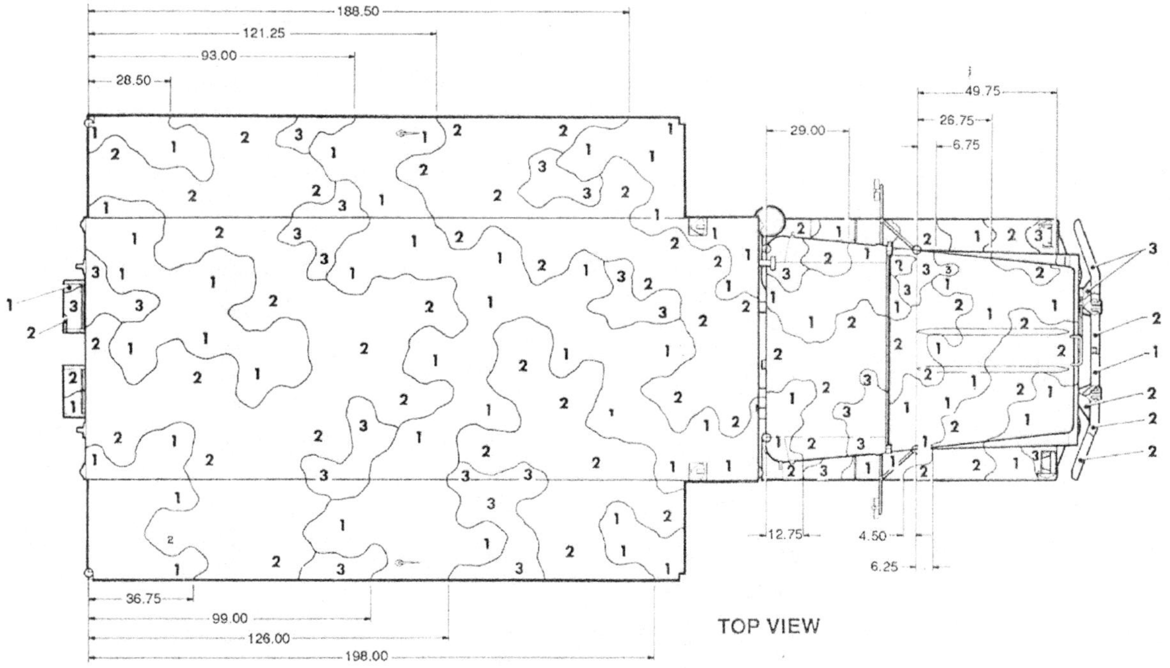

M934A1 M935A1

Figure 72. Truck, van, expansible: 5-ton, M934A1 and M935A1. (3 of 4)

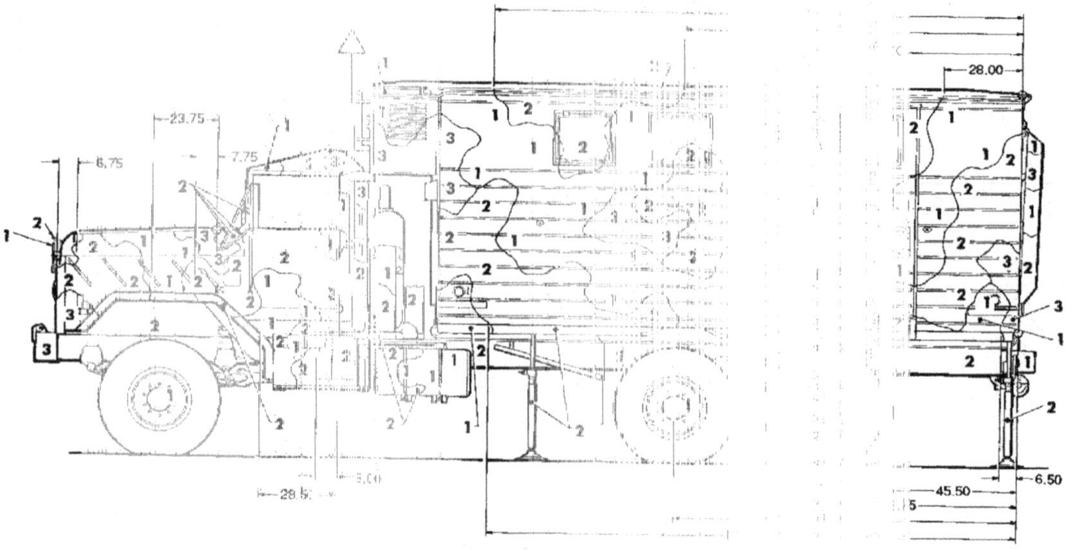

LEFT SIDE VIEW

M934A1 M935A1

Figure 72. Truck, van, expansible: 5-ton, M934A1 and M935A1

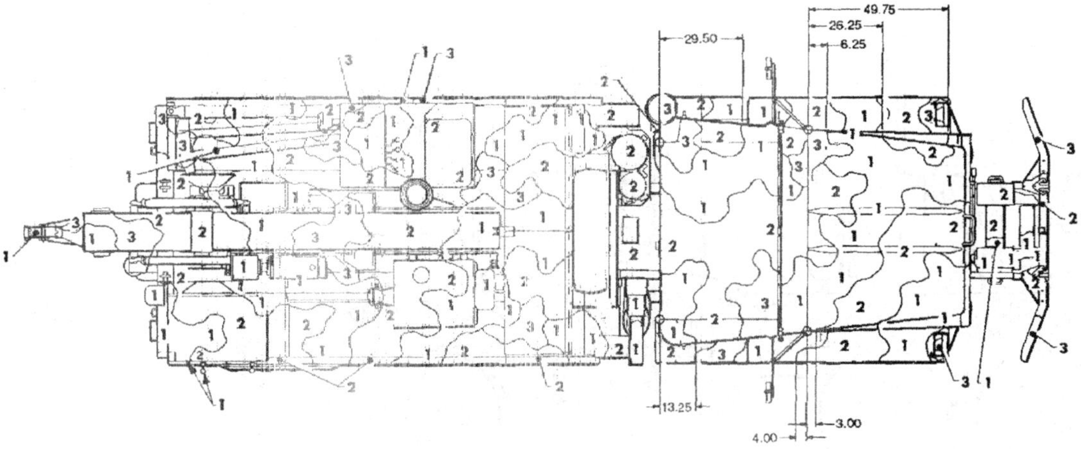

TOP VIEW
M936A1

Figure 73. Truck, wrecker: 5-ton M936A1. (1 of4)

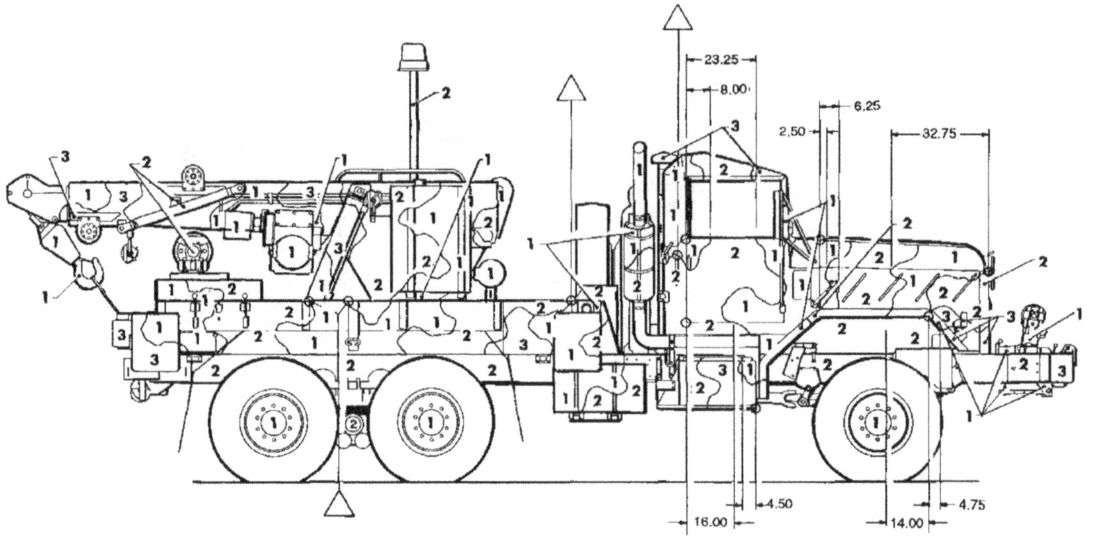

RIGHT SIDE VIEW

M936A1

Figure 73. Truck, wrecker: 5-ton M936AI . (2 of 4)

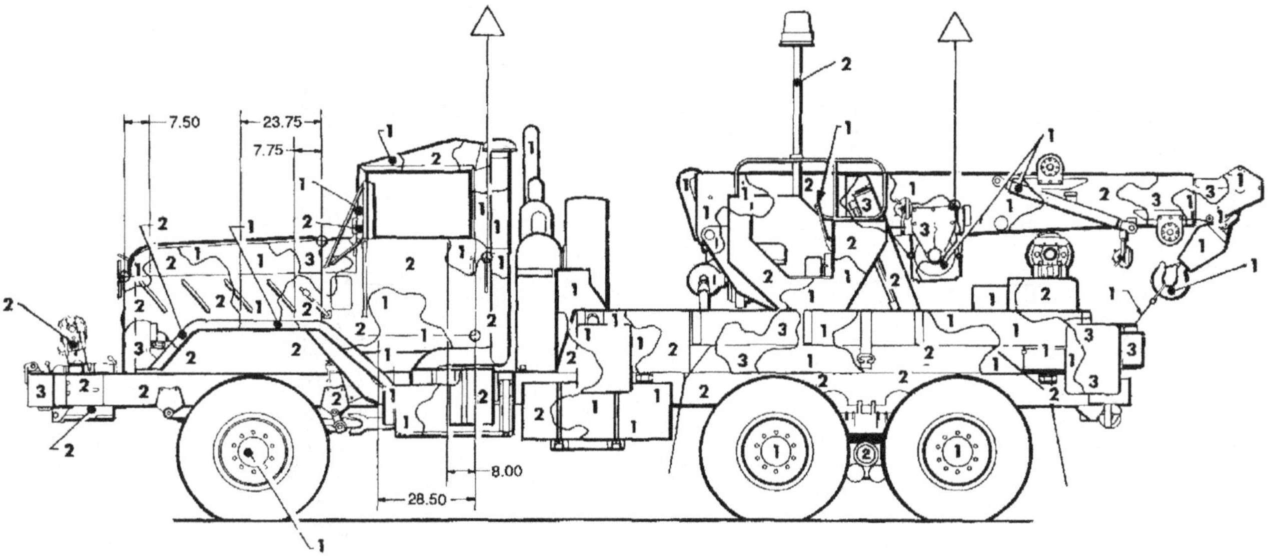

LEFT SIDE VIEW
M936A1
Figure 73. Truck, wrecker: 5-ton M936A1. (3 of 4)

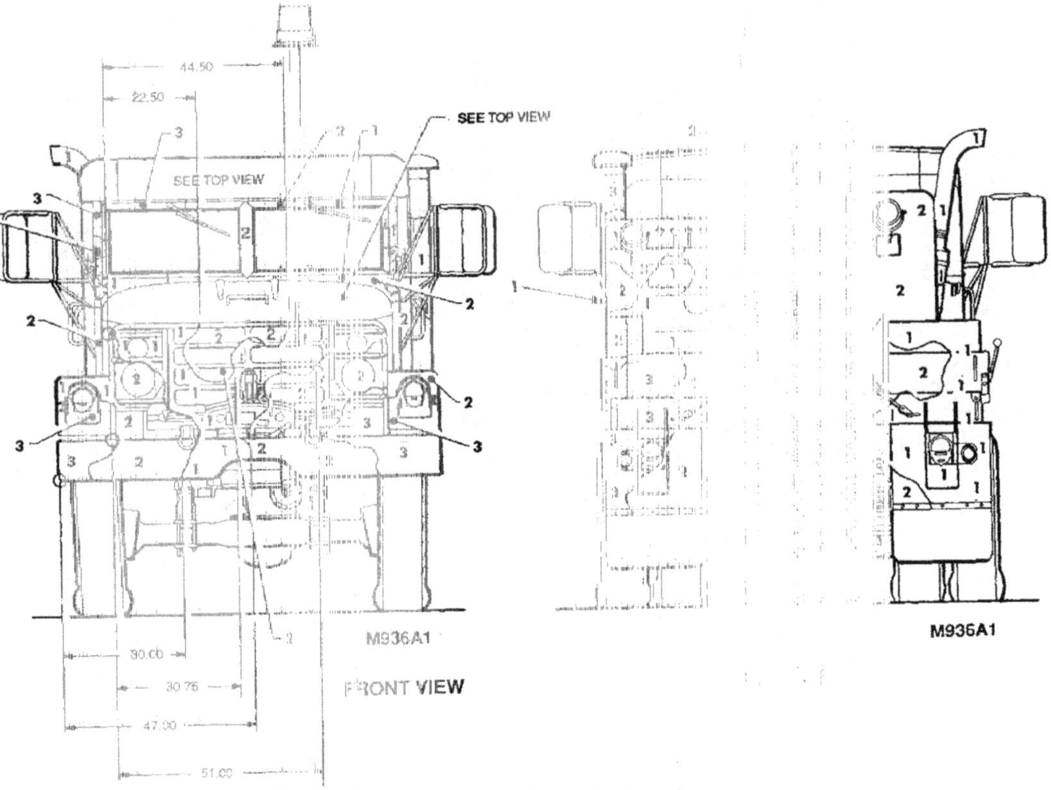

Figure 73. Truck, wrecker: 5-ton M936A1. (4 of

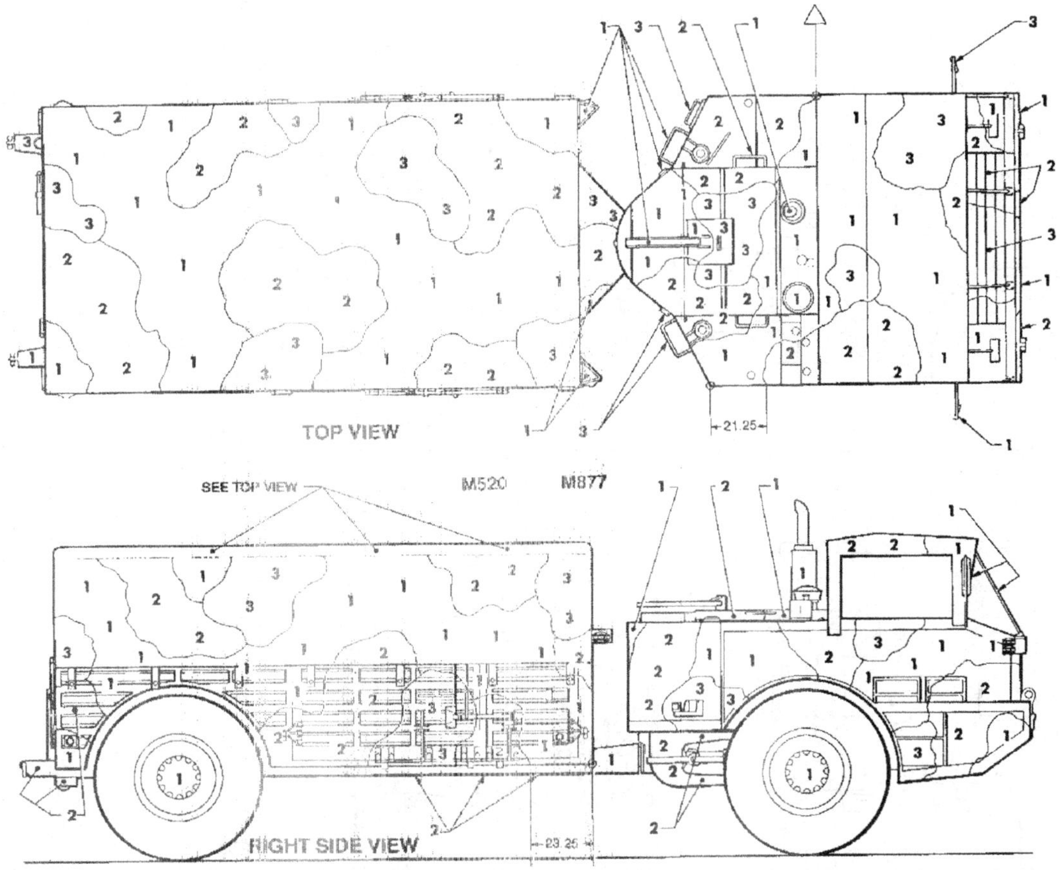

Figure 74. Truck, cargo. 8-ton, 4 x 4, M520 and M877 w/winch. (1 of 2)

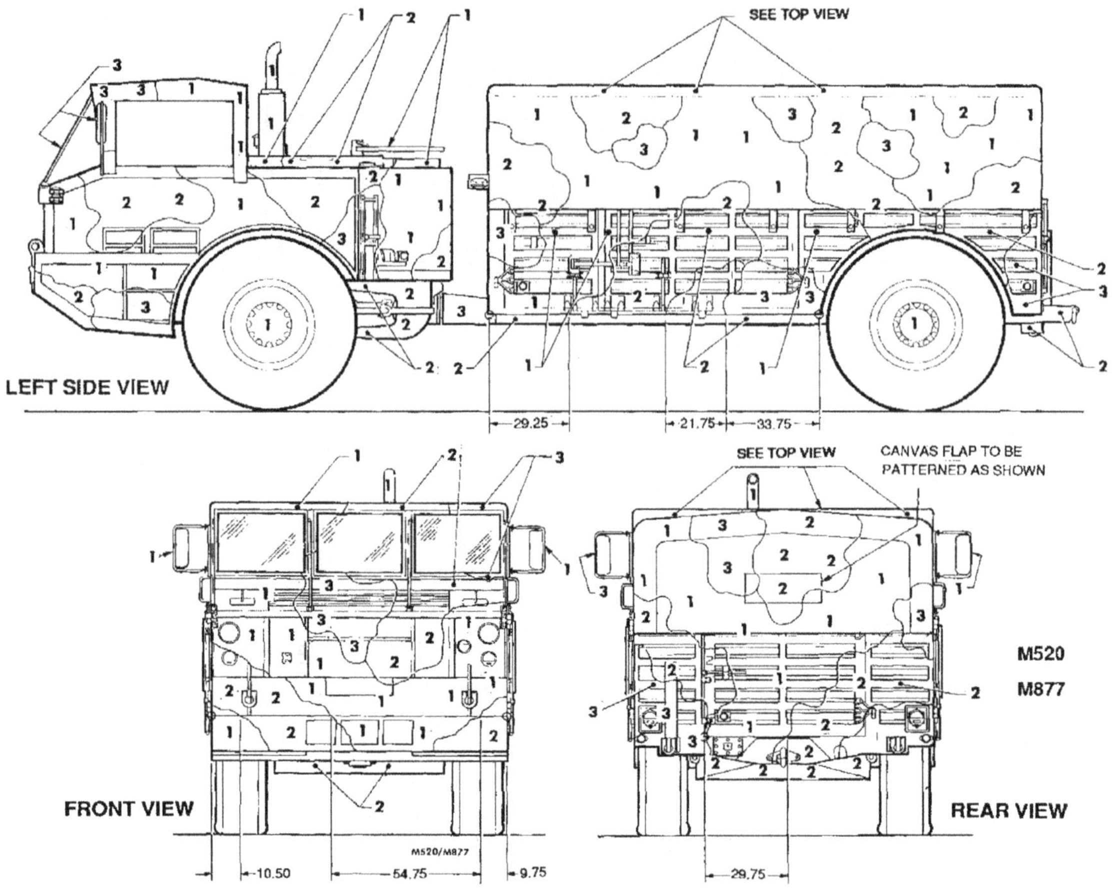

Figure 74. Truck, cargo: 8-ton, 4 x 4, M520 and M877 w/winch. (2 of 2)

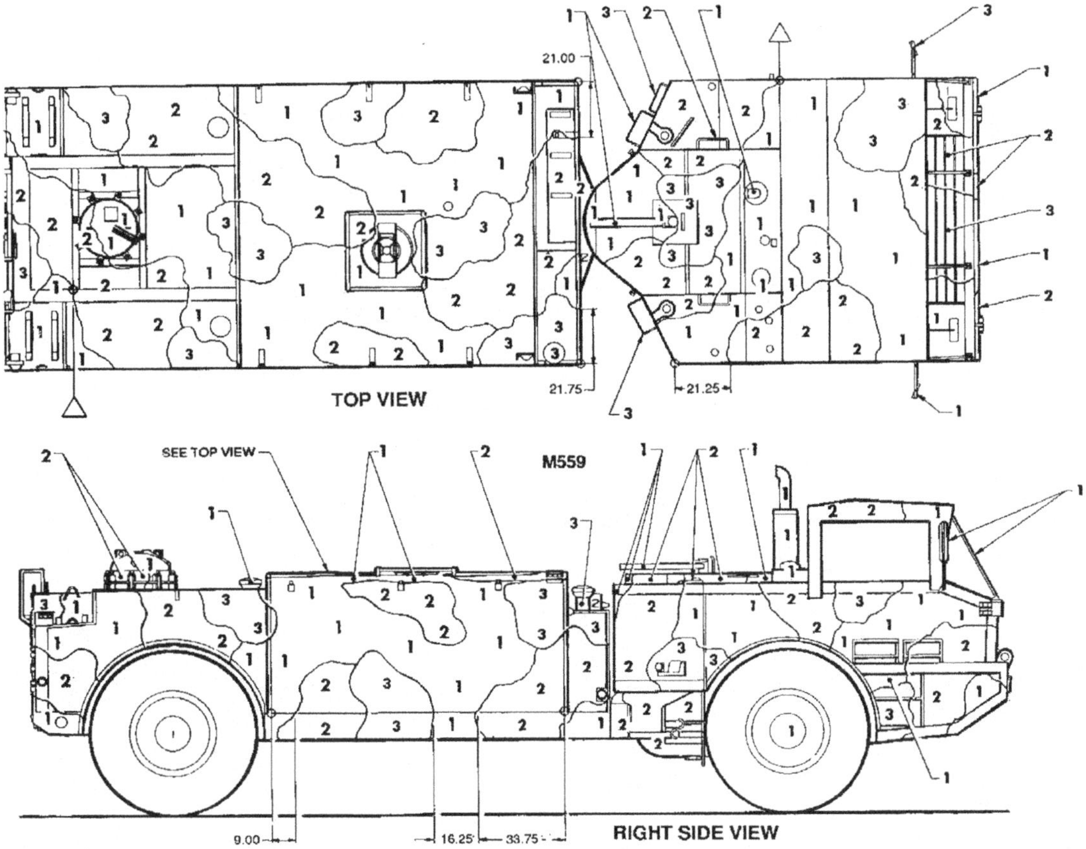

Figure 75. Truck, tank: fuel servicing, 2500 gallon, 4 x 4, M559. (1 of 2)

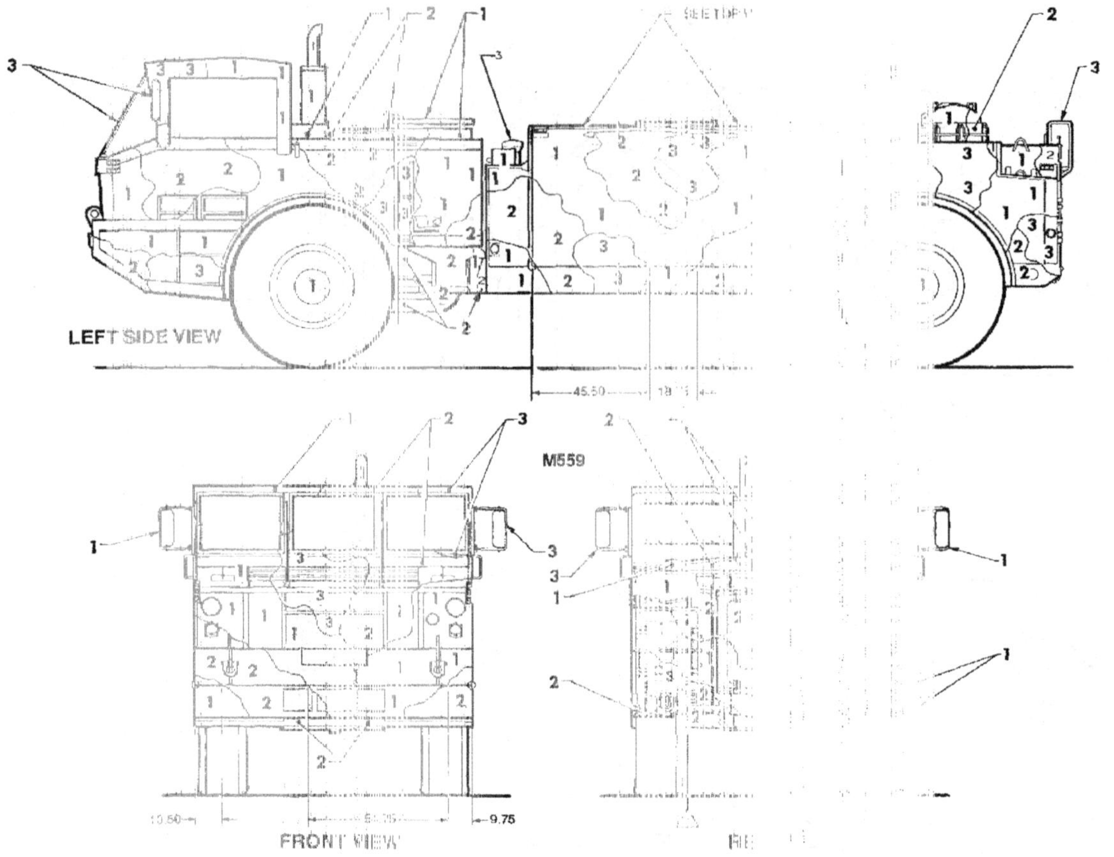

Figure 75. Truck, tank: fuel servicing, 2500 gallon 4 x 4,

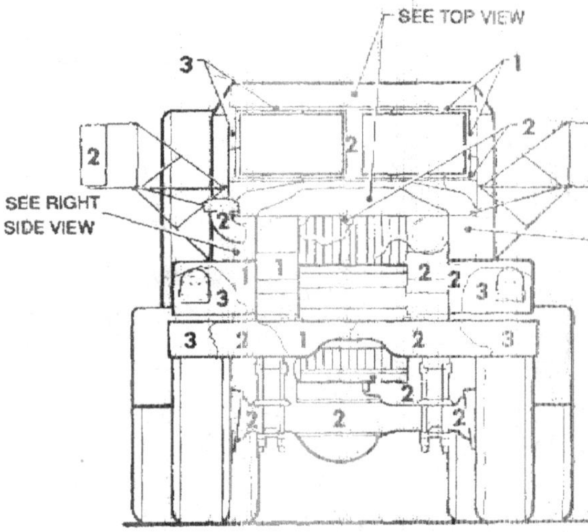

FRONT VIEW

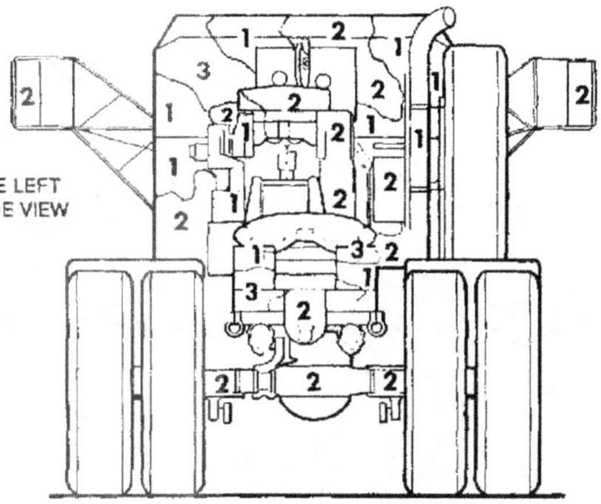

REAR VIEW

M123A2C

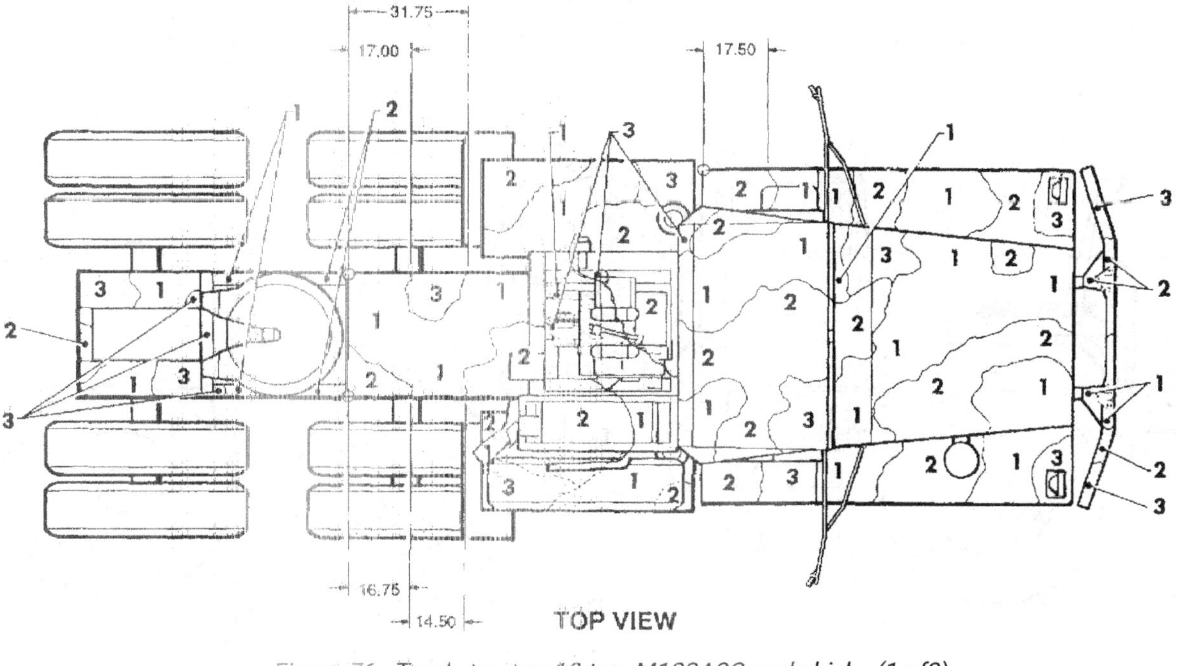

TOP VIEW

Figure 76. Truck, tractor: 10-ton, M123A2C, wolwhich. (1 of2)

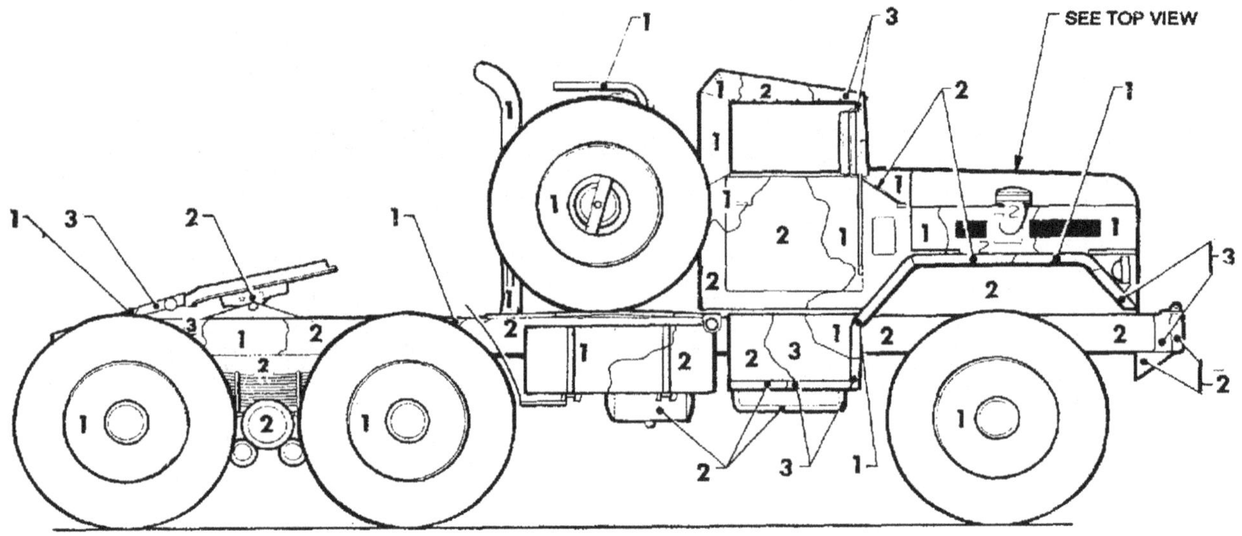

RIGHT SIDE VIEW

M123A2C

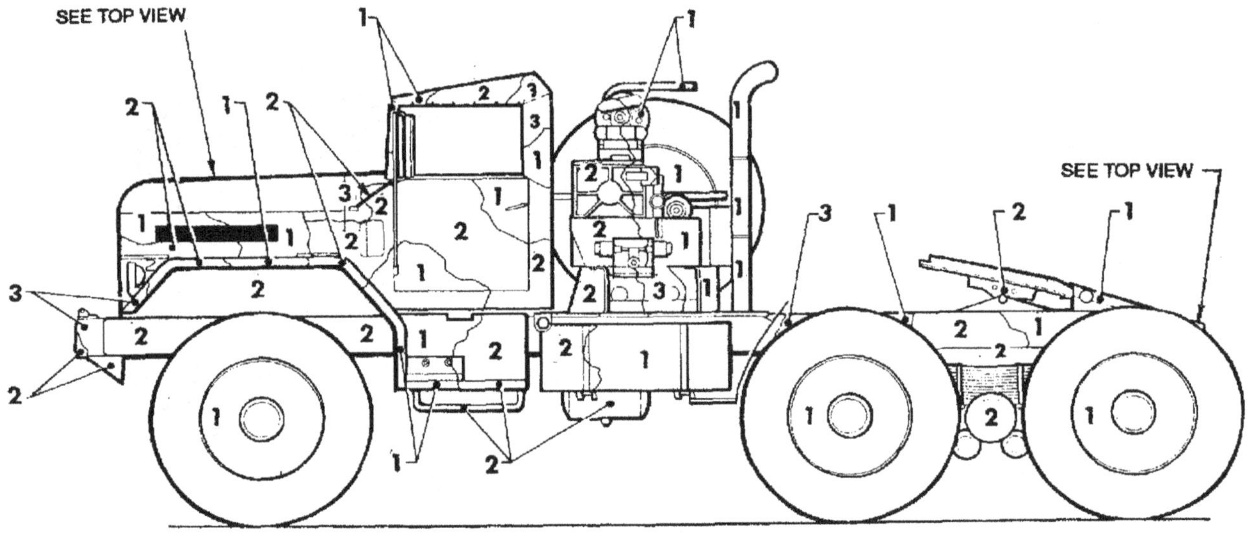

LEFT SIDE VIEW

Figure 76. Truck, tractor: 10-ton, M123A2C, wo/winch. (2 of 2)

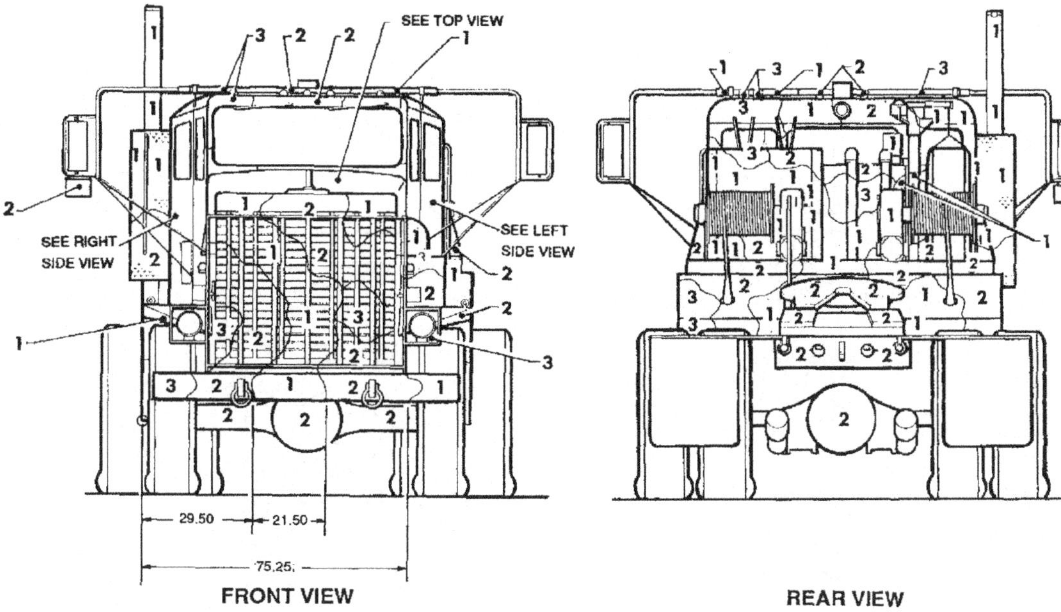

M911

Figure 77. Truck, tractor: 10-ton, M911. (1 of 4)

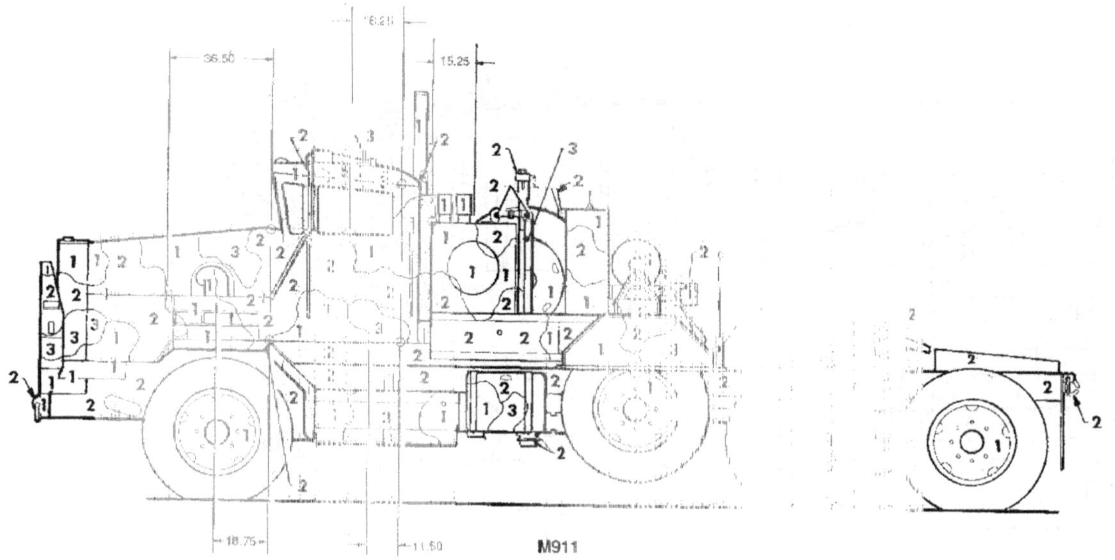

M911

LEFT SIDE VIEW

Figure 27. Truck, tractor: 10-ton, M911. (2 of

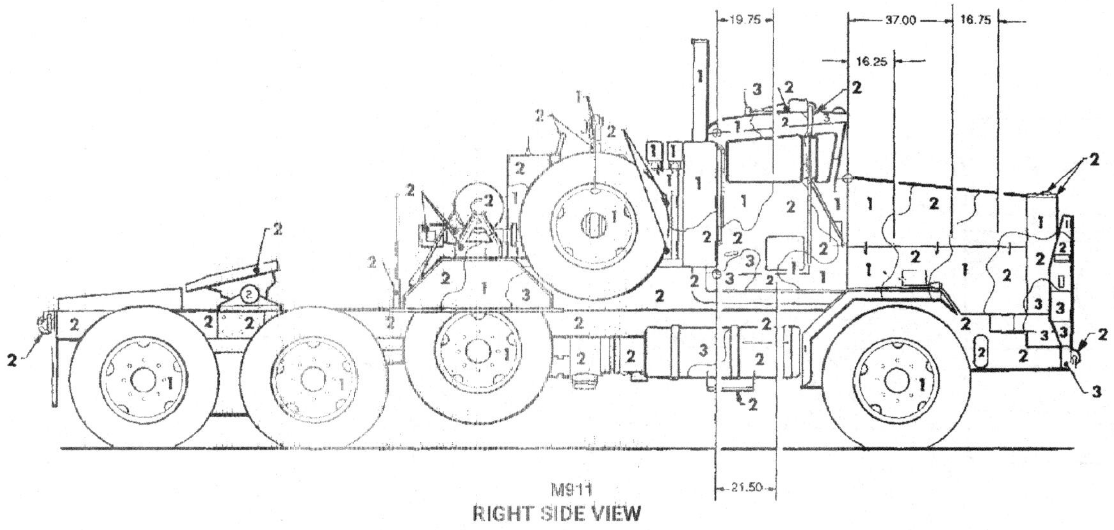

Figure 77. Truck, tractor: 10-ton, M911. (3 of4)

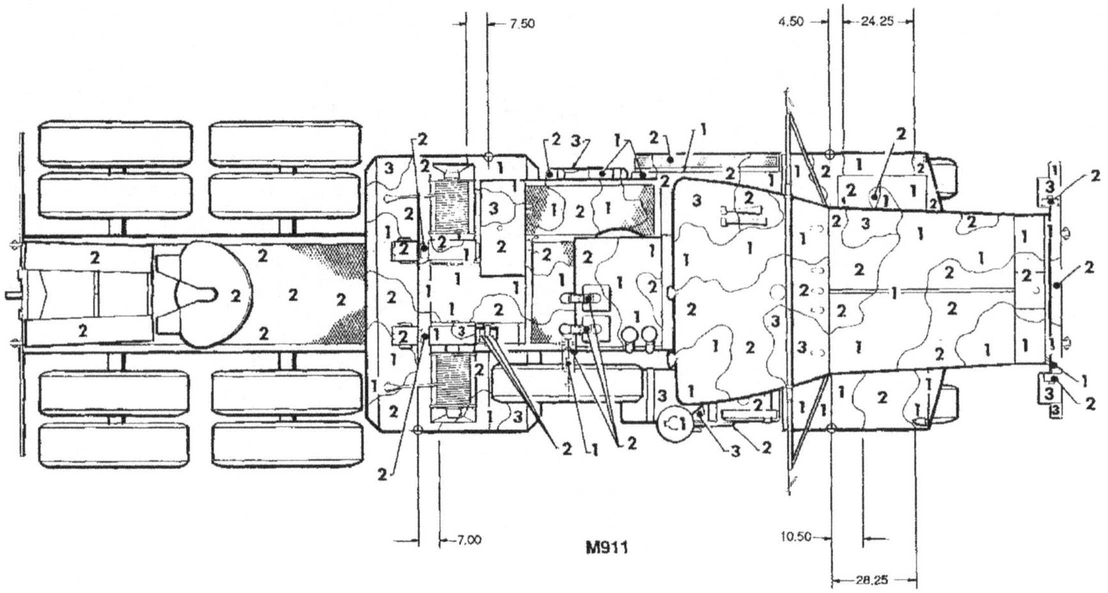

TOP VIEW

Figure 77. Truck, tractor: 10-ton, M911. (4 of 4)

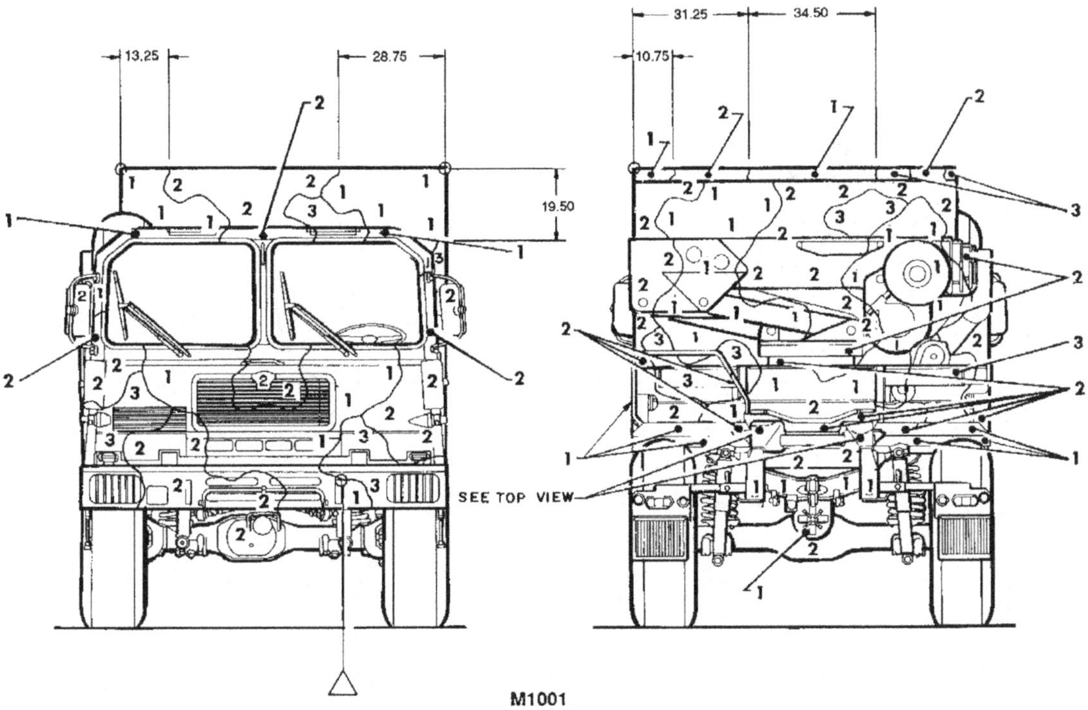

Figure 78. Truck, tractor: 10-ton, M001. (1 of 4)

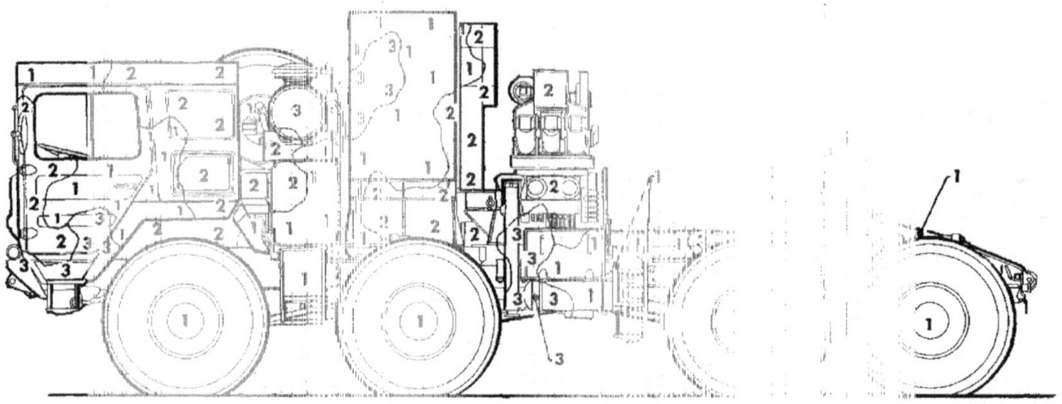

M1 001
LEFT SIDE VIEW

Figure 78. Truck, tractor: 10-ton, M1001. (2 [illegible]

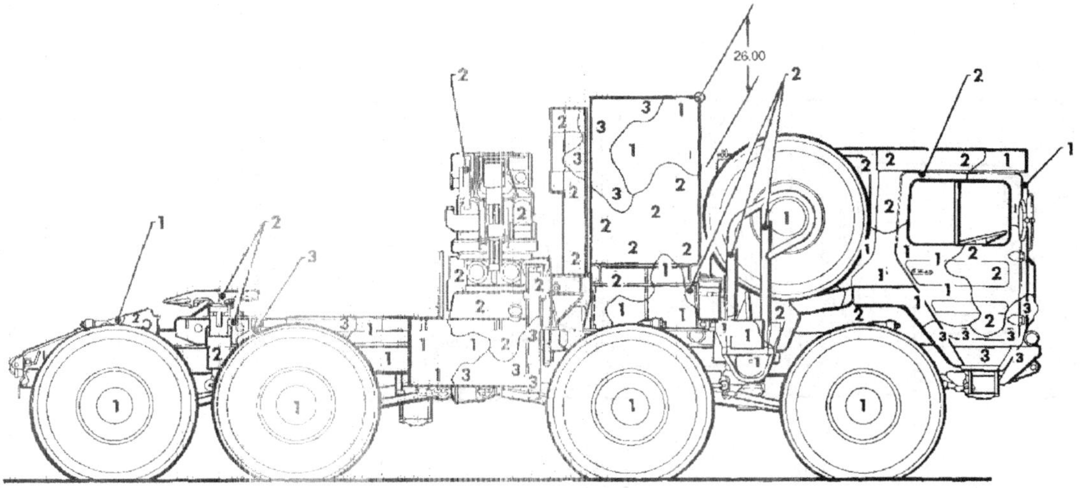

Figure 78. Truck, tractor: 10-ton, M1001. (3 of 4)

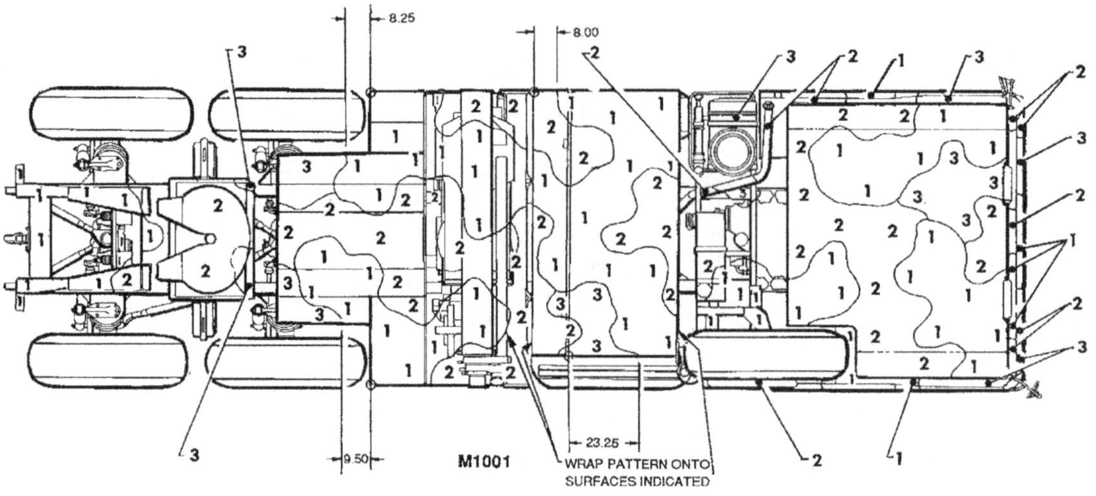

Figure 78. Truck, tractor: 10-ton, M1001. (4 of 4)

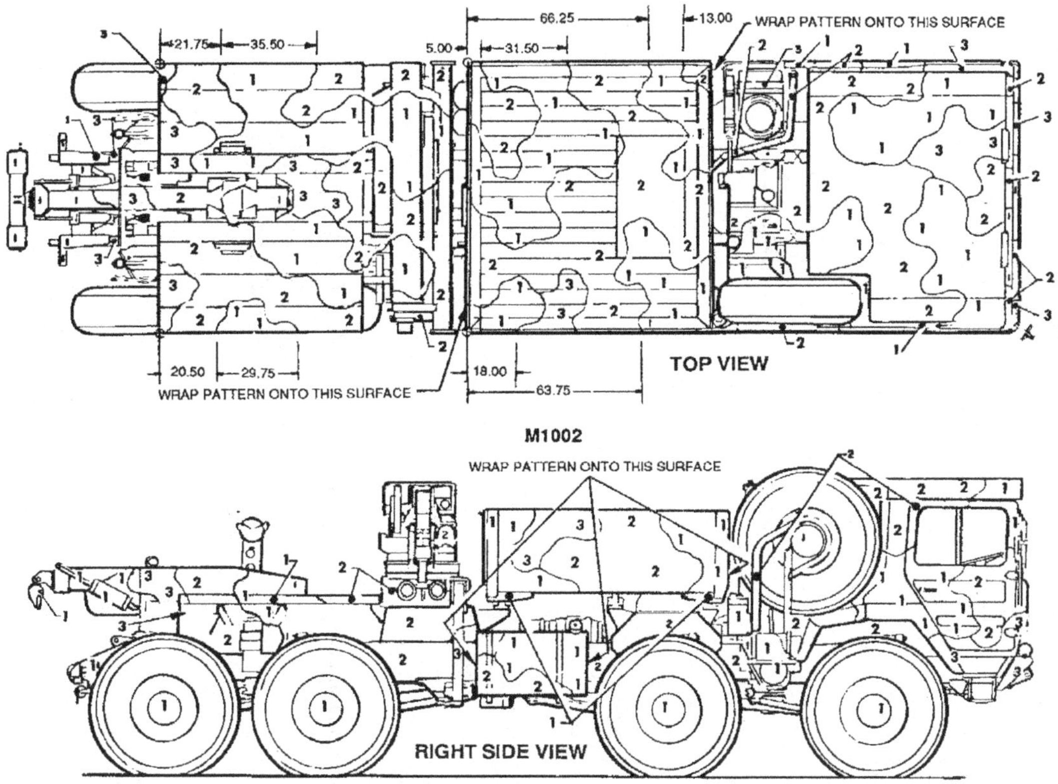

Figure 79. Truck, wrecker: 10-ton, M1002. (1 of 2)

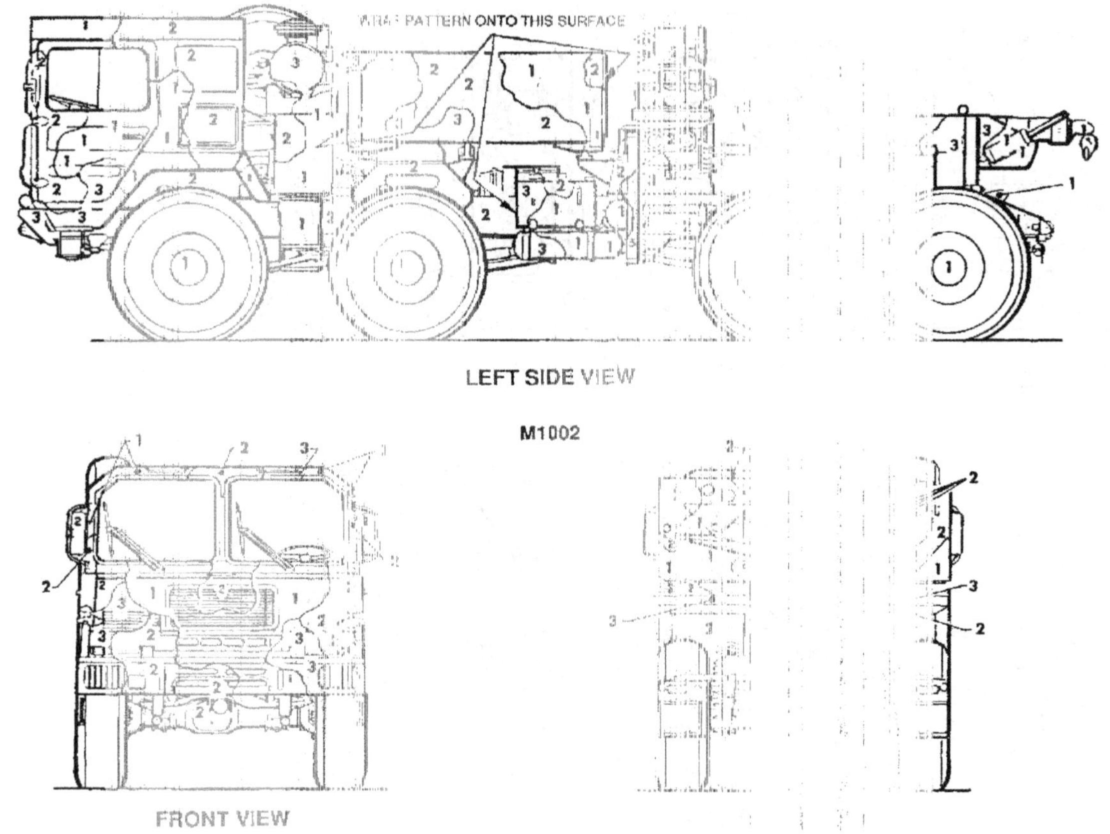

Figure 7[illegible] Truck, wrecker: 10-ton, M1002. [illegible]

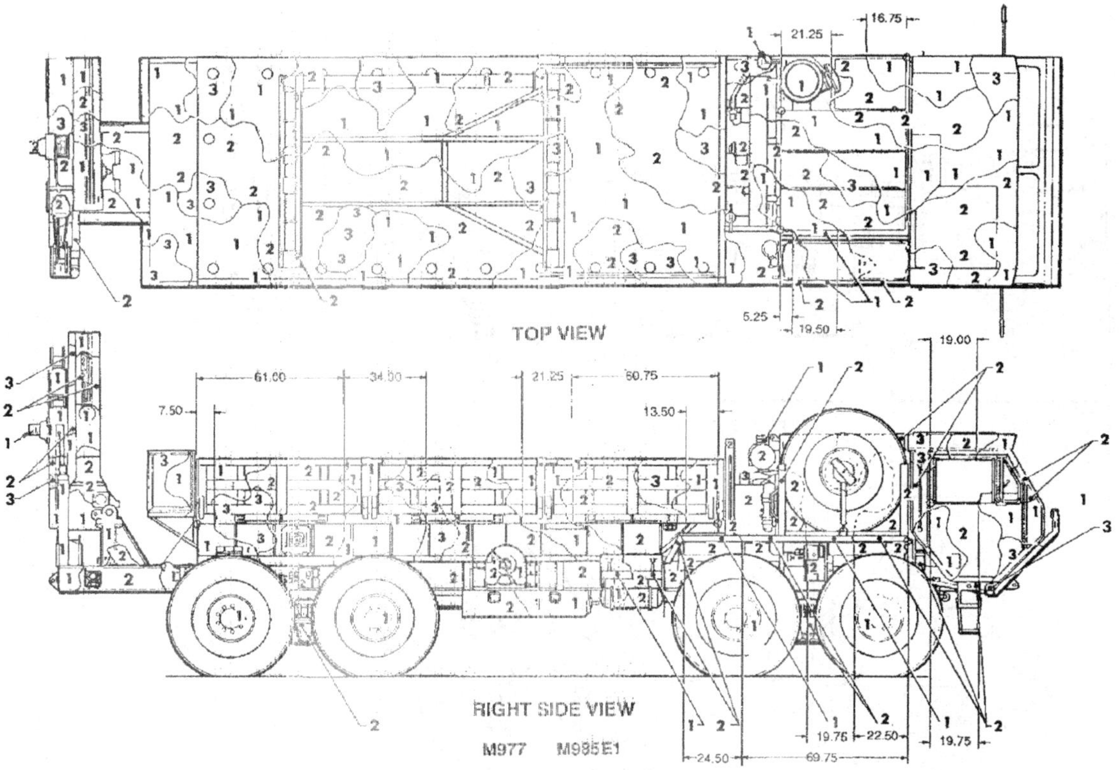

Figure 80. Truck, cargo: 11-ton, M977 and M985E1. (1 of 2)

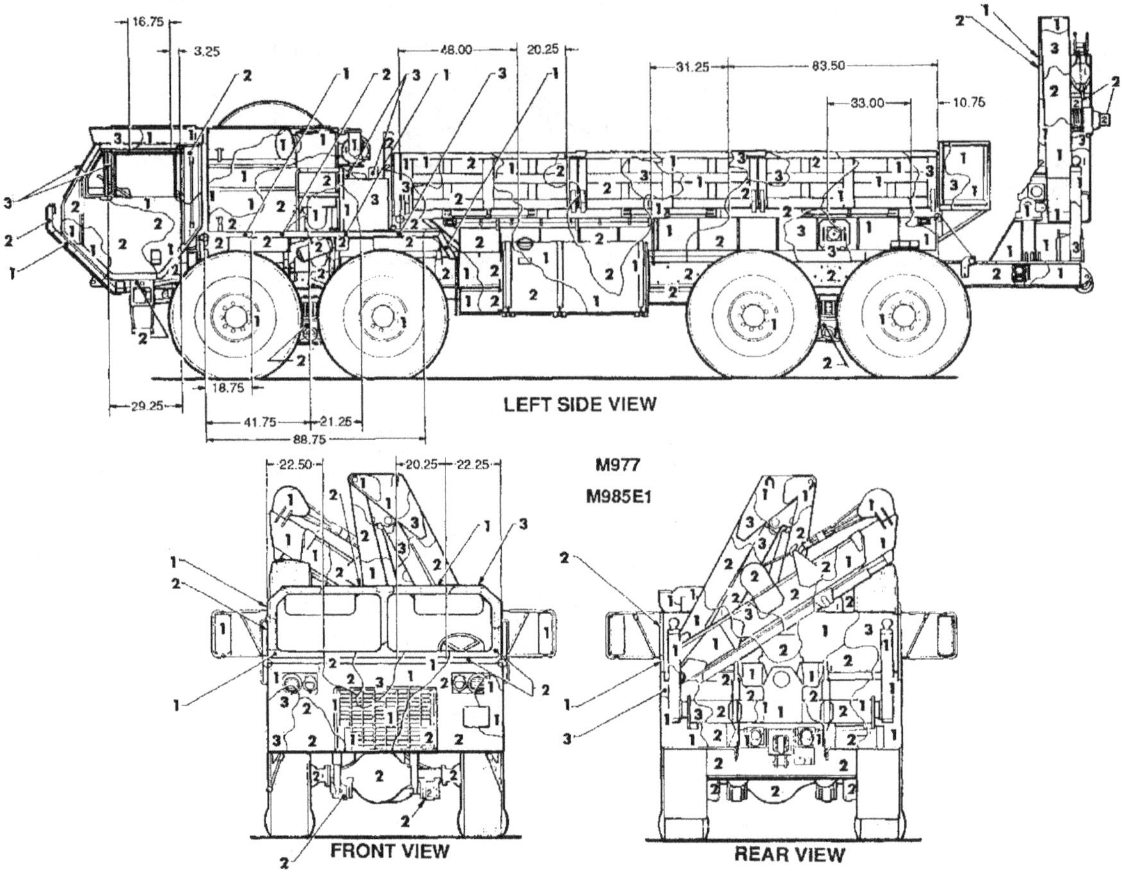

Figure 80. Truck, cargo: 11-ton, M977 and M985E1. (2 o f2)

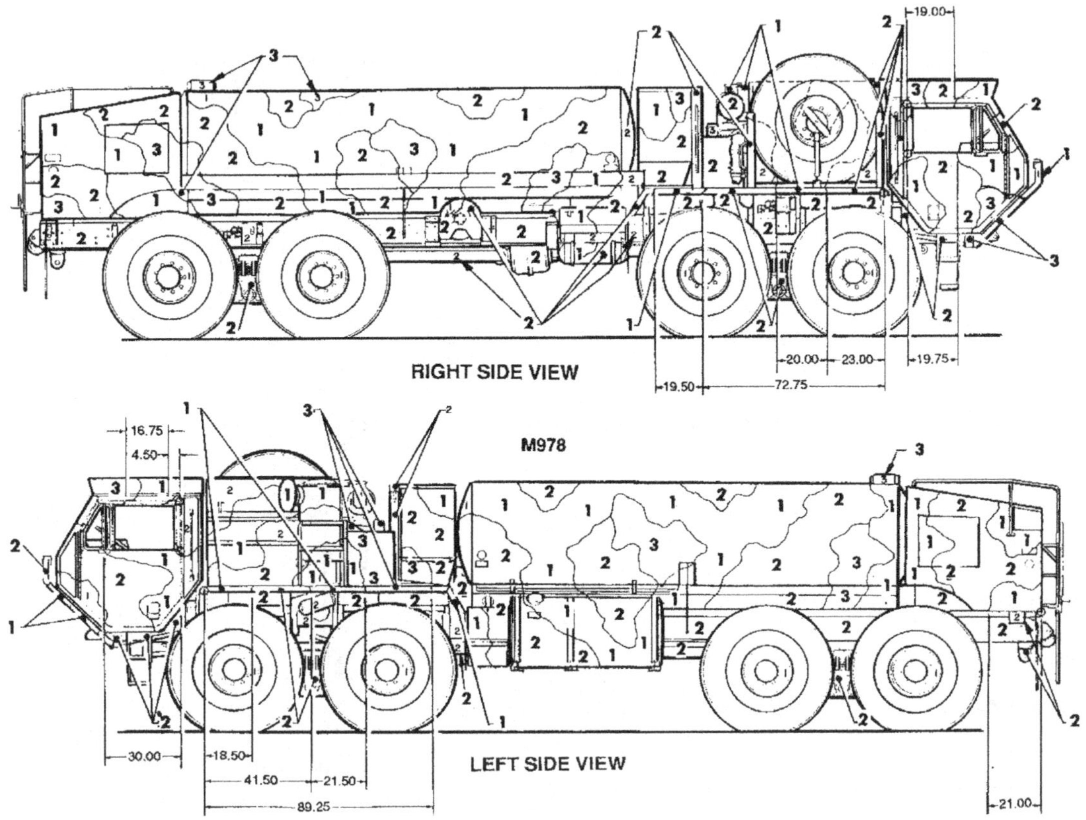

Figure 81. Truck, tanker: 11-ton, M978. (1 of 2)

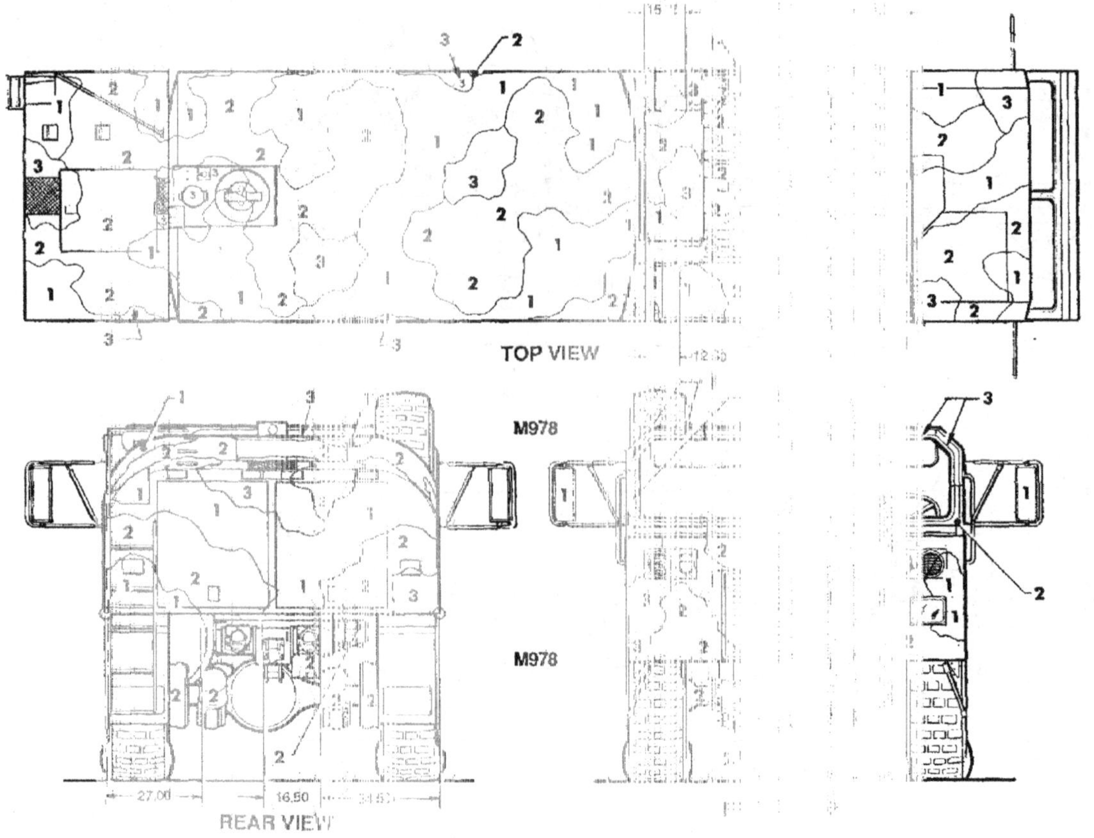

Figure 91. Truck, tanker: 11-ton, M978 (2 of

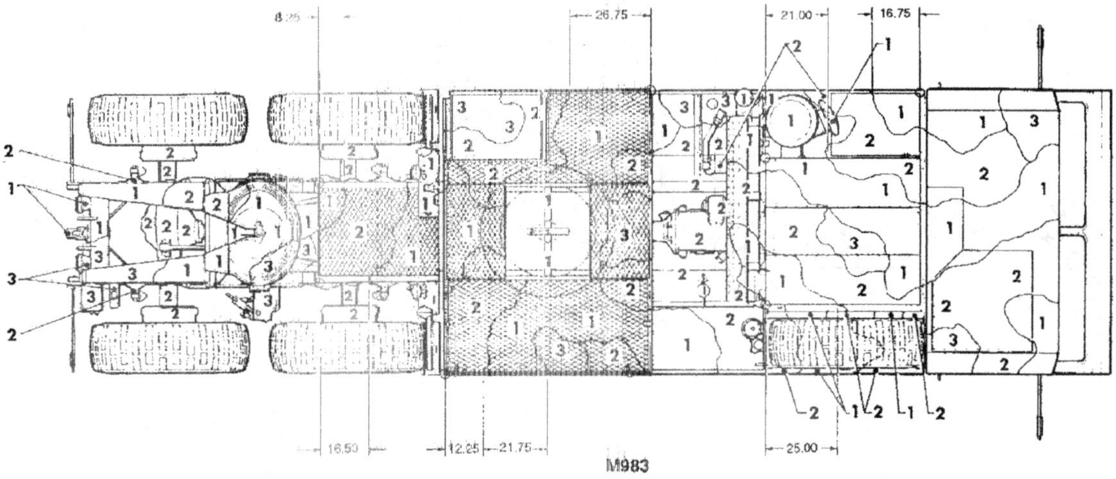

TOP VIEW

Figure 82. Truck, tractor: 11-ton, M983. (1 of 4)

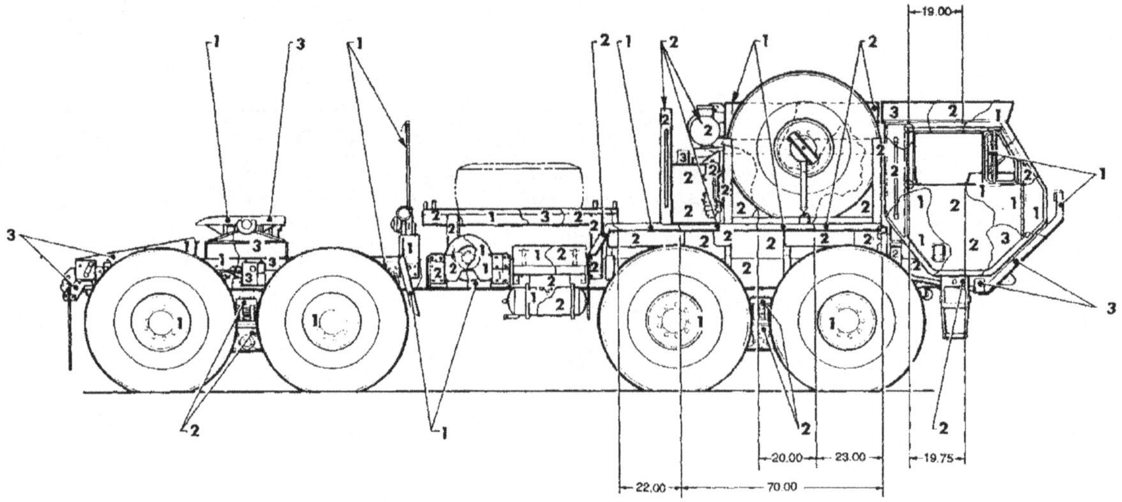

M983

RIGHT SIDE VIEW

Figure 82. Truck, tractor: li-ton, M983. (2 of 4)

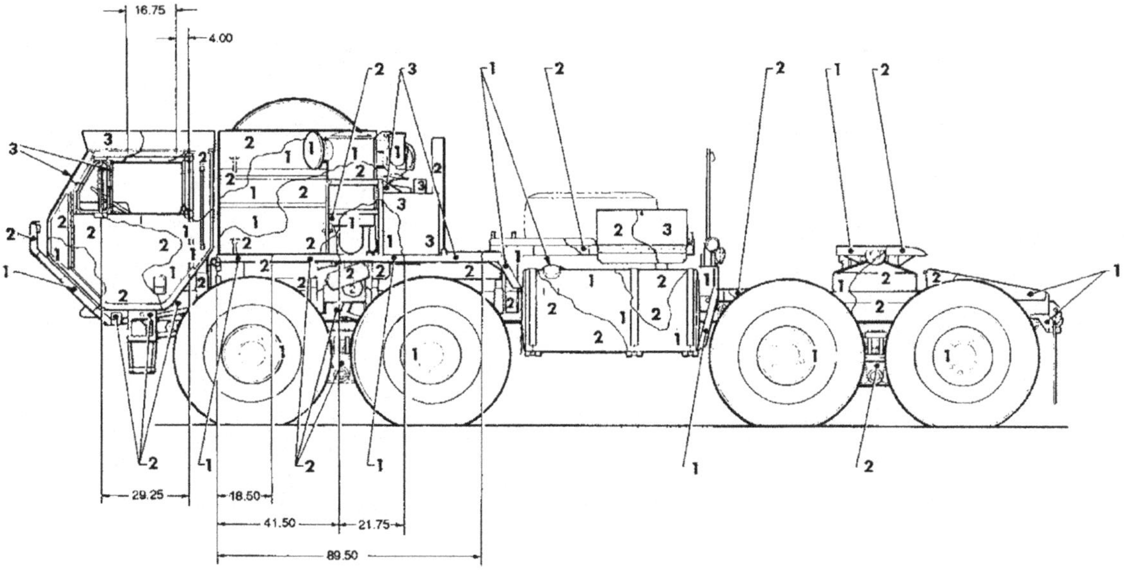

M983

LEFT SIDE VIEW

Figure 82. Truck, tractor: 11-ton, M983. (3 of 4)

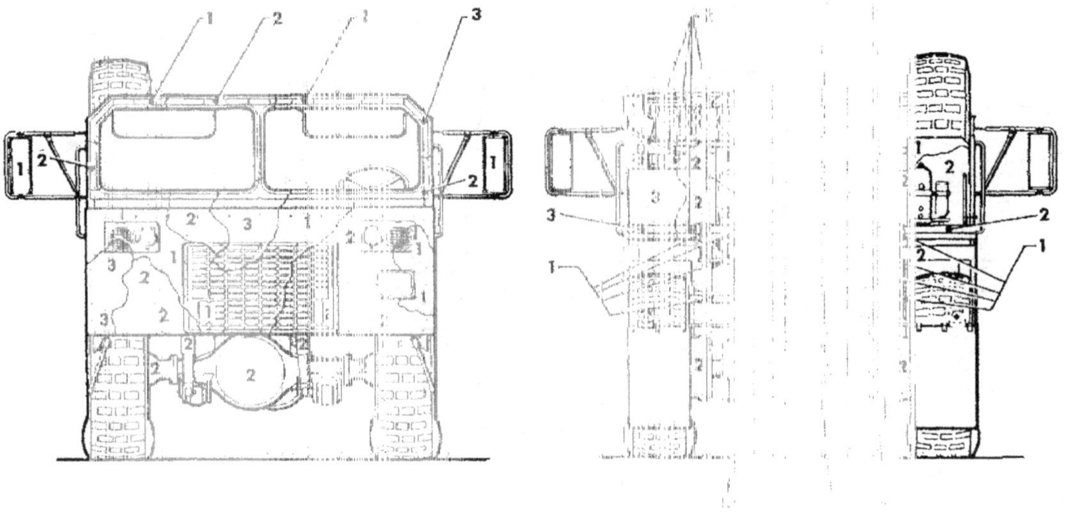

M983

Figure 32. Truck, tractor: 11-ton, M983. (4 of

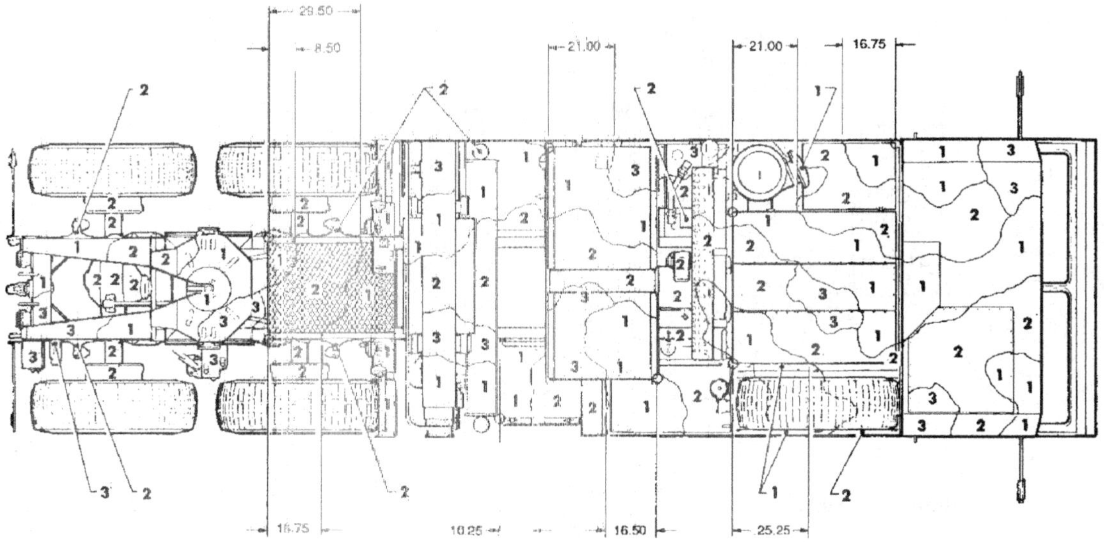

TOP VIEW
M983 W/CRANE

Figure 83. Truck, tractor: 11-ton, w/crane, M983. (1 of 4)

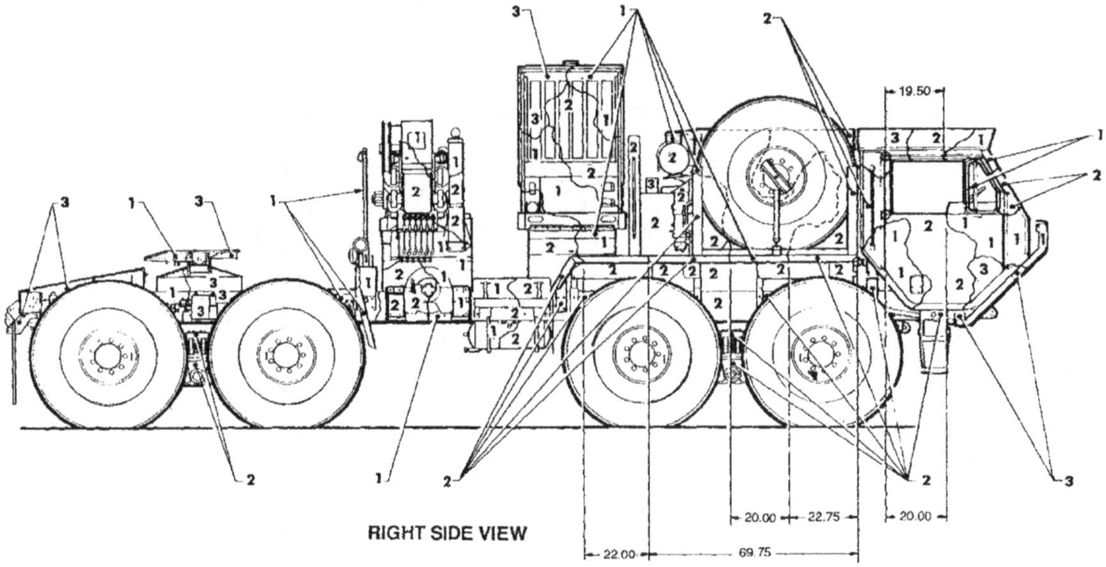

M983 W/CRANE

Figure 83. Truck, tractor: 11 -ton, w/crane, M983. (2 of 4)

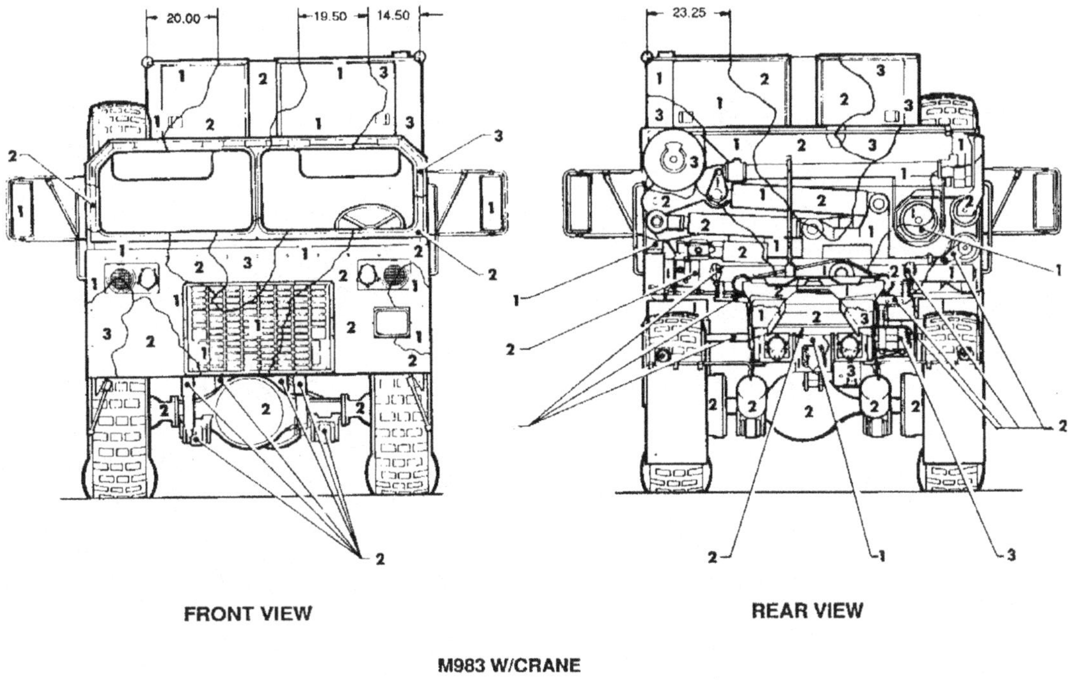

Figure 83. Truck, tractor: 11-ton, w/crane, M983. (3 of 4)

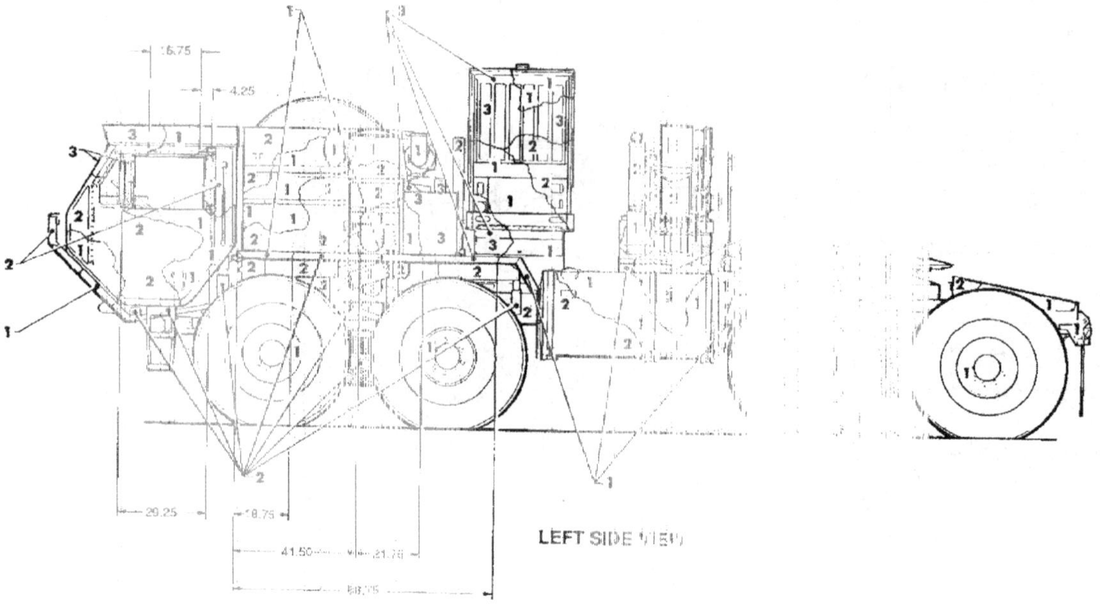

M983 W/CRANE

Figure 93. Truck, tractor: 11-ton, w/crane, M983.

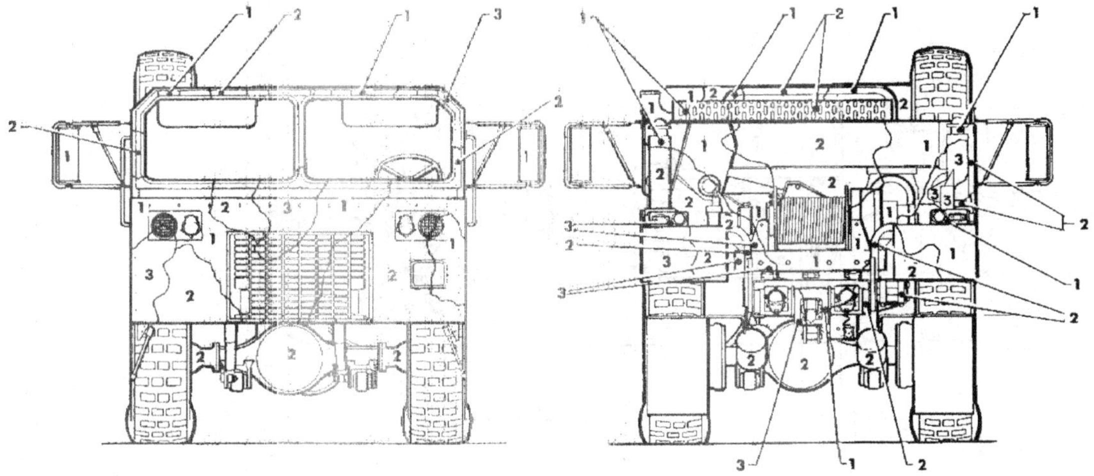

FRONT VIEW

REAR VIEW

M984

Figure 84. Truck, recovery: 11-ton, M984. (1 of 4)

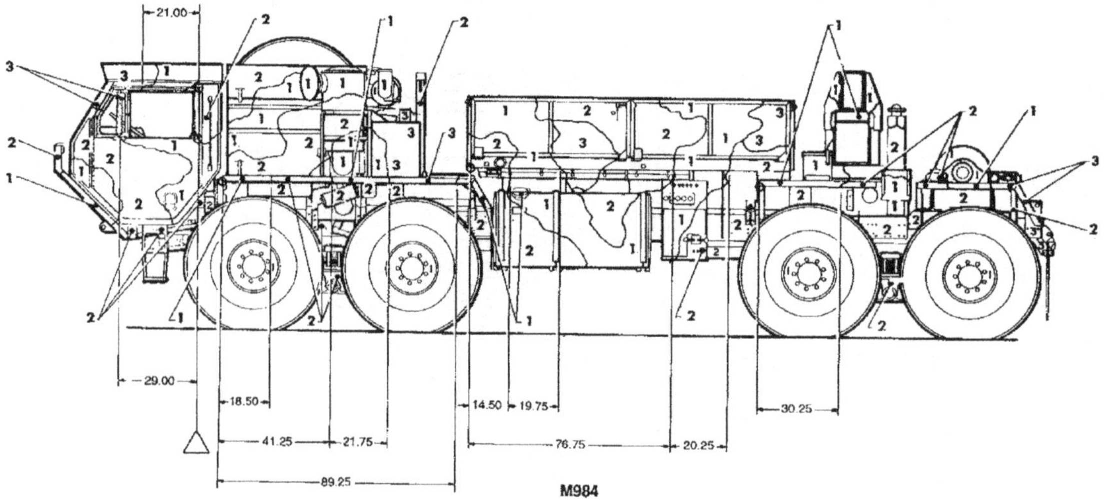

LEFT SIDE VIEW

Figure 84. Truck, recovery: 11-ton, M984. (2 of 4)

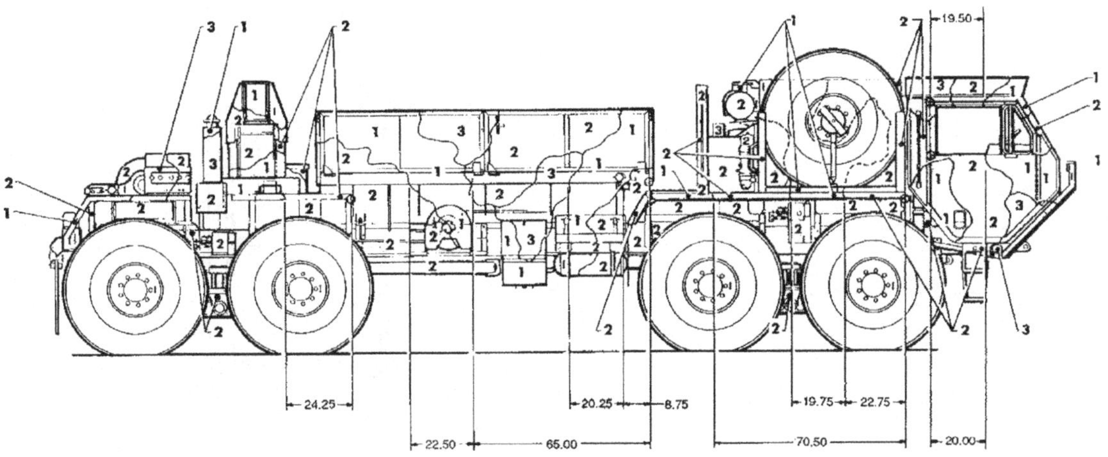

RIGHT SIDE VIEW

M984

Figure 84. Truck, recovery: 11-ton, M984. (3 of 4)

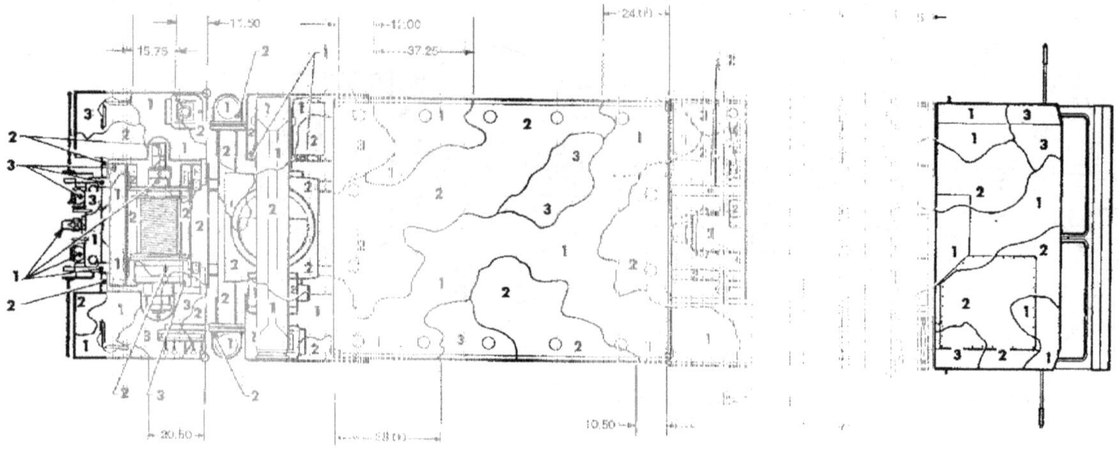

TOP VIEW

M984

Figure 84. Truck, recovery: 11-ton, M984 (cont)

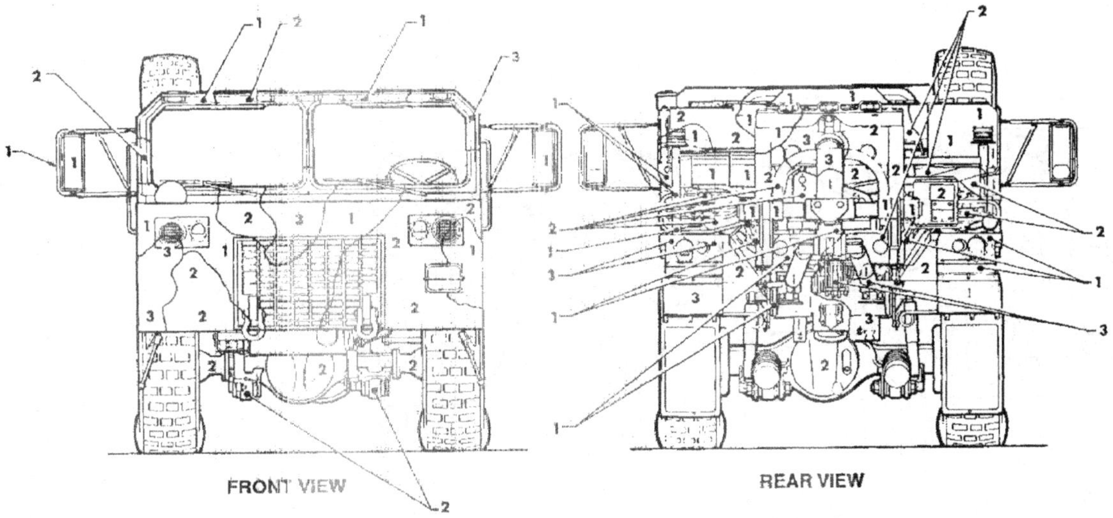

M984E1

Figure 85. Truck, wrecker: 11-ton, M984E1. (1 of 4)

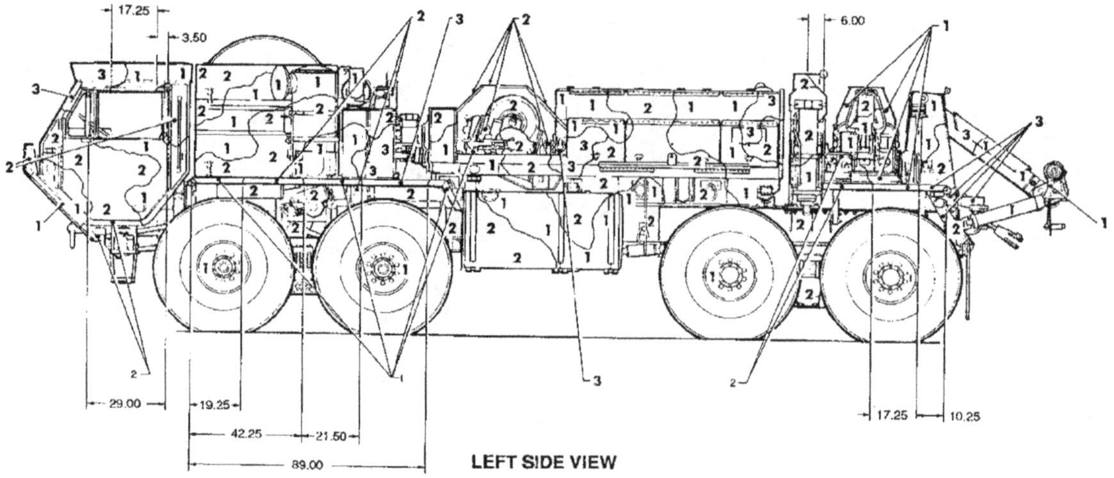

M984E1

Figure 85. Truck, wrecker: 11-ton, M984E1. (2 of 4)

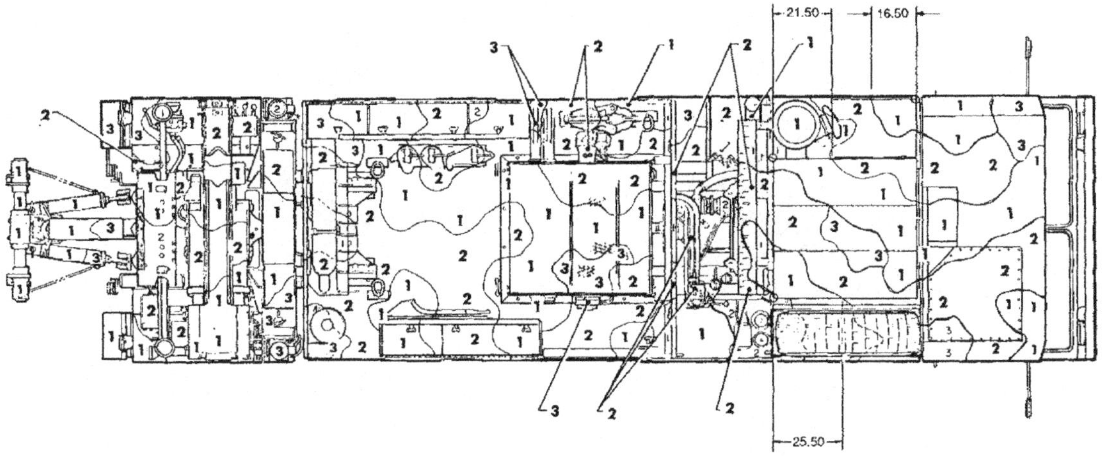

TOP VIEW
M984E1

Figure 85. Truck, wrecker: I1-ton, M984E1. (3 of 4)

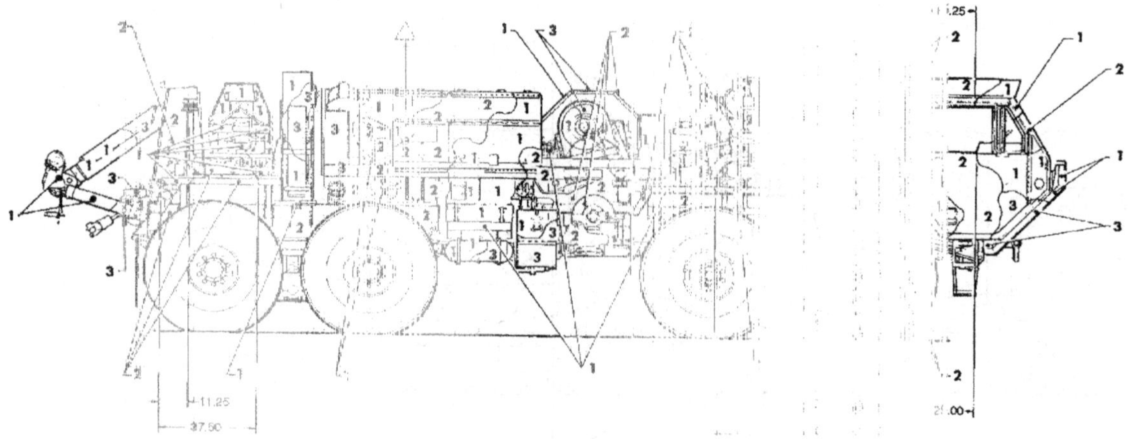

Figure 85. Truck, wrecker: 11-ton, M984E1 (4

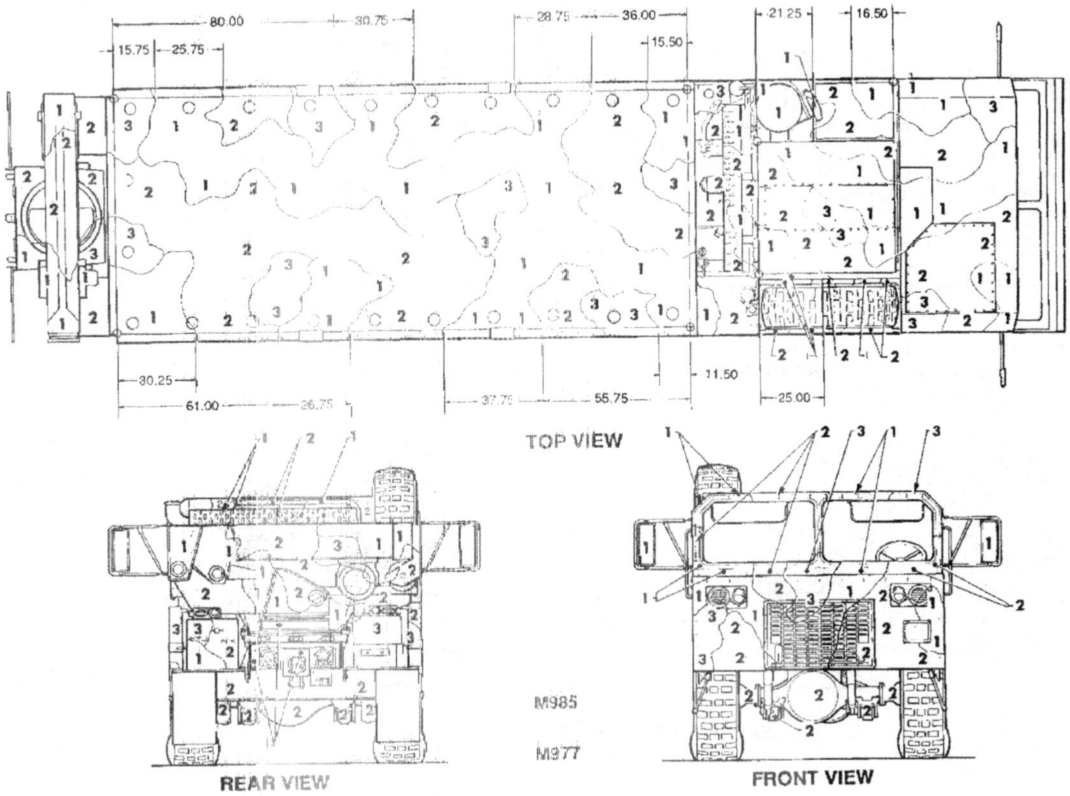

Figure 86. Truck, cargo: 11-ton, M977 and M985. (1 of 2)

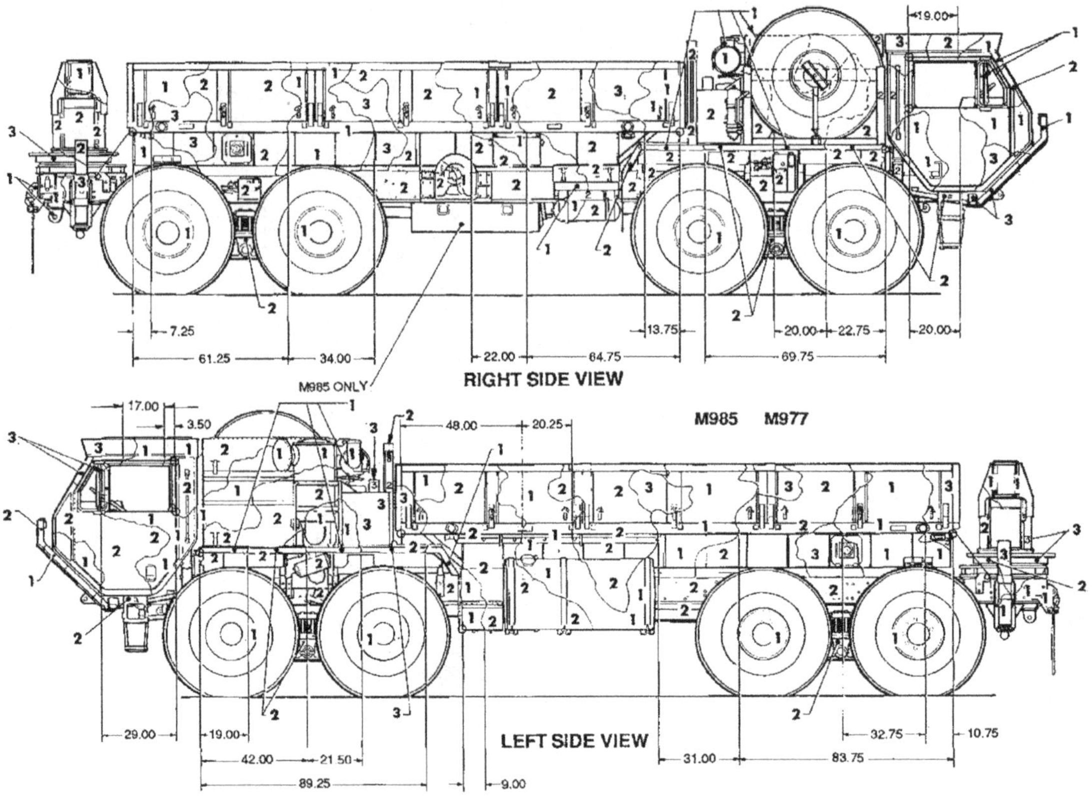

Figure 86. Truck, cargo: 11-ton, M977 and M985. (2 of 2)

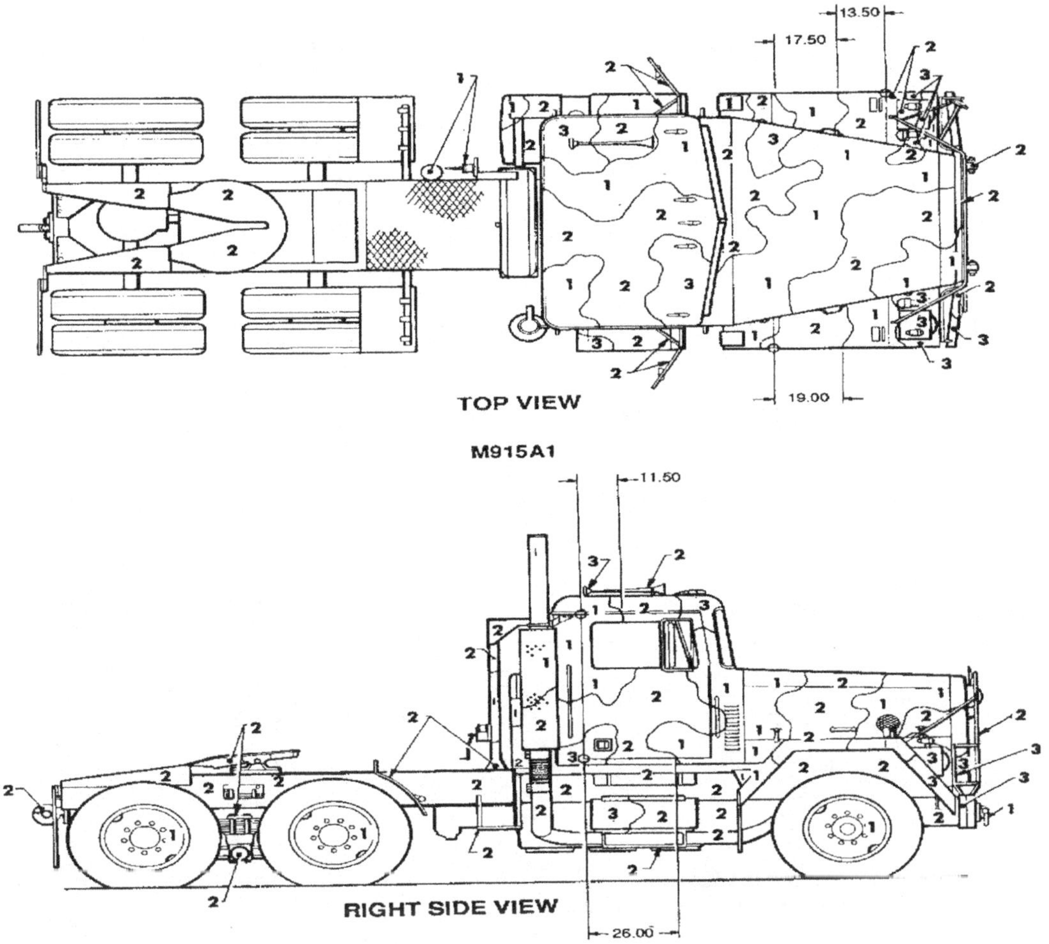

Figure 87. Truck, tractor: 22-1/2-ton linehaul, M915A1. (I of 2)

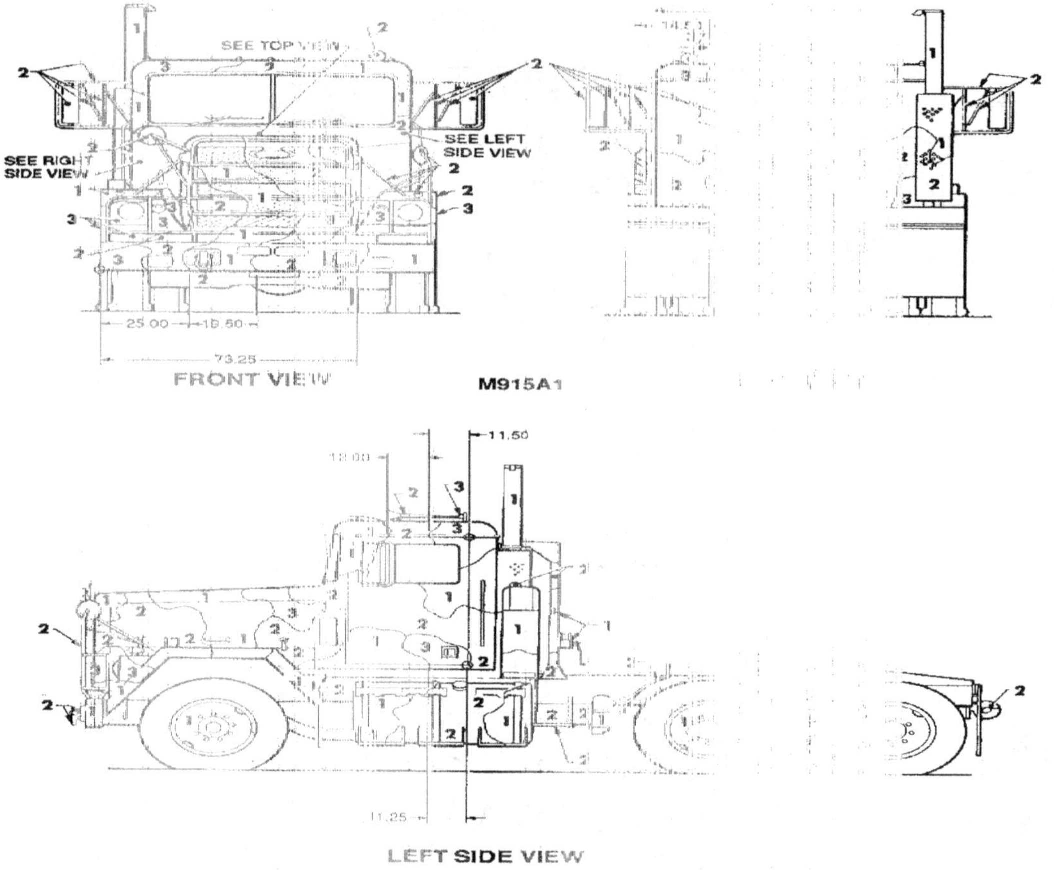

Figure 67. Truck, tractor: 22-1/2-ton linehaul, M915

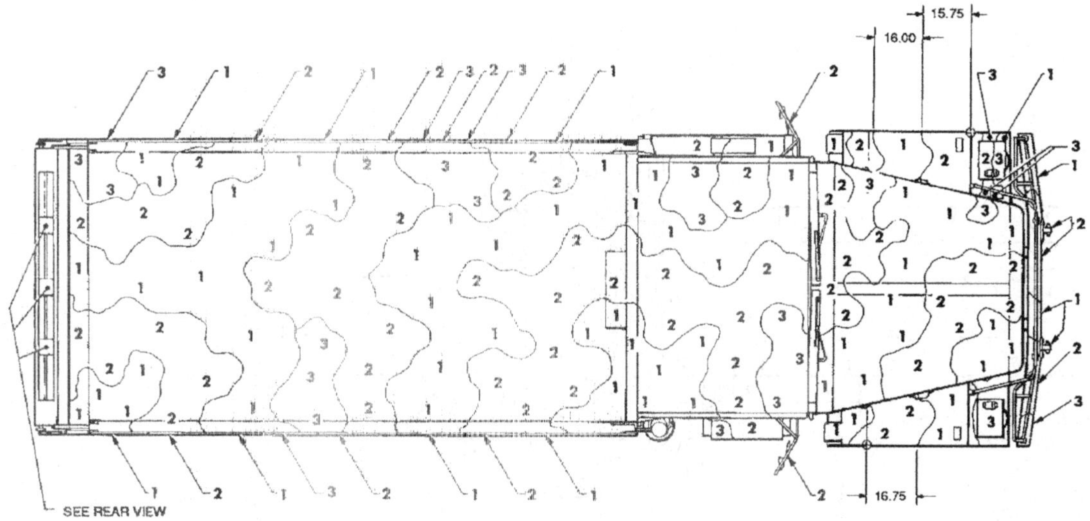

TOP VIEW
M917

Figure 88. Truck, dump: 20-ton, M917. (1 of 4)

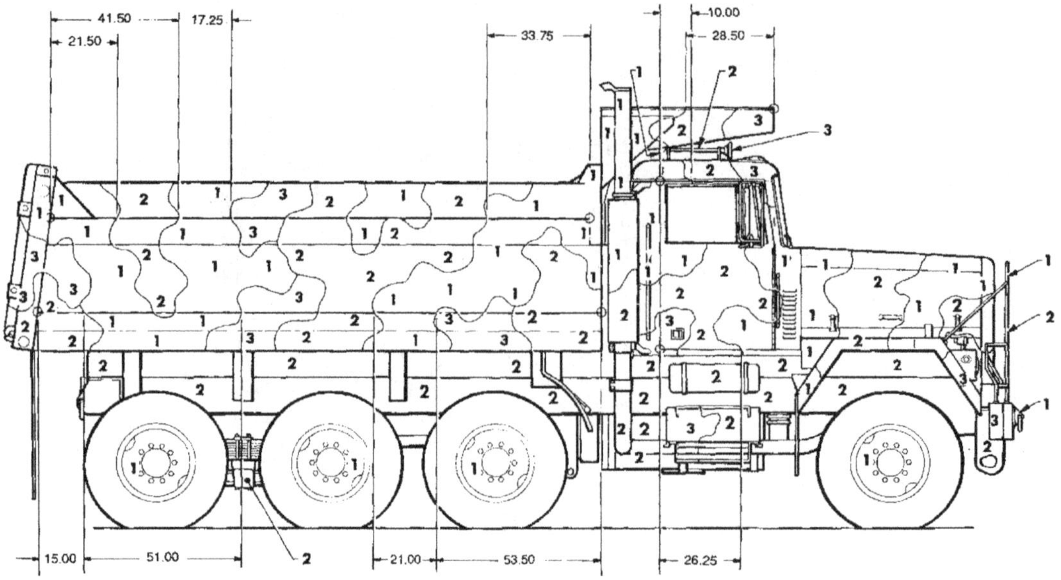

M917

RIGHT SIDE VIEW

Figure 88. Truck, dump: 20-ton, M917. (2 of 4)

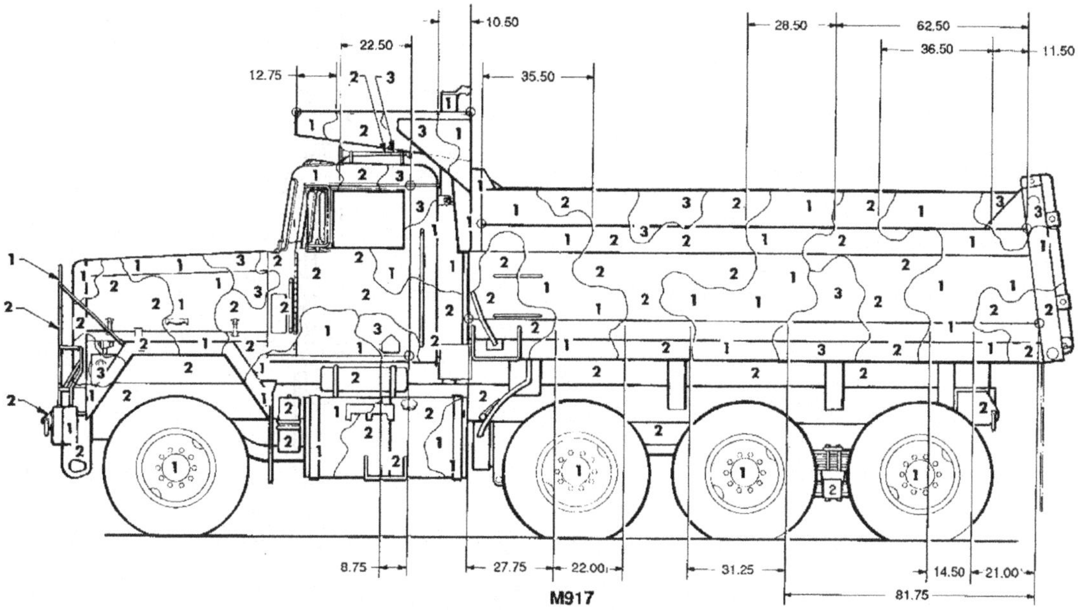

LEFT SIDE VIEW

Figure 88. Truck, dump: 20-ton, M917. (3 of 4)

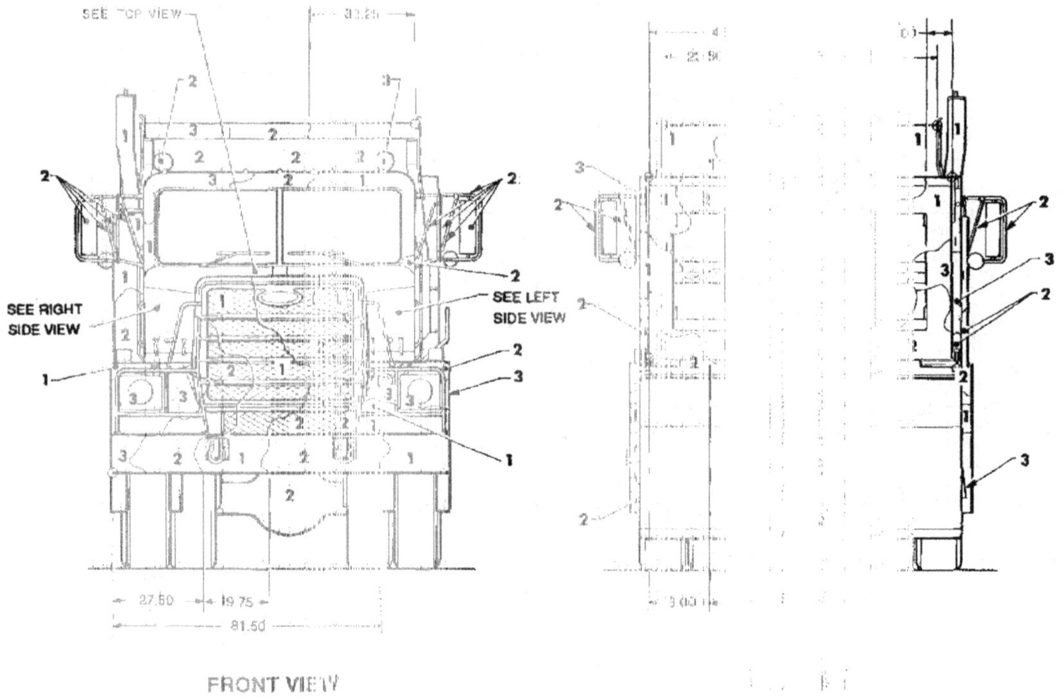

M917

Figure 38. Truck, dump: 20-ton, M917. (4 of

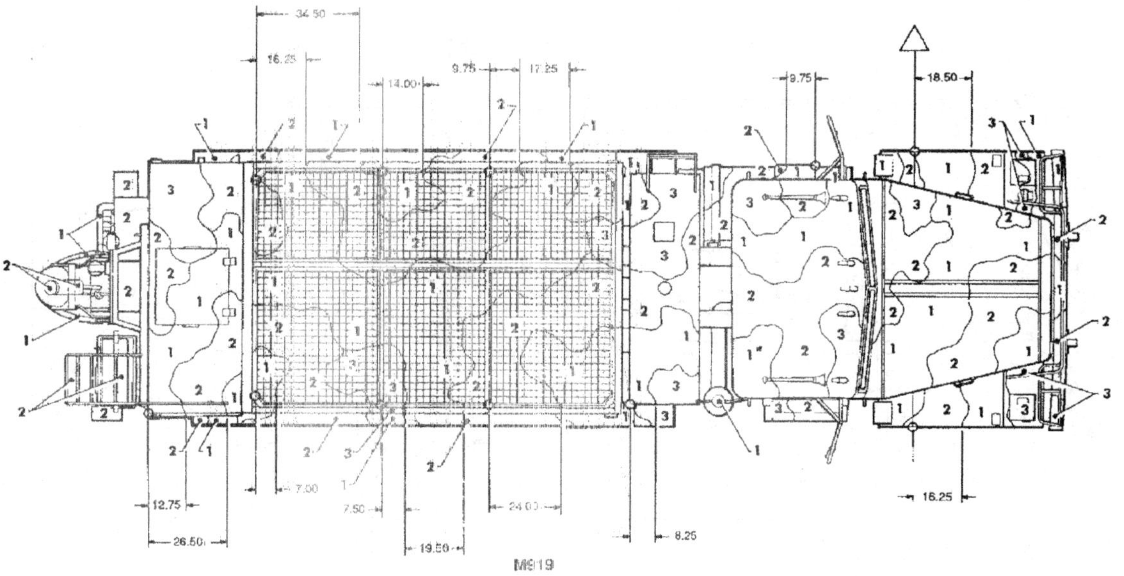

TOP VIEW

Figure 89. Truck, concrete: mobile, M919. (1 of 4)

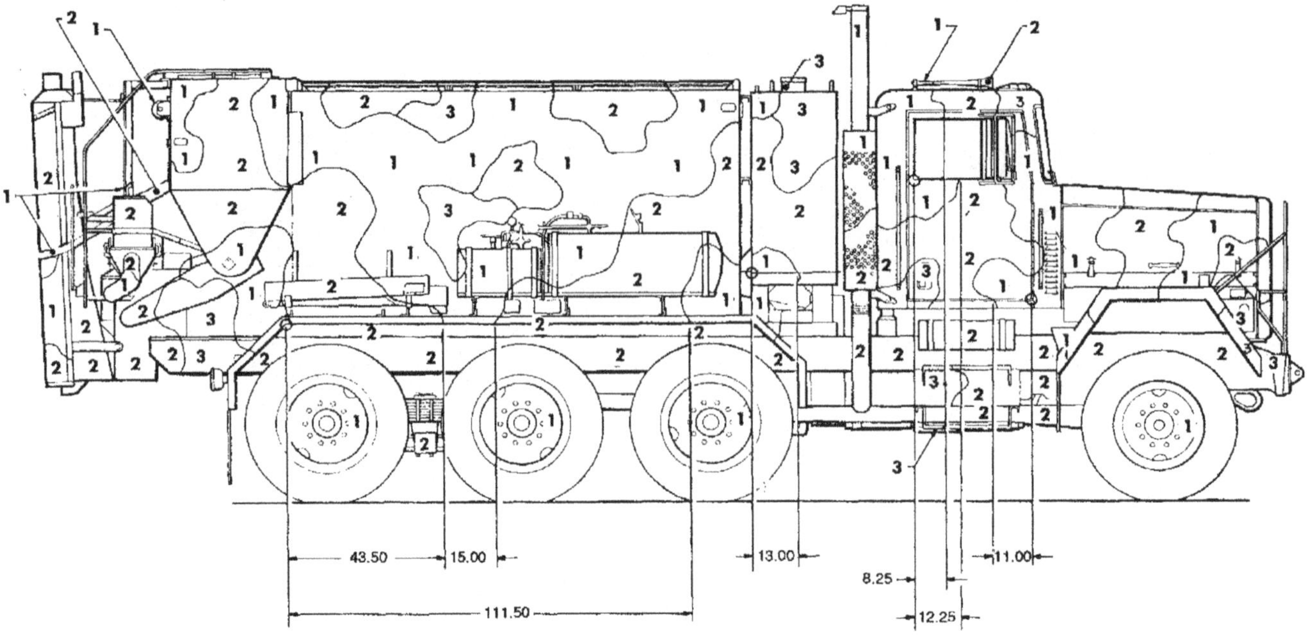

RIGHT SIDE VIEW

M919

Figure 89. Truck, concrete: mobile, M919. (2 of4)

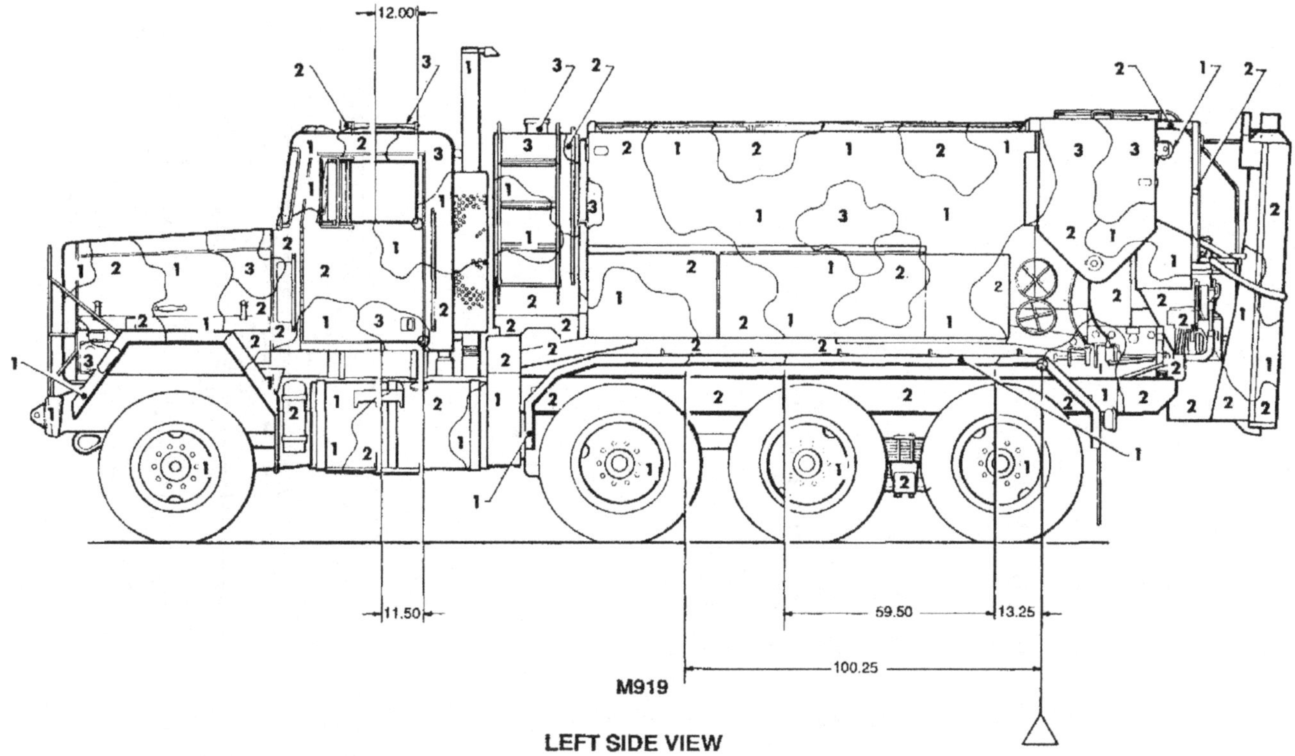

Figure 89. Truck, concrete: mobile, M919. (3 of 4)

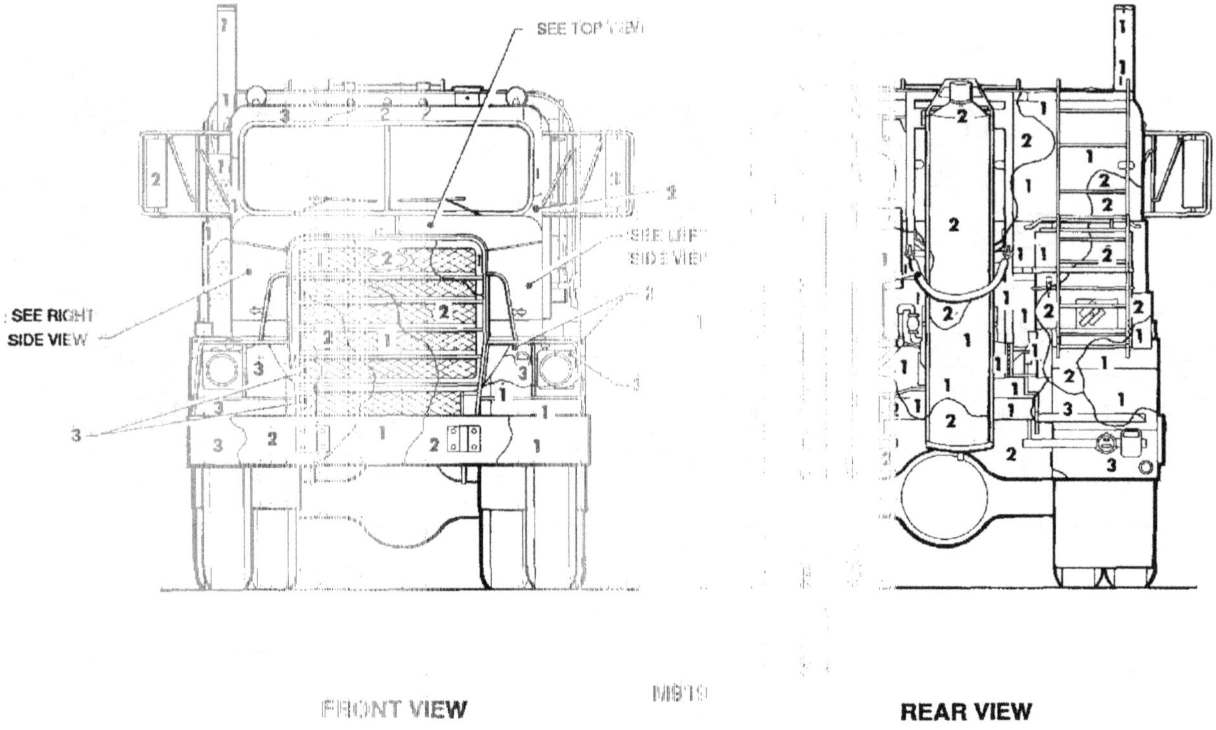

Figure 89. Truck, concrete mobile

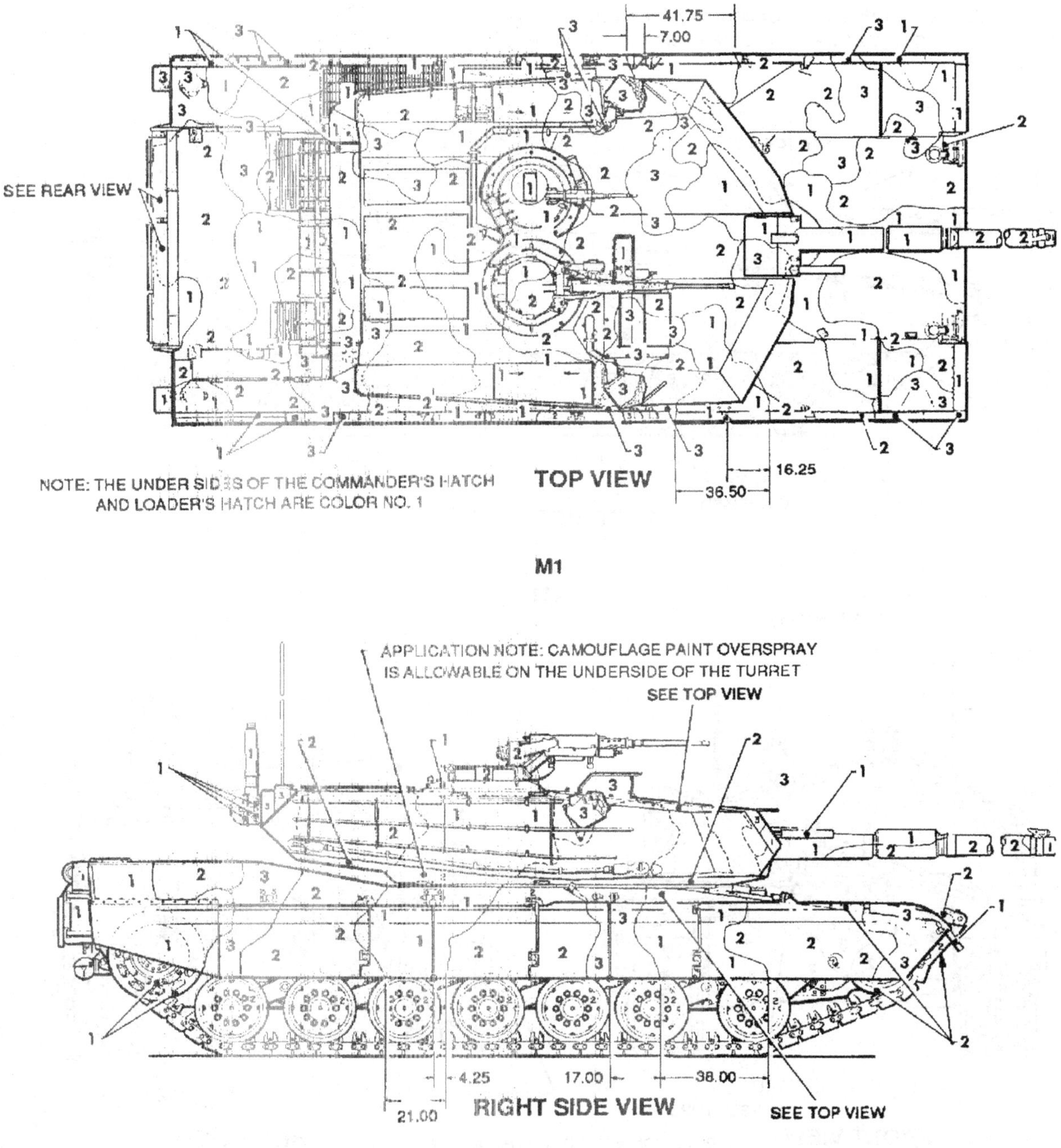

Figure 90. Tank, combat, full tracked: 105-mm gun, M1/M1 chassis. (1 of 2)

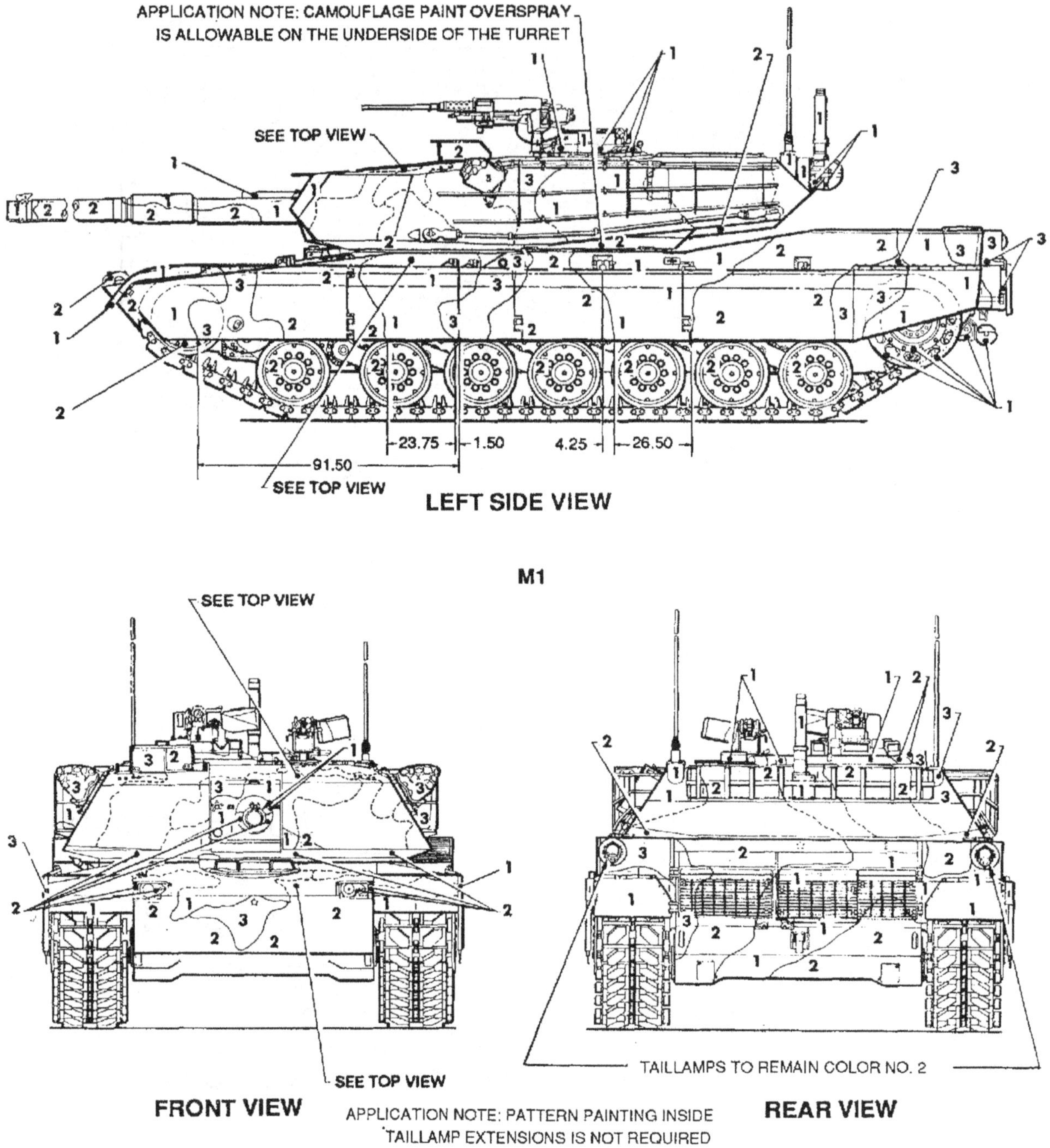

Figure 90. Tank, combat, full tracked: 105-mm gun, M1IM1 chassis. (2 of2)

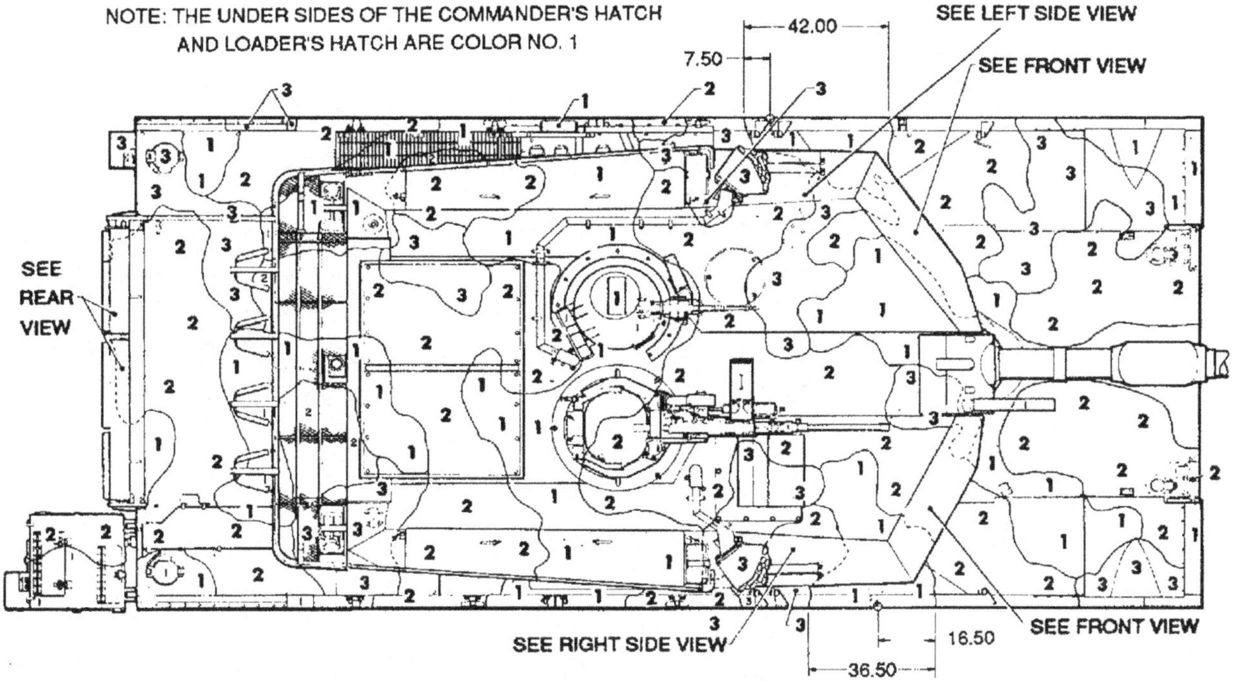

TOP VIEW

M1A1

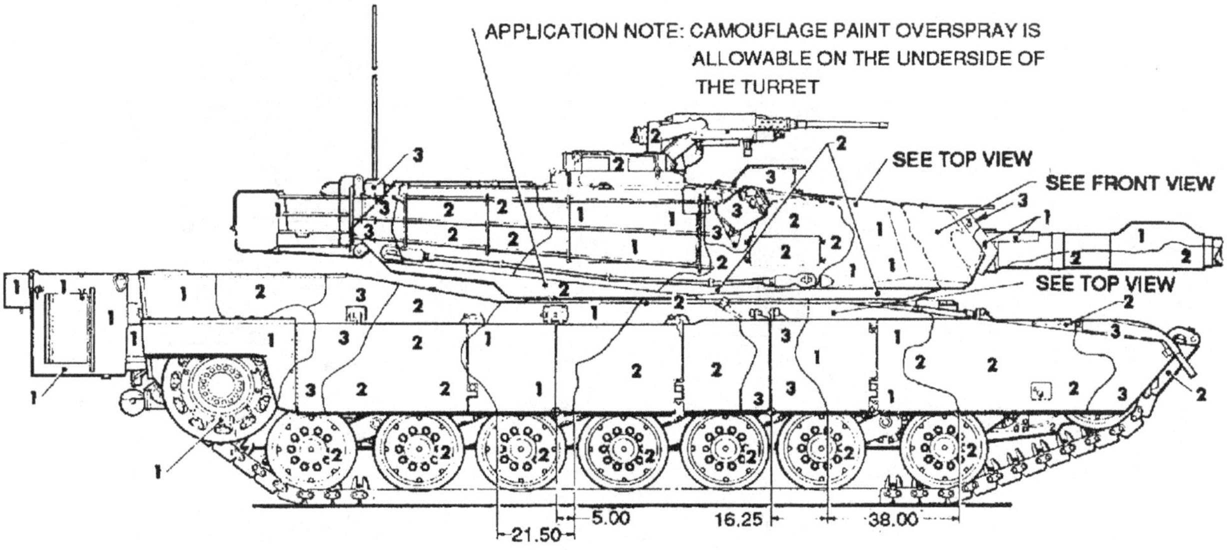

RIGHT SIDE VIEW

Figure 91. Tank, combat, full- tracked: 120-mm gun, M1A1. (1 of2)

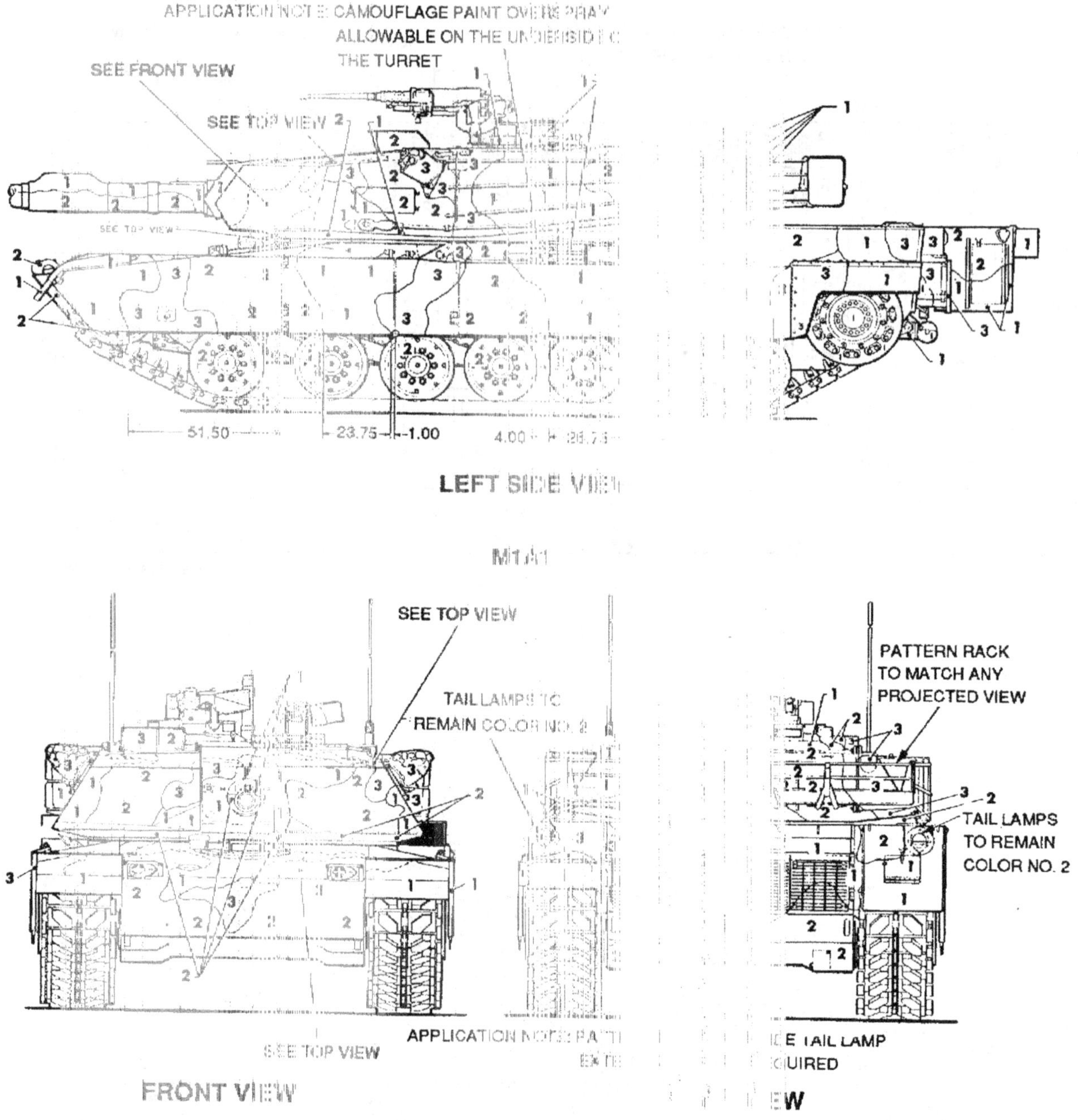

Figure 91. Tank, combat, full tracked: 120- ... (2 of 2)

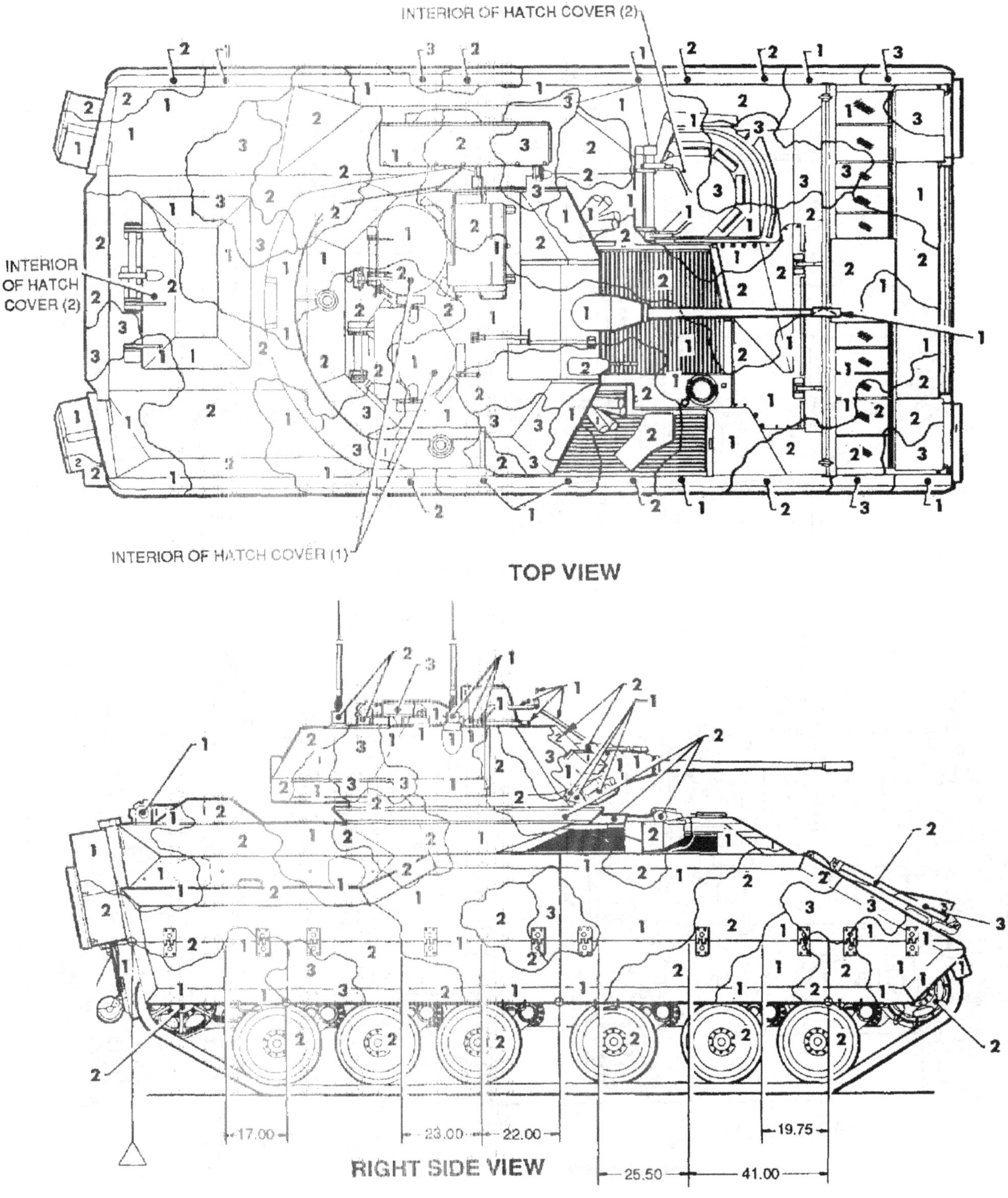

Figure 92. Fighting Vehicle, infantry M2/M2A1, cavalry M3/M3A1. (1 of2)

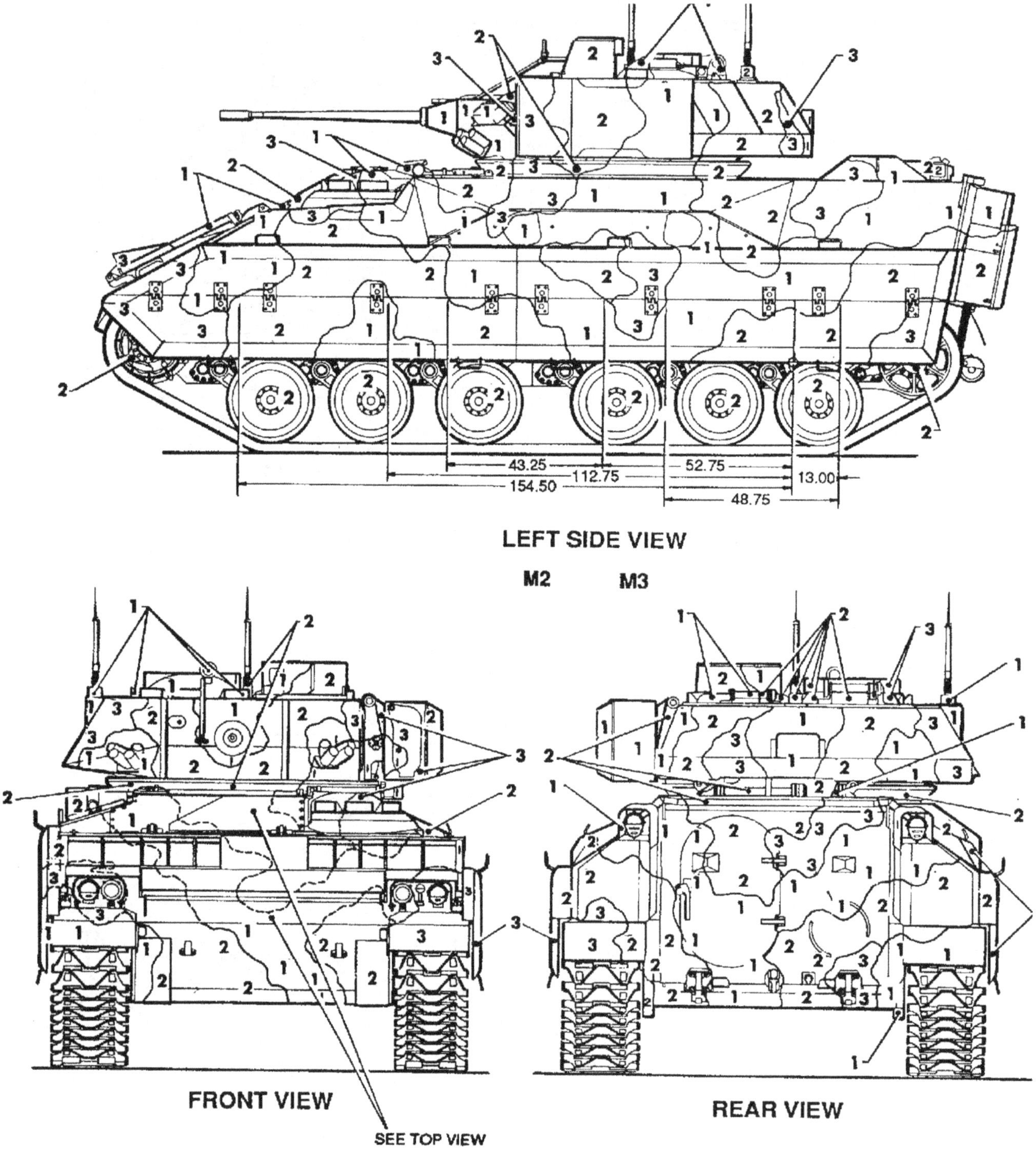

Figure 92. Fighting Vehicle, infantry M2/M2A1, cavalry M3/M3A1. (2 of 2)

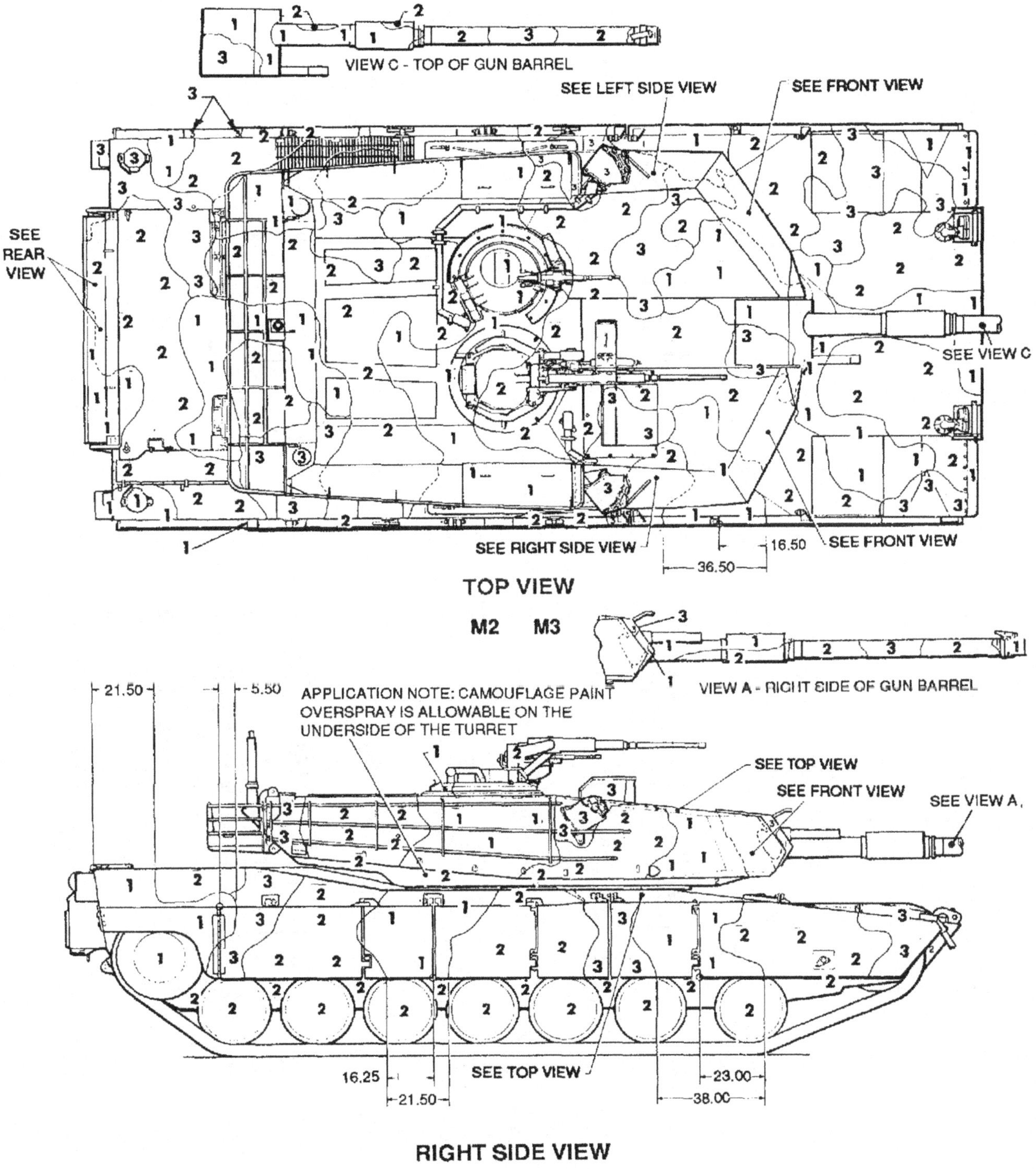

Figure 93. Tank, combat, full tracked: 105-mm gun, 1PM1. (1 of 2)

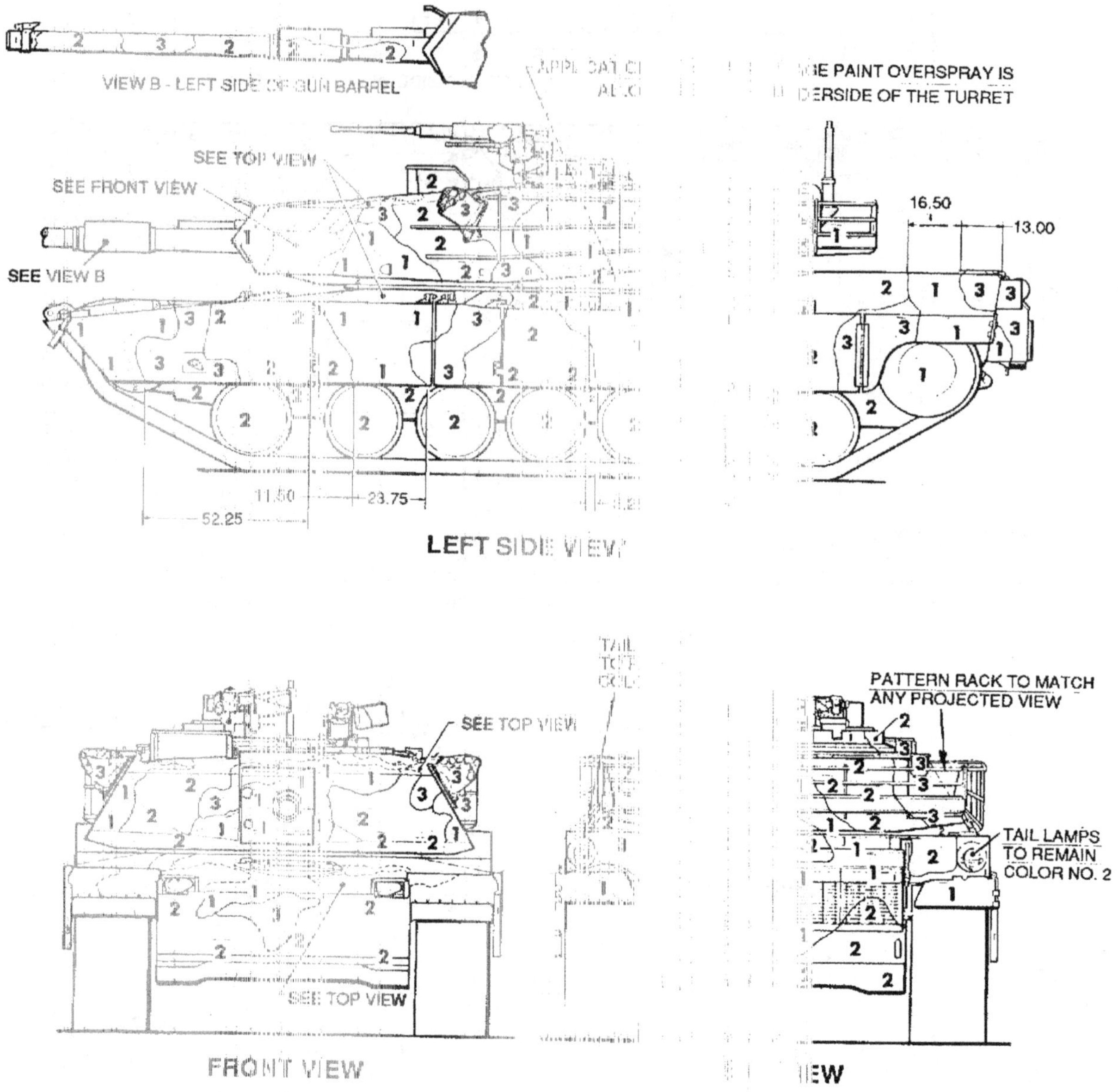

Figure 93. Tank, combat, full tracked: 105 (2 of2)

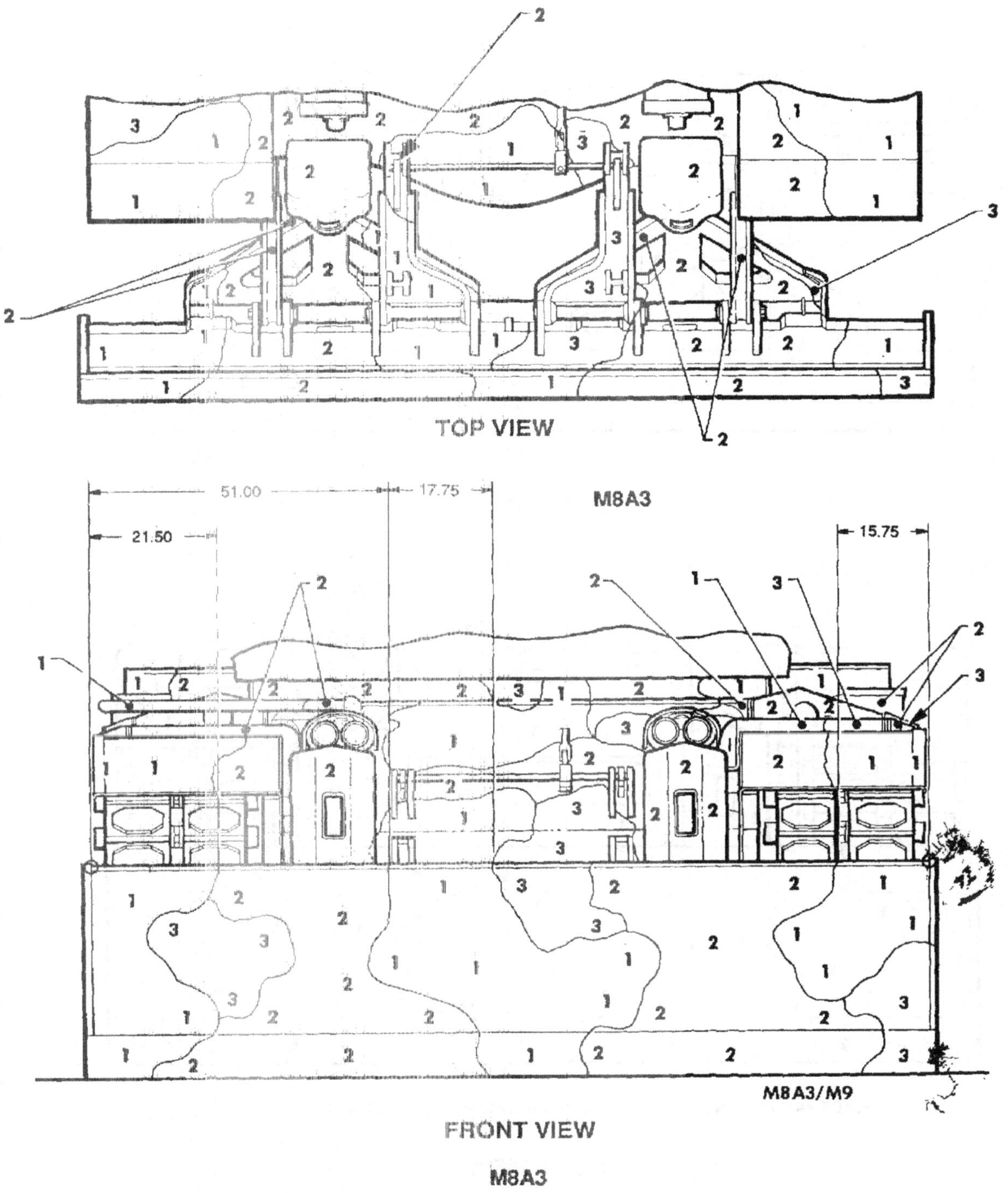

Figure 94. Blade, bulldozer, earthmoving: M8A3/M9. (1 of 2)

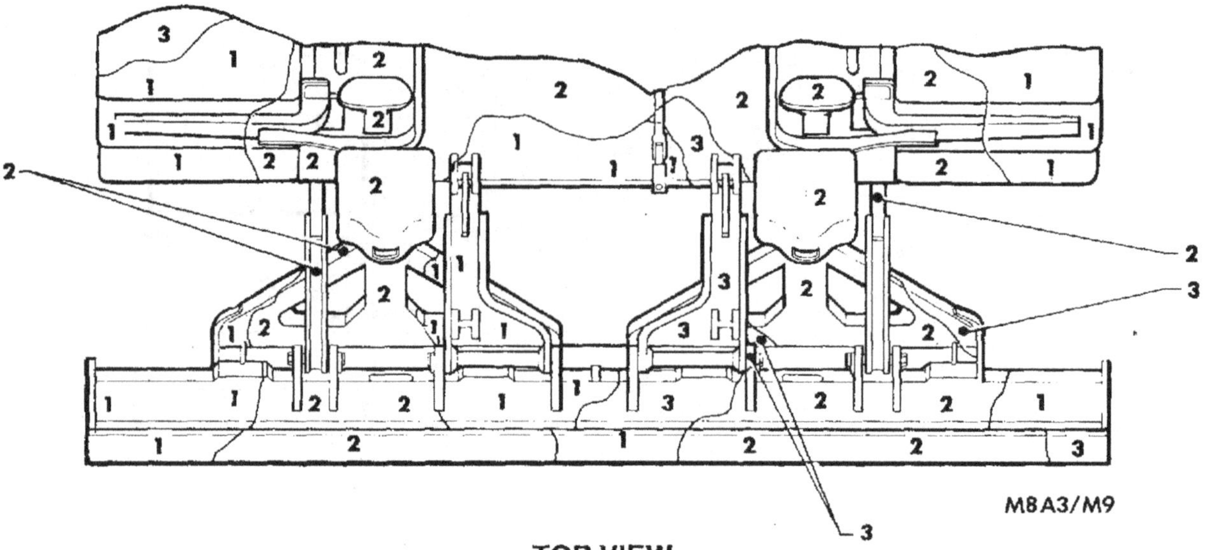

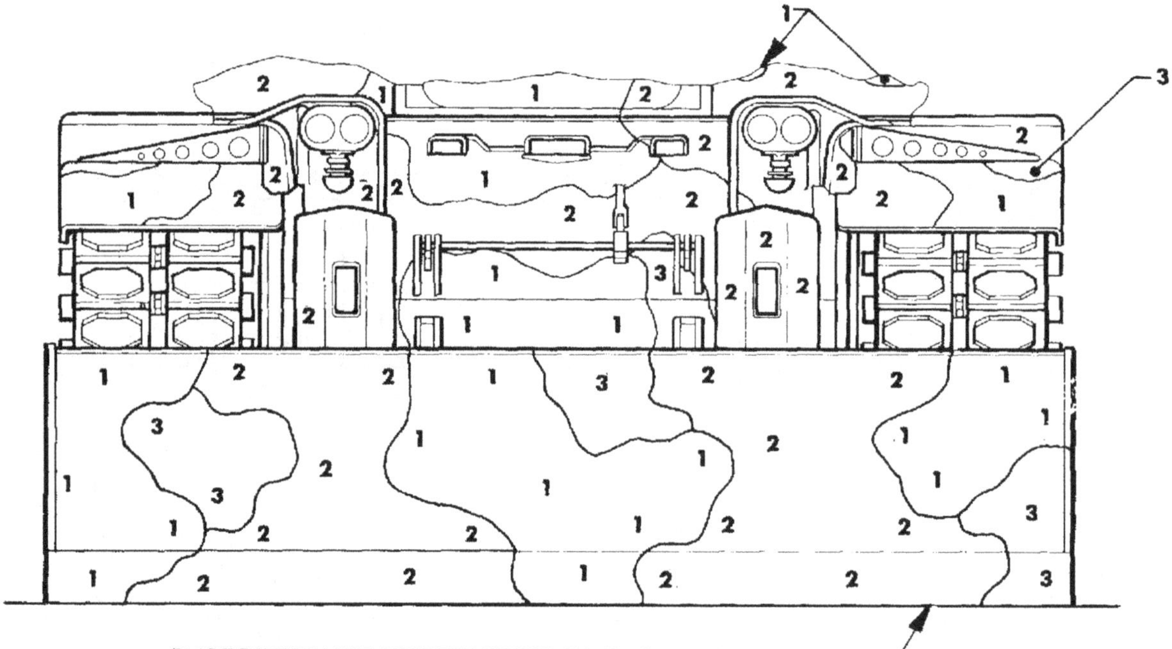

Figure 94. Blade, bulldozer, earthmoving: M8A3/M9. (2 of 2)

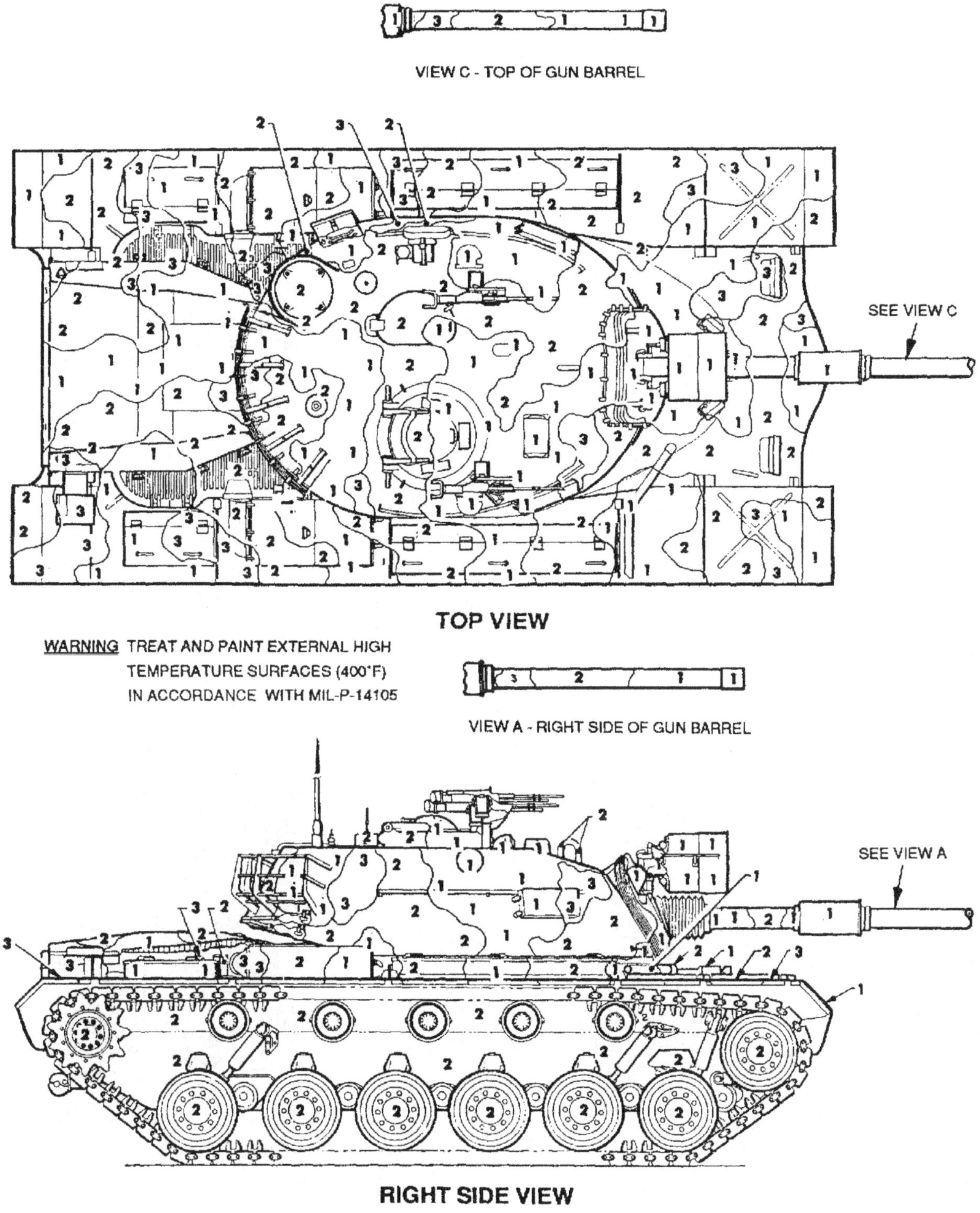

Figure 95. Tank, combat, full tracked: 105-mm gun, M48A5. (1 of 2)

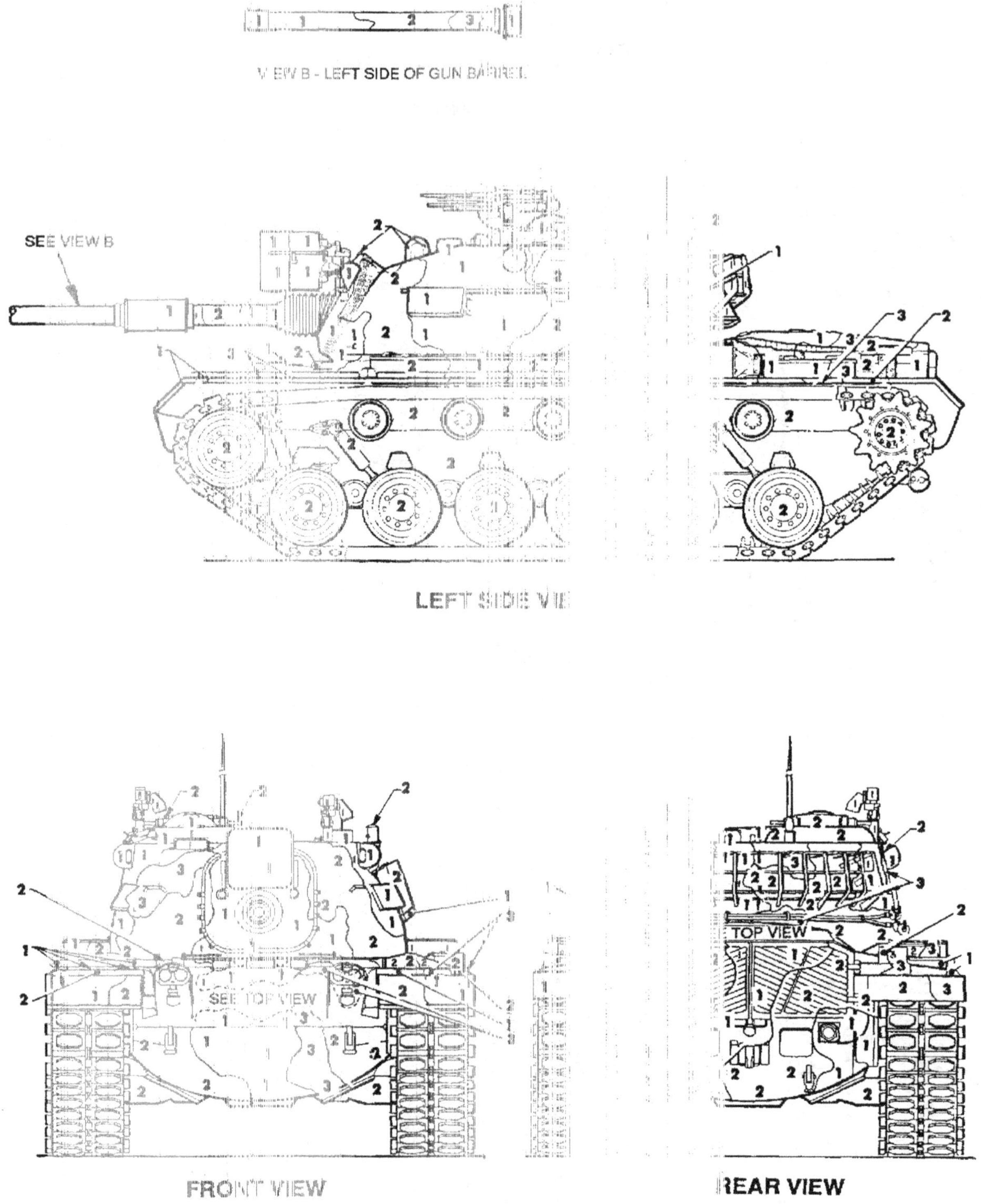

Figure 95. Tank, combat, full tracked: 105- (2 of 2)

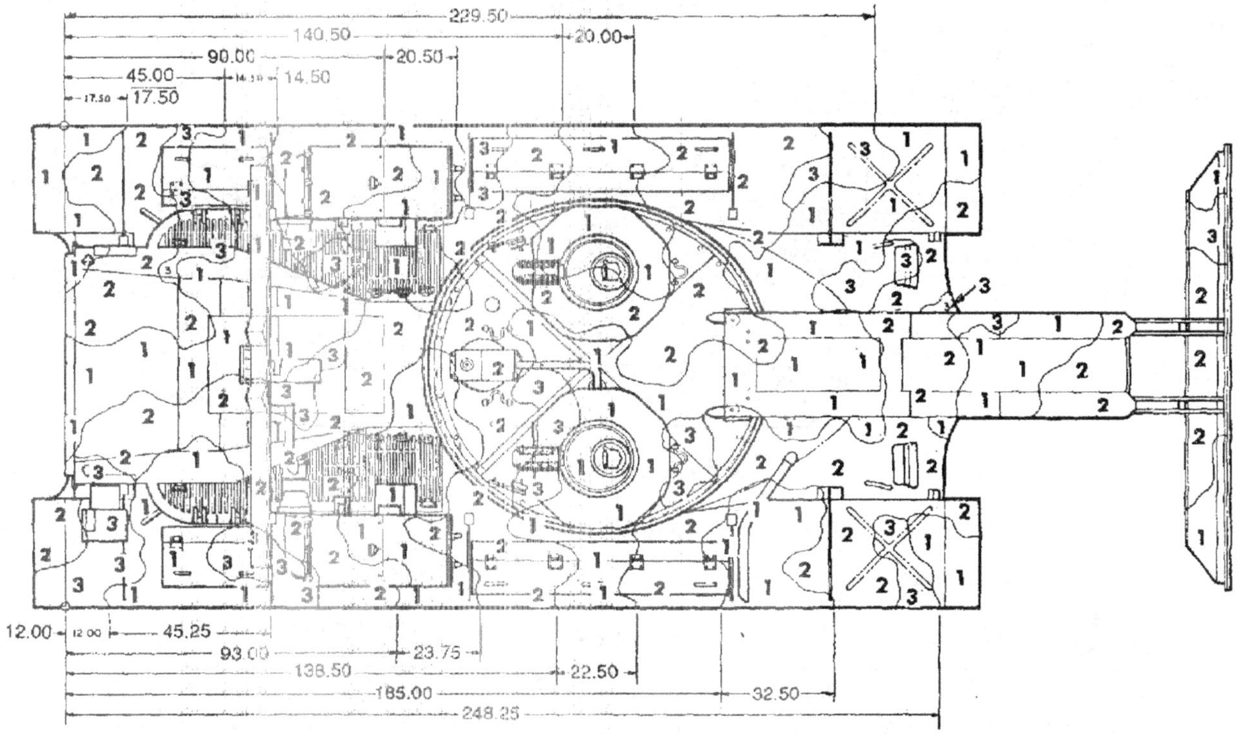

TOP VIEW

M48A5

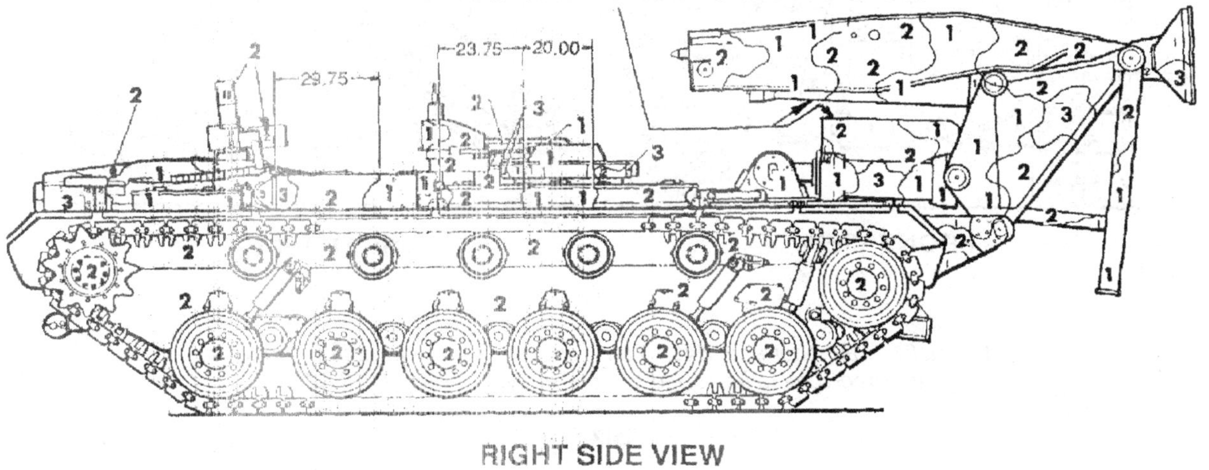

RIGHT SIDE VIEW

Figure 96. Launcher and M48A5 tank chassis, transporting:for Bridge, armored-vehicle-launched scissoring type, class 60. (1 of2)

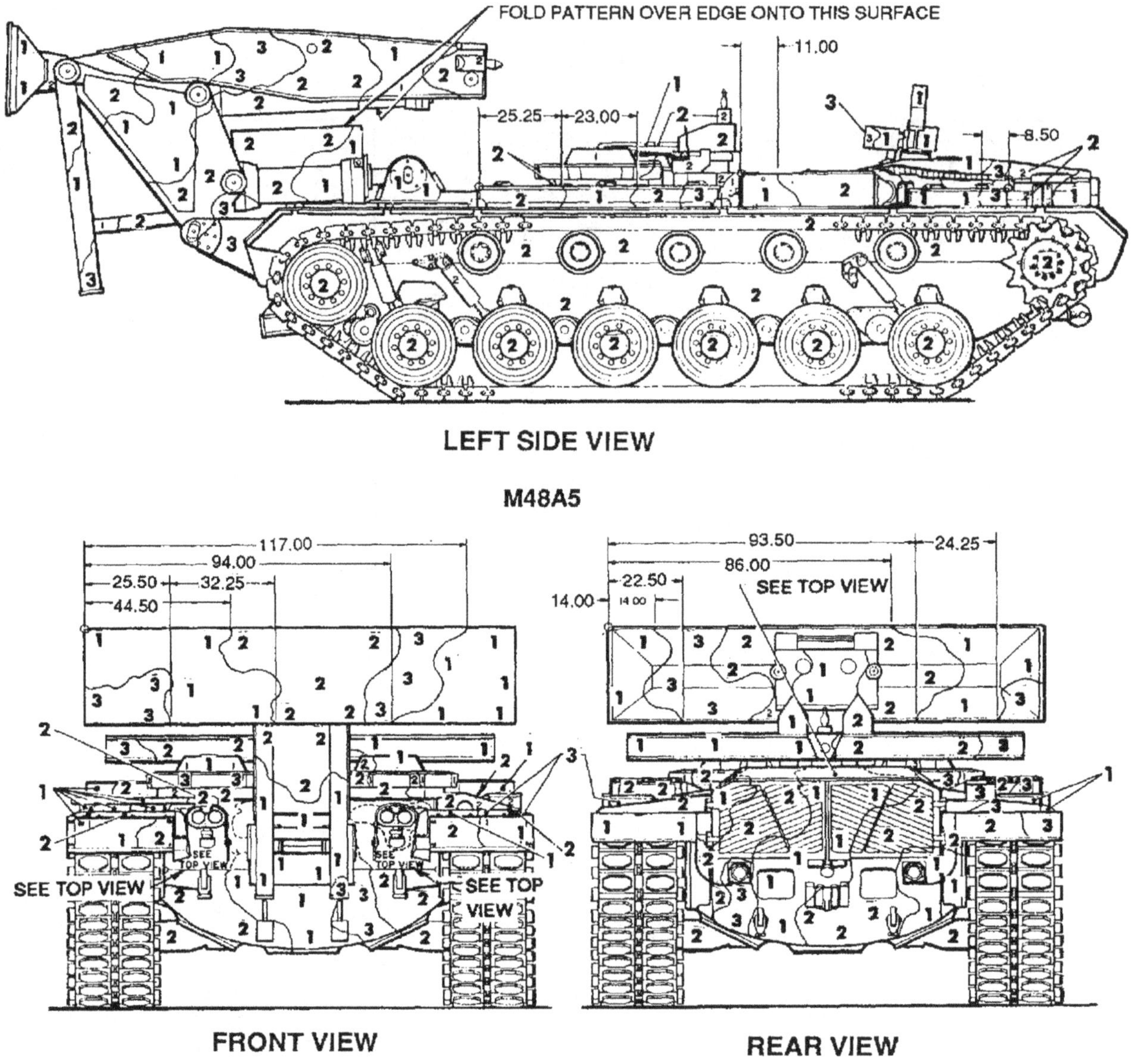

Figure 96. Launcher and M48A5 tank chassis, transporting:for Bridge, armored-vehicle-launched scissoring type, class 60. (2 of2)

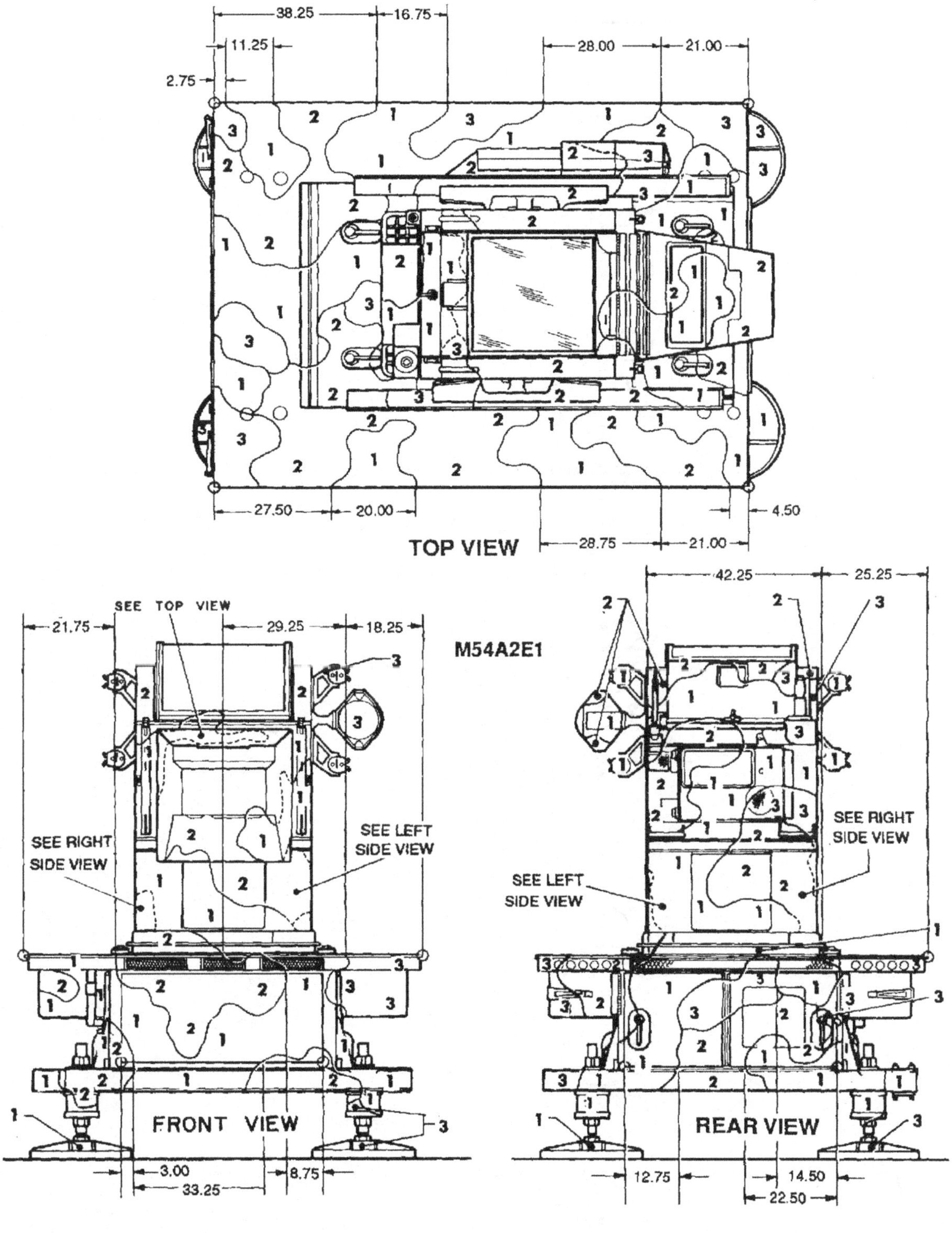

Figure 97. Launcher, platform, M54A2E1. (1 of 2)

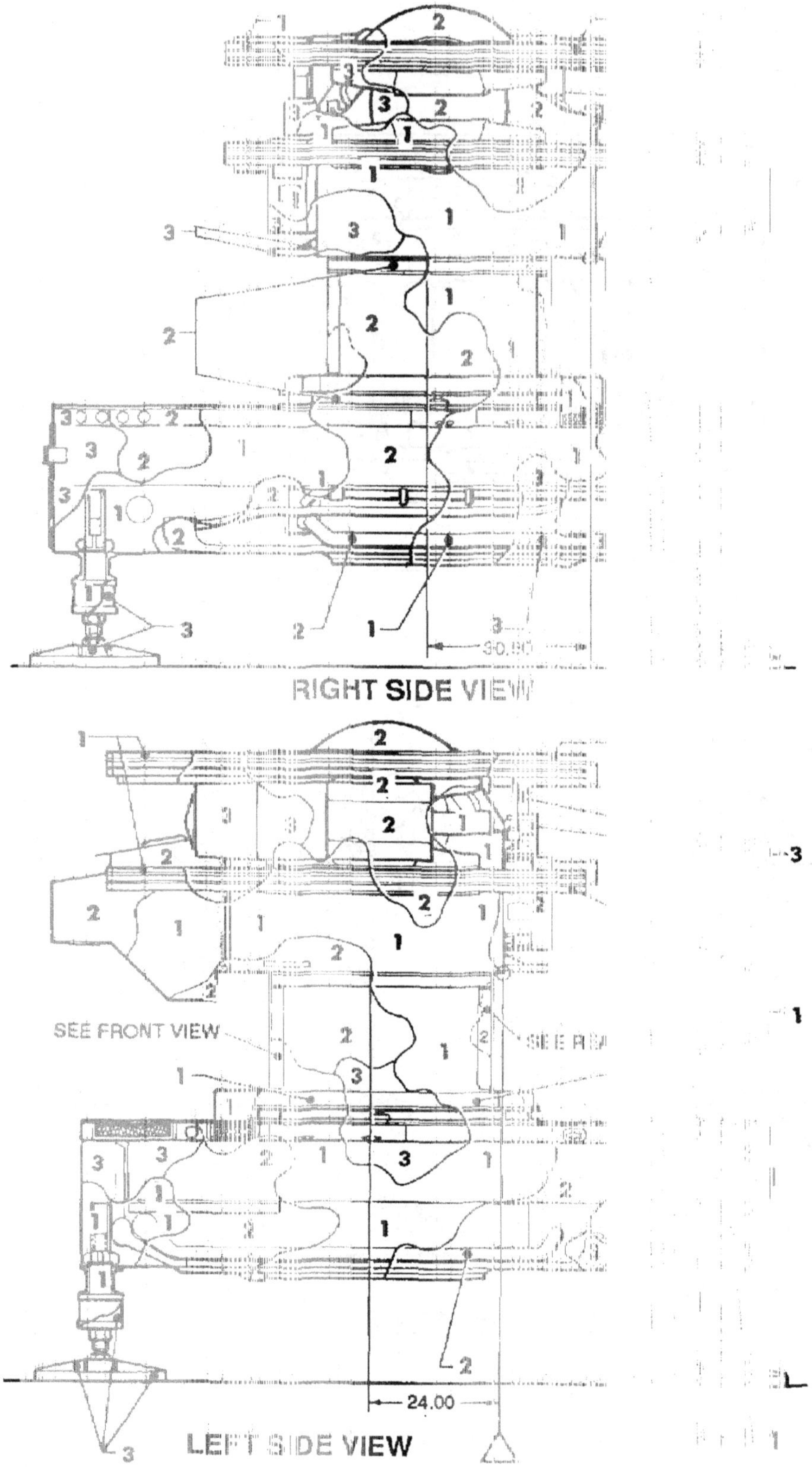

Figure 97. Launcher, platform, M54A2E1.

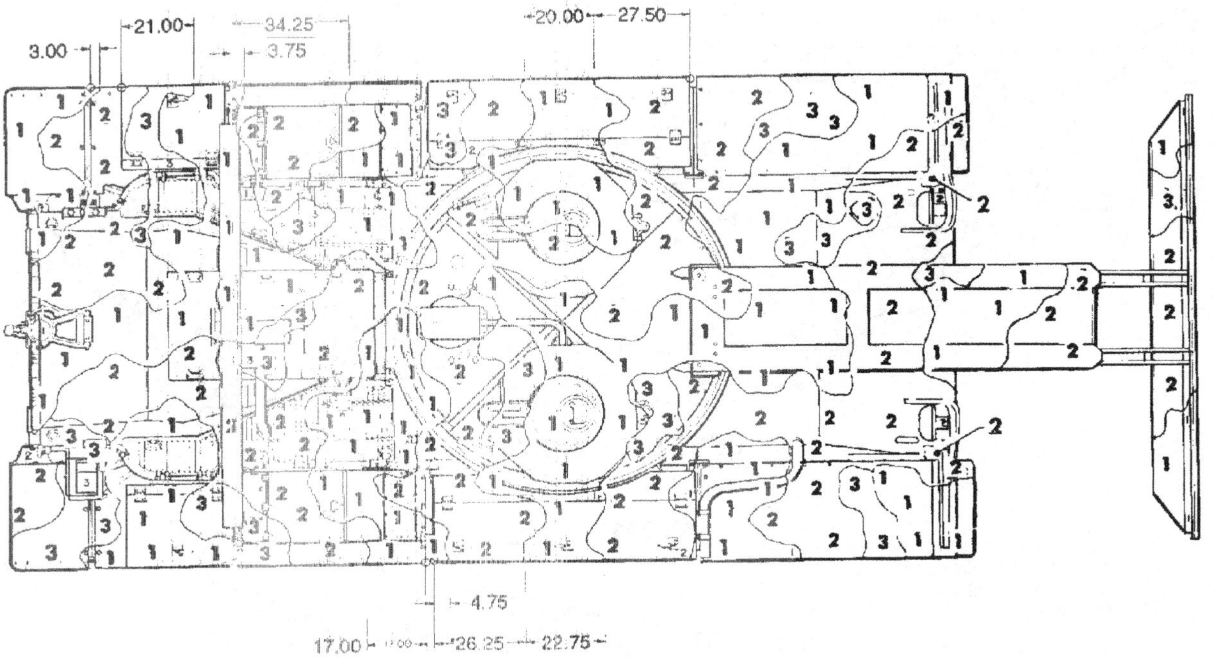

TOP VIEW

M60A1

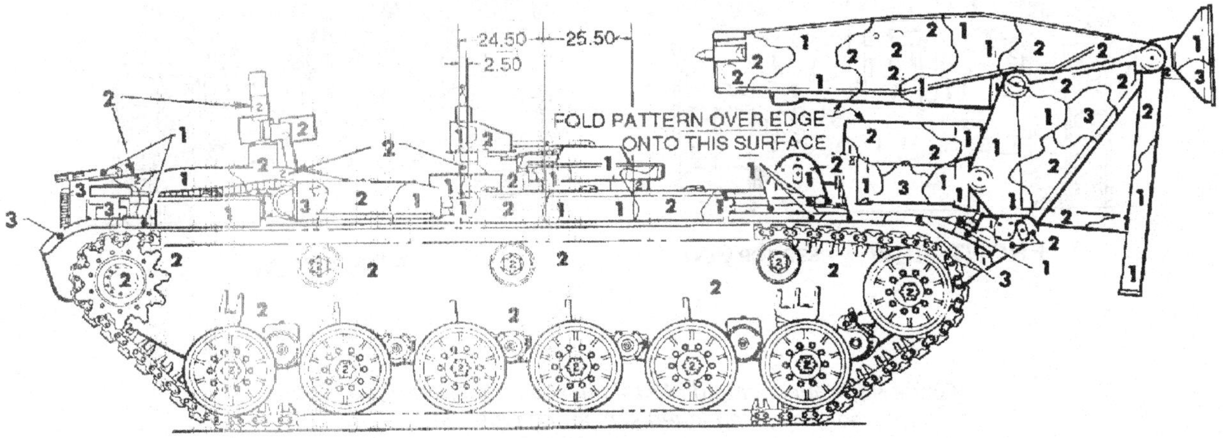

RIGHT SIDE VIEW

Figure 98. Launcher and M60A1 chassis, transporting: for Bridge, armored-vehicle-launched scissoring type, class 60. (1 of 2)

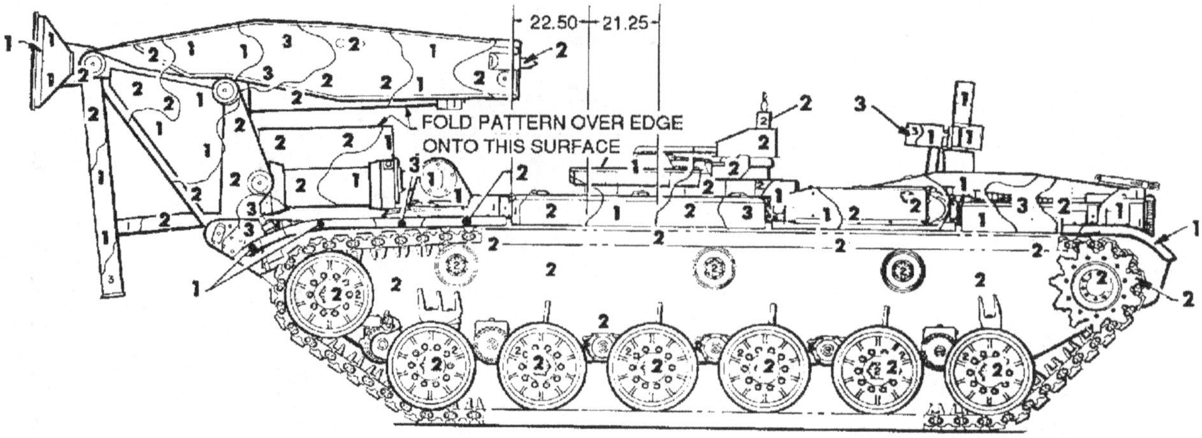

LEFT SIDE VIEW

M60A1

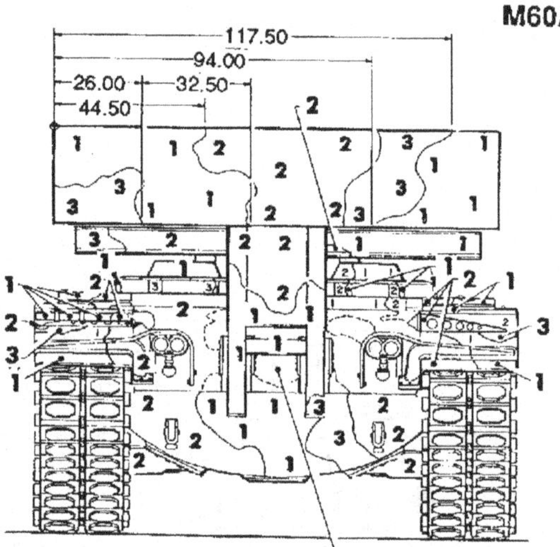

FRONT VIEW

SEE TOP VIEW

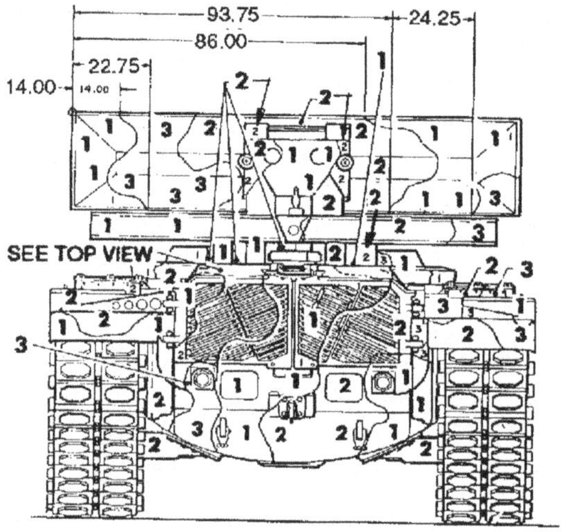

REAR VIEW

Figure 98. Launcher and M60AI chassis, transporting:for Bridge, armored-vehicle-launched scissoring type, class 60. (2 of2)

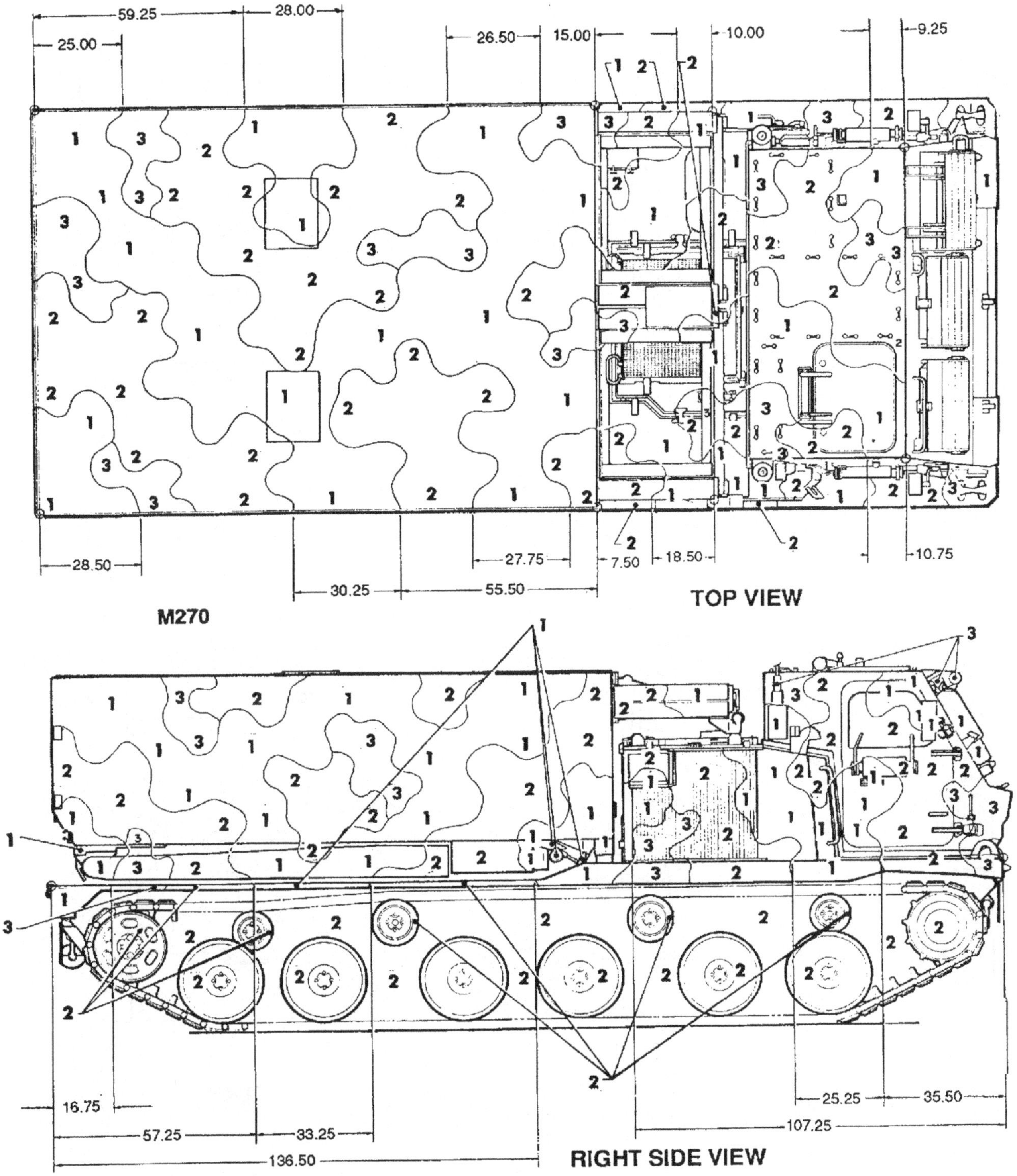

Figure 99. Self-propelled Loader/Launcher, M270. (1 of 2)

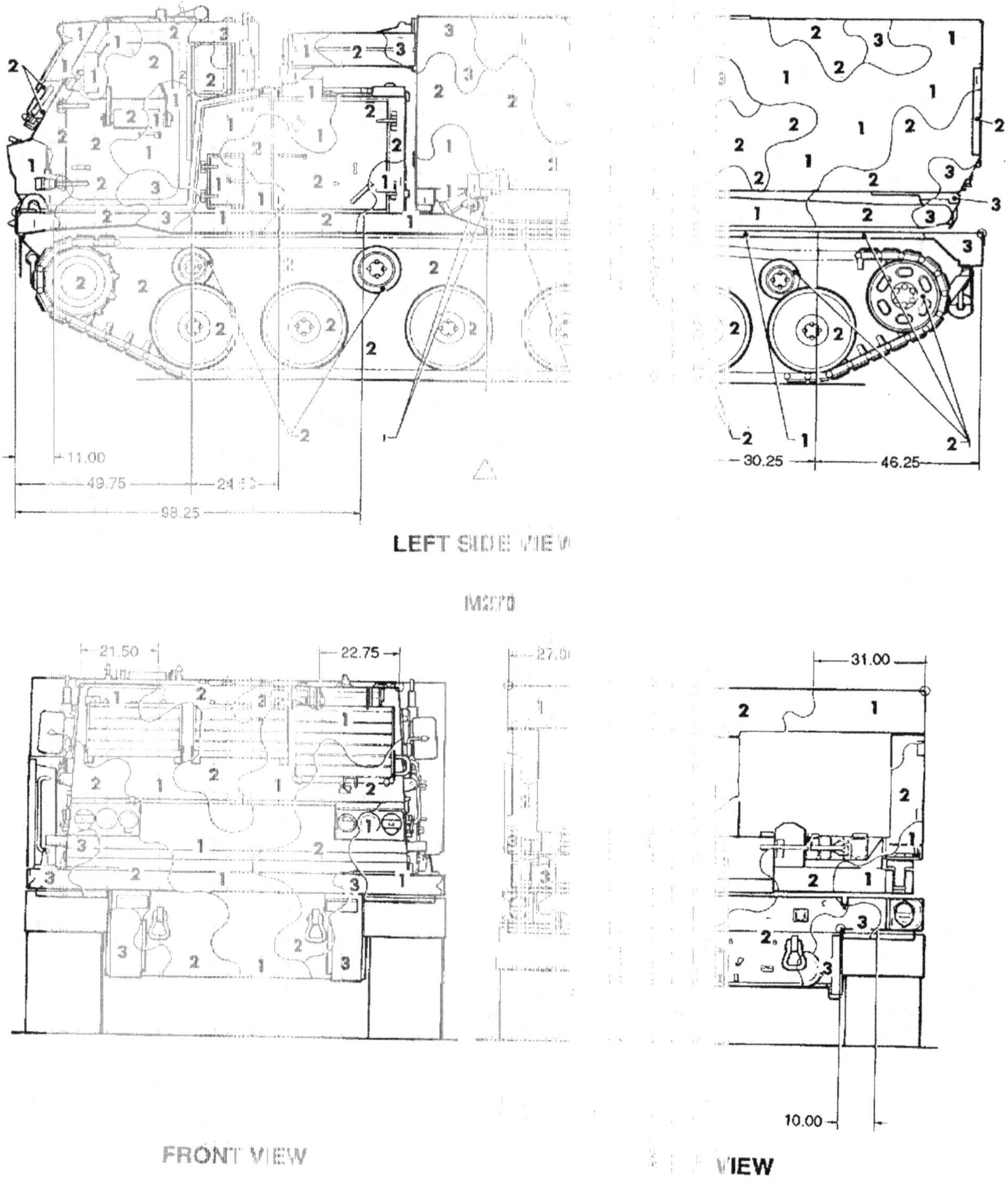

Figure 99. Self-propelled Loader/Launcher (Sheet 2 of 2)

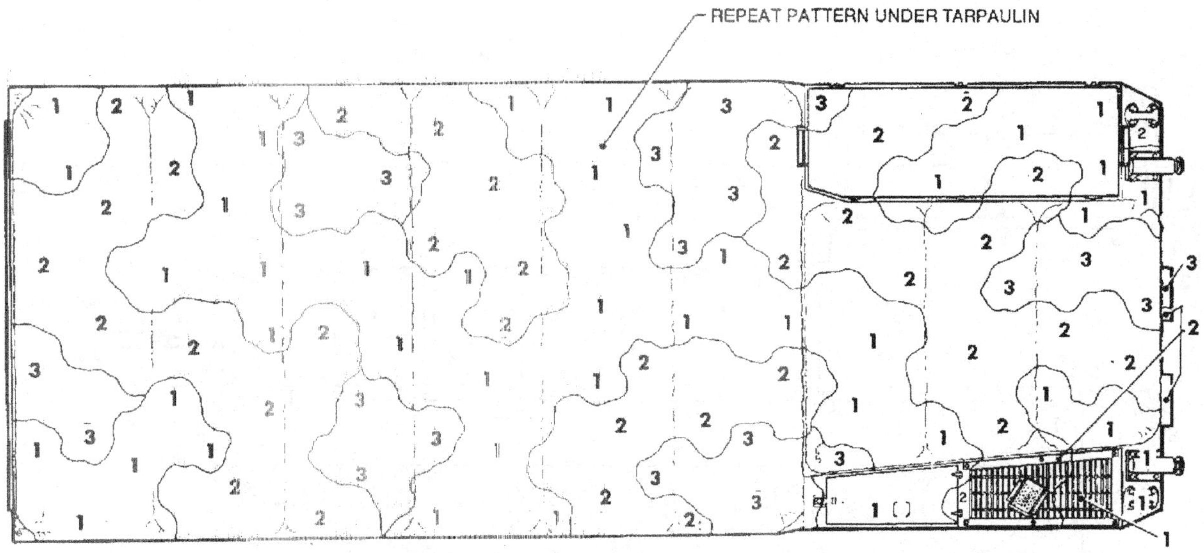

TOP VIEW

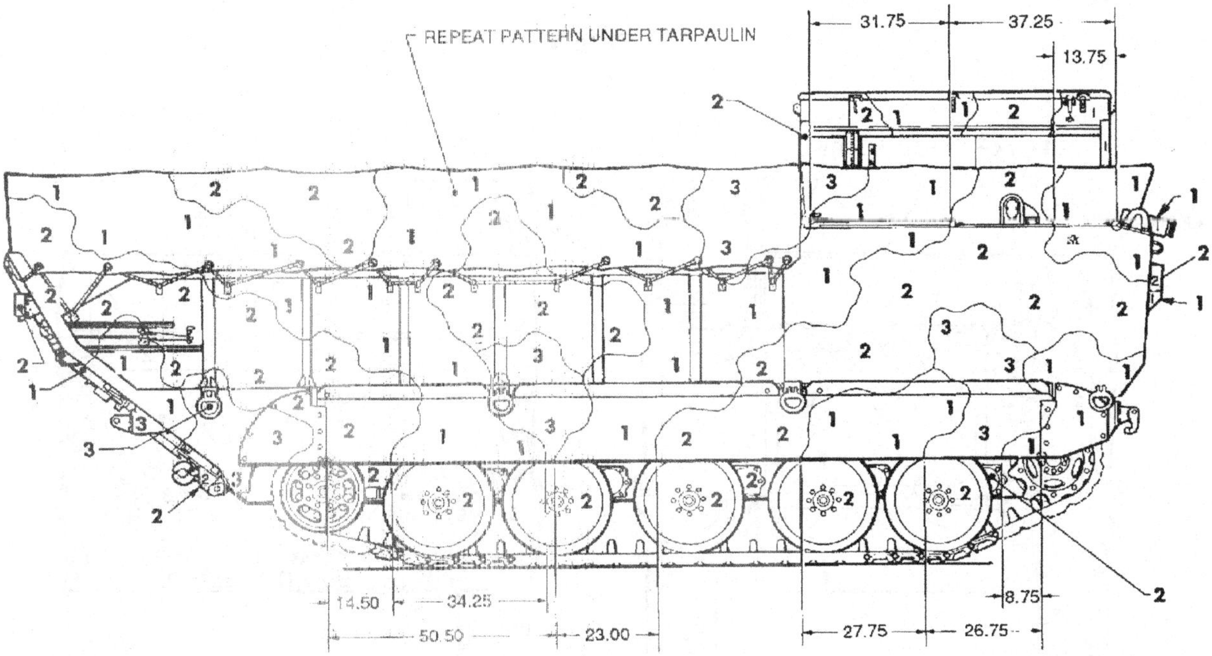

RIGHT SIDE VIEW

Figure 100. Self-propelled Launcher (SPL) M752, Loader-transporter (LT) M688. (1 of 2)

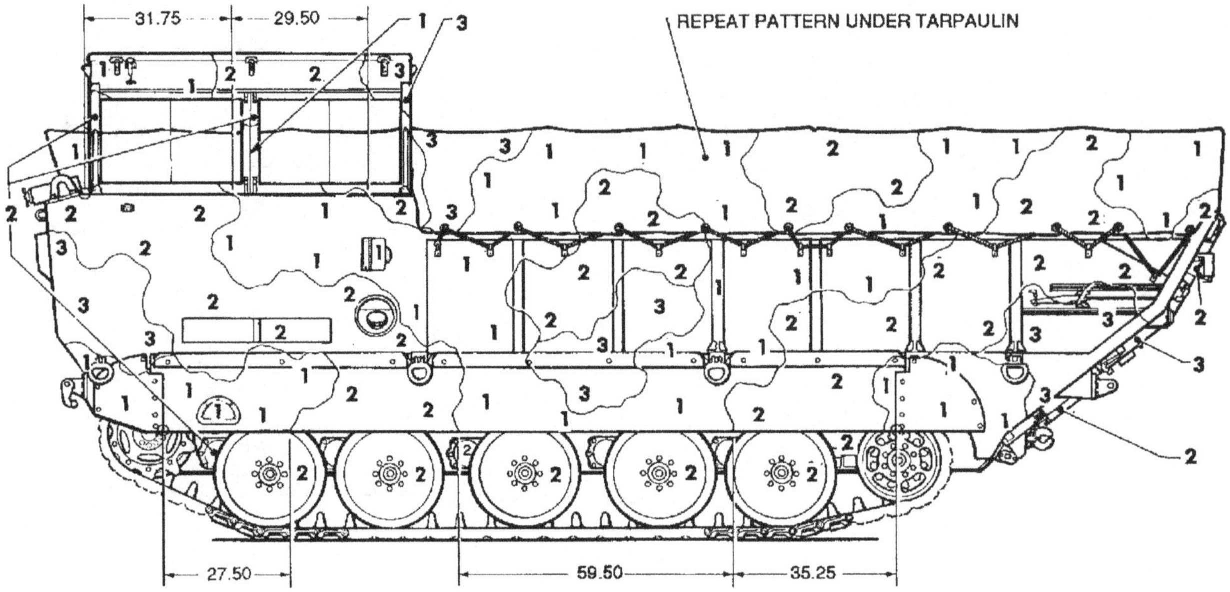

LEFT SIDE VIEW

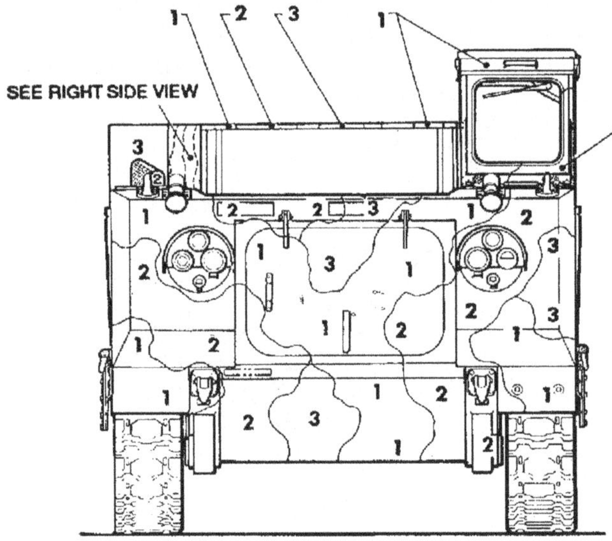

FRONT VIEW

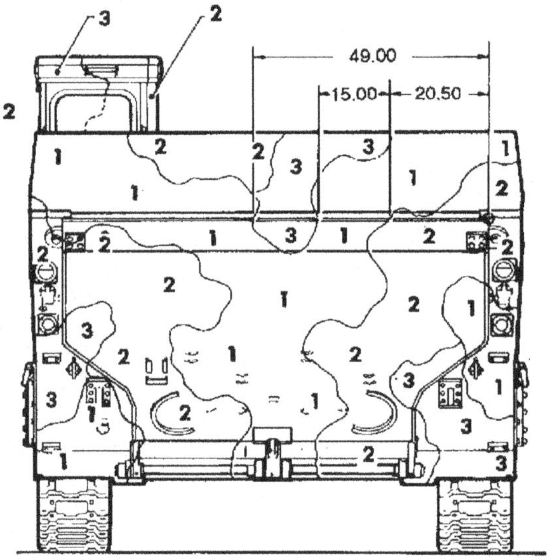

REAR VIEW

Figure 100. Self-propelled Launcher (SPL) M752, Loader-transporter (LT) M688. (2 of2)

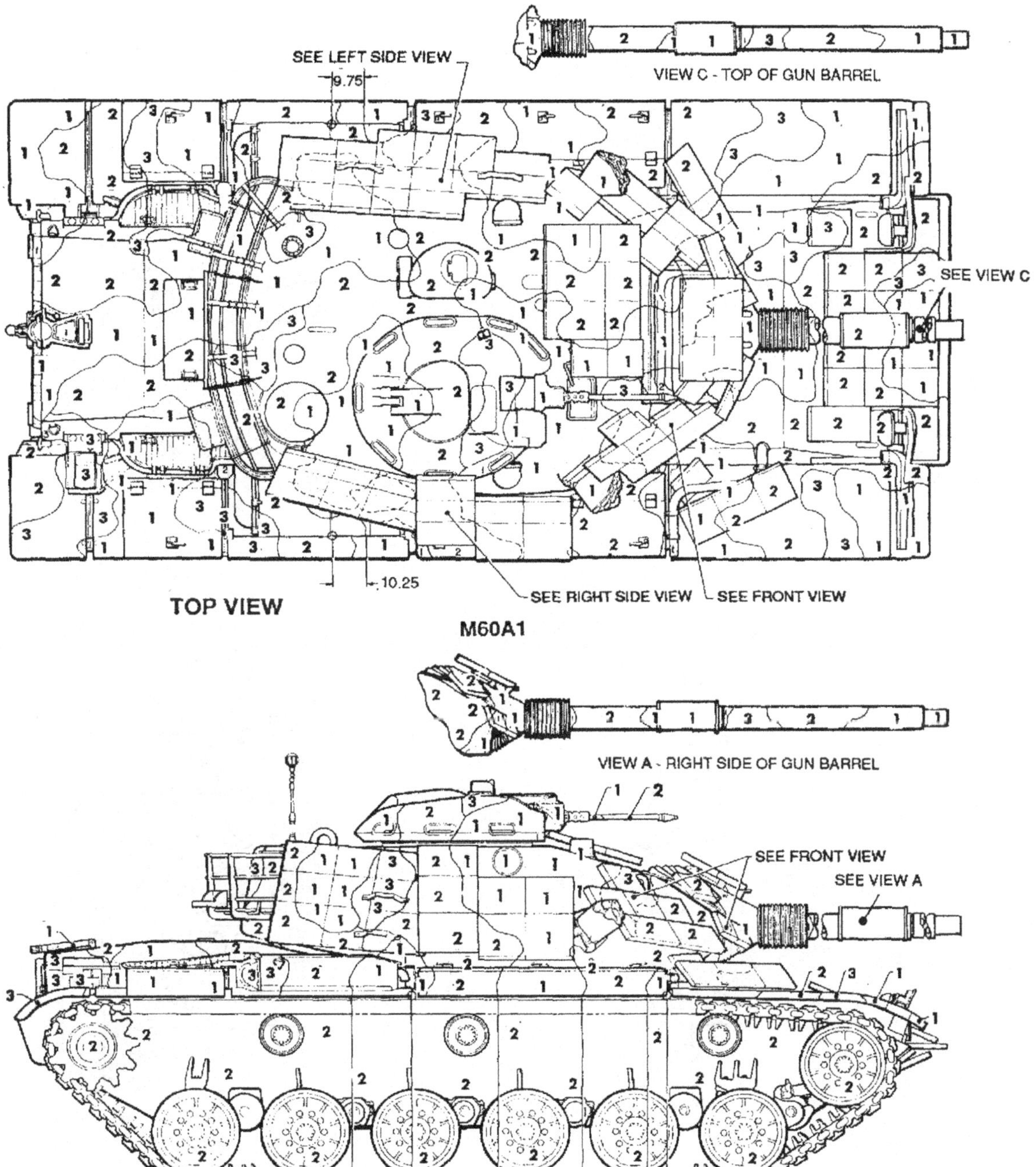

Figure 101. Tank, combat, full tracked: 105-mm gun, M60A1. (1 of 2)

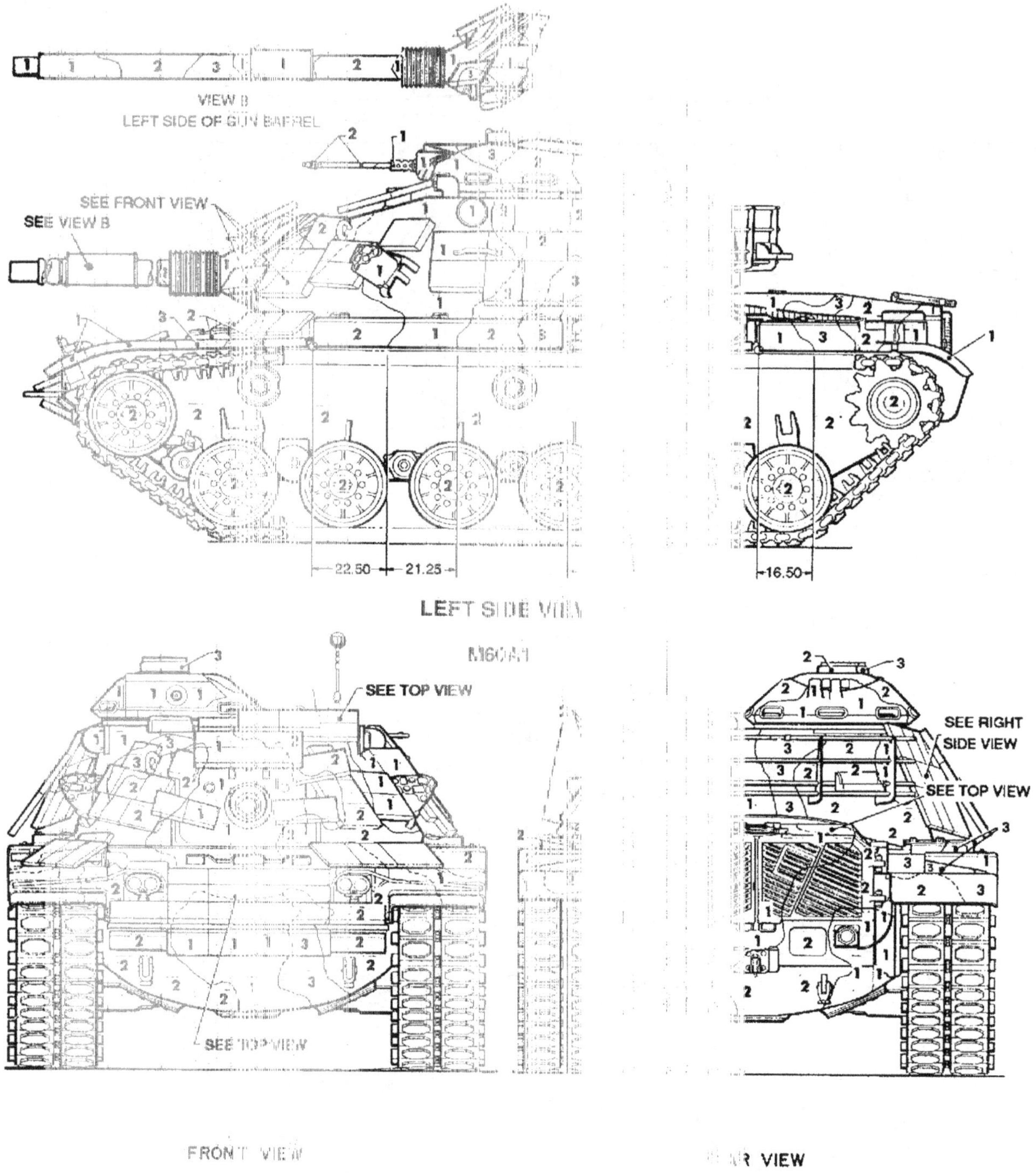

Figure 101. Tank, combat, full tracked: 105 [illegible]. (2 of2)

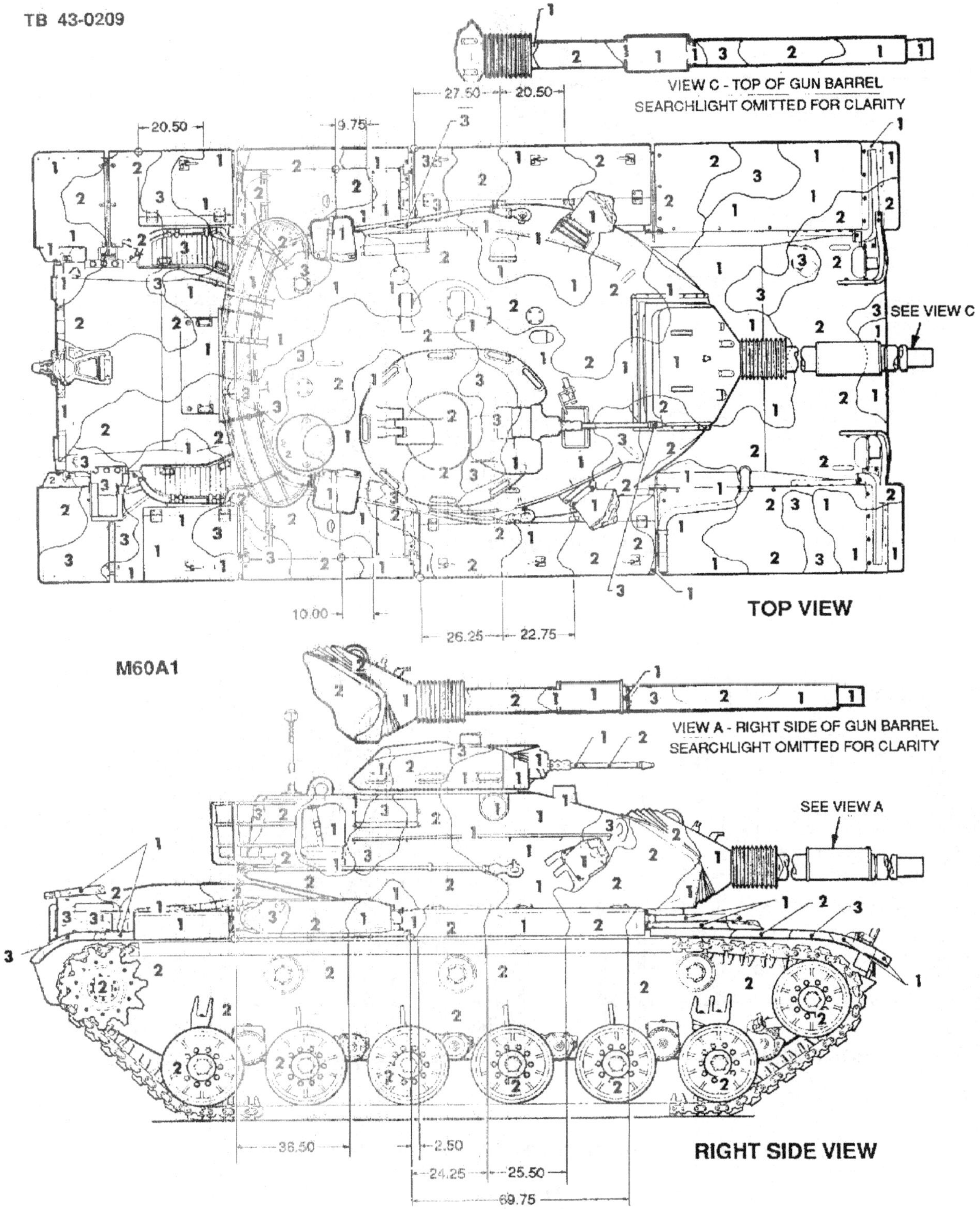

Figure 102. Tank, combat, full tracked: 105-mm gun, M60A3. (1 of 2)

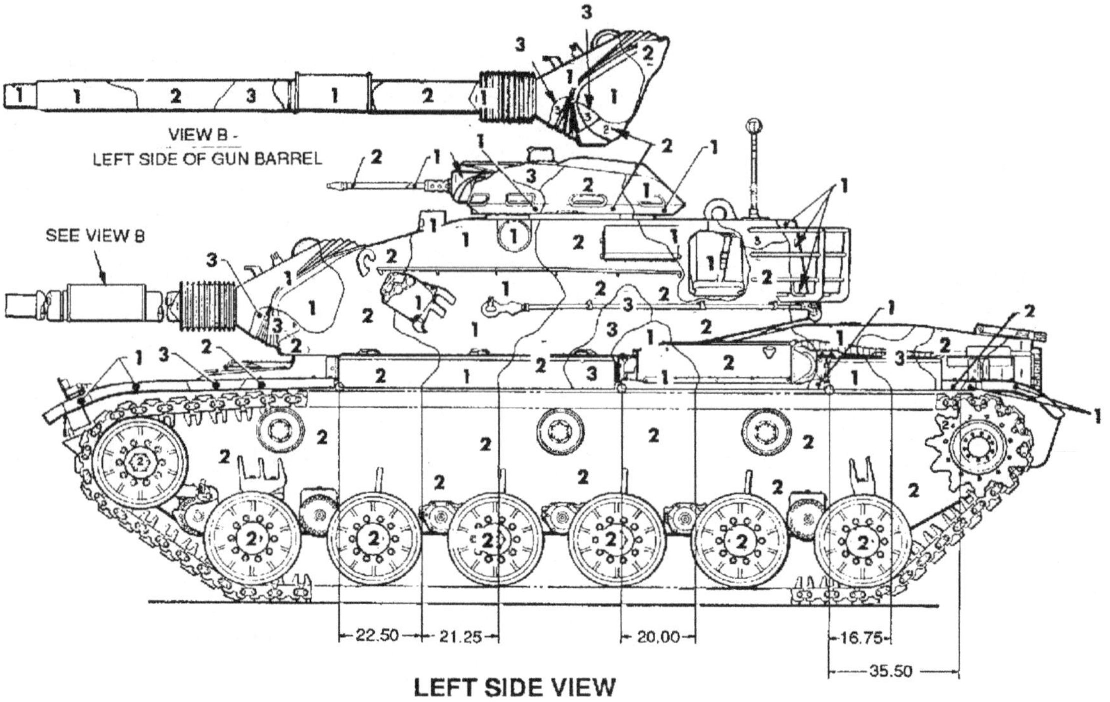

LEFT SIDE VIEW

M60A3

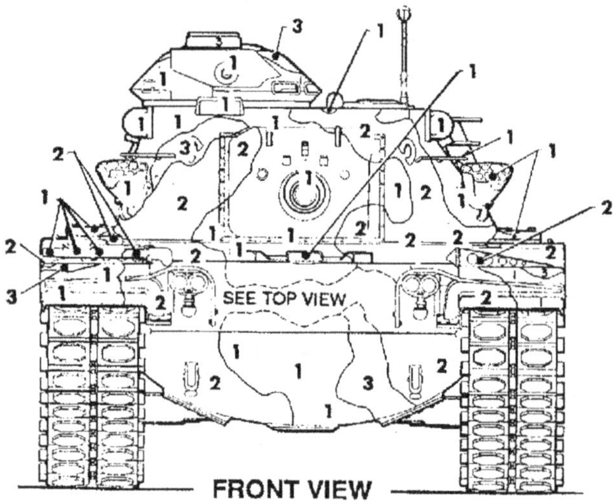

FRONT VIEW

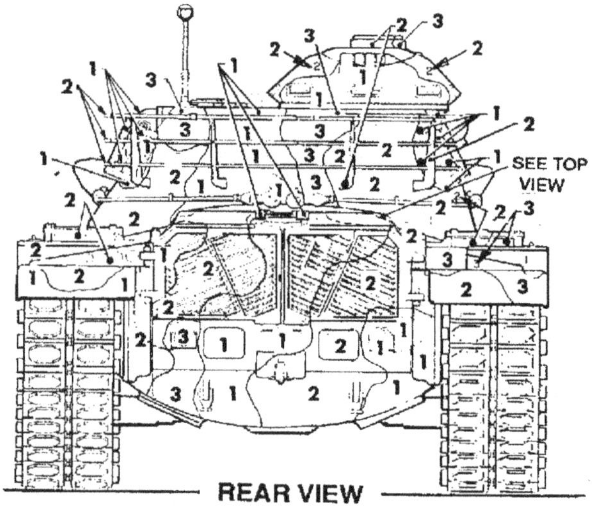

REAR VIEW

Figure 102. Tank, combat, full tracked: 105-mm gun, M60A3. (2 of2)

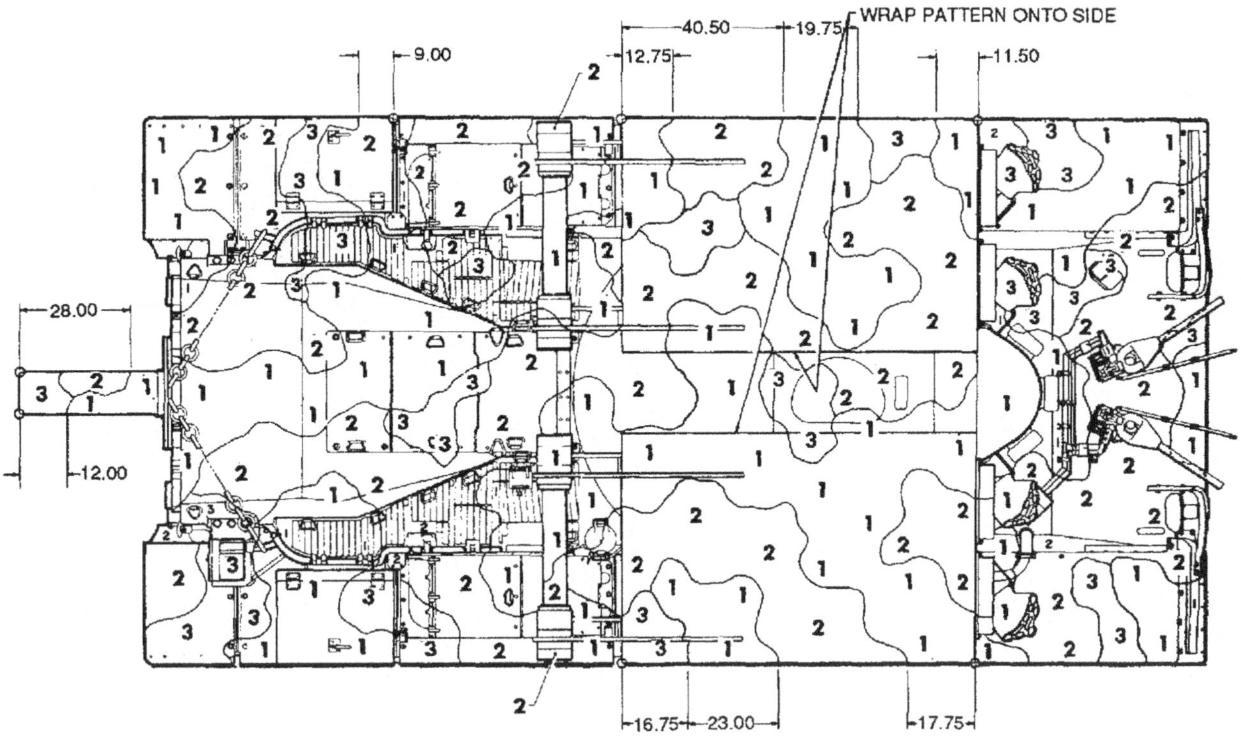

TOP VIEW

M60A3

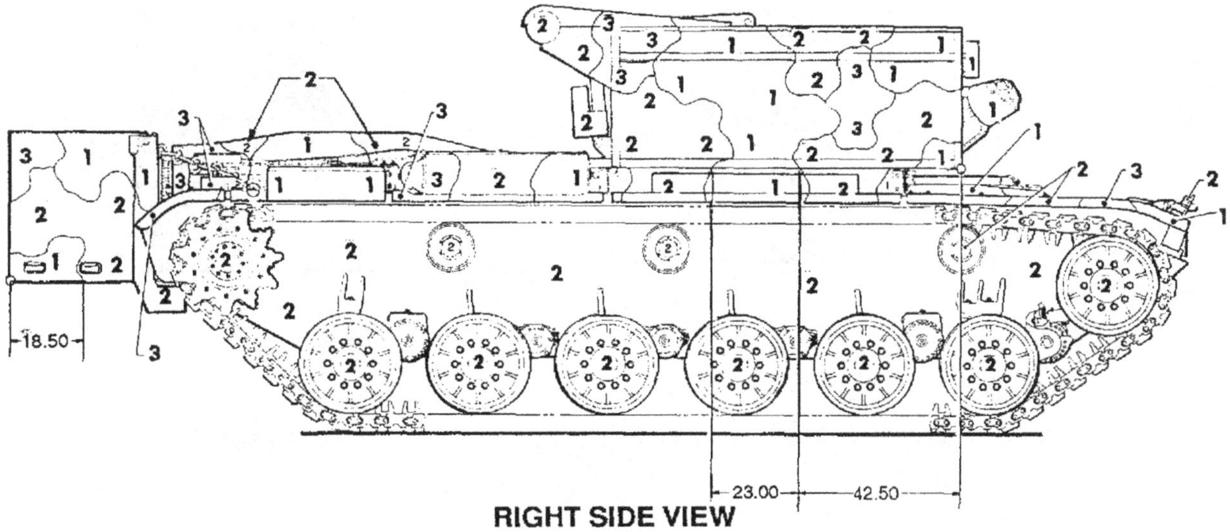

RIGHT SIDE VIEW

Figure 103. Tank, mine-clearing: robot-line-charge, roller and clams, M60A3. (1 of 2)

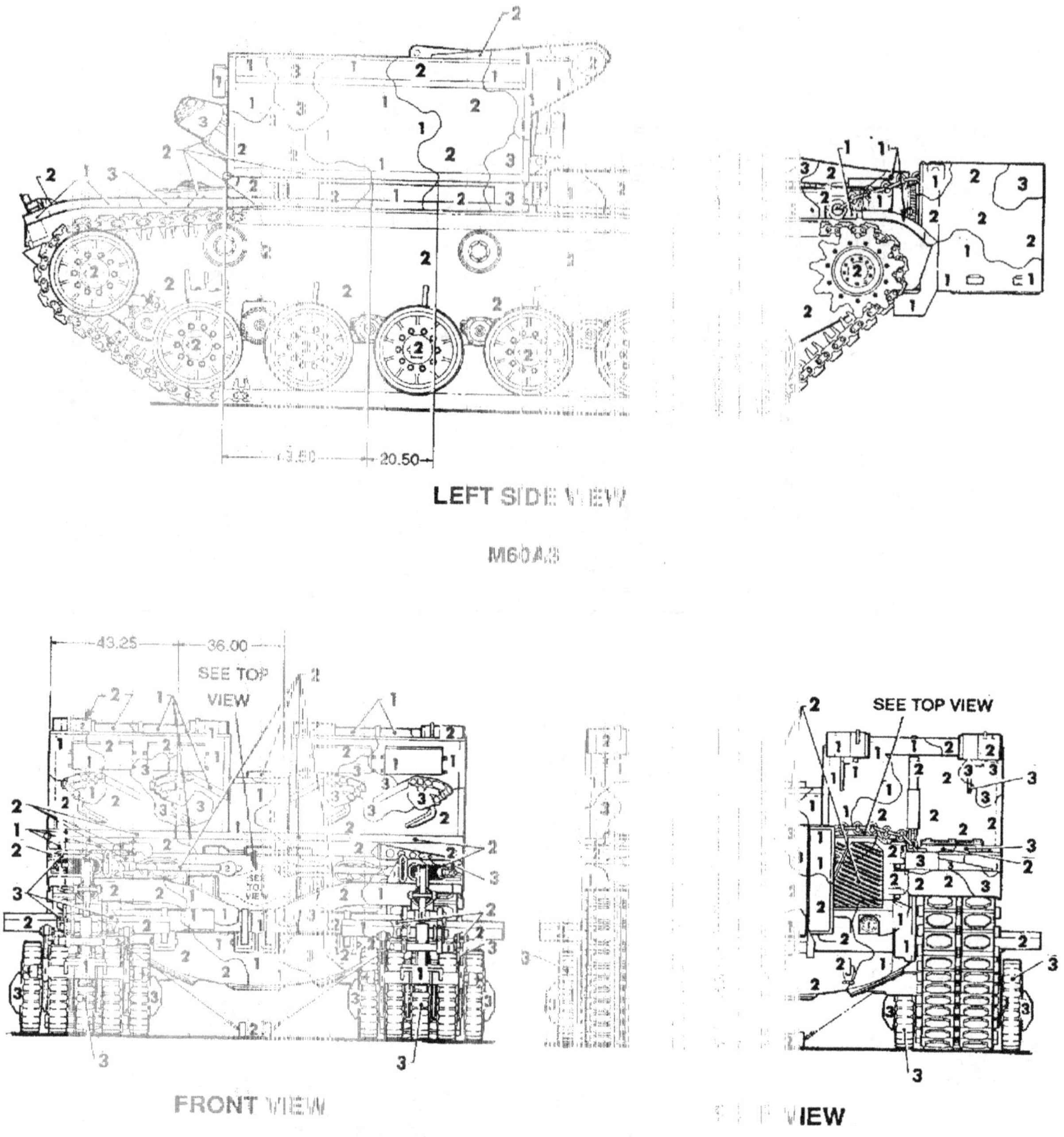

Figure 103. Tank, mine-clearing: robot-line-charge, [illegible] M60A3. (2 of 2)

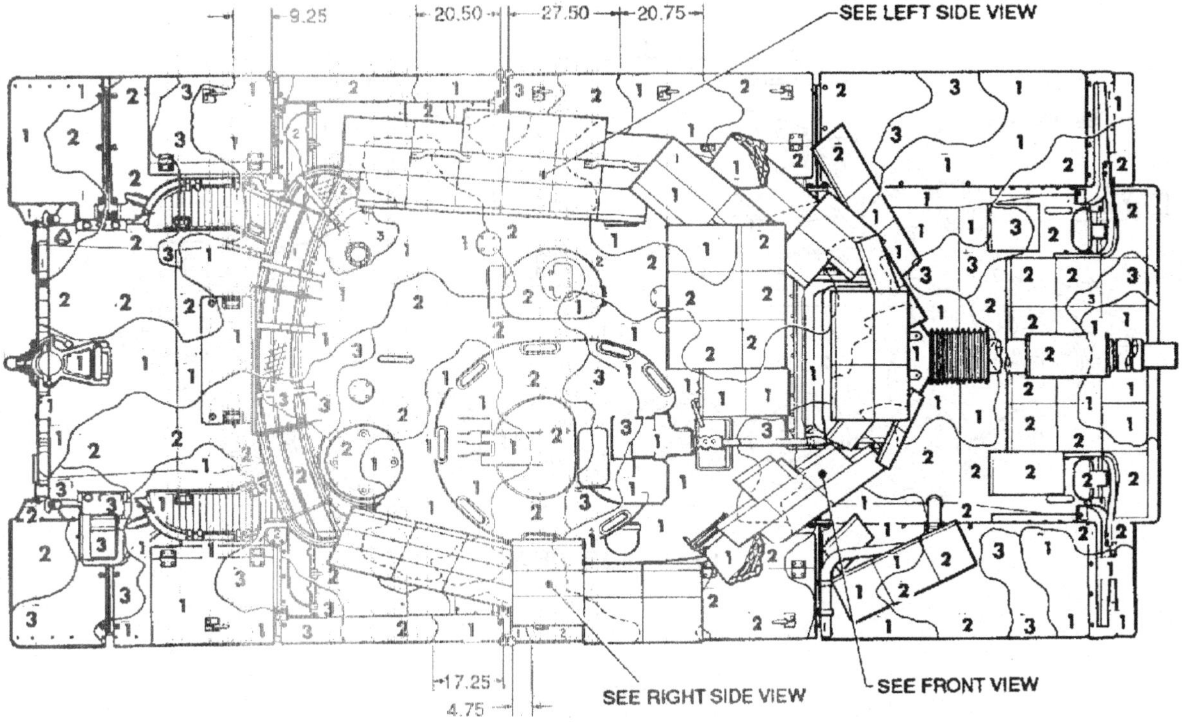

TOP VIEW

M60

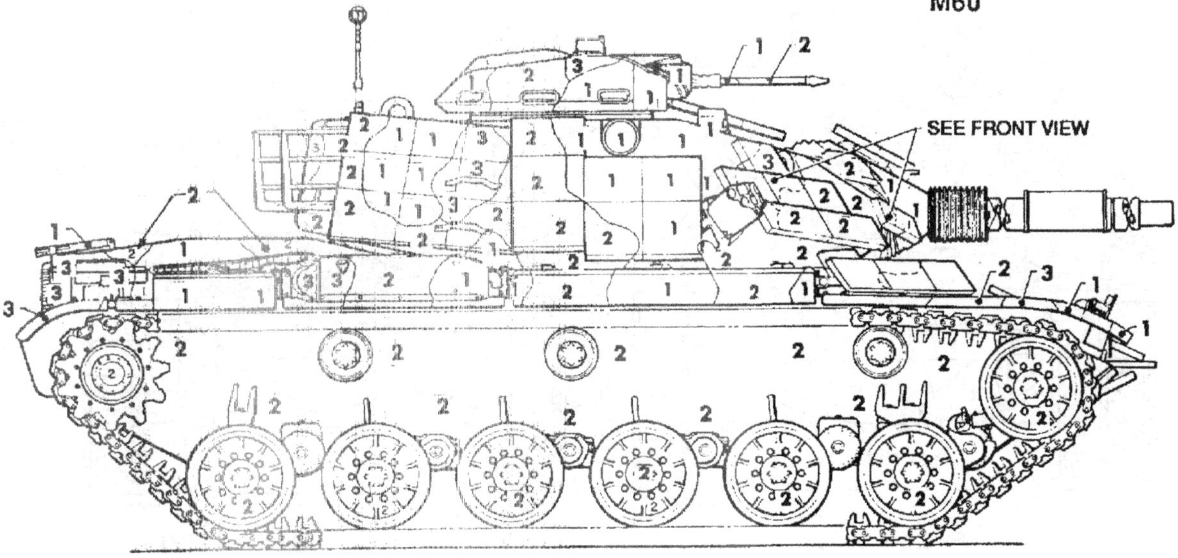

RIGHT SIDE VIEW

Figure 104. Tank, combat, full tracked: 105-mm gun, M60A3TTS. (1 of2)

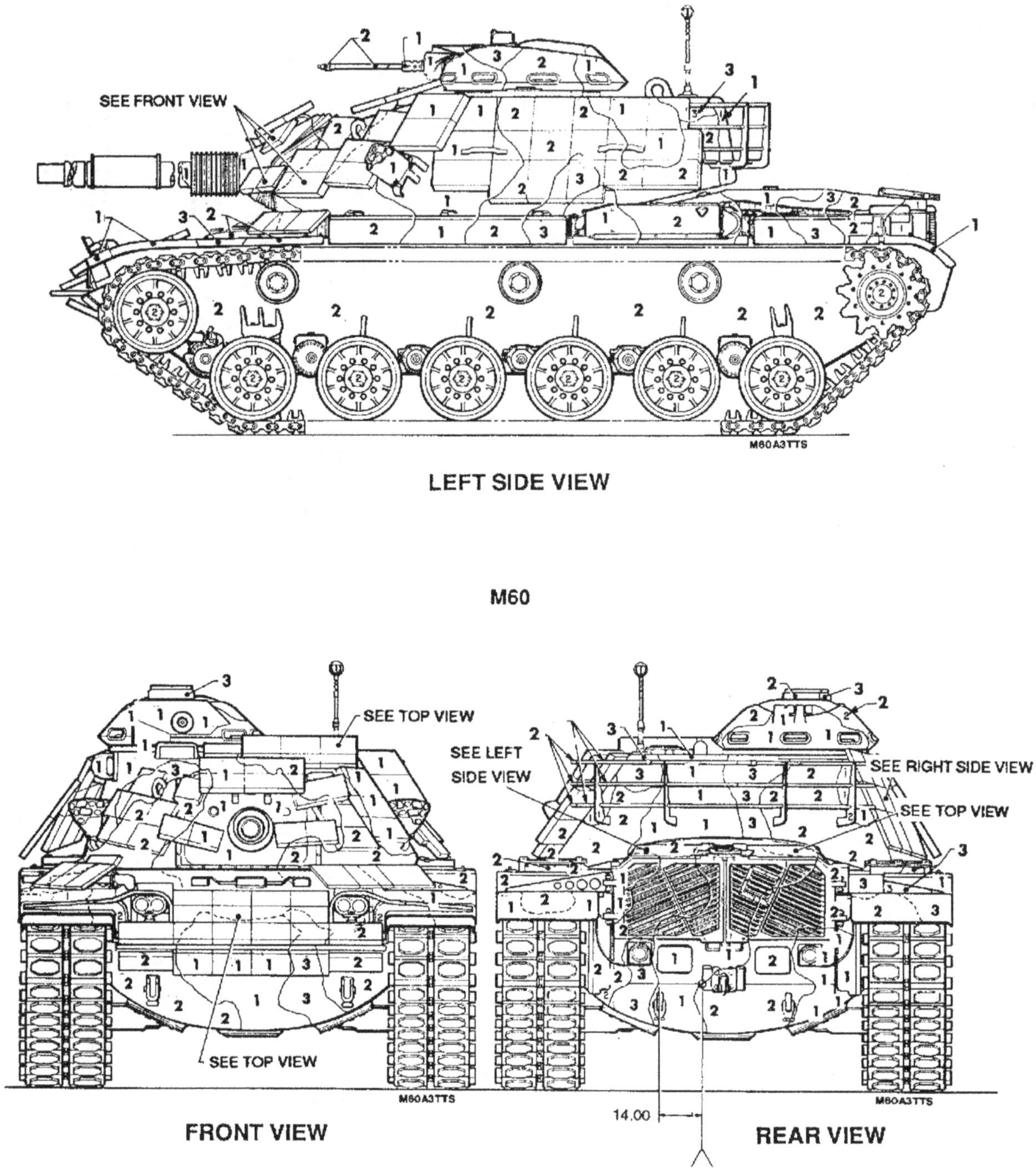

Figure 104. Tank, combat, full tracked: 105-mm gun, M60A3TTS. (2 of 2)

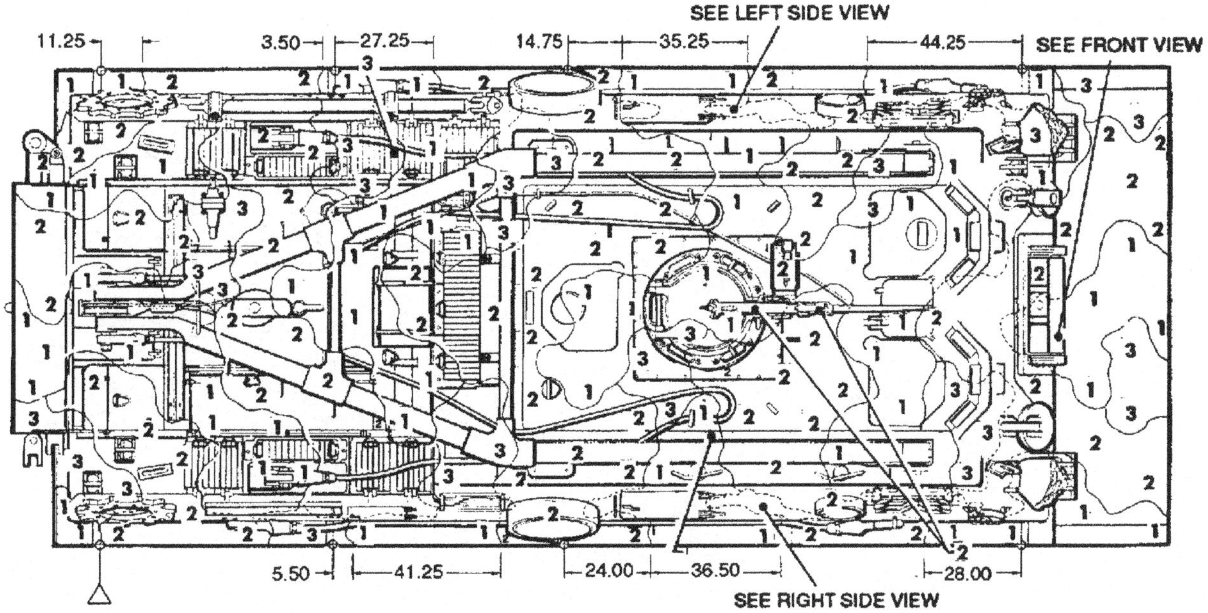

TOP VIEW

M88A1

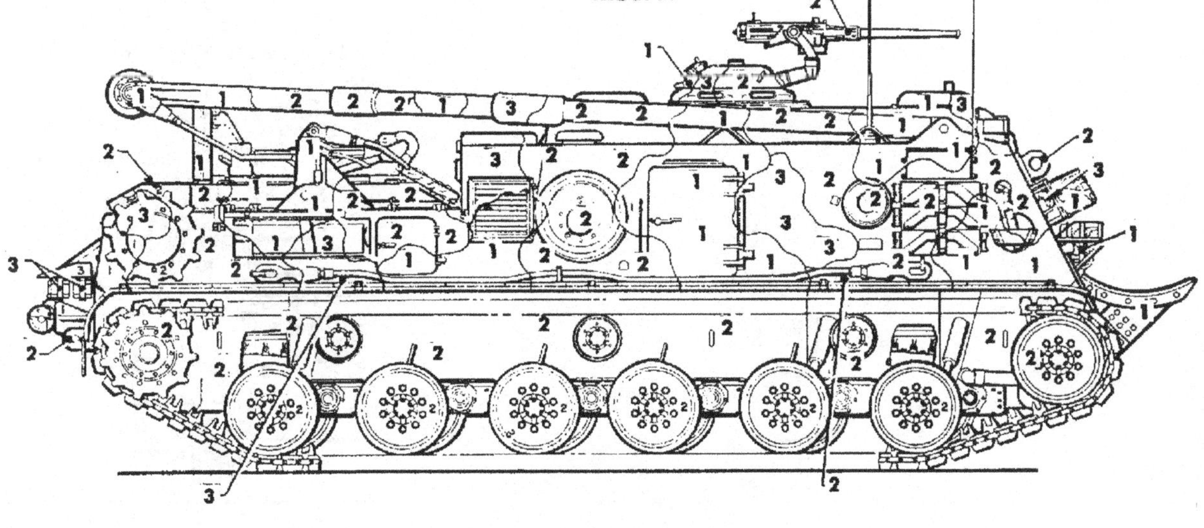

RIGHT SIDE VIEW

Figure 105. Tank, recovery vehicle, full tracked: medium, M88A1. (1 of 2)

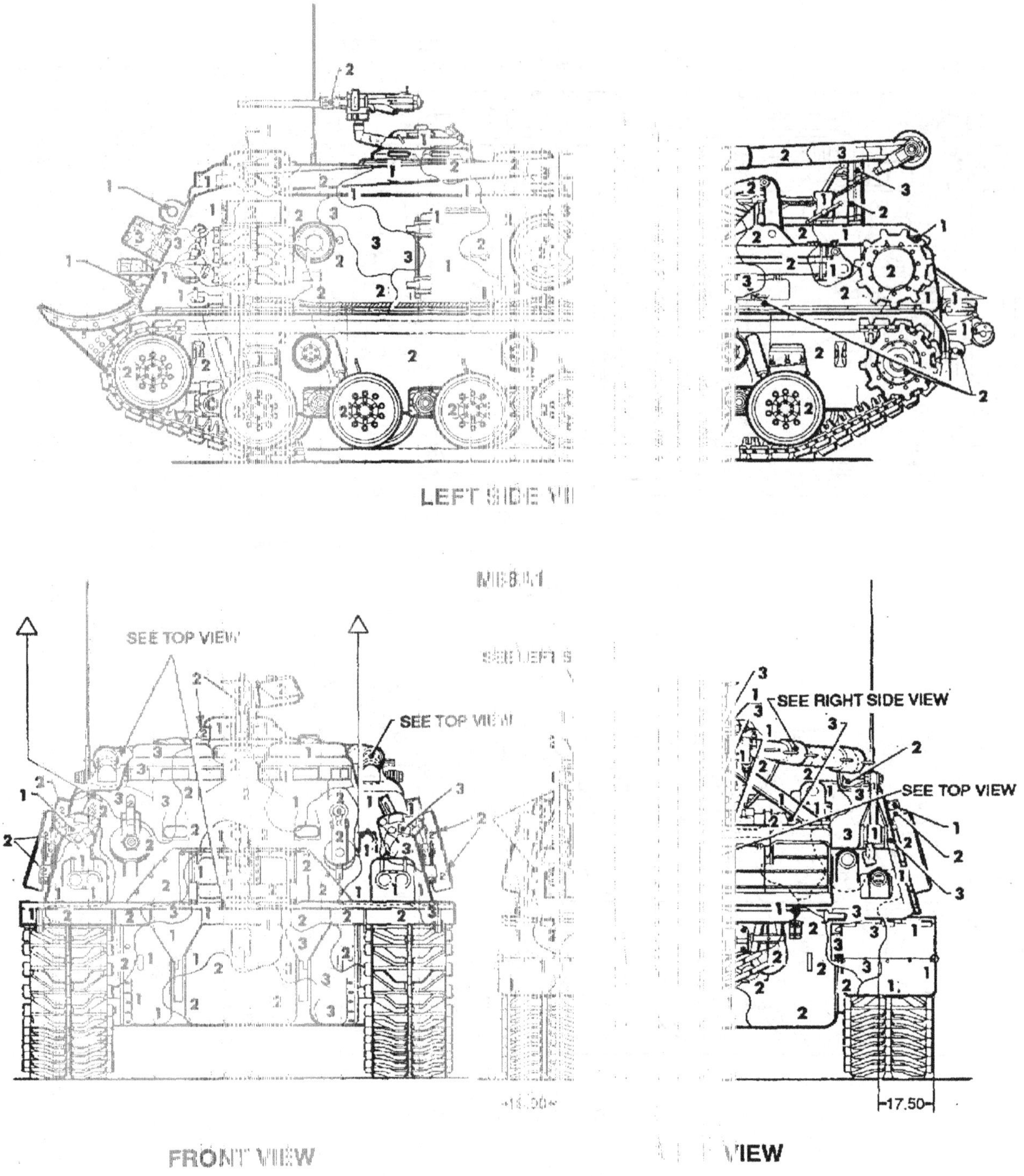

Figure 105. Tank, recovery vehicle ,full tracke[illegible]A1. (2 of2)

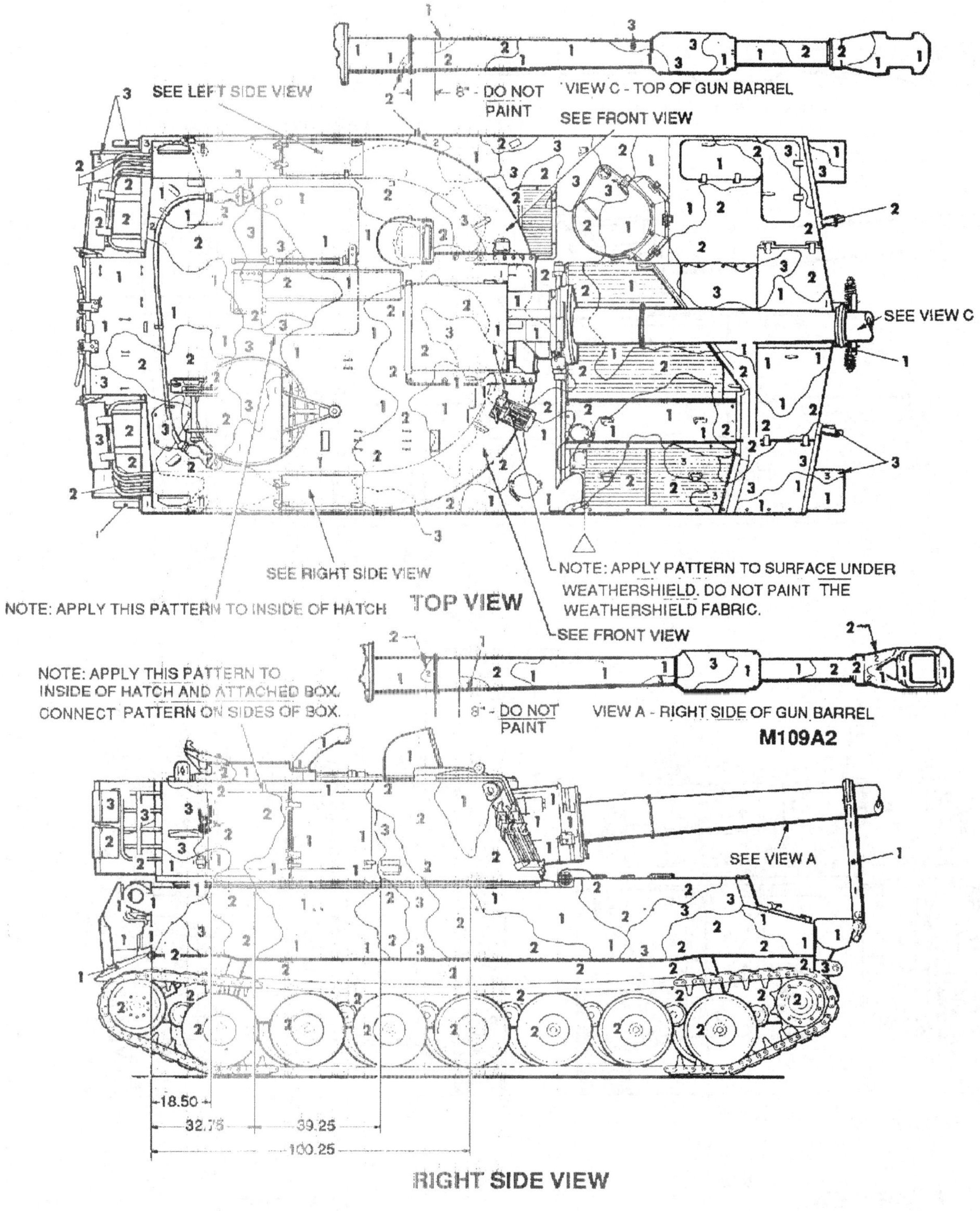

Figure 106. Howitzer, medium, self-propelled: 155-mm, M109A2. (1 of2)

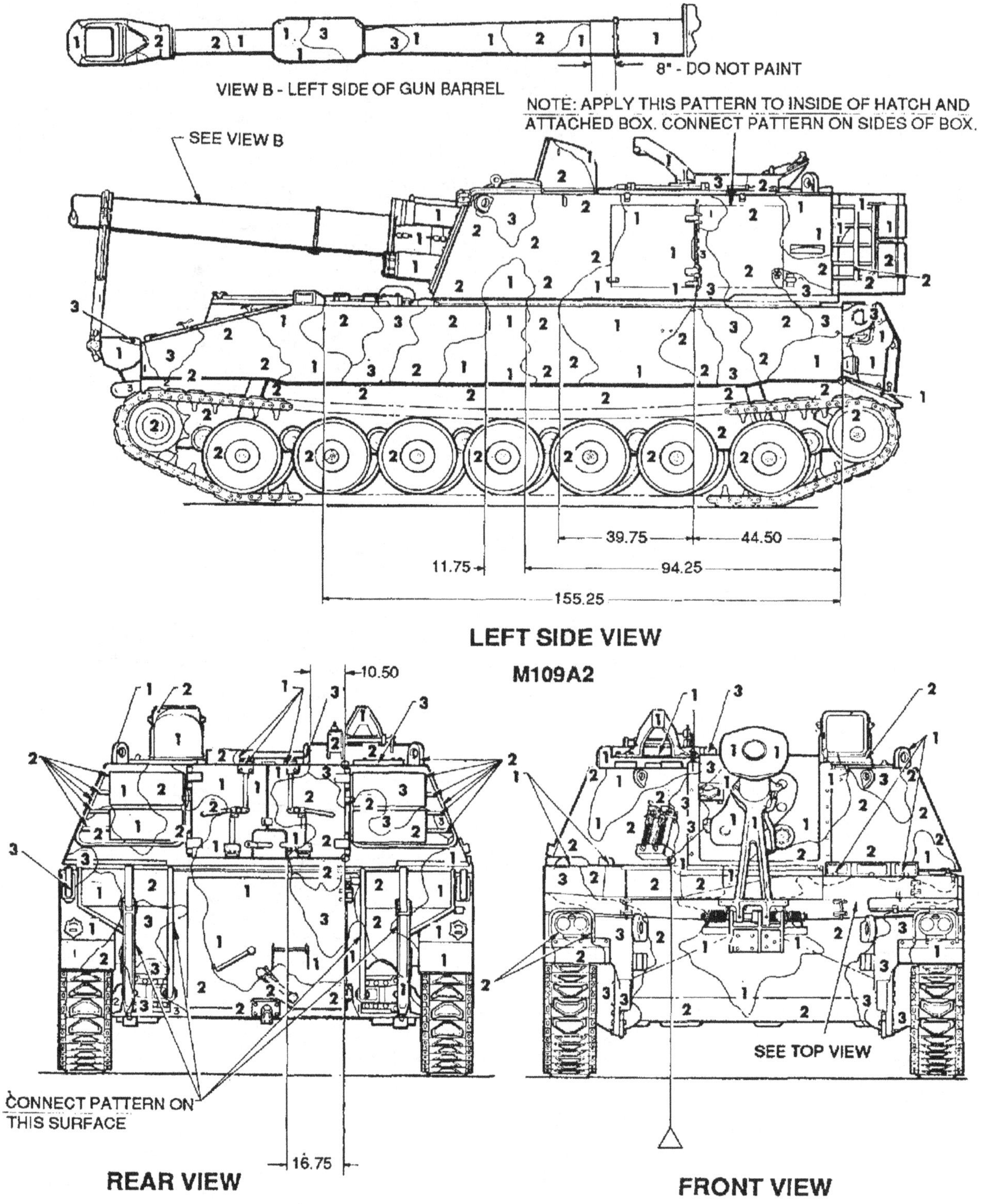

Figure 106. Howitzer, medium, self-propelled: 155-mm, M109A2. (2 of 2)

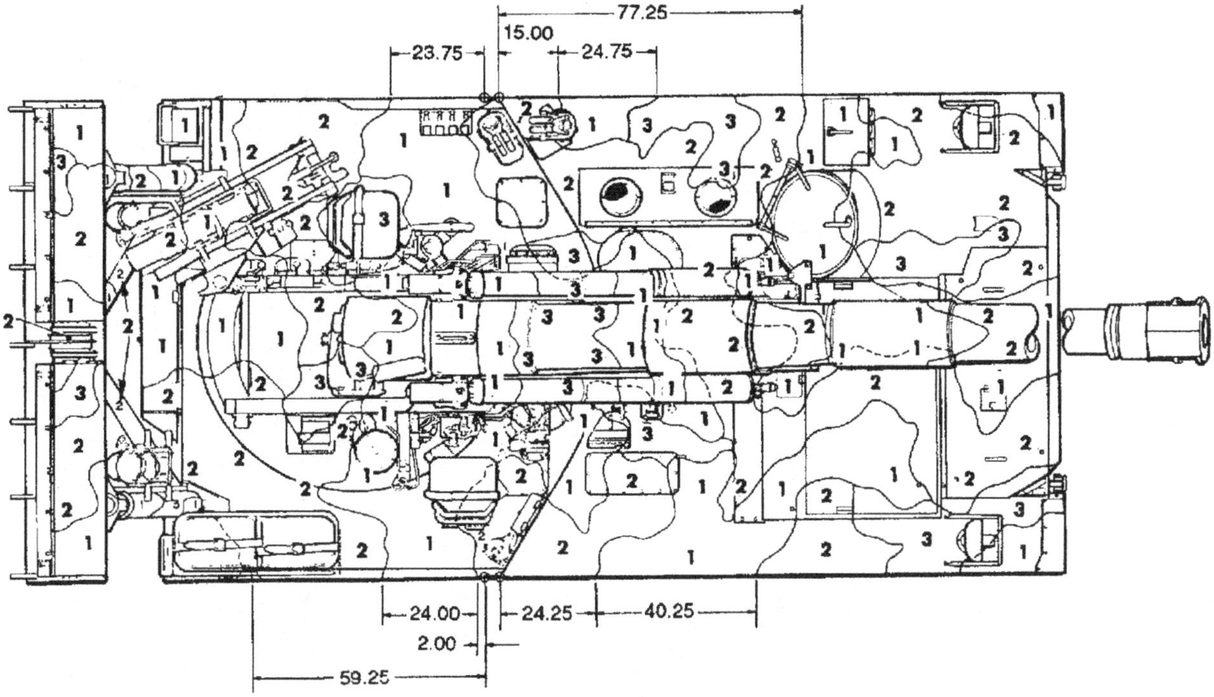

TOP VIEW

M110A2

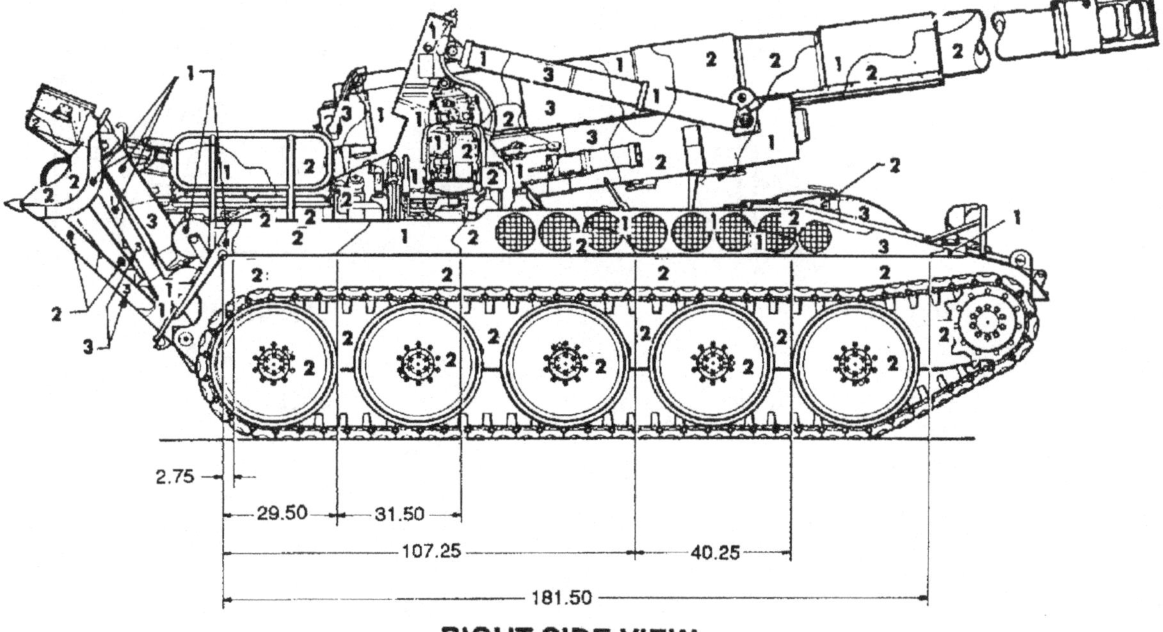

RIGHT SIDE VIEW

Figure 107. Howitzer, self-propelled: 8-inch, M110A2. (1 of 2)

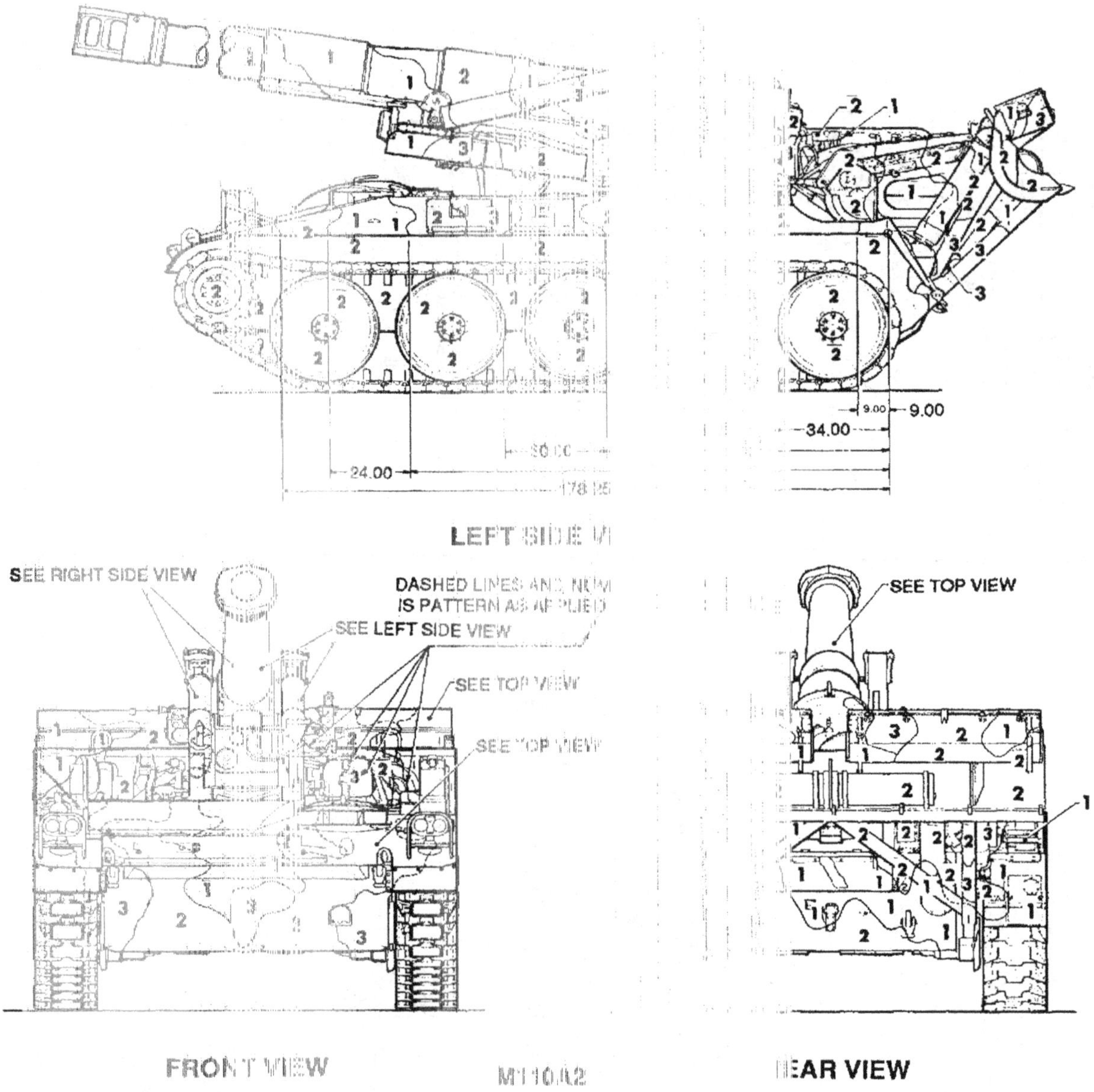

Figure 107. Howitzer, self-propelled: 8 ... 2 of 2)

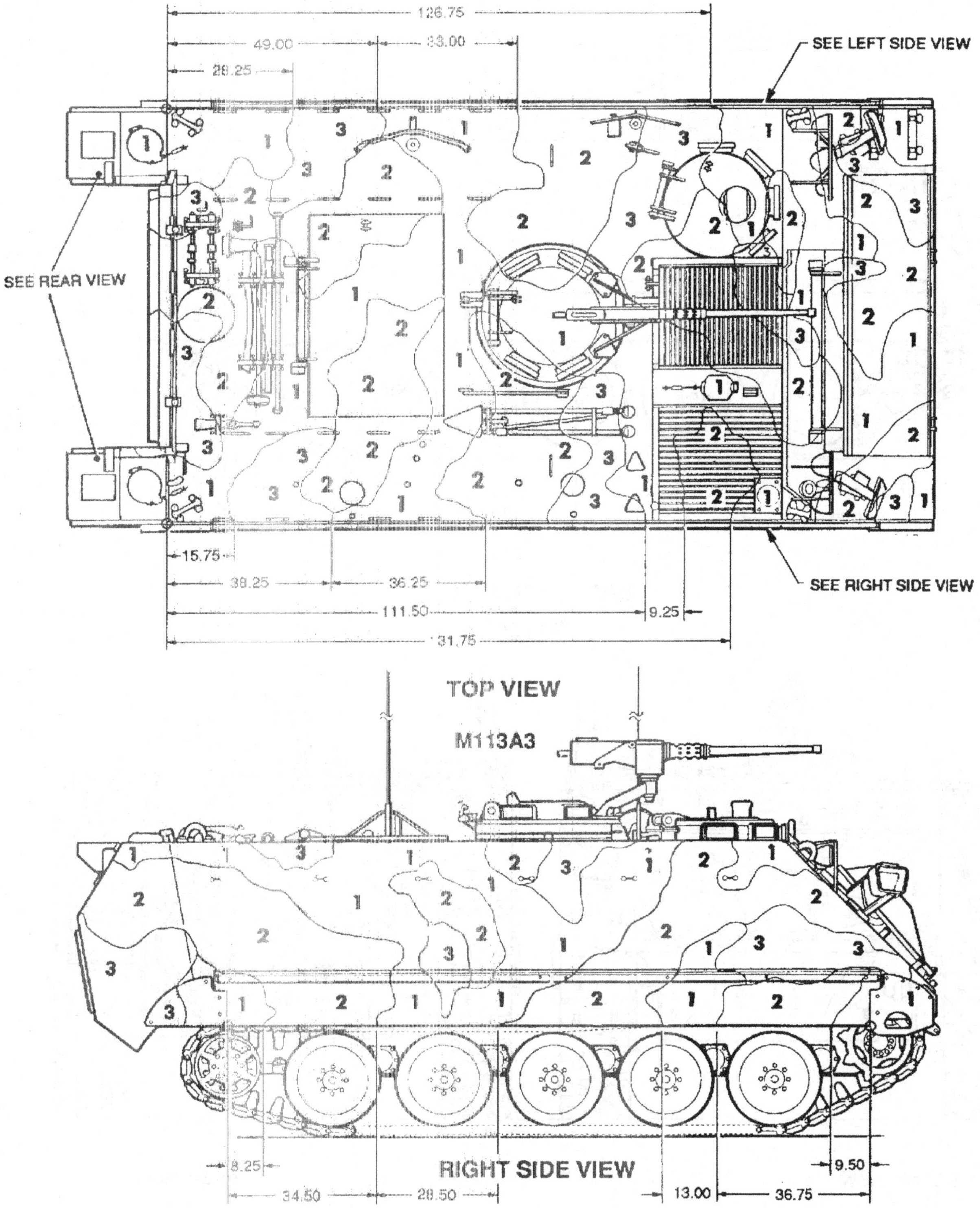

Figure 108. Carrier, personnel: M113A3. (1 of 2)

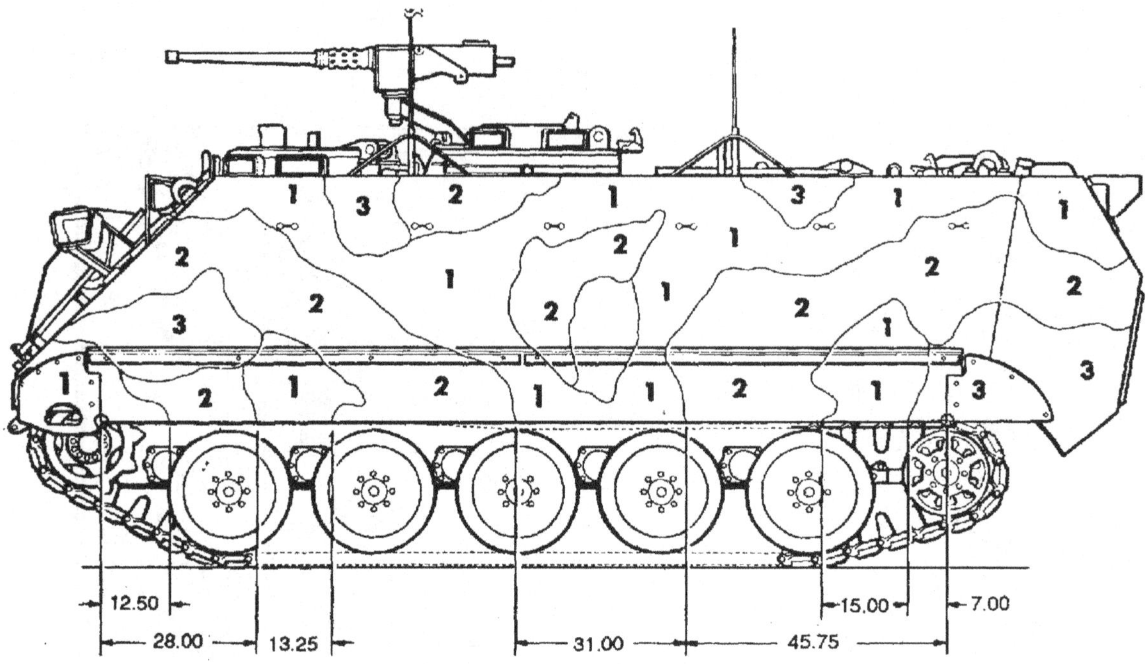

LEFT SIDE VIEW

M113A3

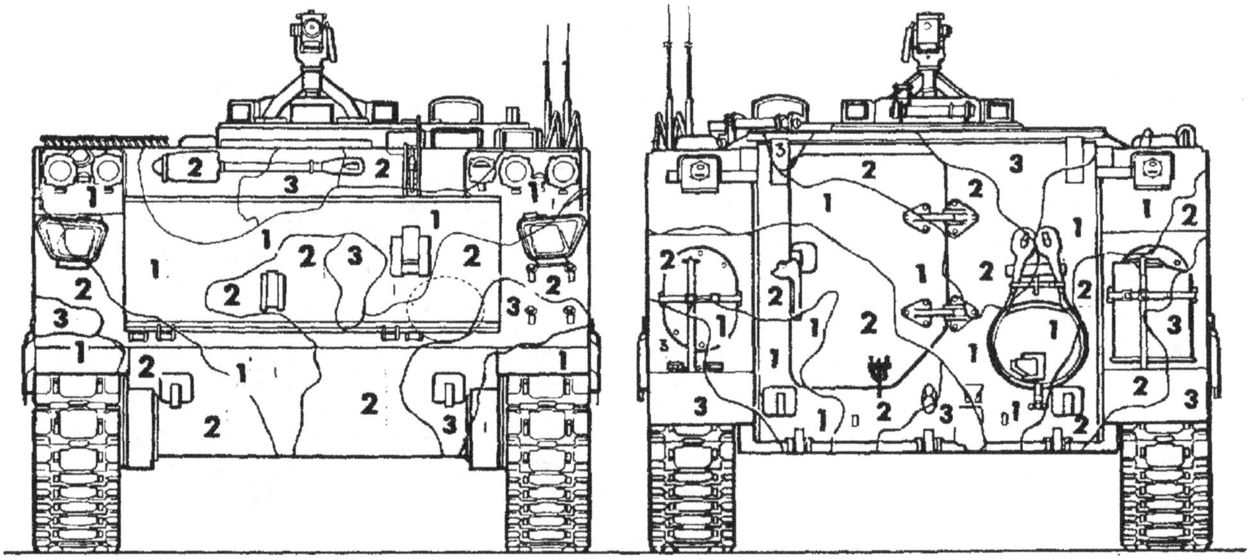

FRONT VIEW **REAR VIEW**

Figure 108. Carrier, personnel: M113A3. (2 of2)

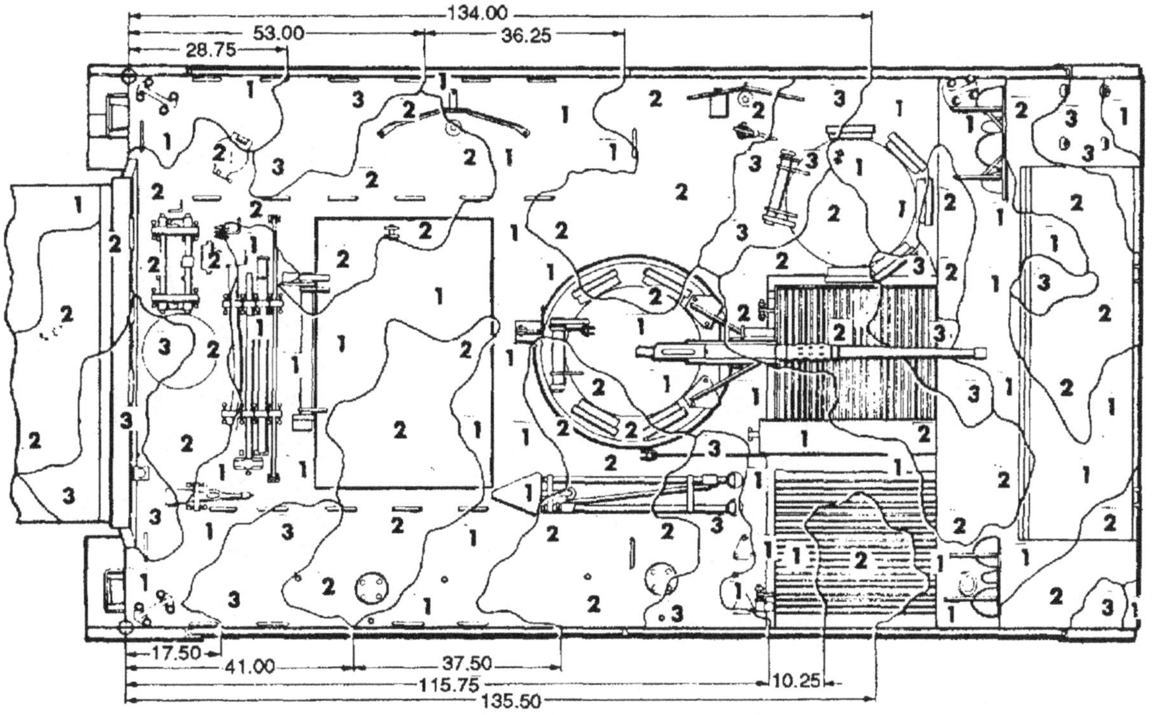

TOP VIEW

M113A2

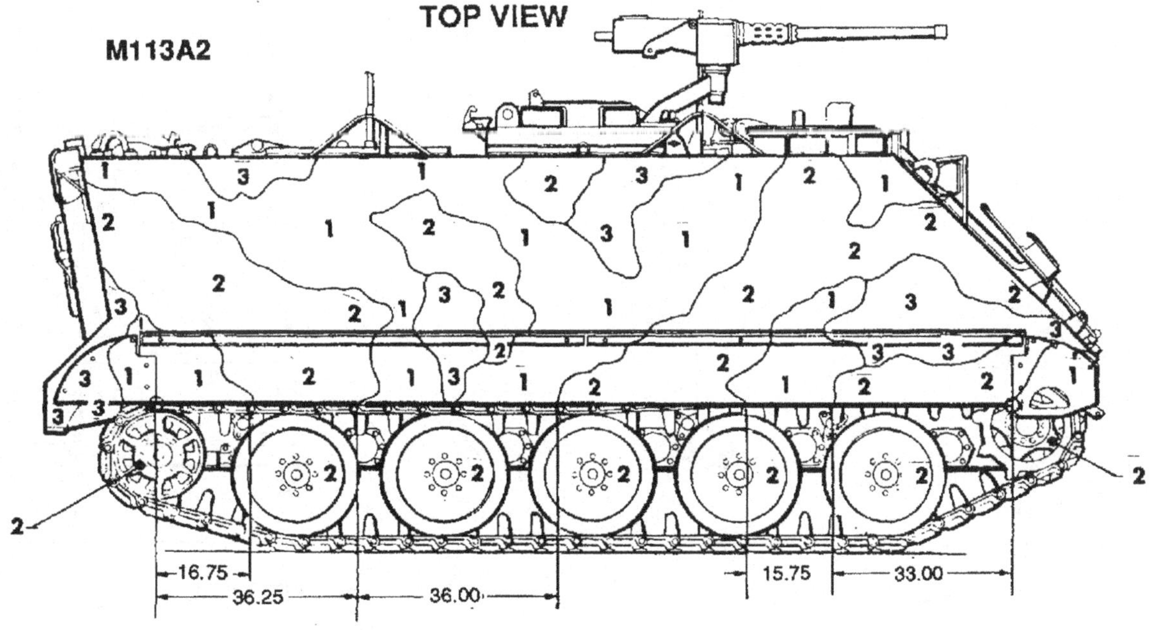

RIGHT SIDE VIEW

Figure 109. Carrier, personnel, full tracked, armored: M113A2. (I of2)

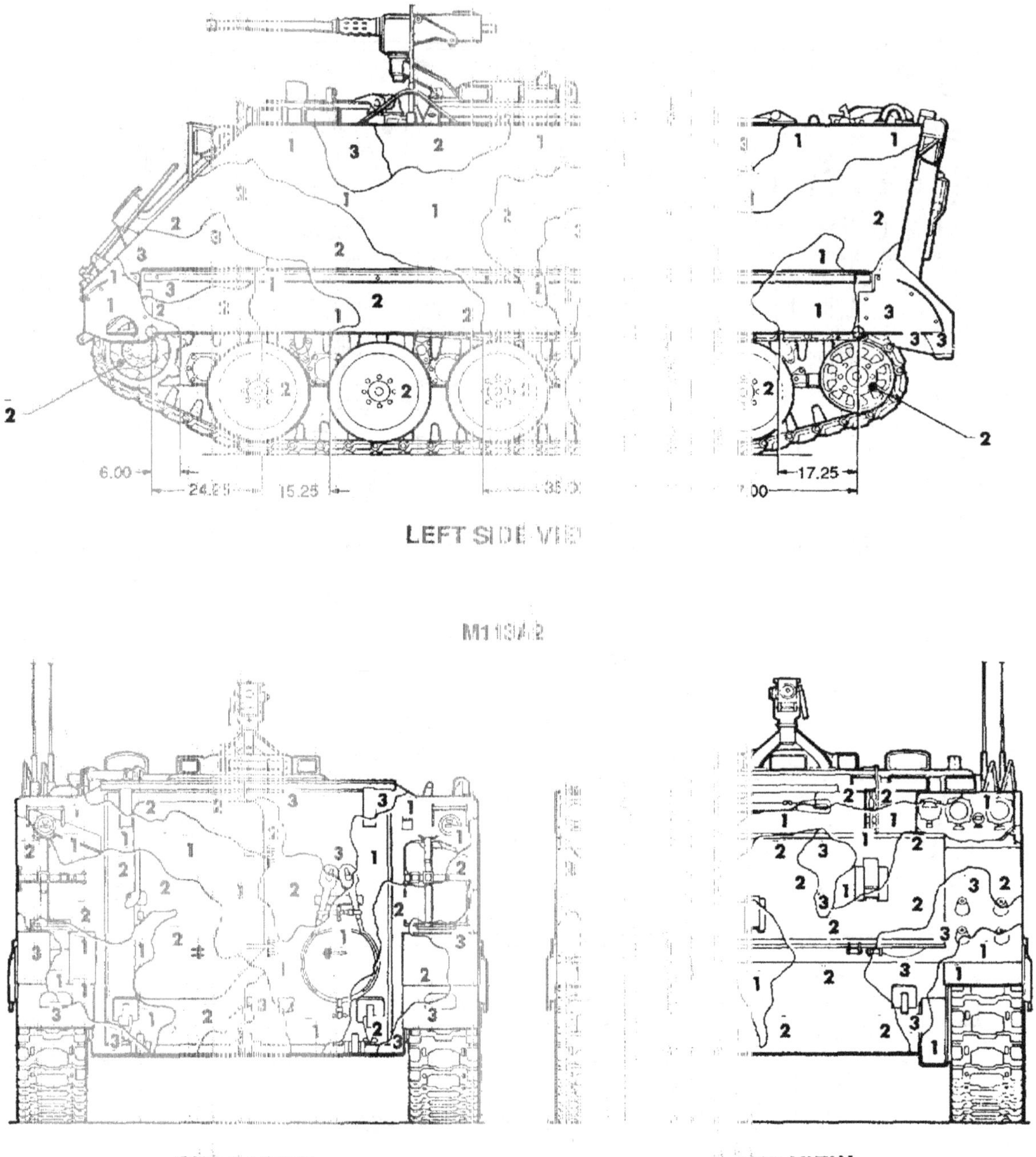

Figure 109. Carrier, personnel full tracked, armored: M113A2 (2 of 2)

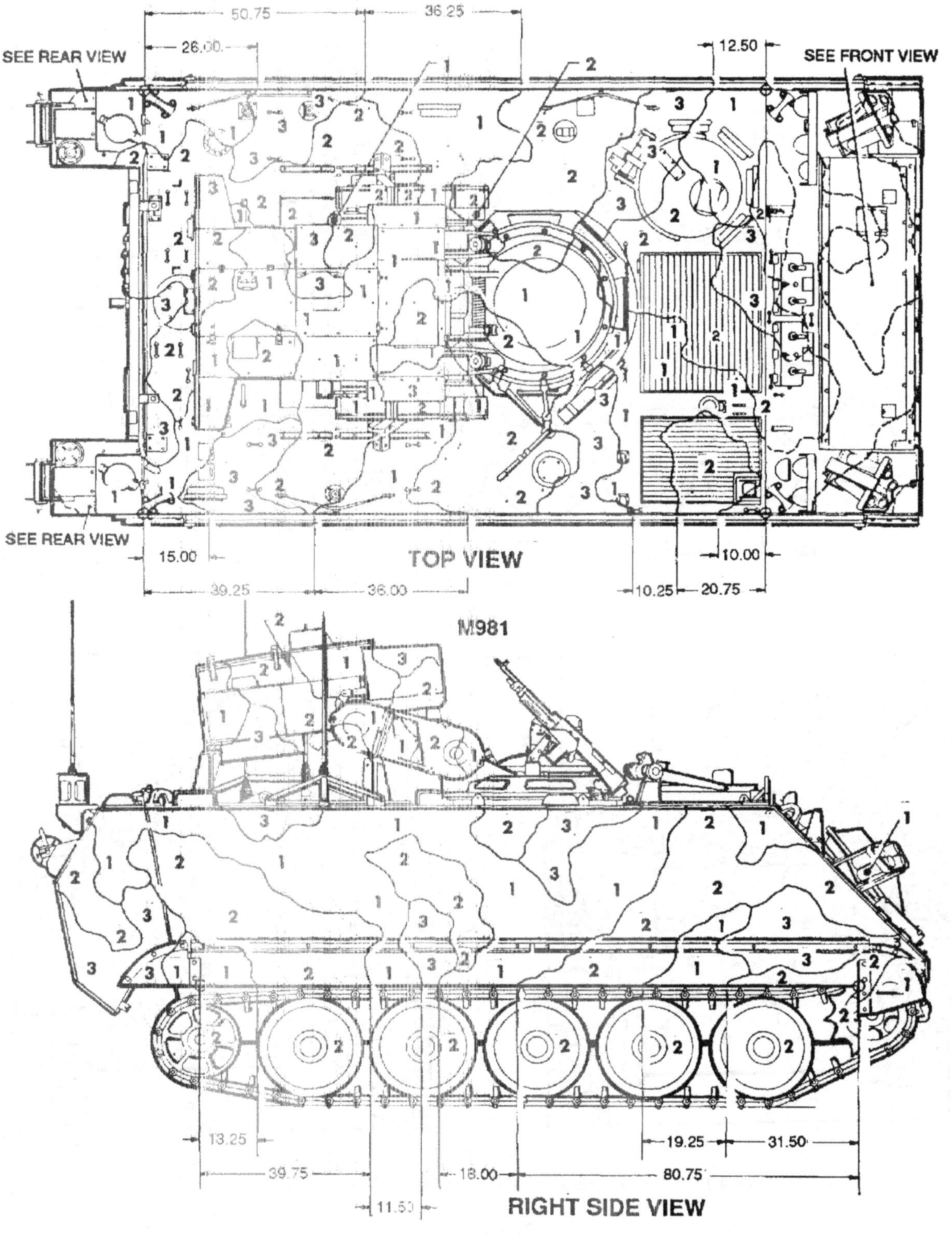

Figure 110. Carrier, personnel, fire support, M981. (1l of2)

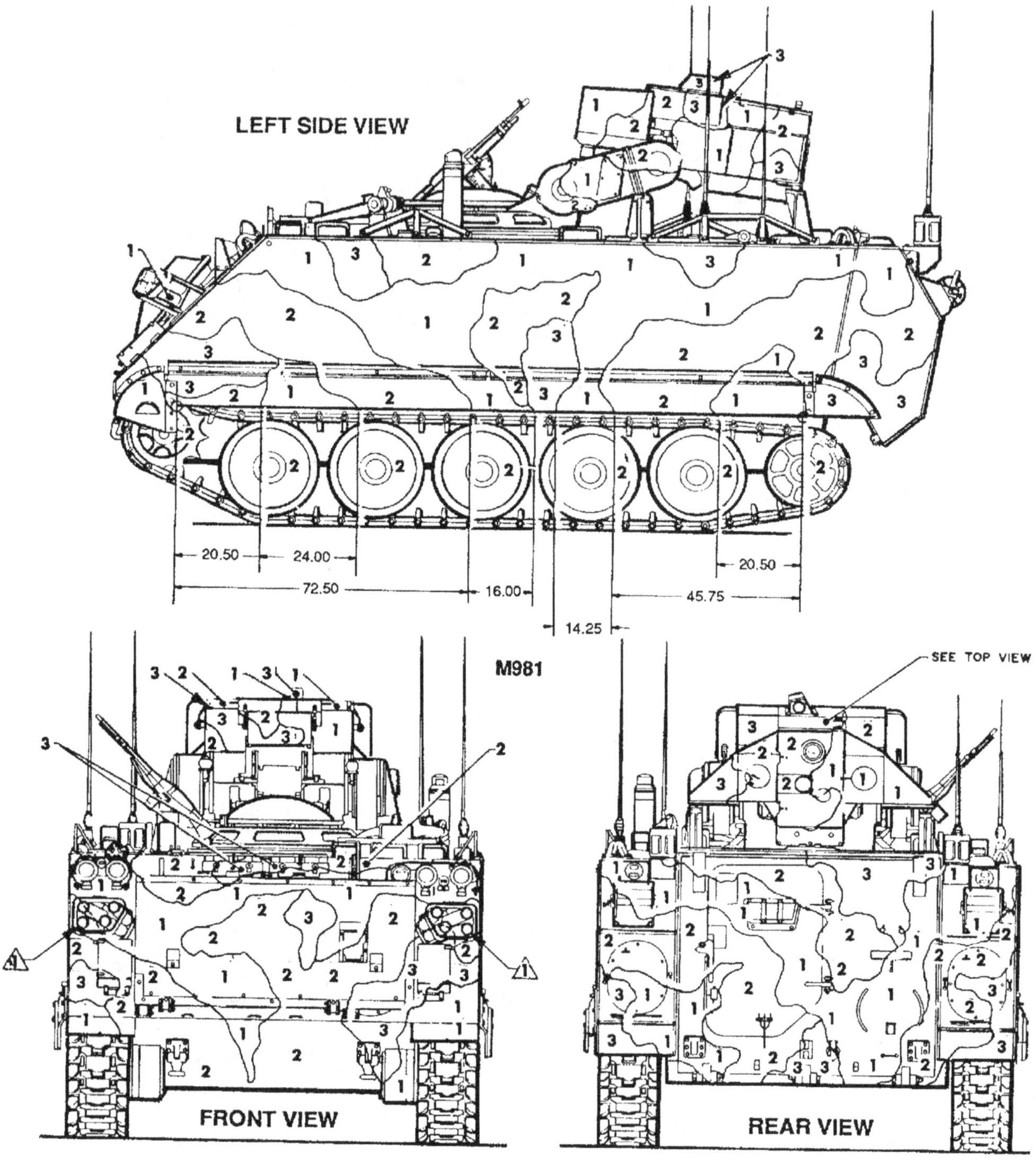

Figure 110. Carrier, personnel, fire support, M981. (2 of 2)

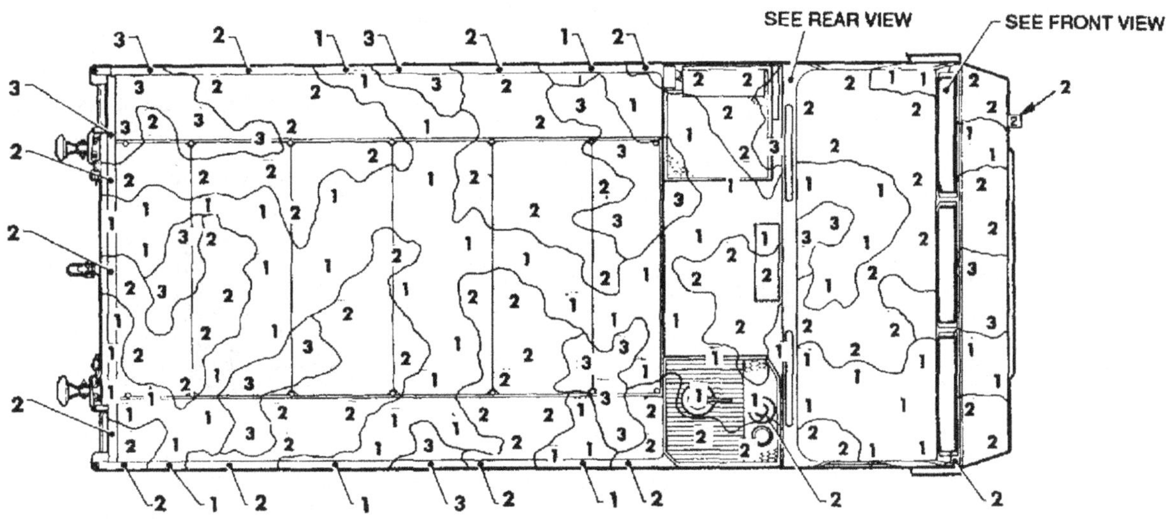

TOP VIEW

M548

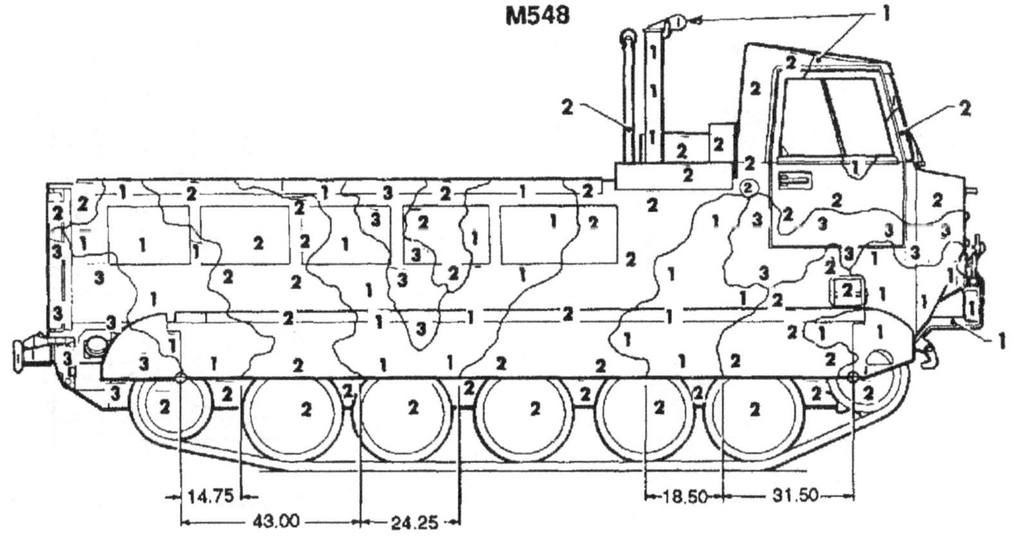

RIGHT SIDE VIEW

Figure 111. Carrier, cargo, full tracked: M548. (1 of 2)

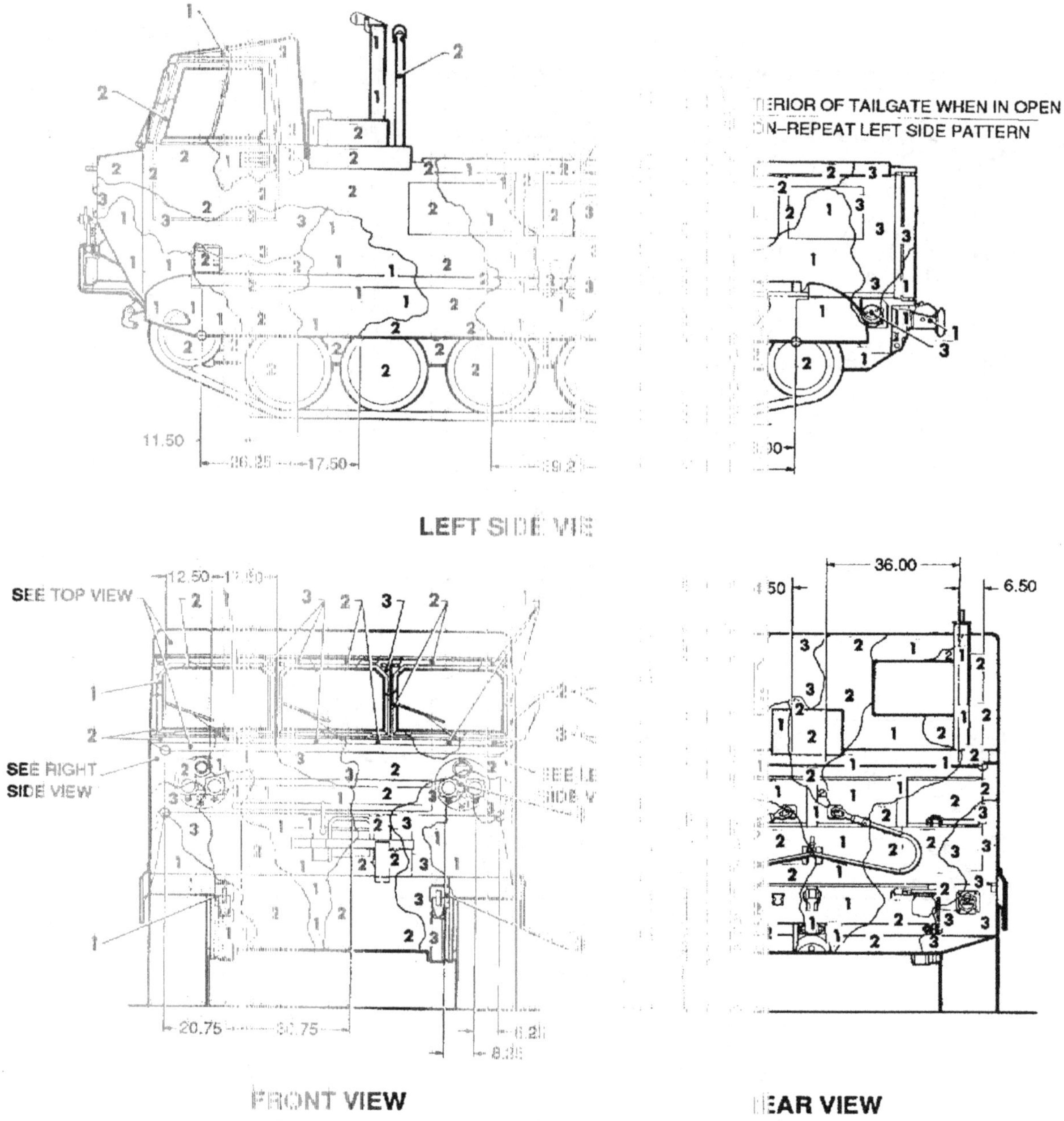

M548

Figure 111. Carrier, cargo, full track … 2)

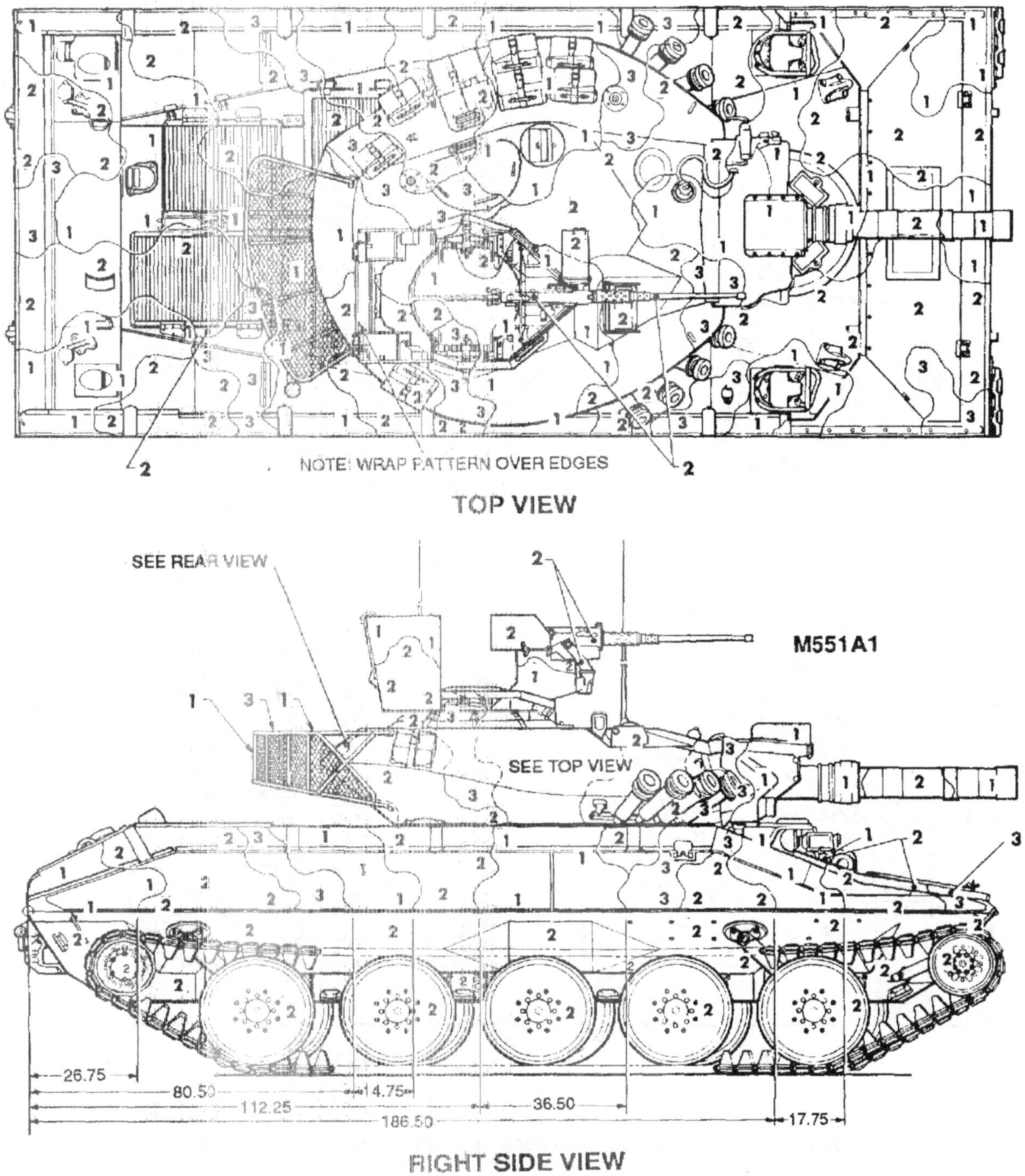

Figure 112. Armored, reconnaissance/airborne vehicles full tracked, M551AI. (I of2)

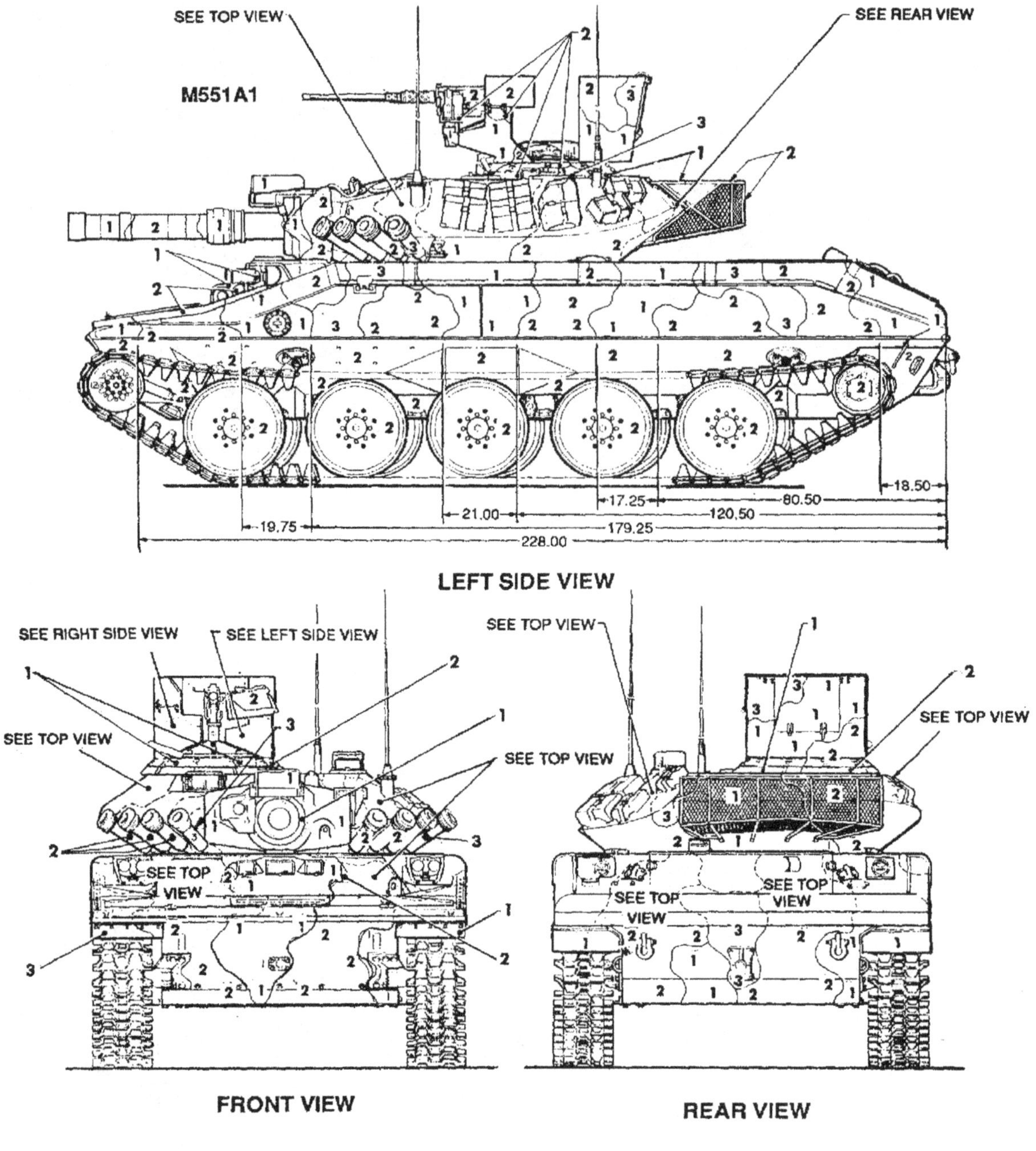

Figure 112. Armored, reconnaissance/airborne vehicles full tracked, M551AI. (2 of 2)

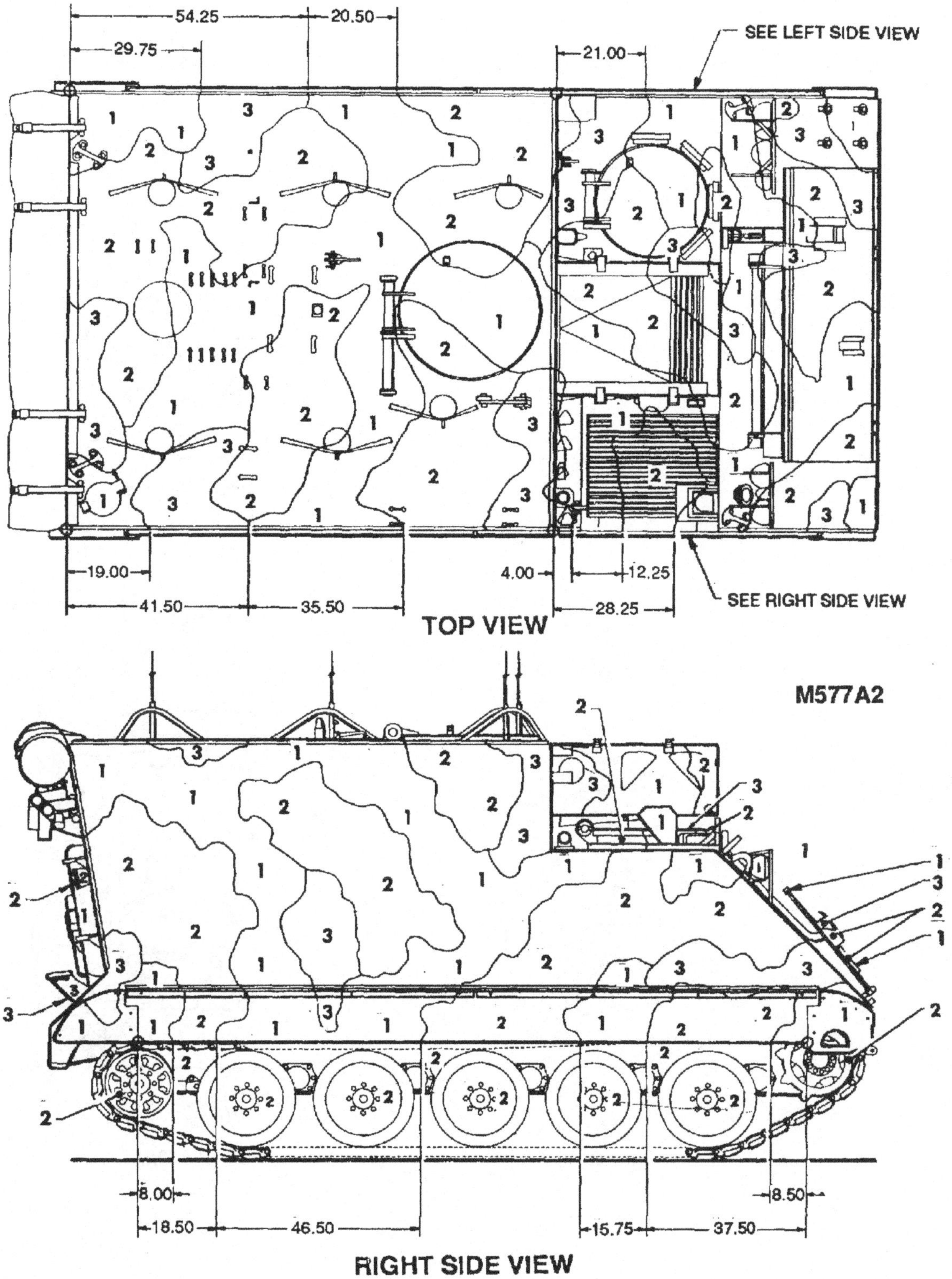

Figure 113. Carrier, command post, full tracked, armored, M577A2. (1 of 2)

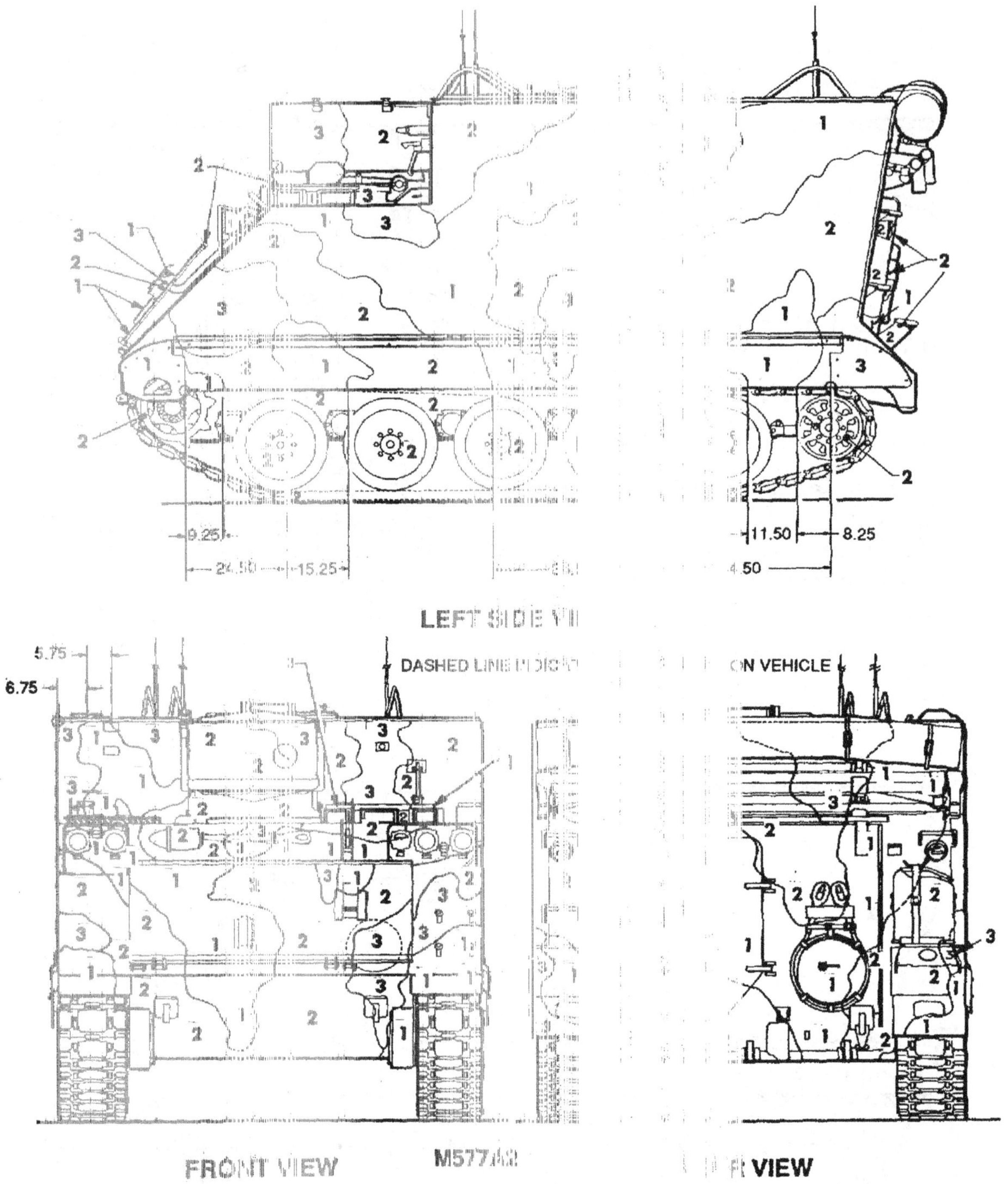

Figure 113. Carrier, command post, full tracke[illegible]7A2. (2 of 2)

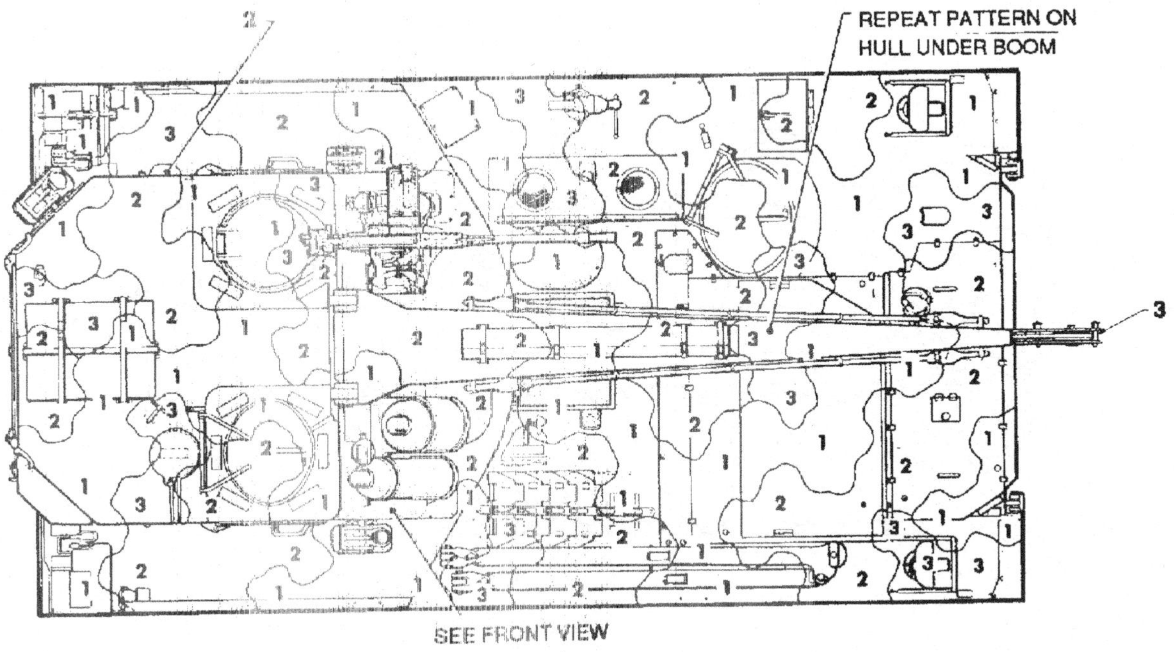

TOP VIEW

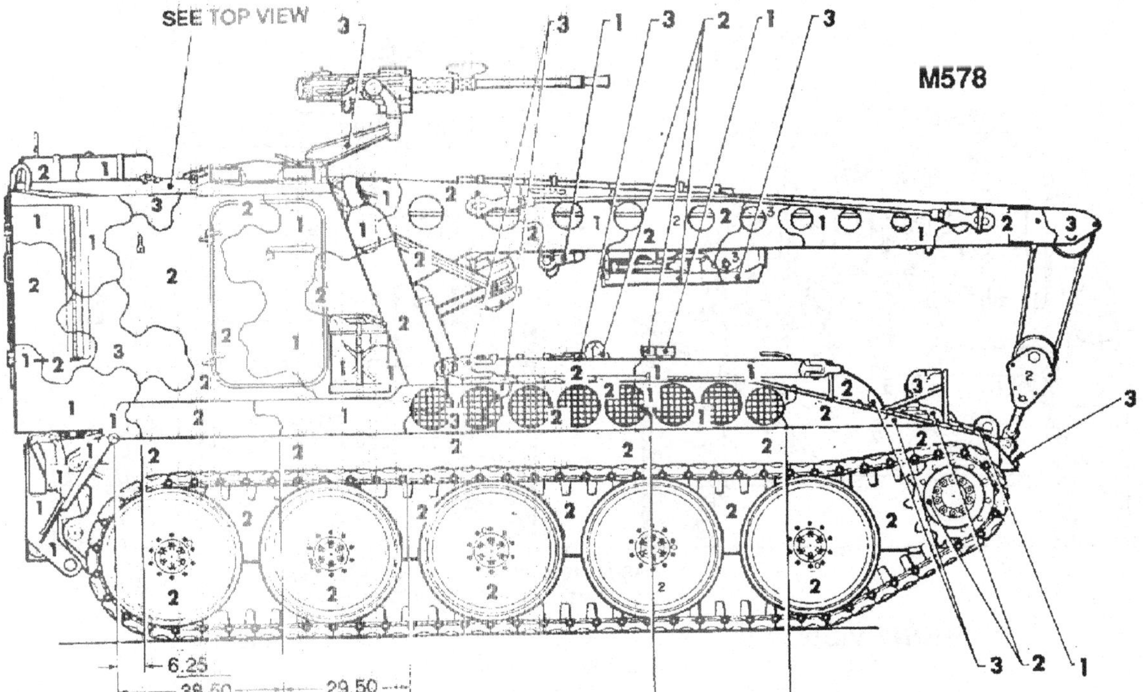

RIGHT SIDE VIEW

Figure 114. Recovery vehicle, armored: M578. (1 of 2)

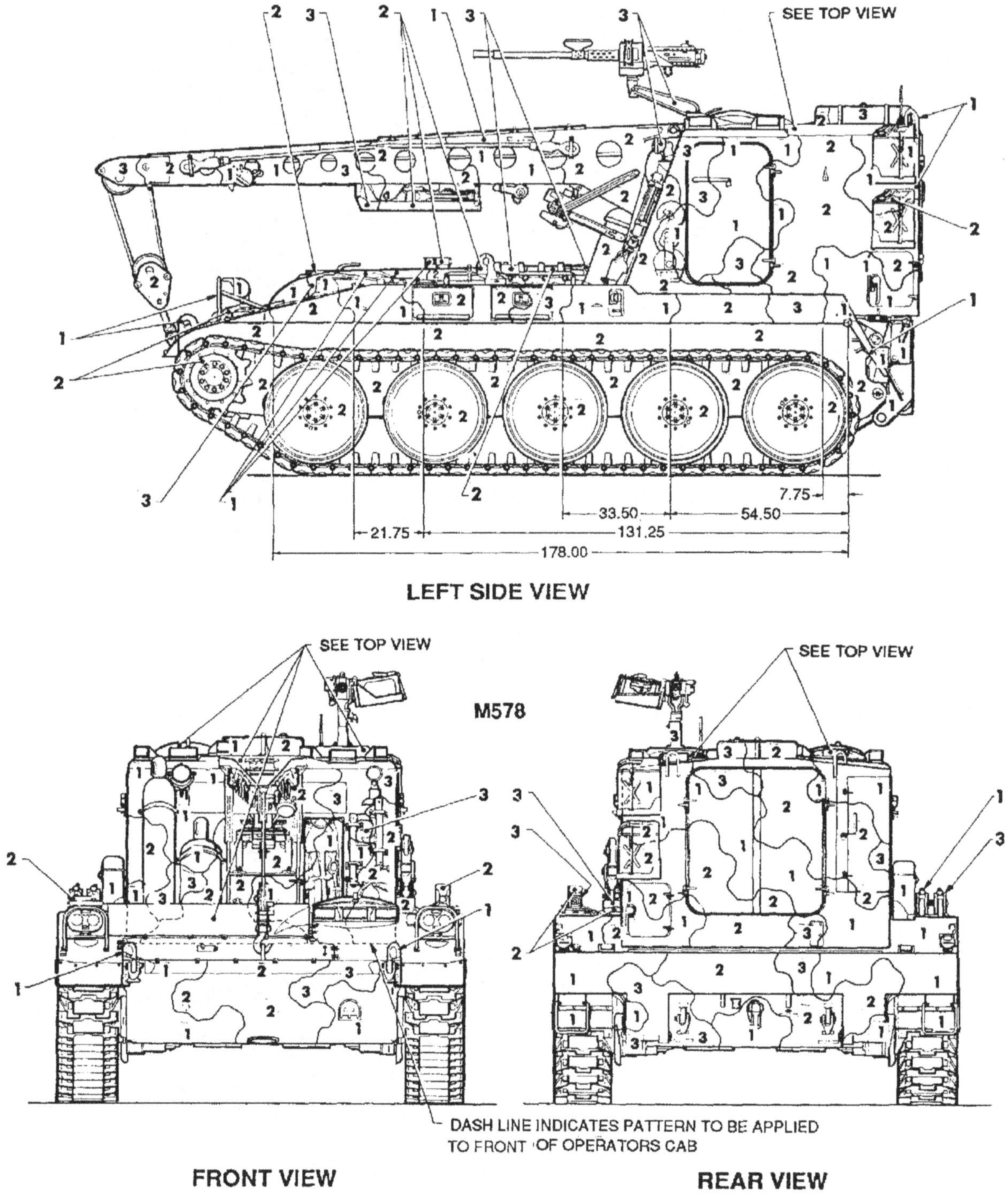

Figure 114. Recovery vehicle, armored: M578. (2 of2)

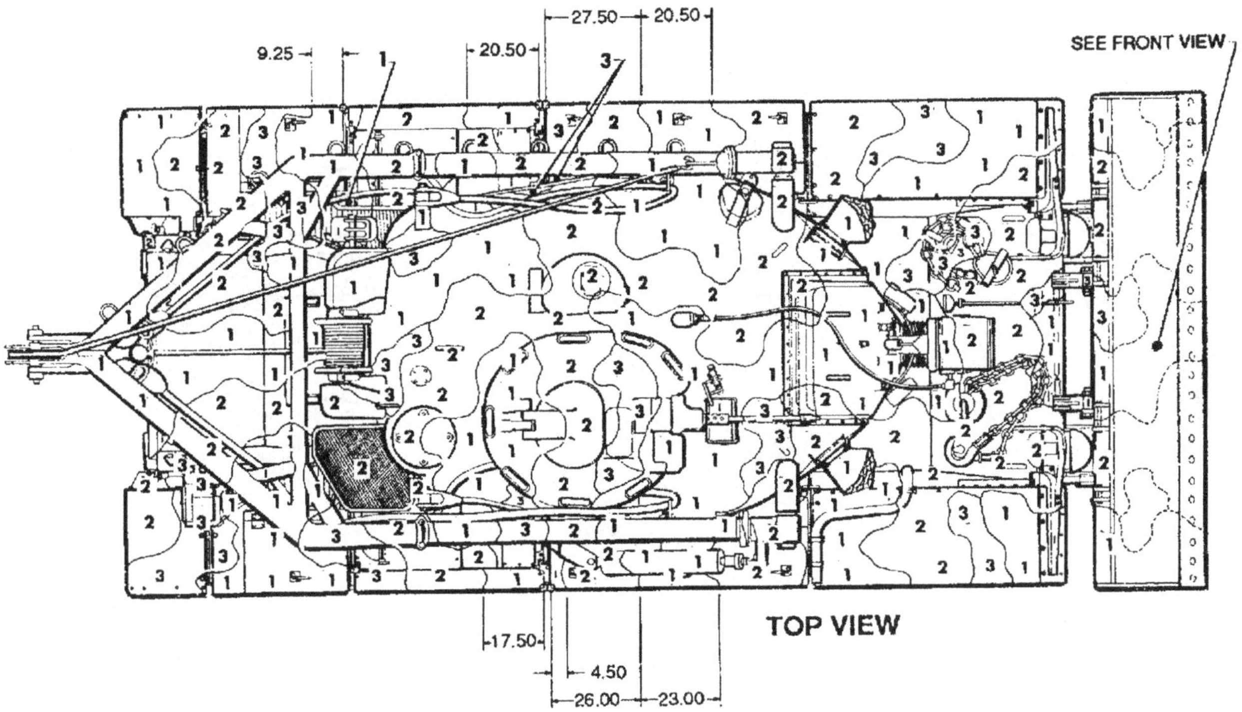

TOP VIEW

M728

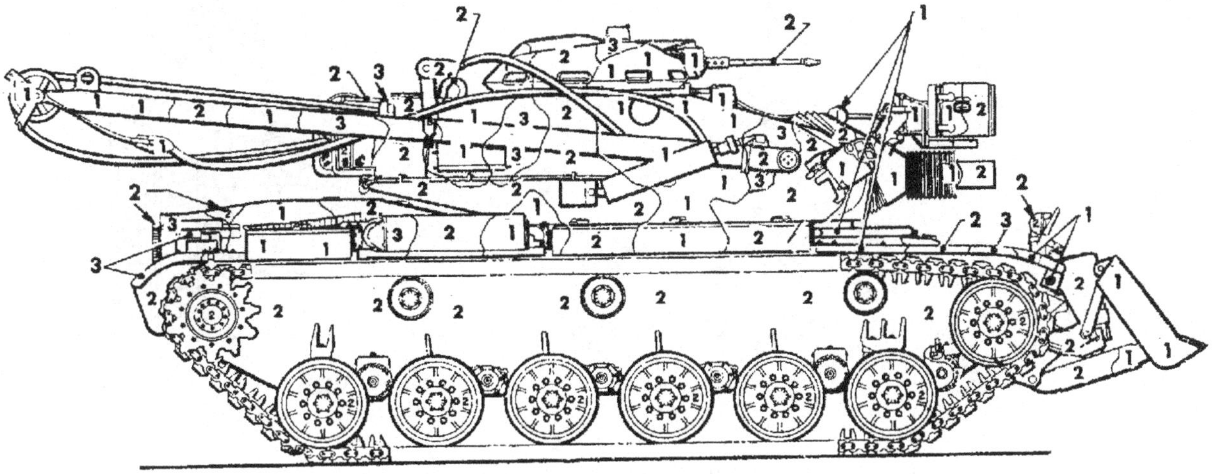

RIGHT SIDE VIEW

Figure 115. Tank, combat, engineers vehicle, full tracked, M728. (I of 2)

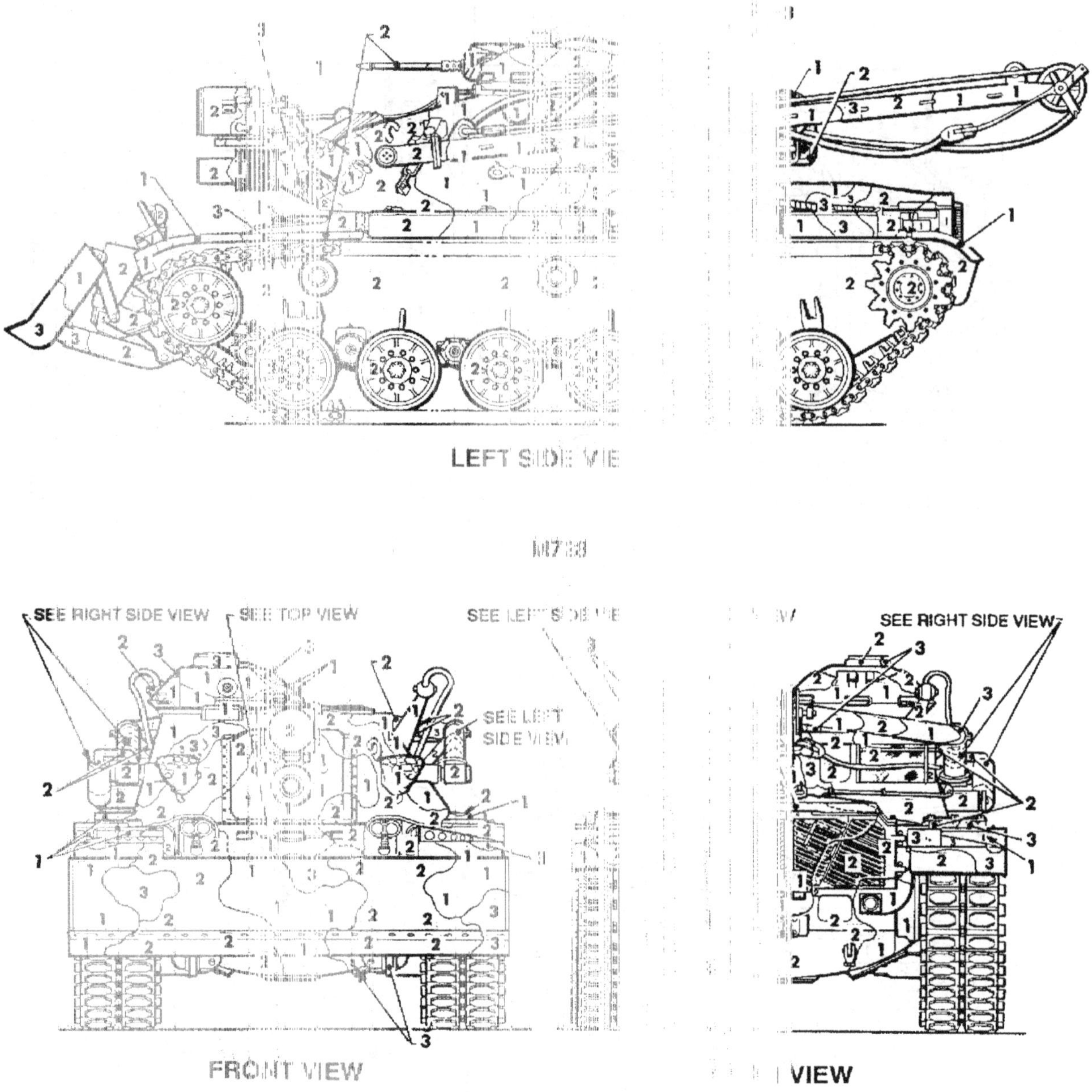

Figure 115. Tank, combat, engineers vehicle M728. (2 of2)

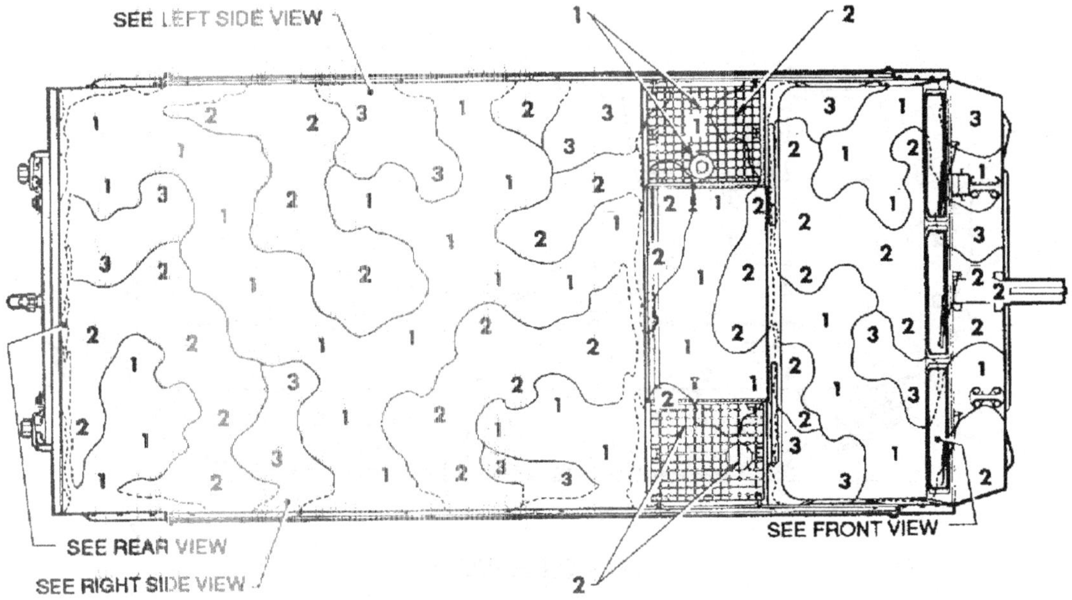

TOP VIEW

M730A1E1

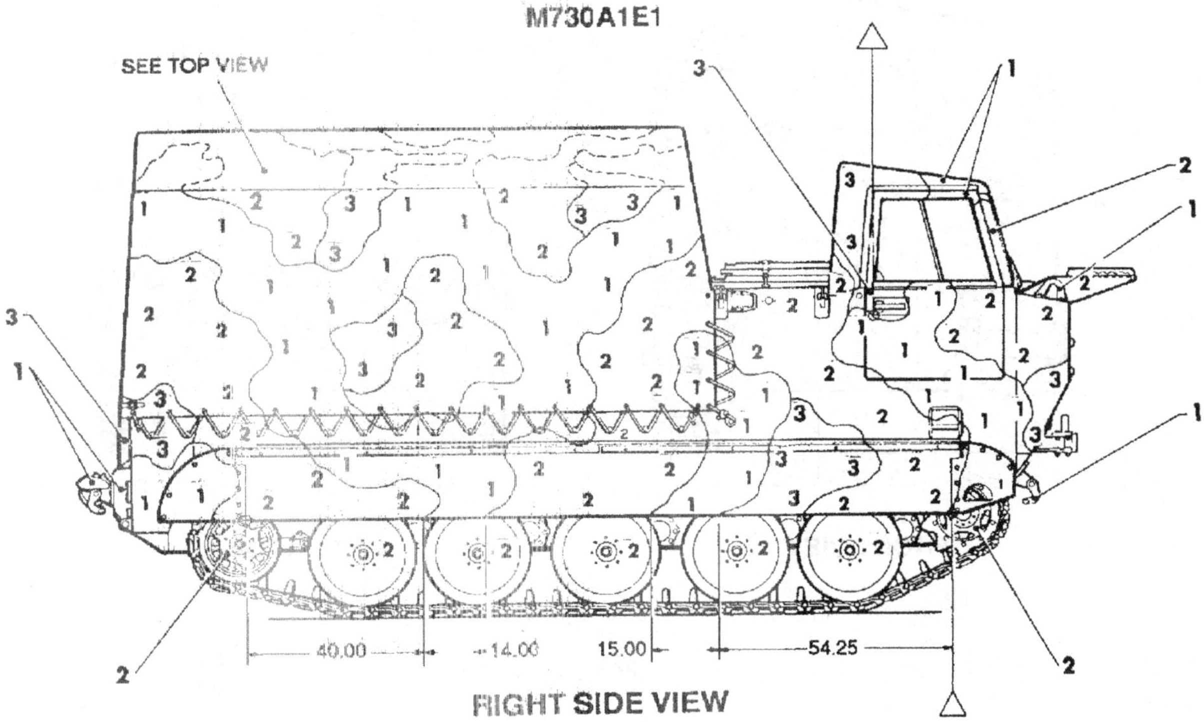

RIGHT SIDE VIEW

Figure 116. Carrier, guided missile equipment, M730AIEI. (I of 2)

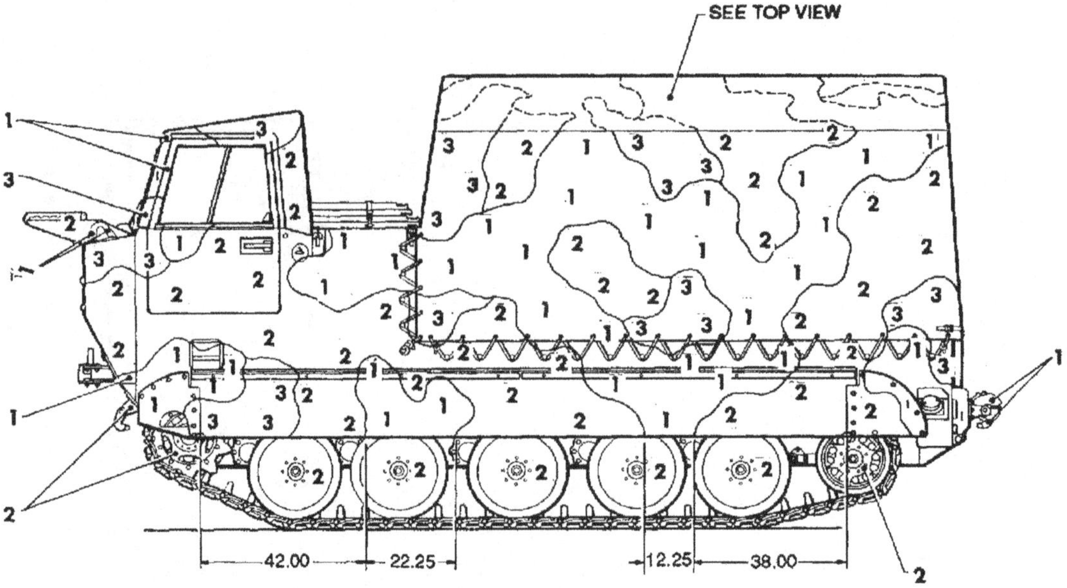

LEFT SIDE VIEW

M730A1E1

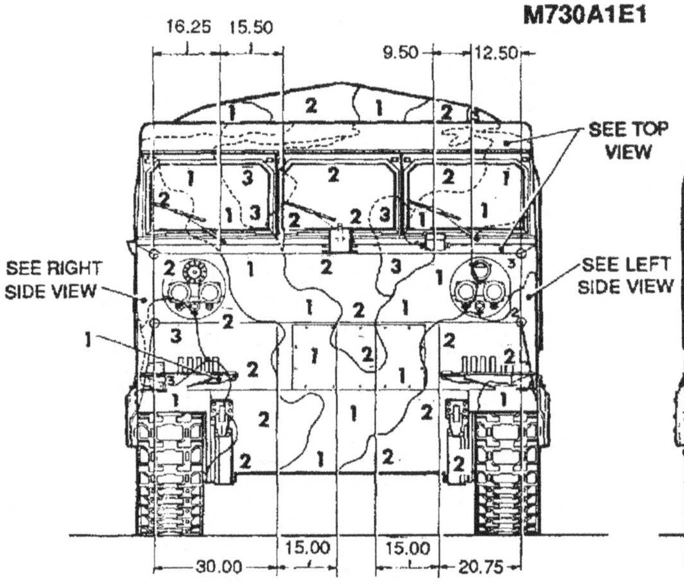

FRONT VIEW

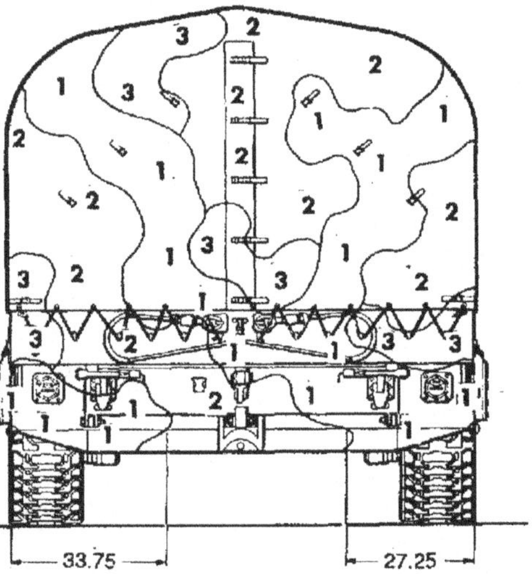

REAR VIEW

Figure 116. Carrier, guided missile equipment, M730AIEI. (2 of2)

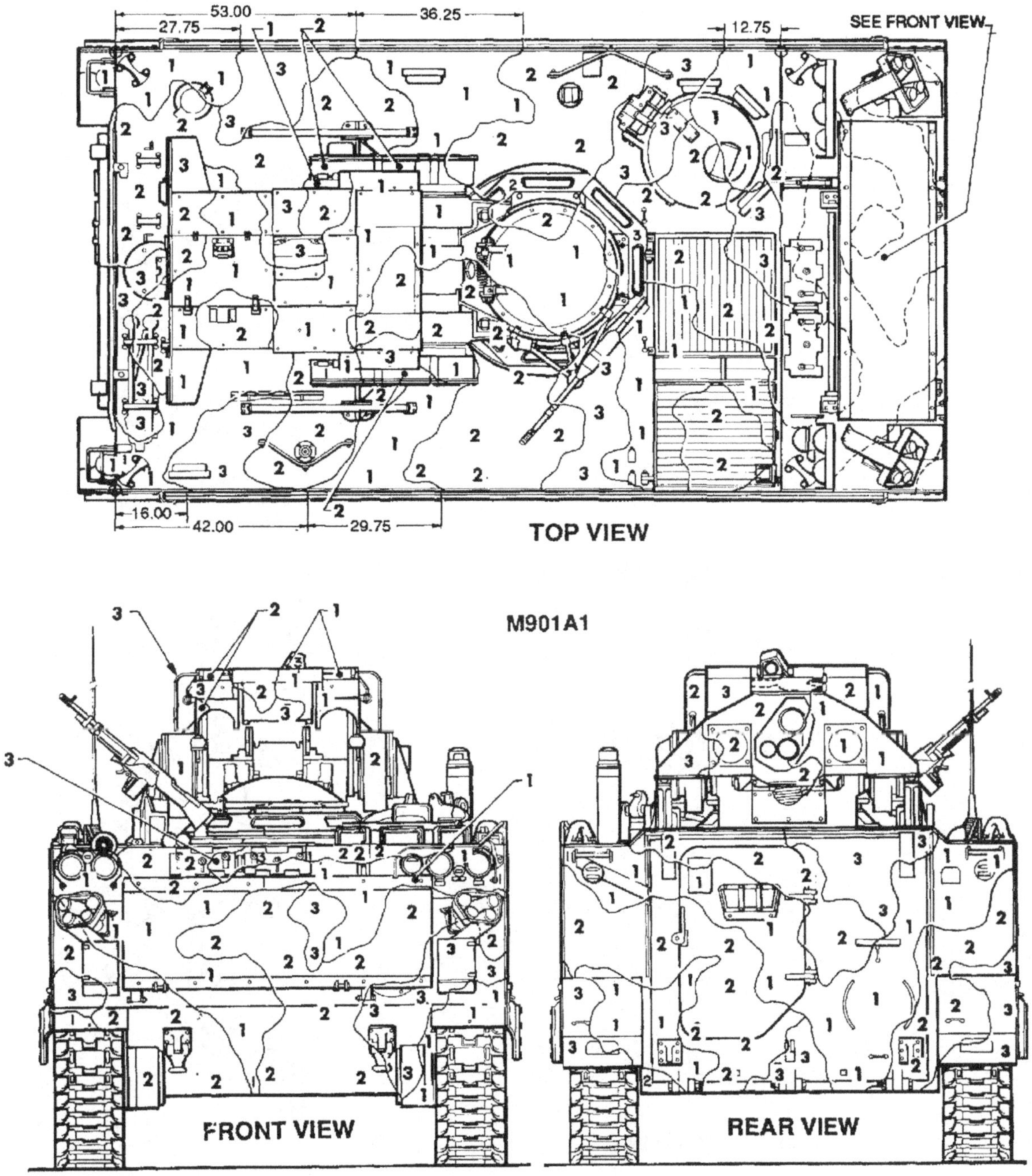

Figure 117. APC, improved tow vehicle, M901AI. (1 of2)

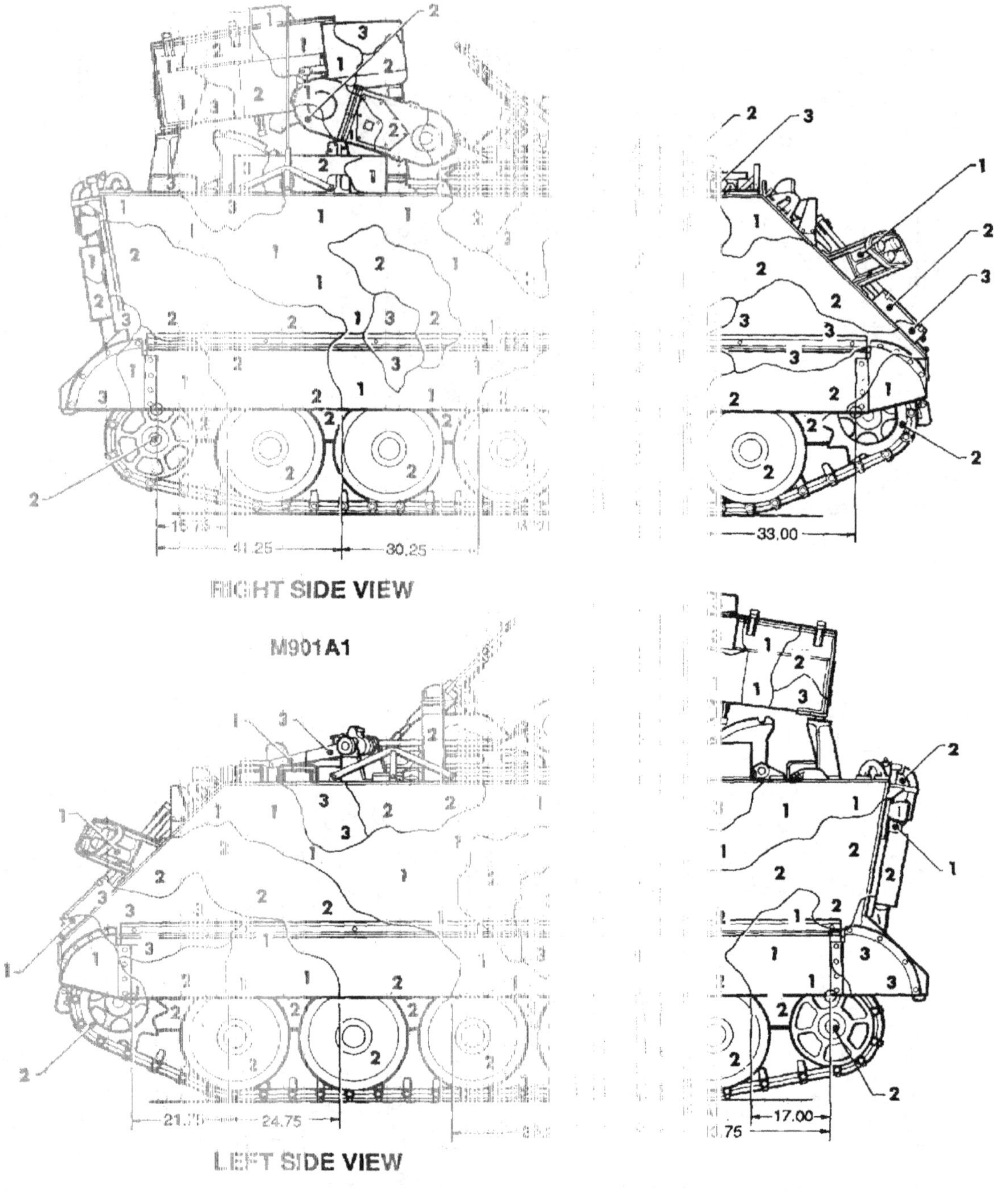

Figure 117. APC, improved tow vehic... of2)

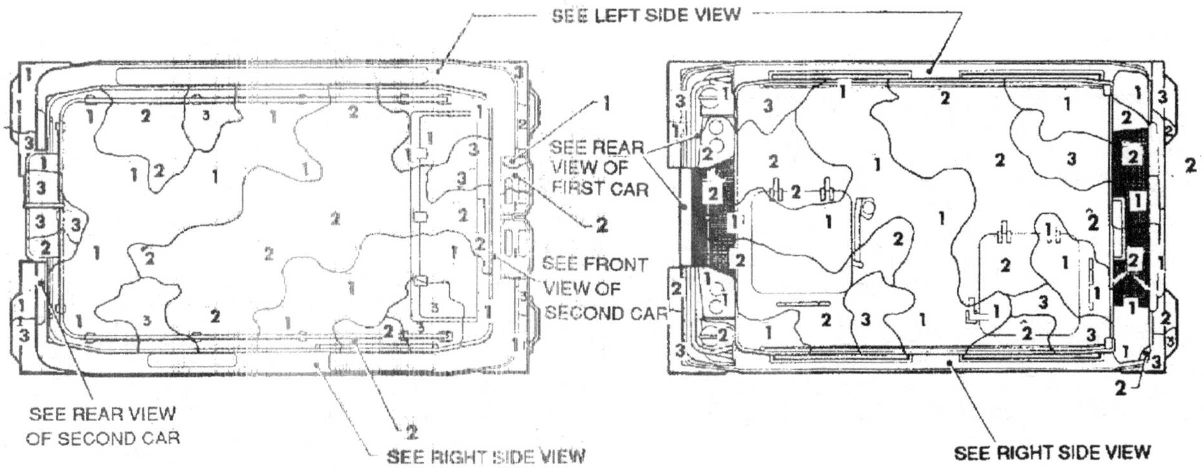

TOP VIEW

M973

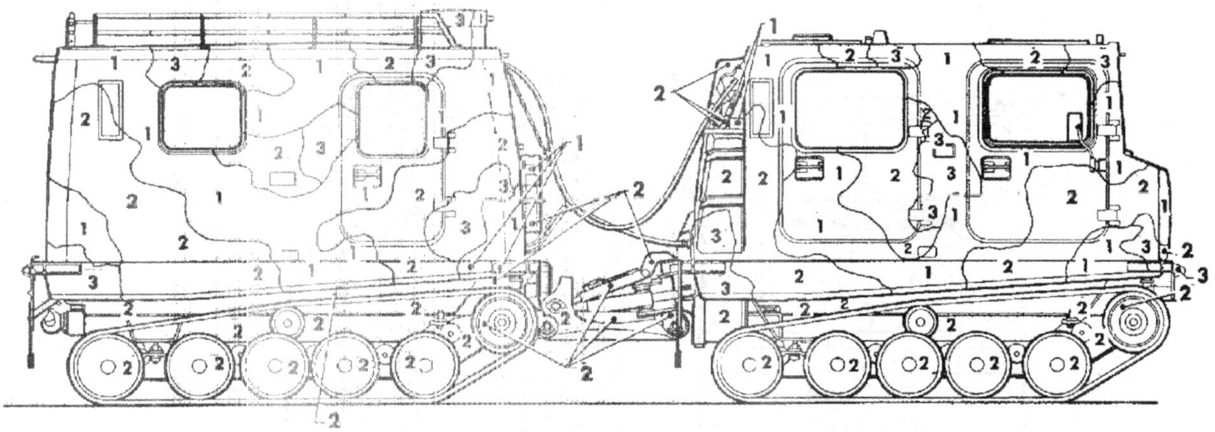

RIGHT SIDE VIEW

Figure 118. Carrier, cargo, tracked, 1-1/2-ton, M973. (I of 2)

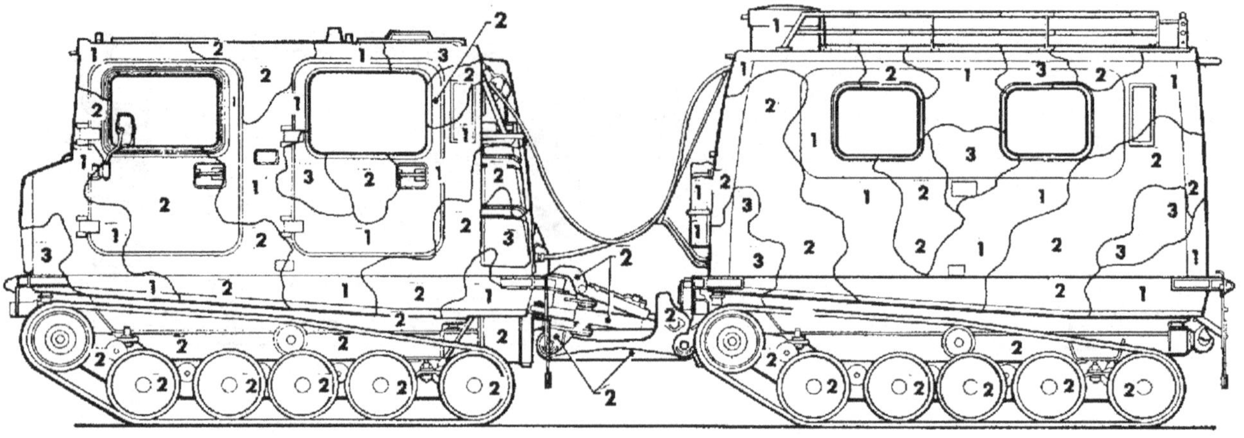

LEFT SIDE VIEW

M973

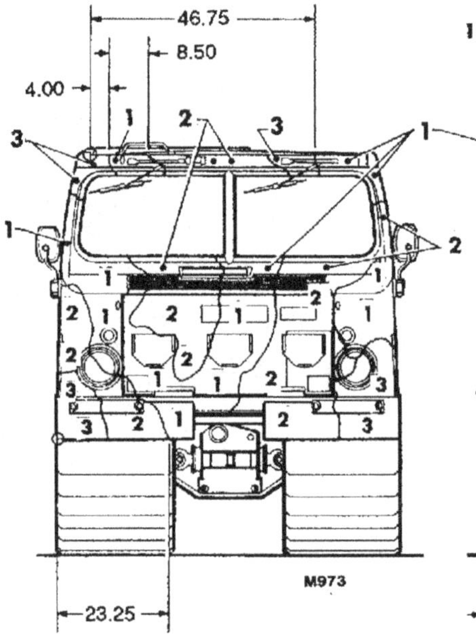

FRONT VIEW

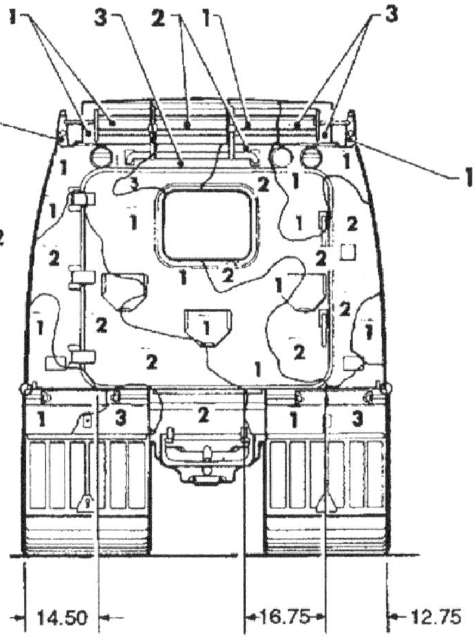

REAR VIEW

Figure 118. Carrier, cargo, tracked, 1-1/2-ton, M973. (2 of 2)

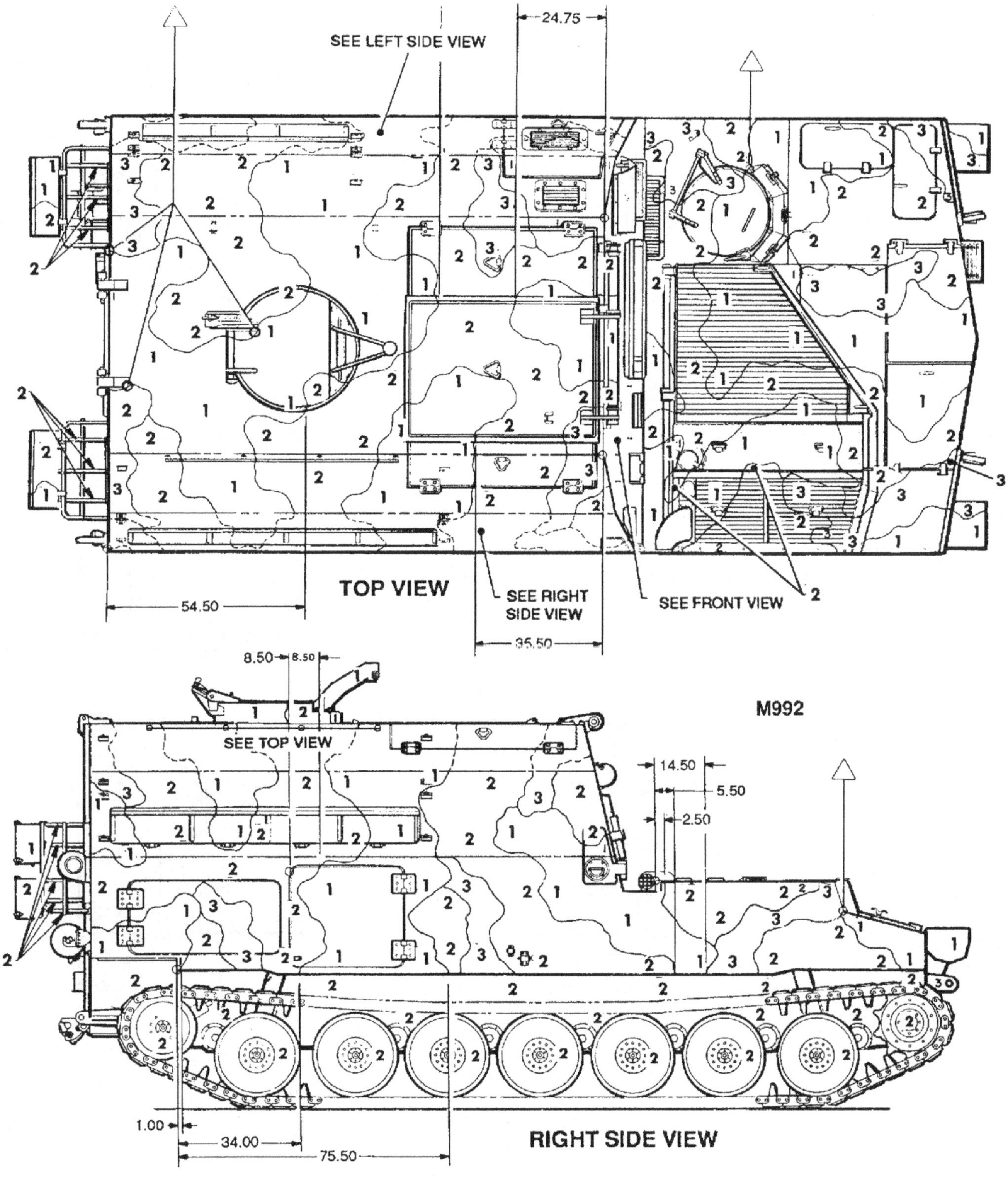

Figure 119. Carrier, ammunition, tracked: M992. (1 of 2)

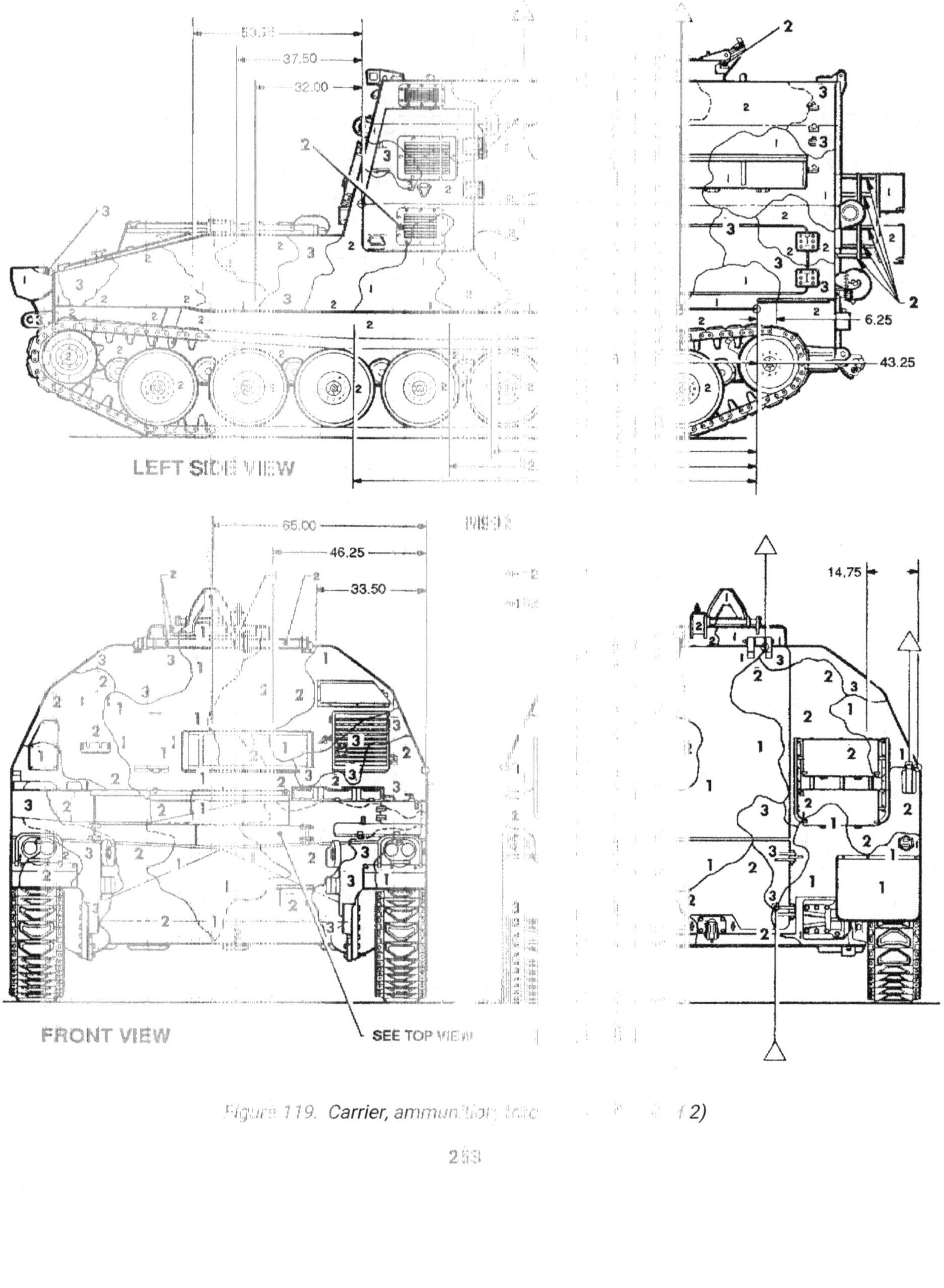

Figure 119. Carrier, ammunition, tracked ... 2)

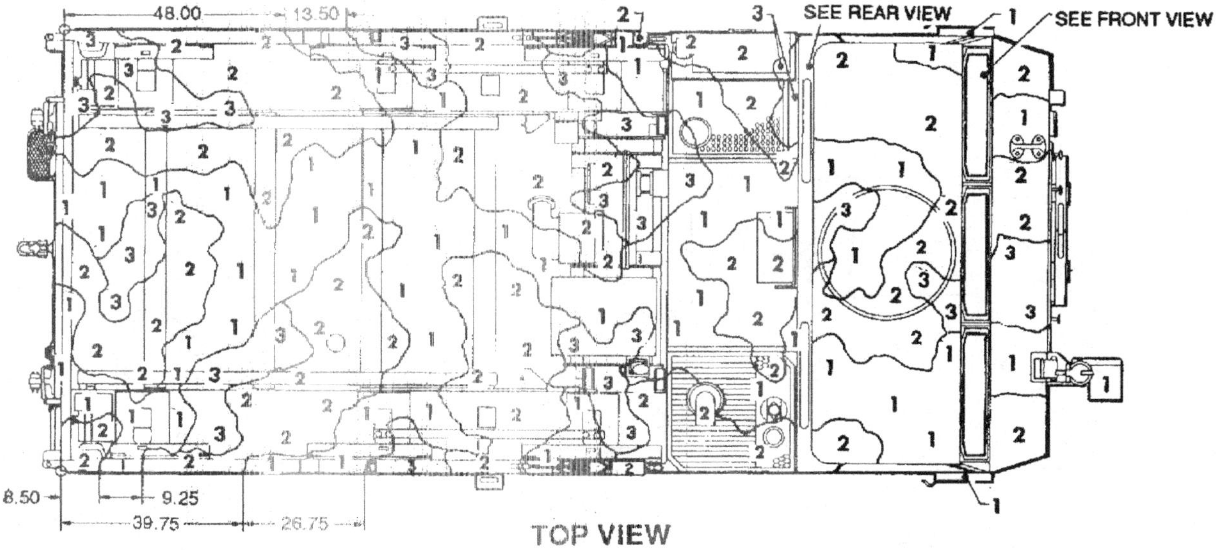

TOP VIEW

M1015

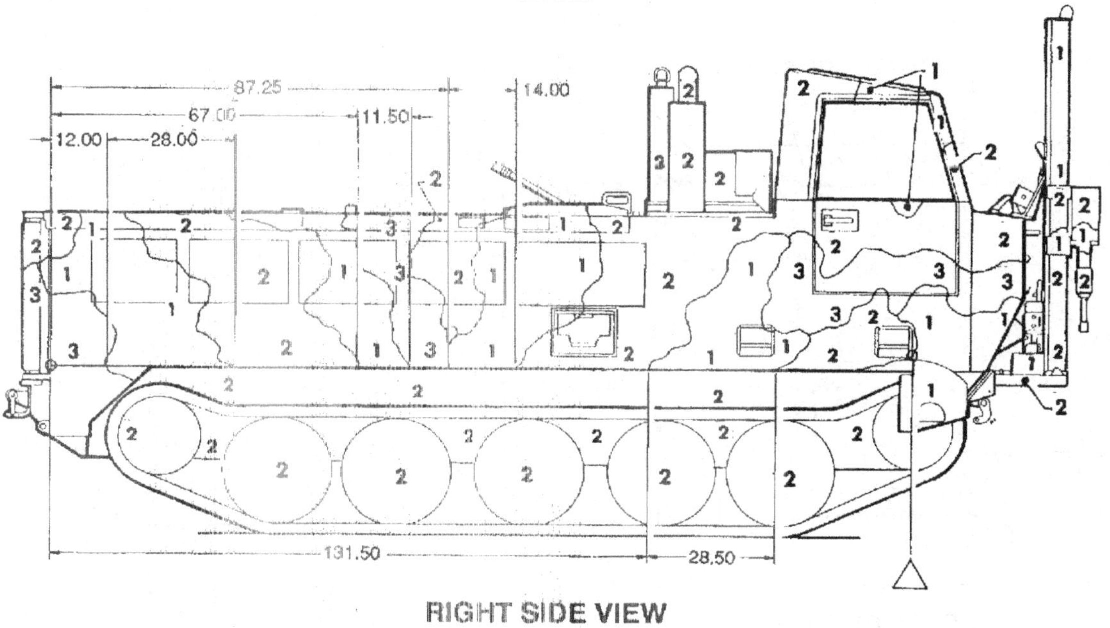

RIGHT SIDE VIEW

Figure 120. Carrier, electronic warfare, tracked: M1015. (1 of 2)

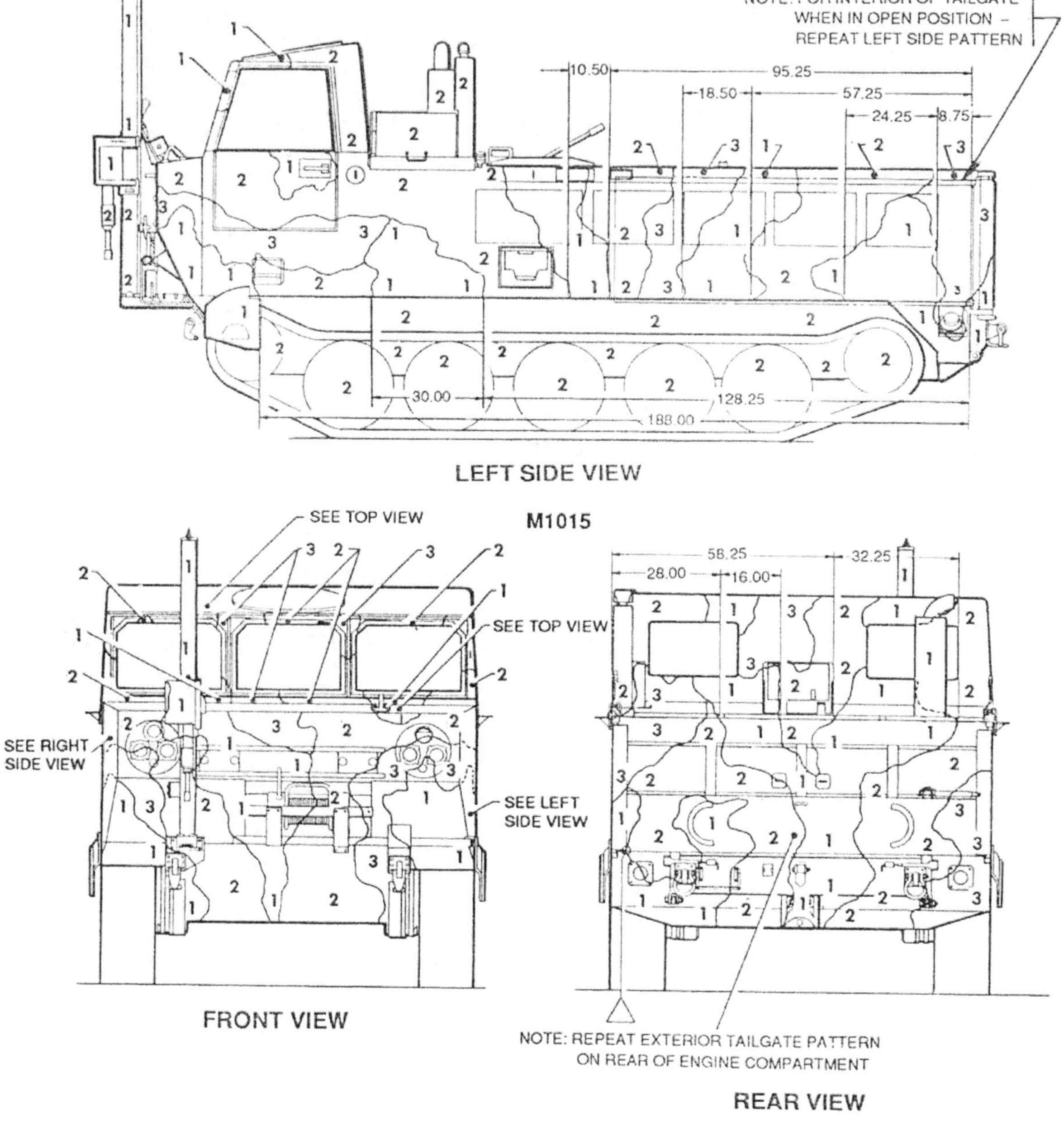

Figure 120. Carrier, electronic warfare, tracked: M1015. (2 of 2)

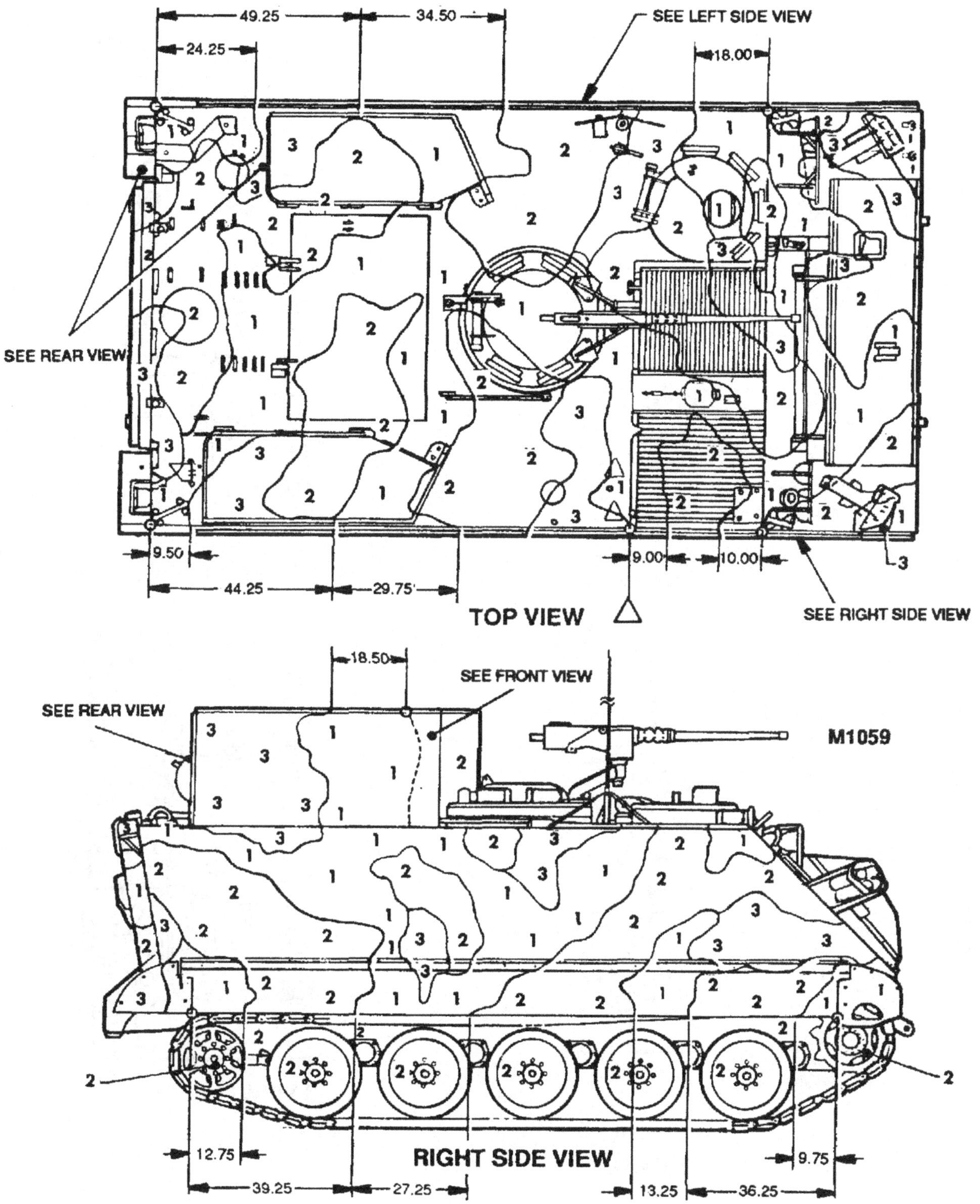

Figure 121. Carrier, smoke generator M1059 (1 of 2)

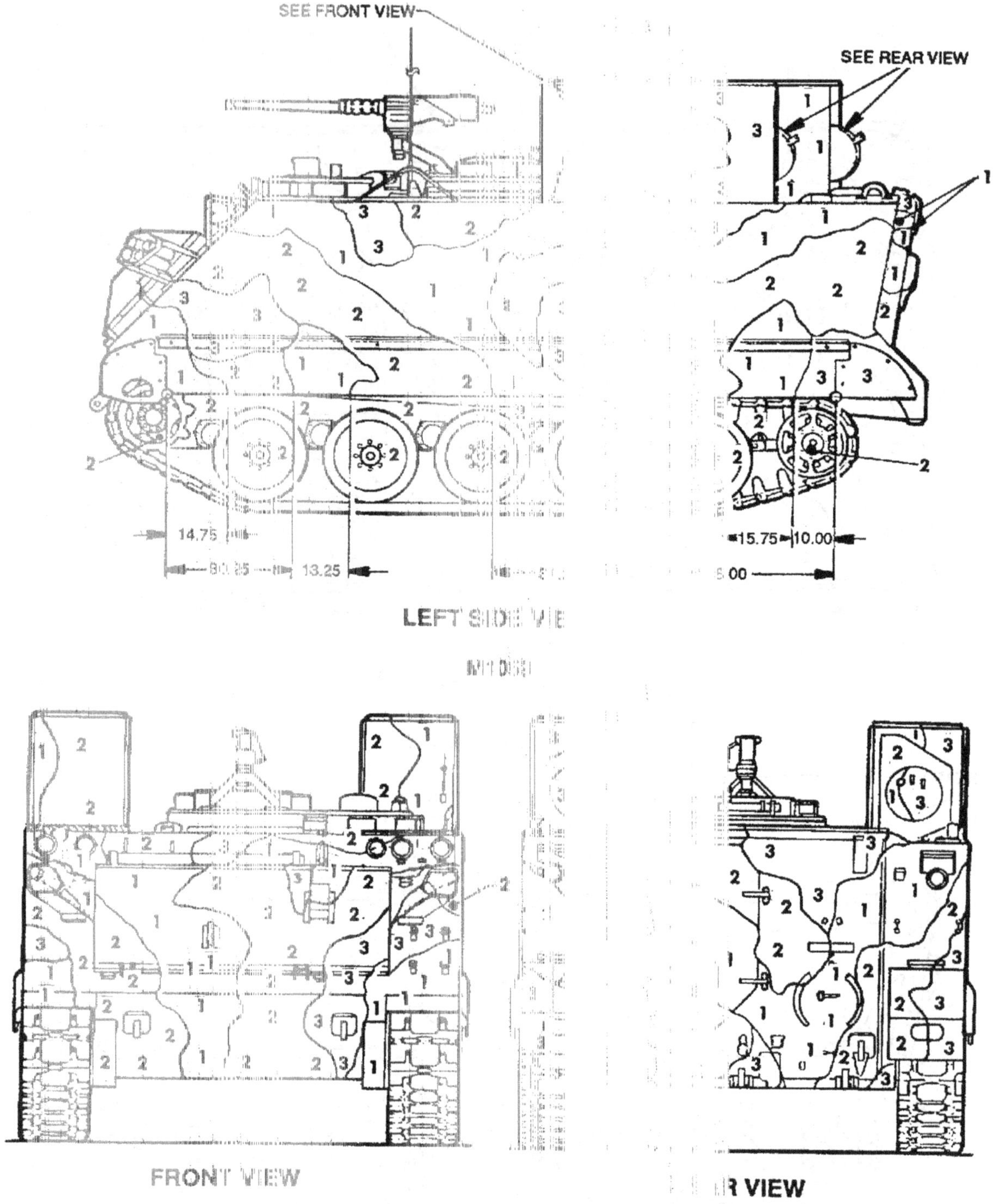

Figure 121. Carrier, smoke genera[illegible] 2)

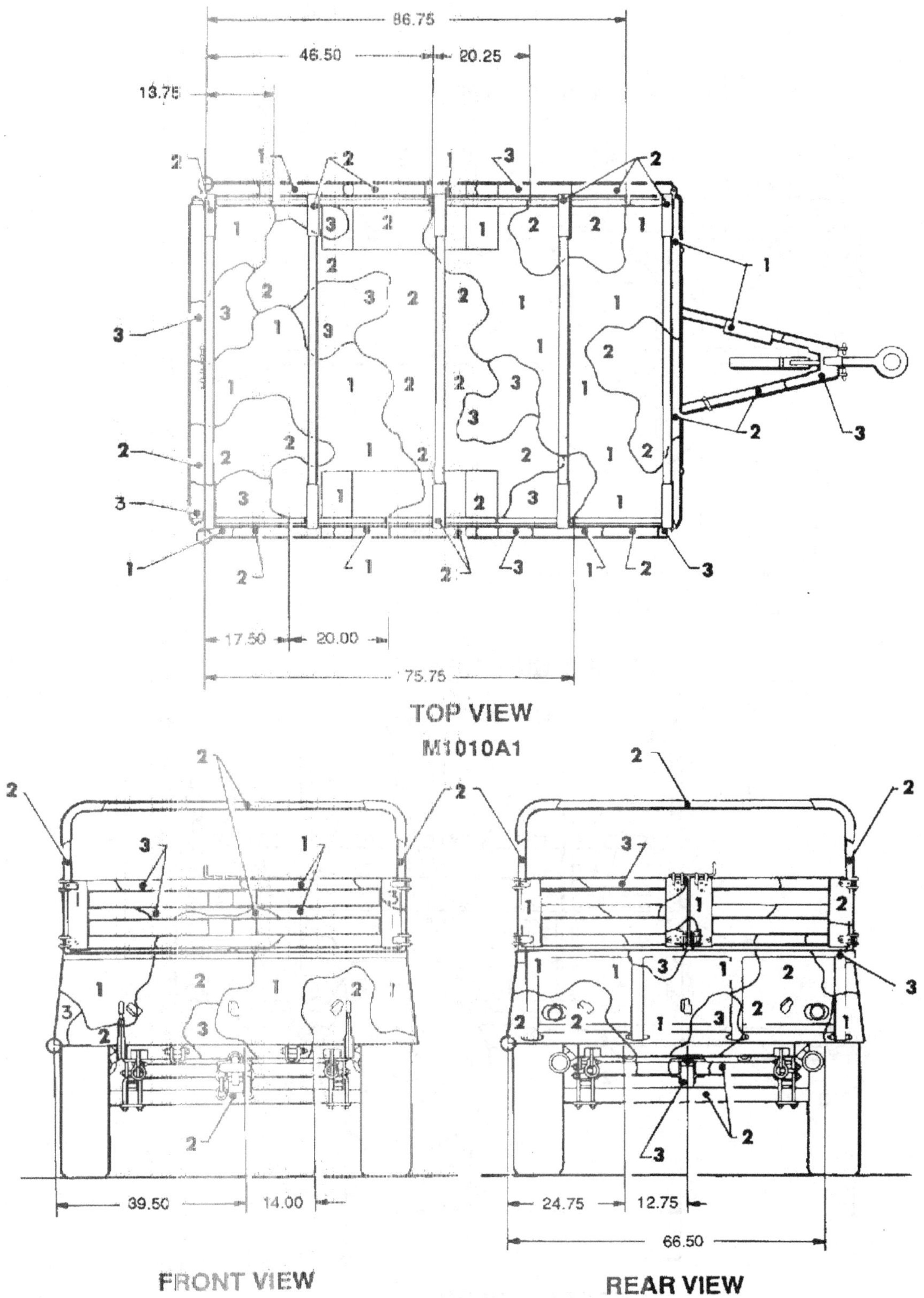

Figure 122. Trailer, cargo: 3/4-ton, 2-wheel, M101A1. (1 of 2)

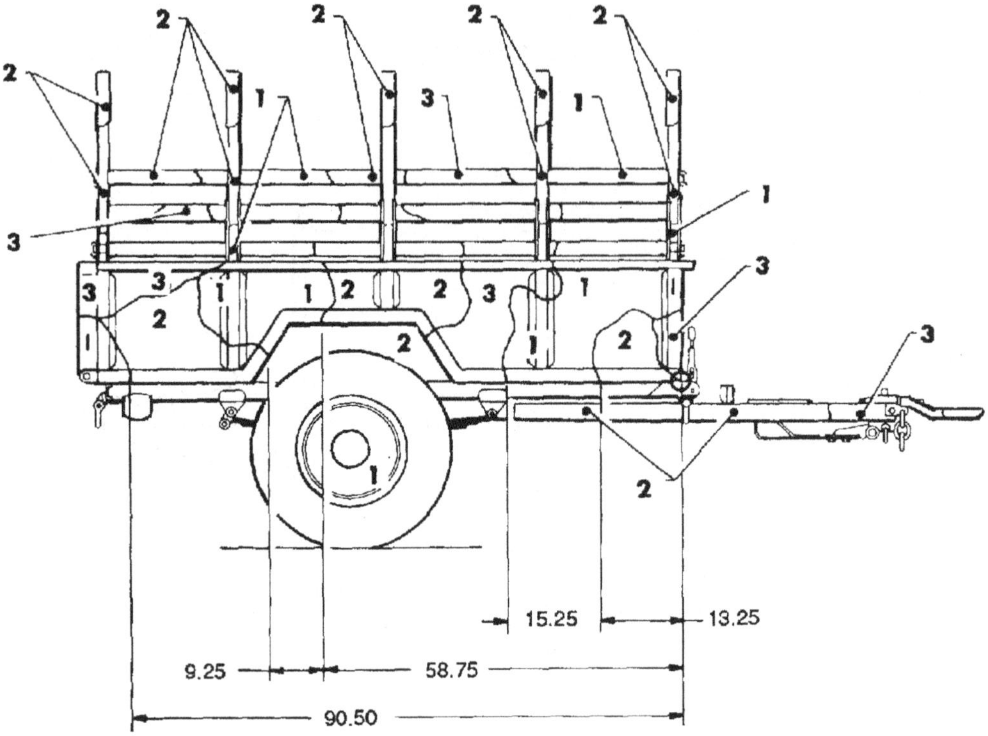

RIGHT SIDE VIEW

M1010A1

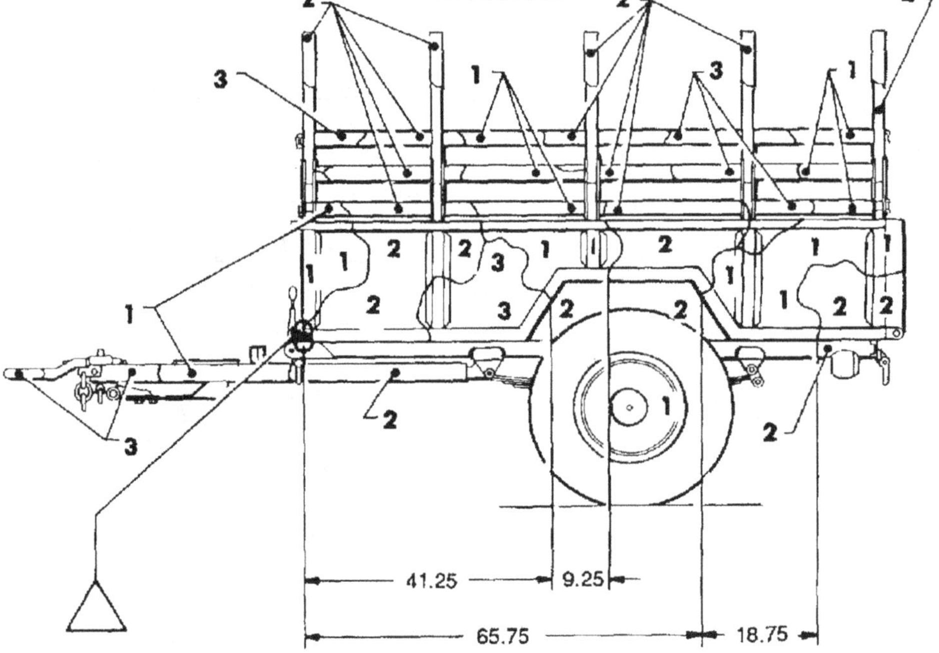

LEFT SIDE VIEW

Figure 122. Trailer, cargo: 3/4-ton, 2-wheel, M101A1. (2 of 2)

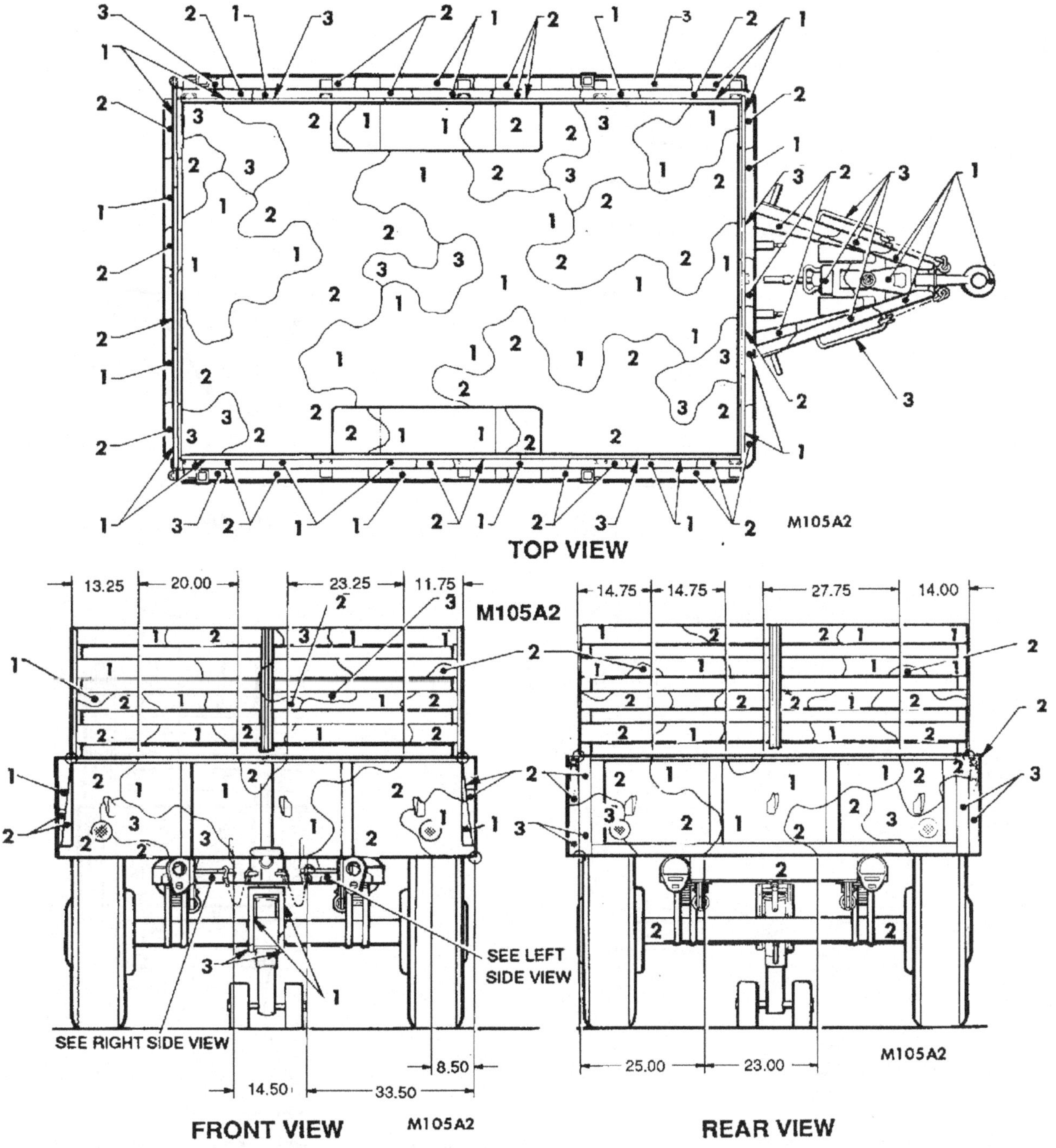

Figure 123. Trailer, cargo: 1-1/4 ton, 2-wheel, M105A2. (1 of 2)

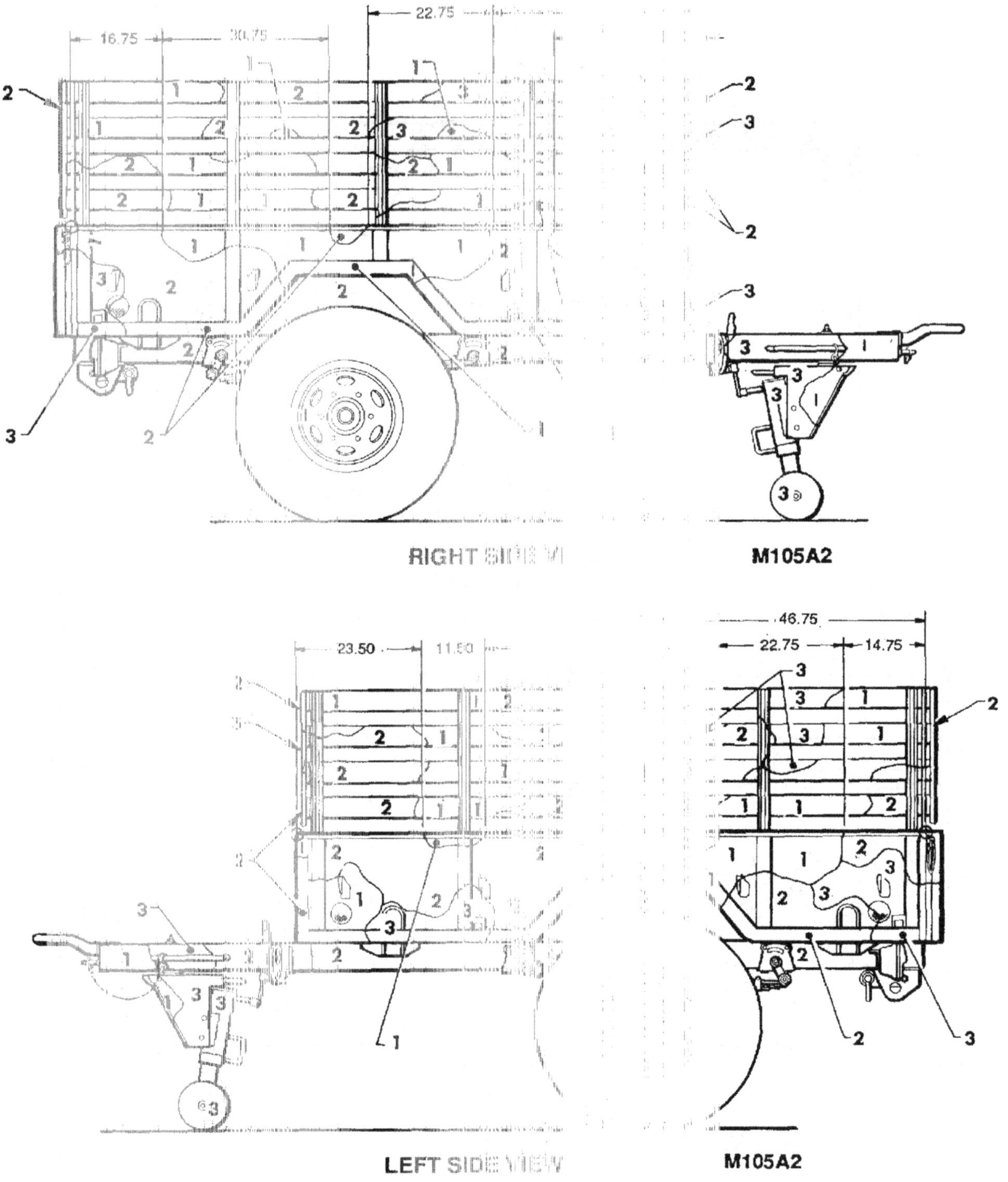

Figure 123. Trailer, cargo: 1-1/4 ton, 2-w [illegible] 2 of 2)

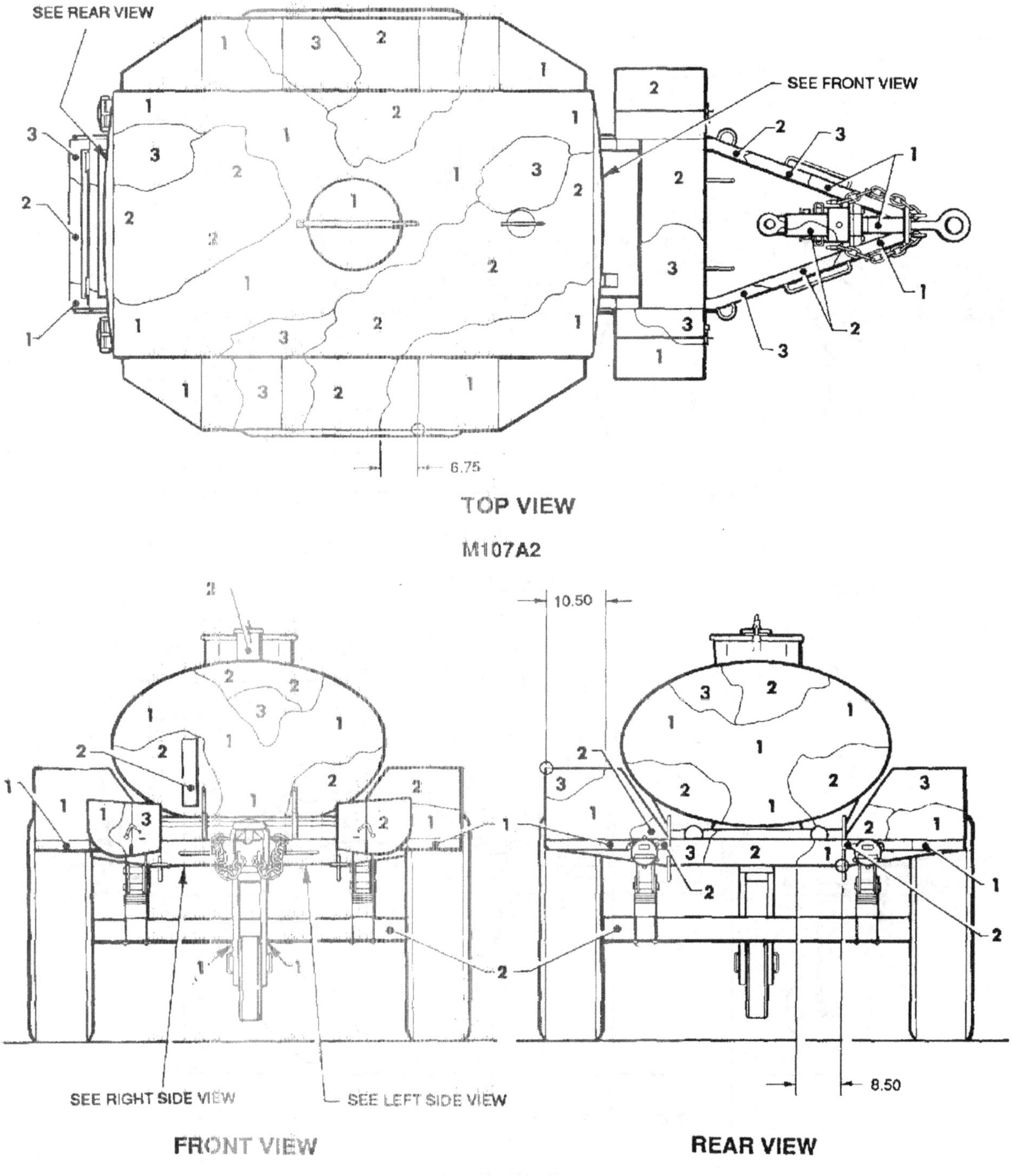

Figure 124. Trailer, tank: water 1-1/4-ton, 2-wheel, M107A2. (1 of 2)

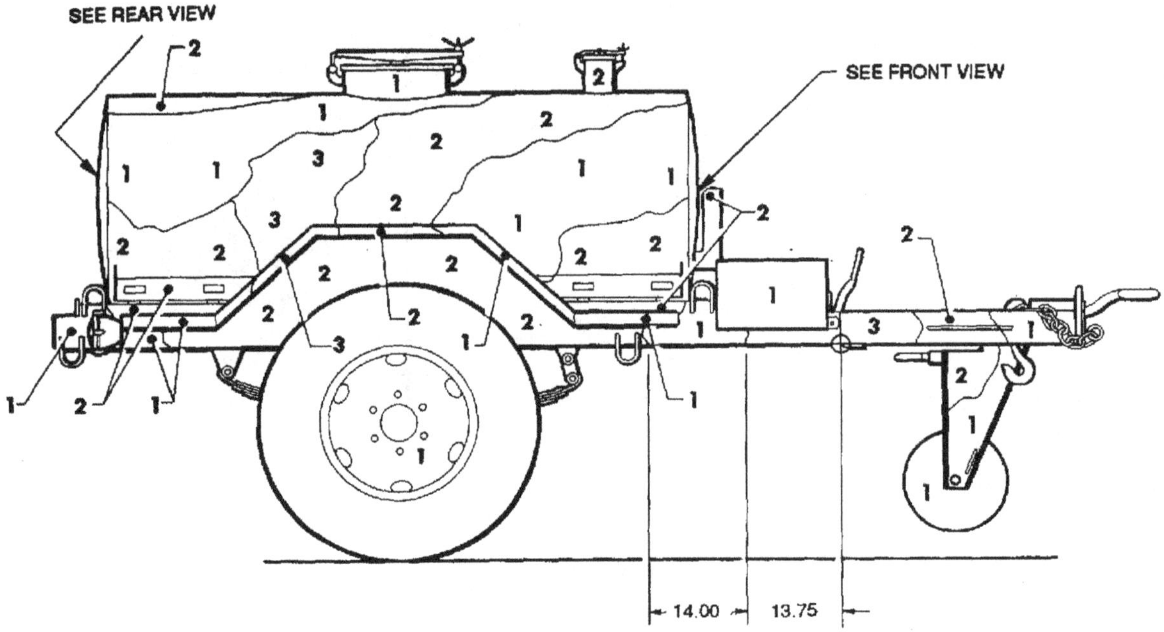

M107A2

RIGHT SIDE VIEW

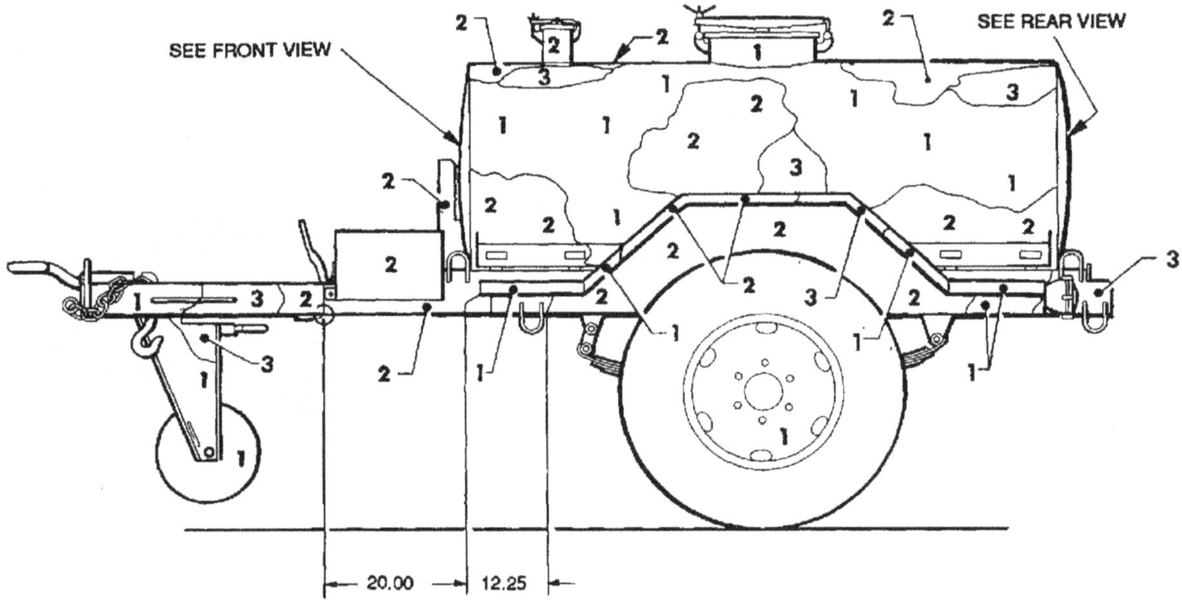

LEFT SIDE VIEW

Figure 124. Trailer, tank: water 1-1/4-ton, 2-wheel, M107A2. (2 of 2)

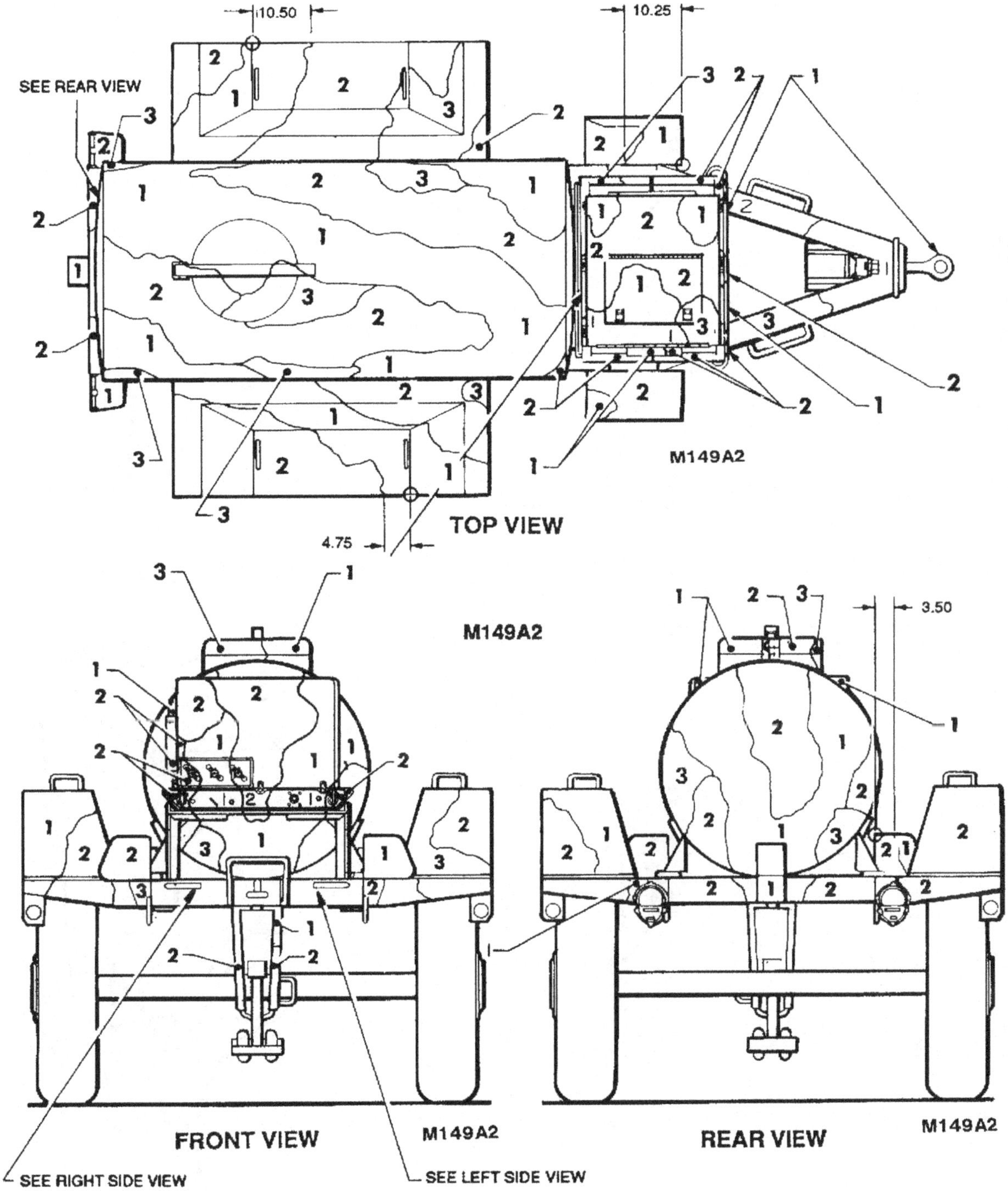

Figure 125. Trailer, tank: water, 400 gallon, 1-1/2-ton, 2-wheel, M149A2. (1 of 2)

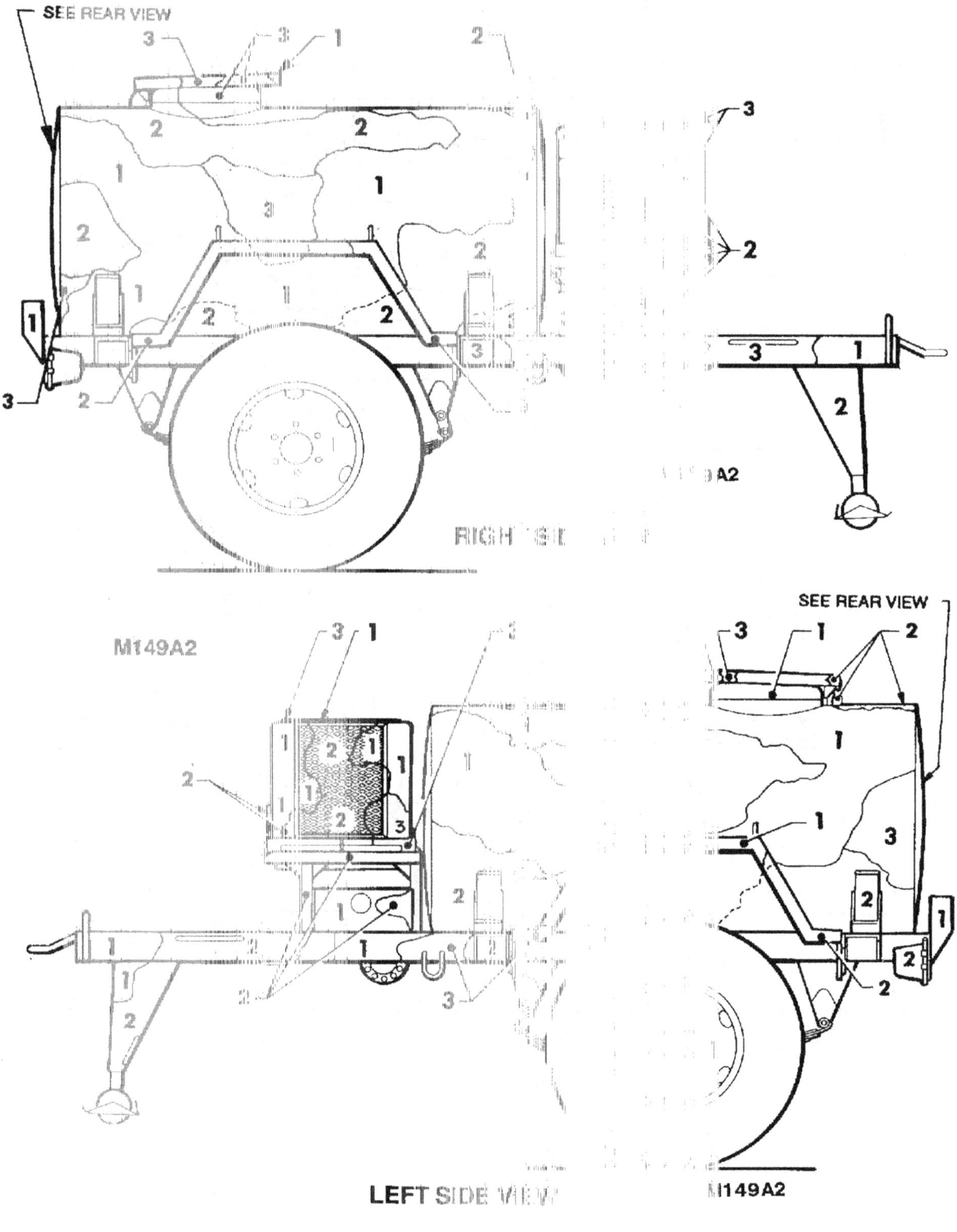

Figure 125. Trailer, tank: water, 400 gallon, ... 49A2. (2 of 2)

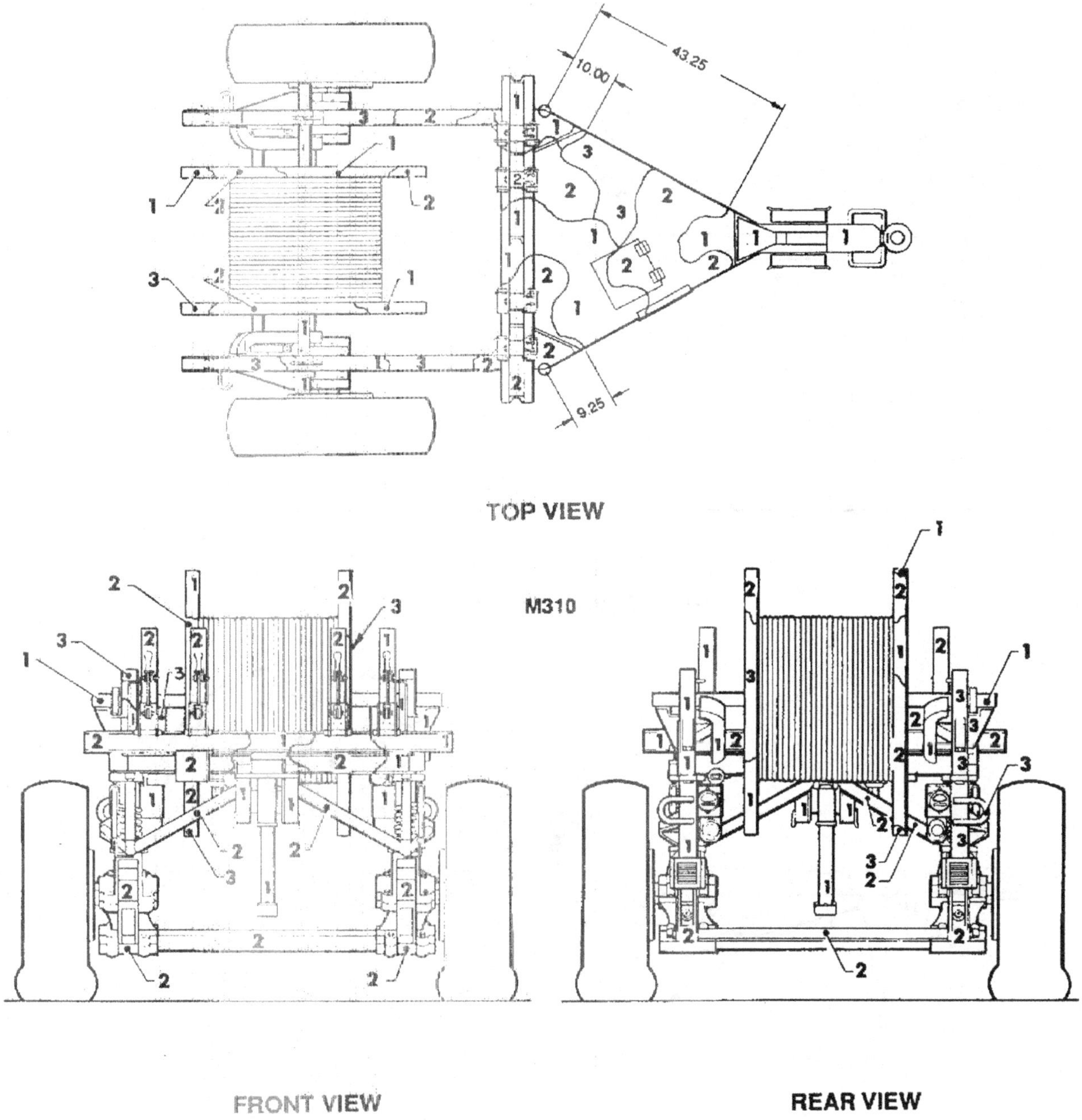

Figure 126. Trailer, cable reel: M310. (1 of 2)

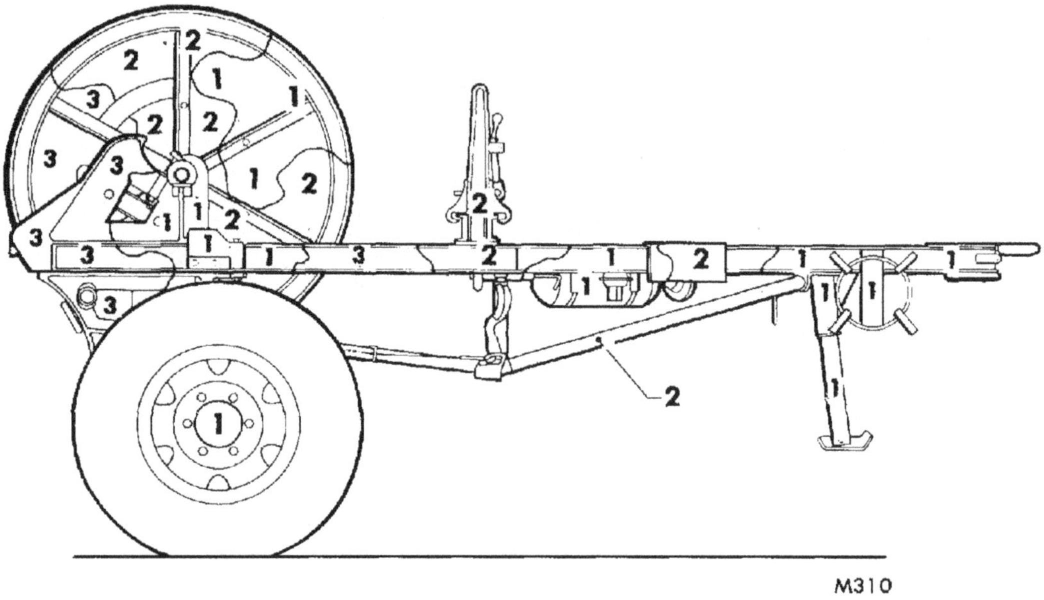

RIGHT SIDE VIEW

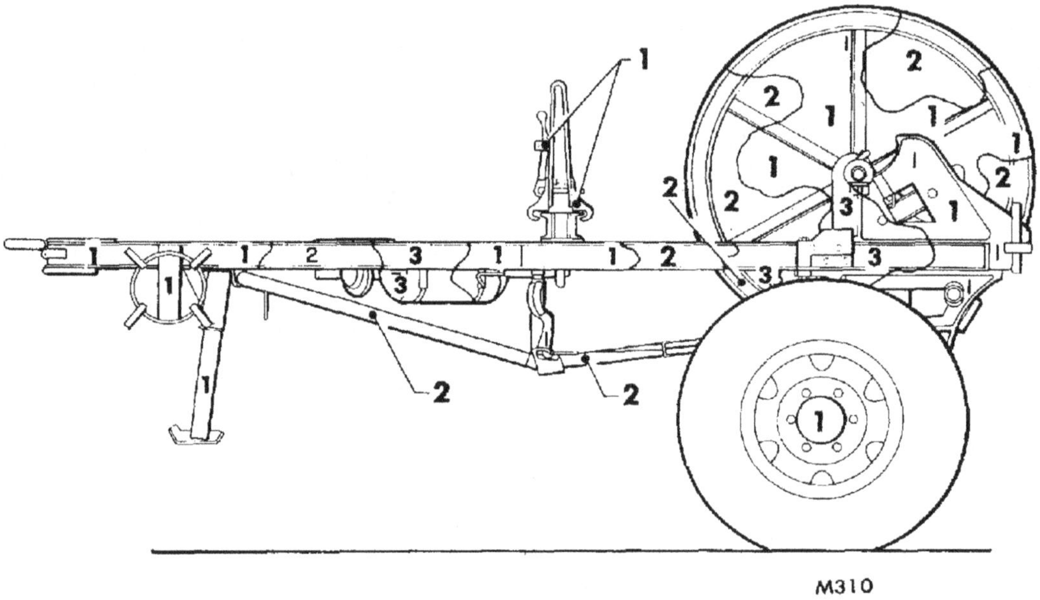

LEFT SIDE VIEW

M310

Figure 126. Trailer, cable reel: M310. (2 of 2)

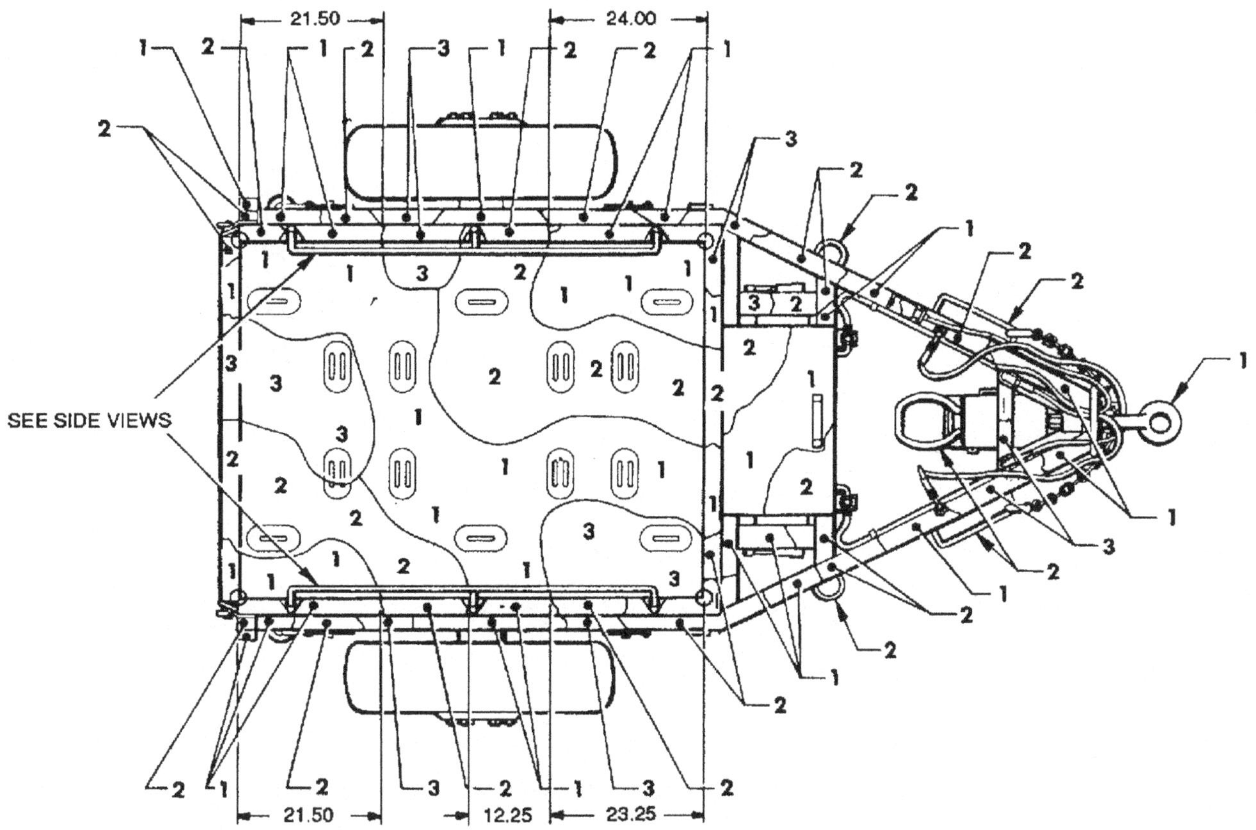

TOP VIEW

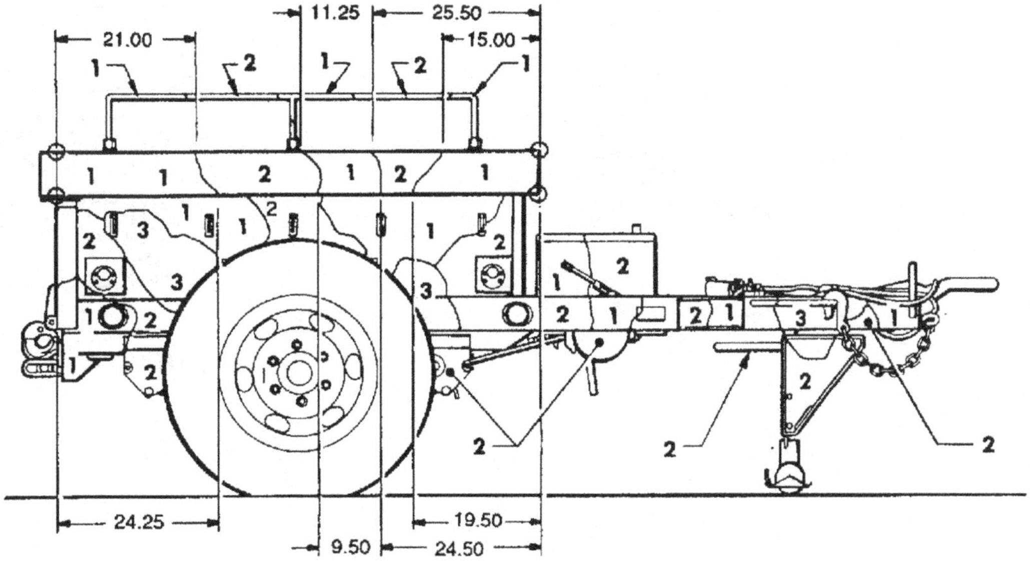

RIGHT SIDE VIEW

Figure 127. Trailer, ammunition: 1-1/2-ton, 2-wheel, M332. (1 of 2)

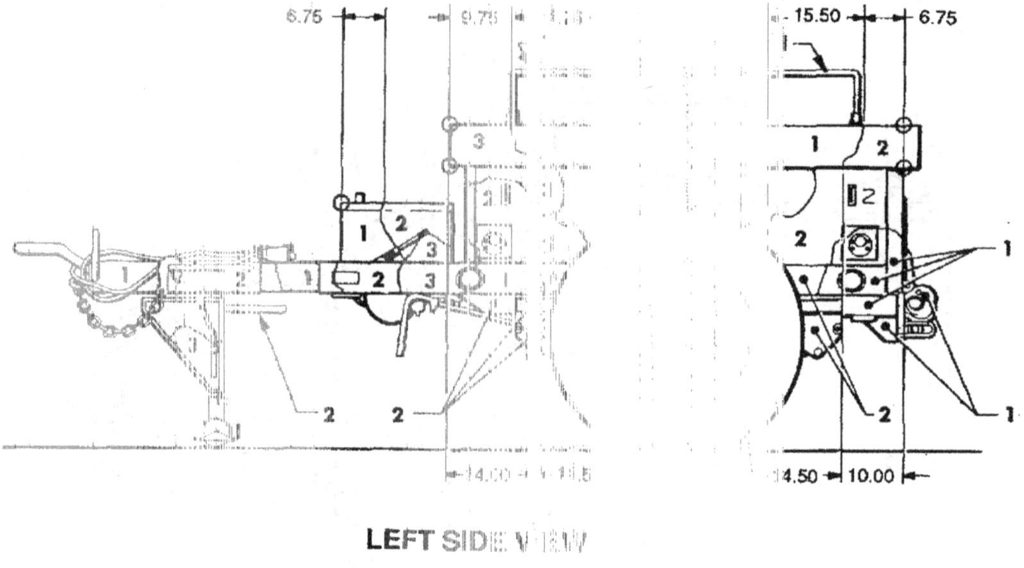

M332

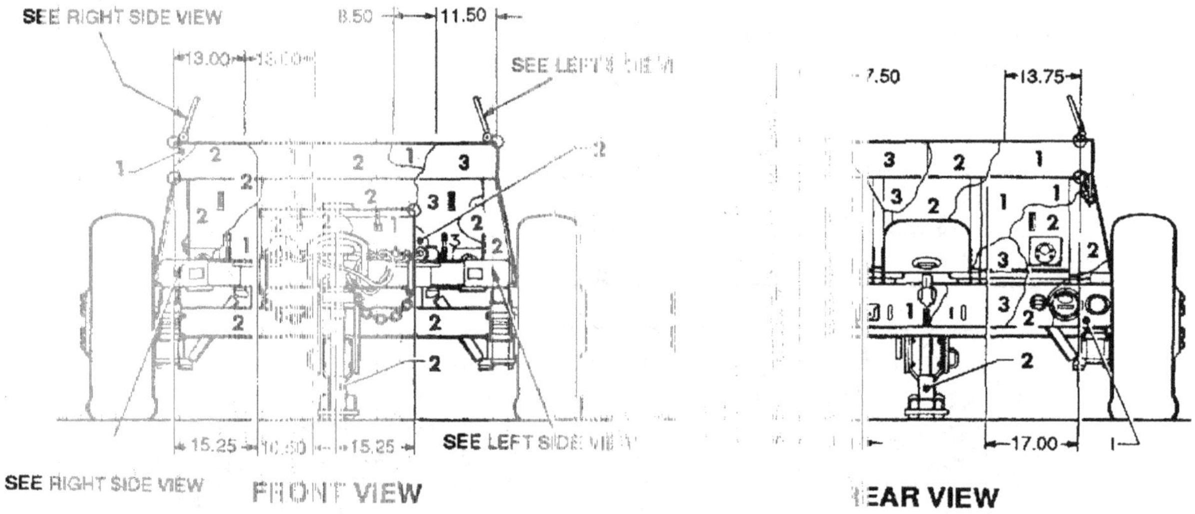

Figure 127. Trailer, ammunition: 1-1/2-ton (2 of 2)

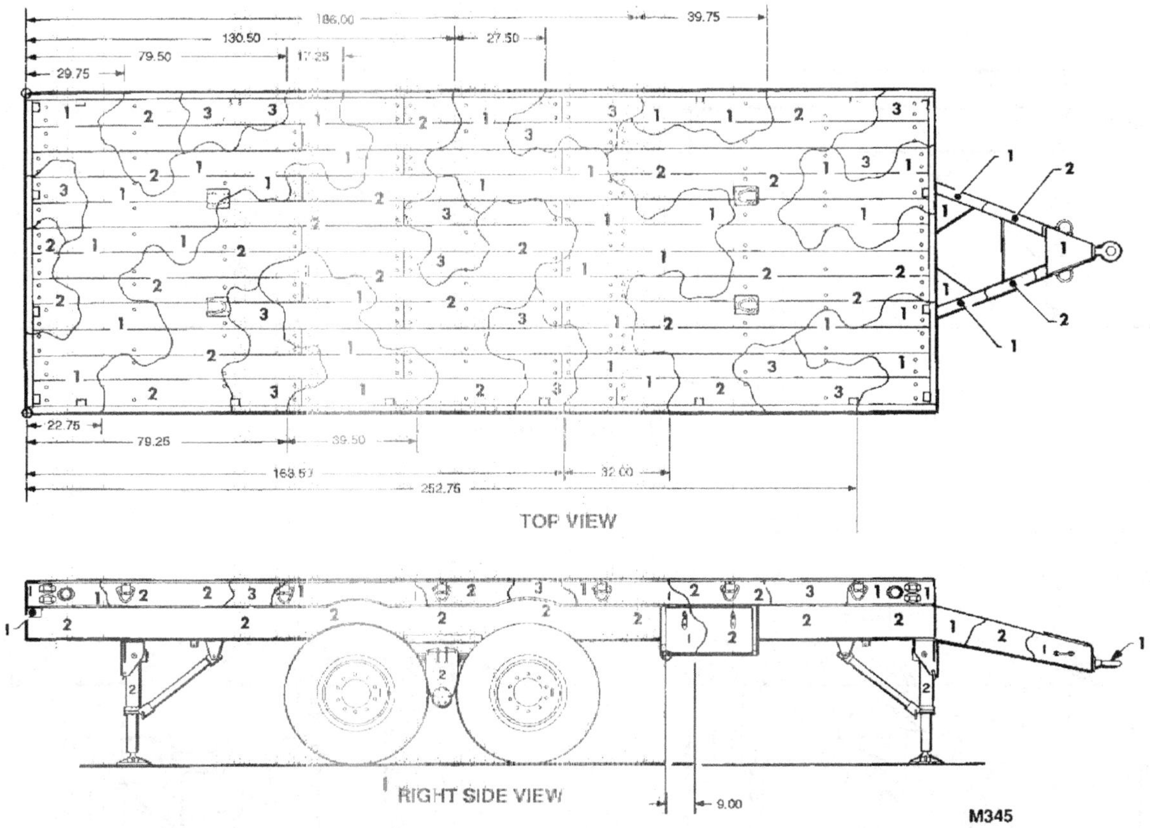

Figure 128. Trailer, flatbed: 10-ton, M345. (1 of 2)

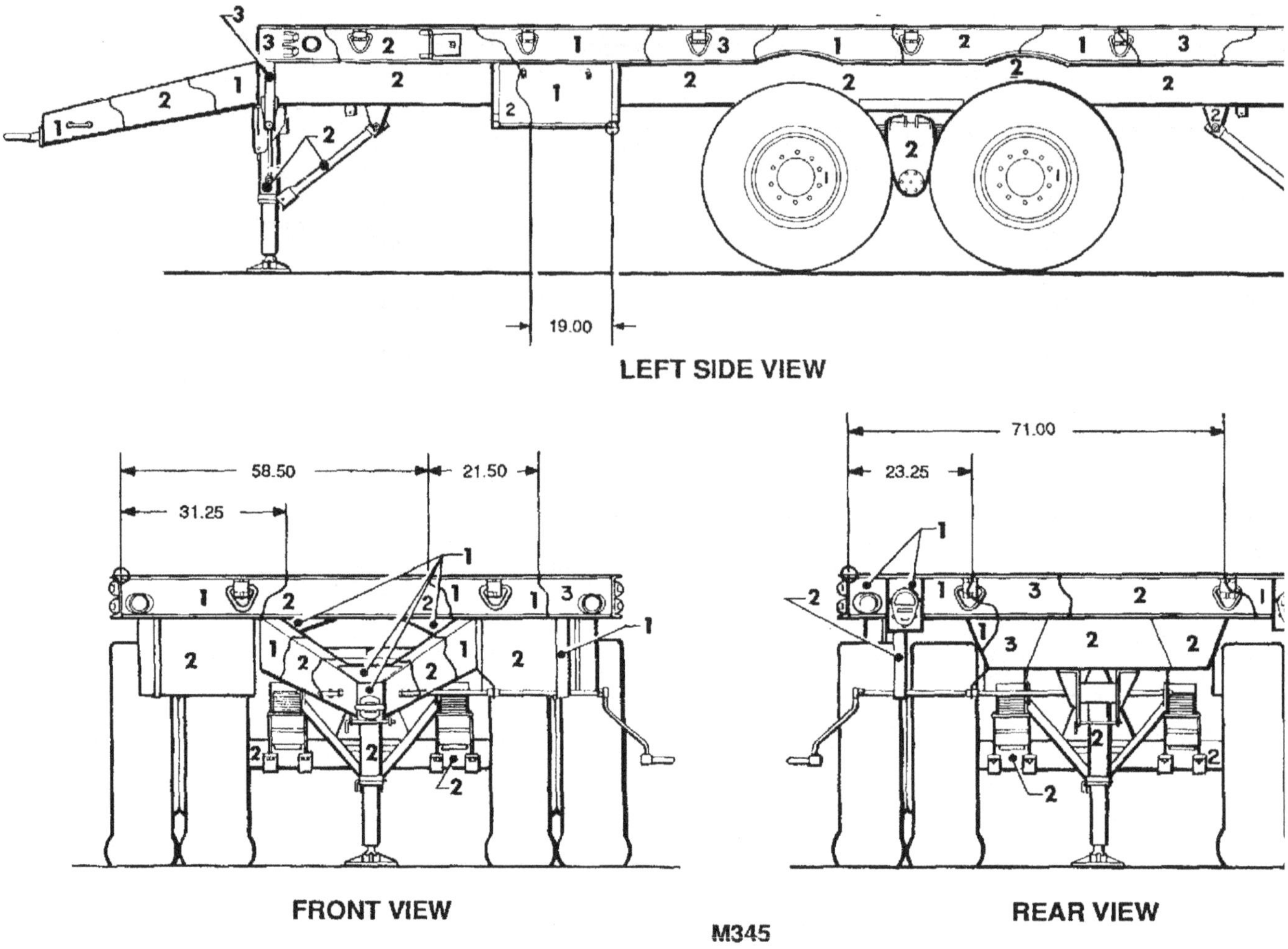

Figure 128. Trailer, flatbed: 10-ton, M345. (2 of 2)

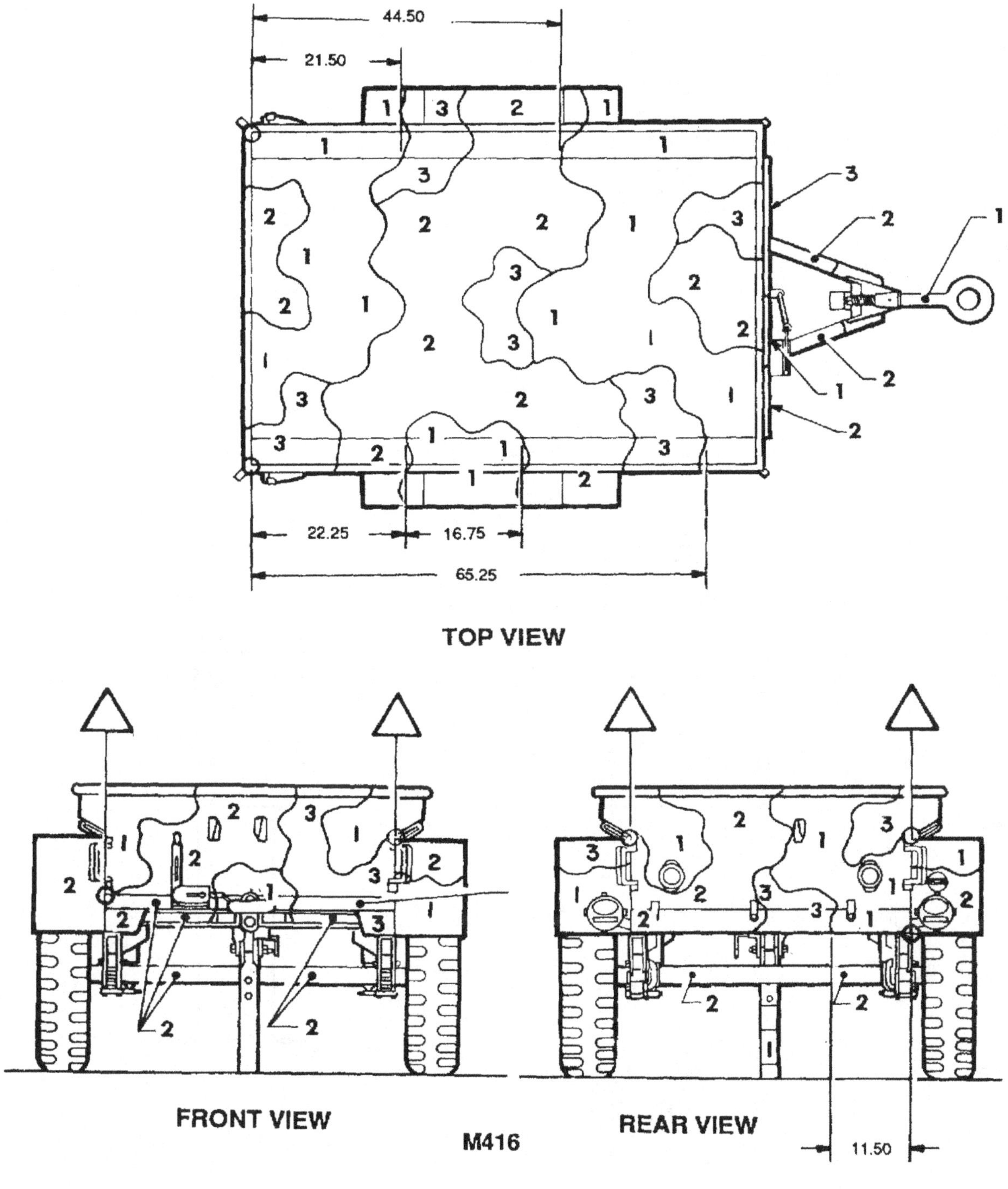

Figure 129. Trailer, cargo: 1/4-ton, 2-wheel, general purpose, M16. (1 of 2)

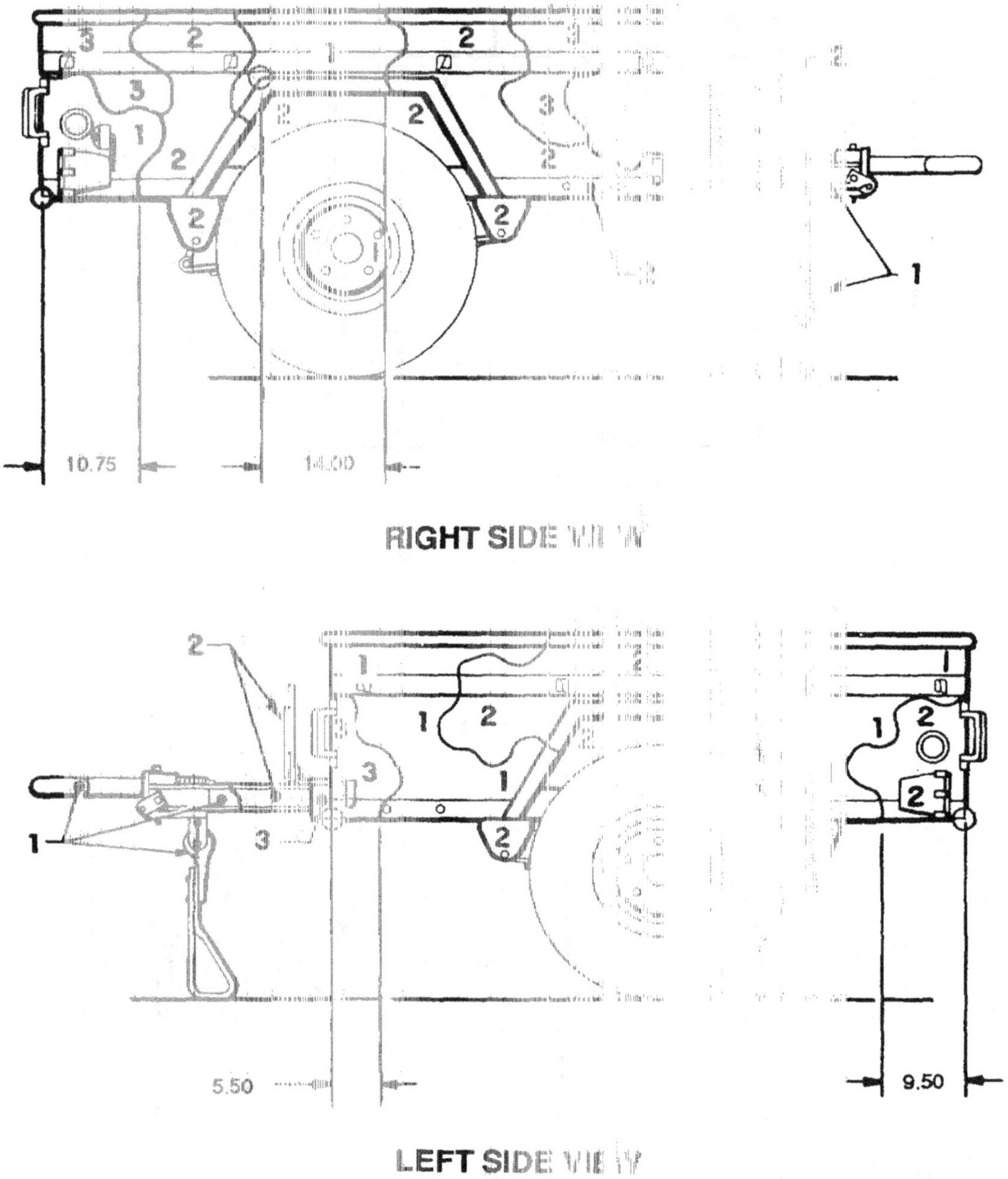

M416

Figure 129. Trailer, cargo: 1/4-ton, 2-wheel, general purpose, M416. (2 of 2)

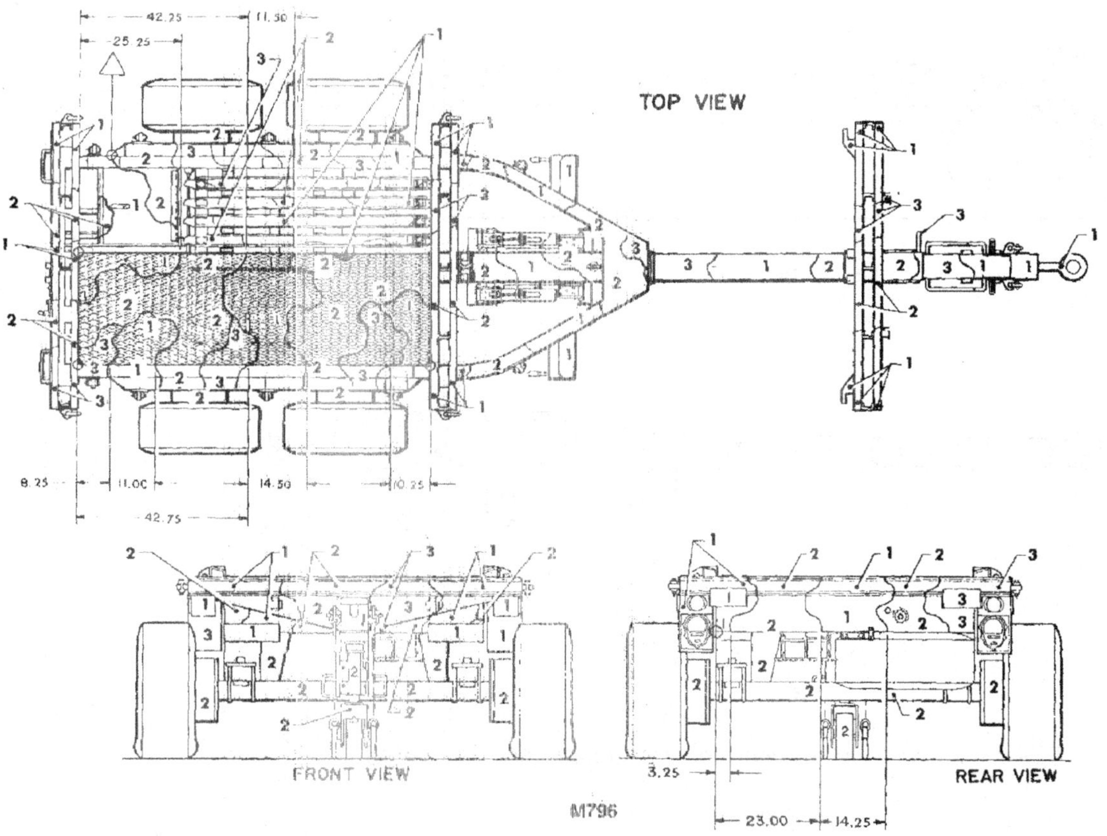

Figure 130. Trailer, bolster: 4-ton, 4-wheel, general purpose, M796. (1 of 2)

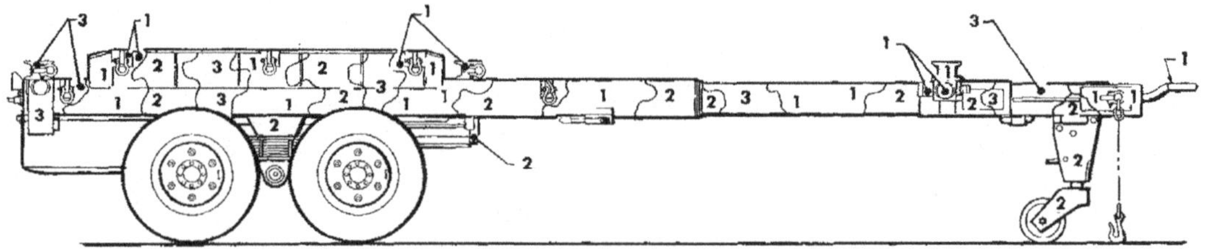

RIGHT SIDE VIEW

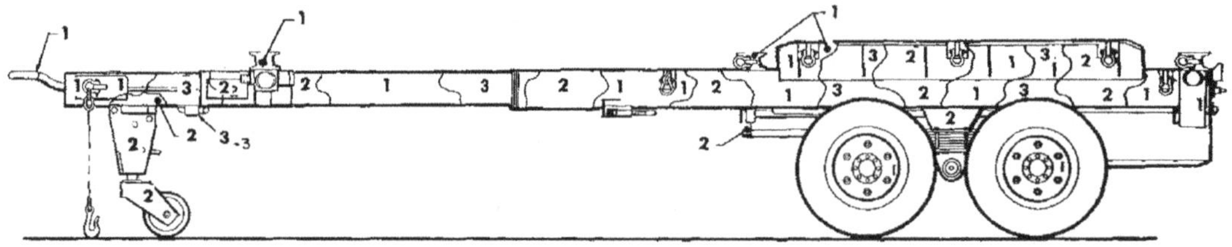

LEFT SIDE VIEW

M796

Figure 130. Trailer, bolster: 4-ton, 4-wheel, general purpose, M796. (2 of 2)

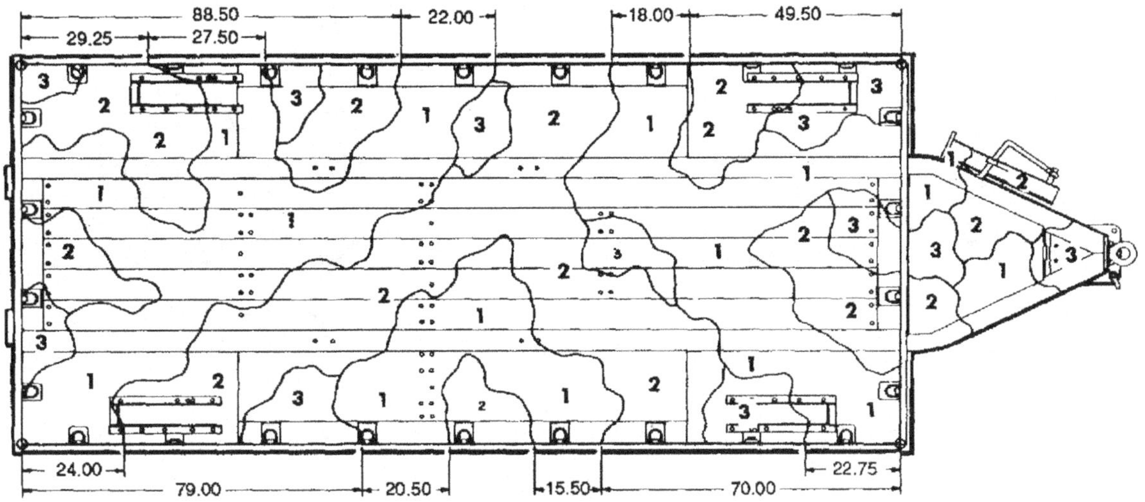

TOP VIEW

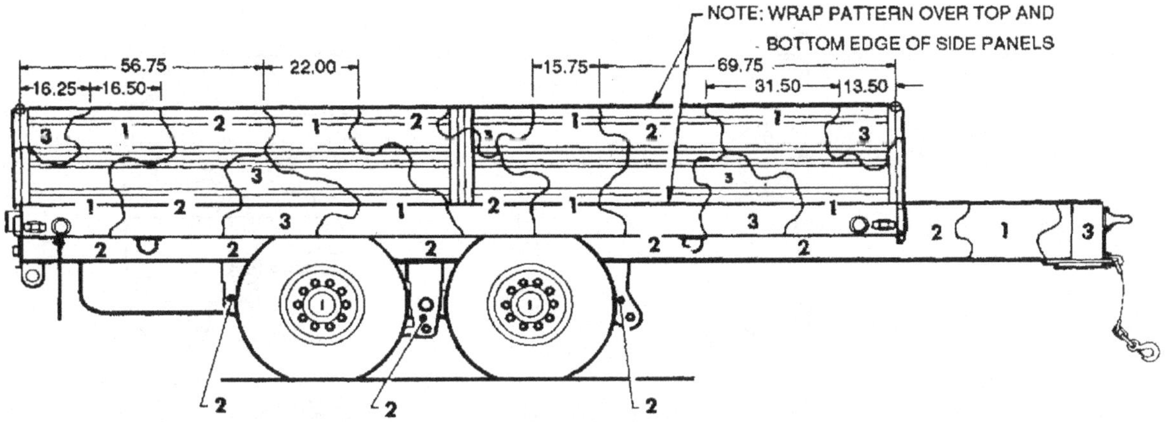

RIGHT SIDE VIEW

M989

Figure 131. Trailer, ammunition: 11-ton, heavy expanded mobility, M989. (1 of 2)

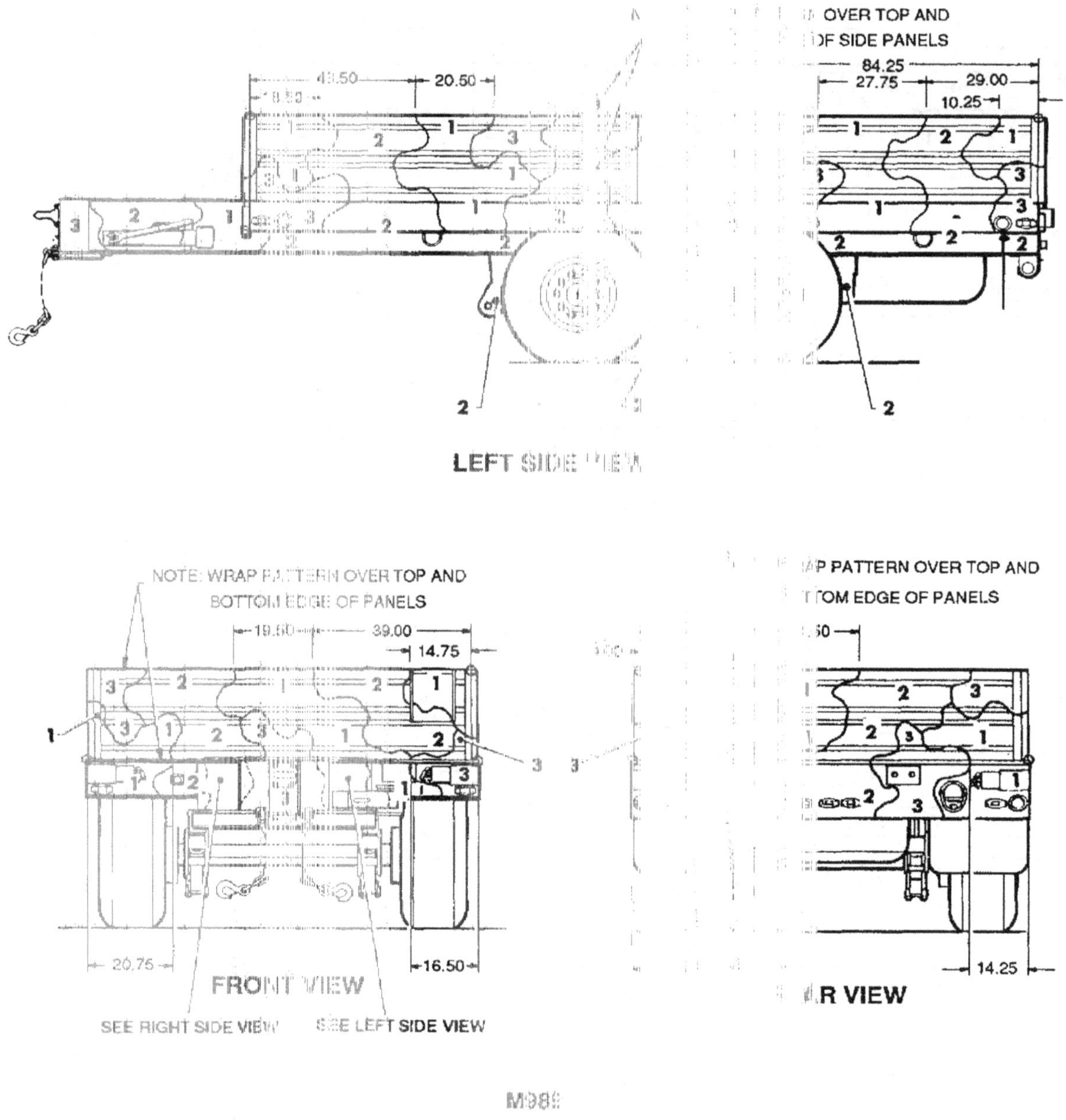

Figure 131. Trailer, ammunition: 11-ton, heavy ... ity, M989. (2 of 2)

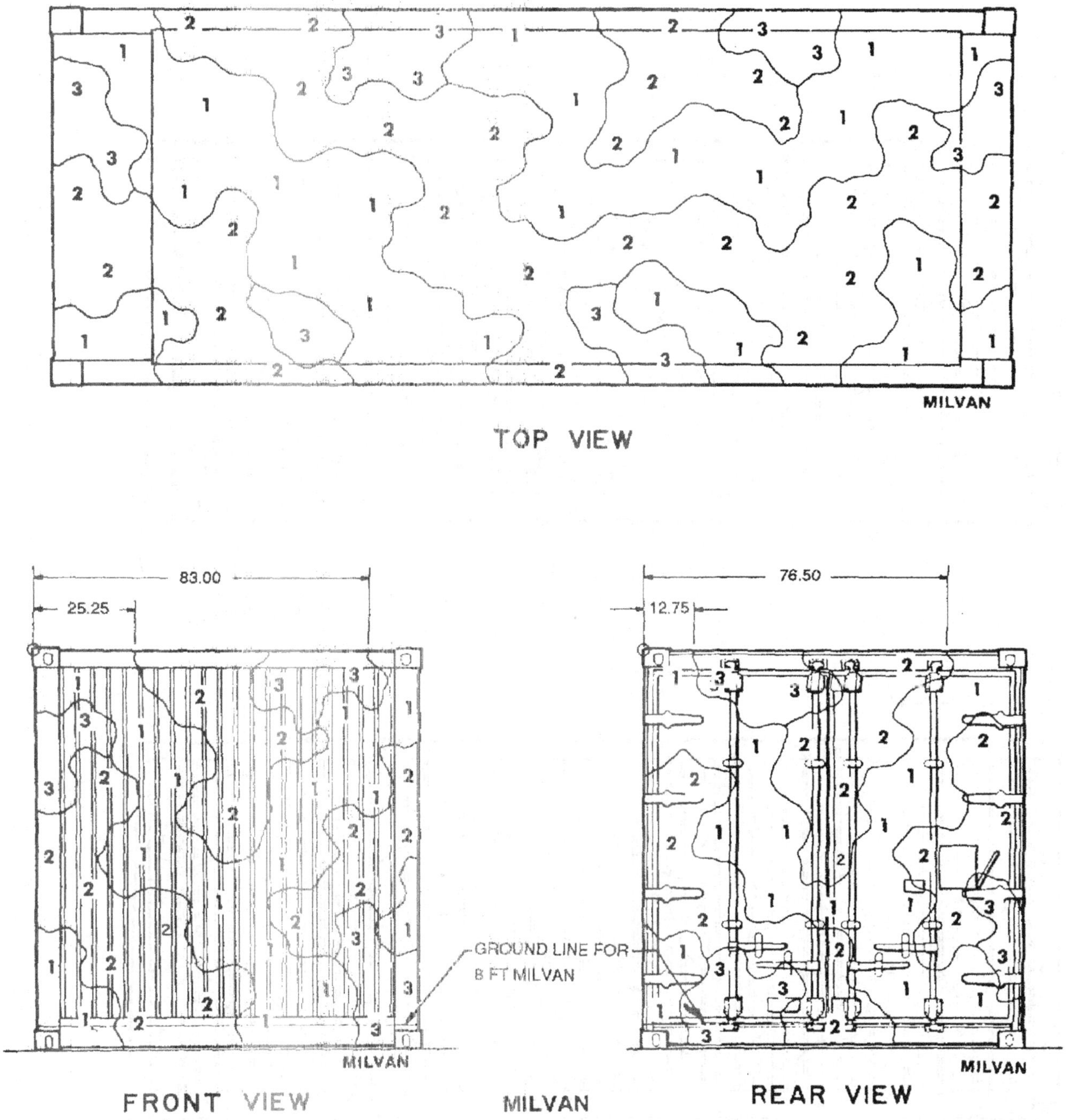

Figure 132. Container, cargo: general, MILVAN. (1 of 2)

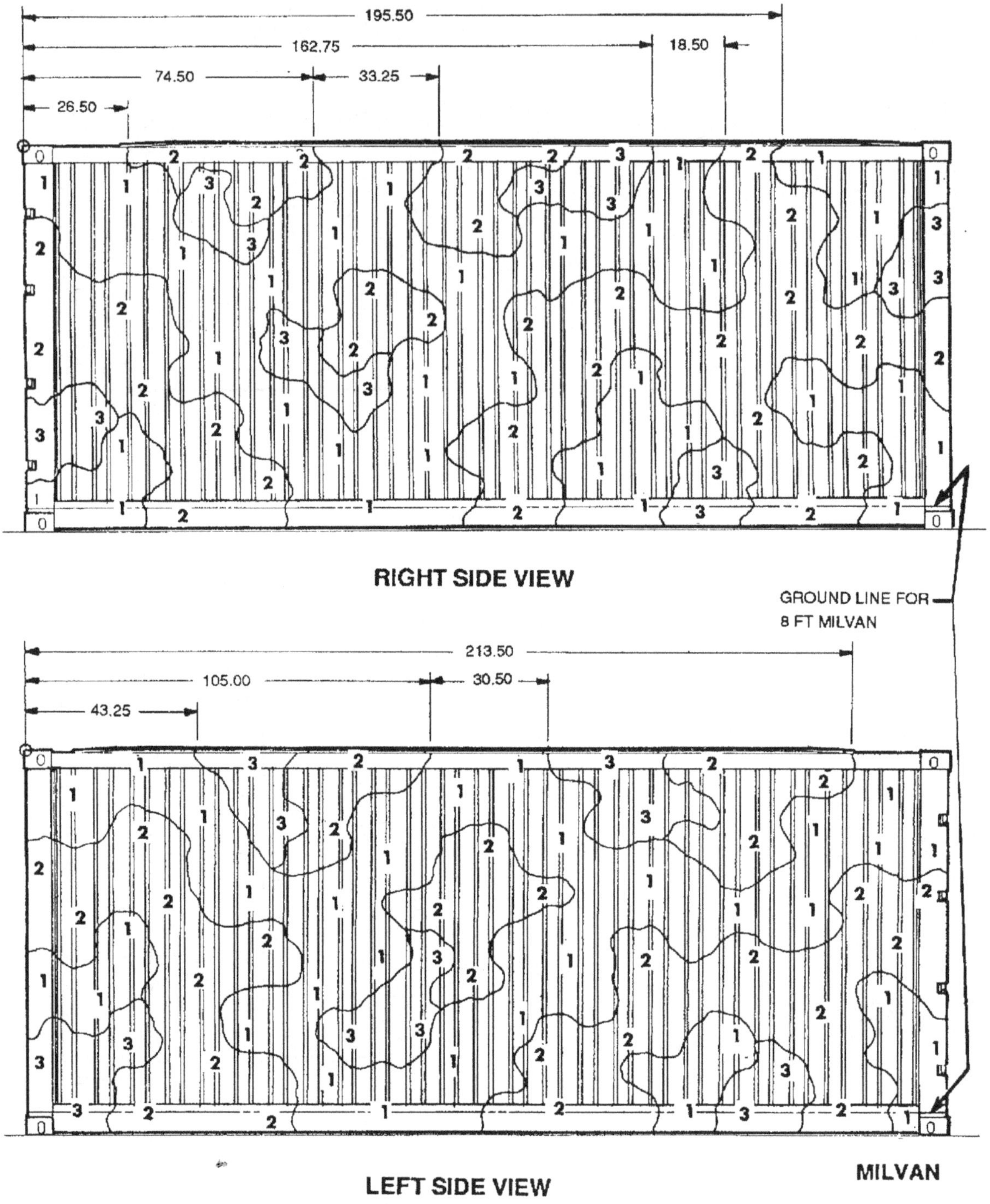

Figure 132. Container, cargo: general, MILVAN. (2 of 2)

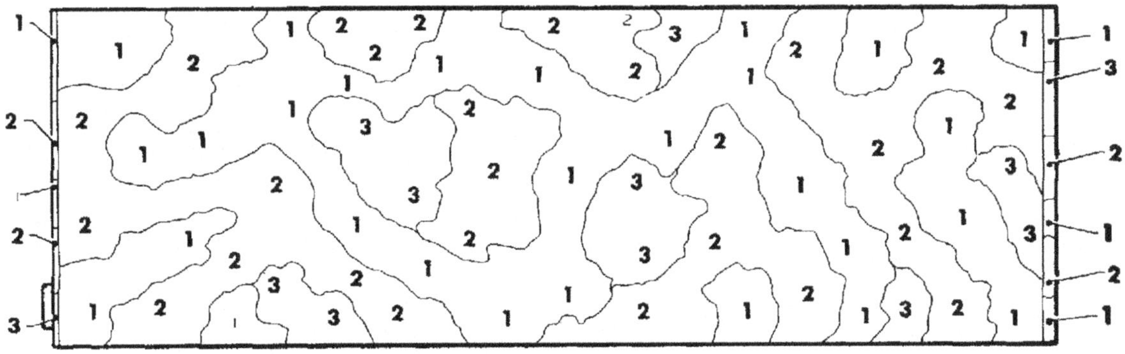

TOP VIEW

M119A1

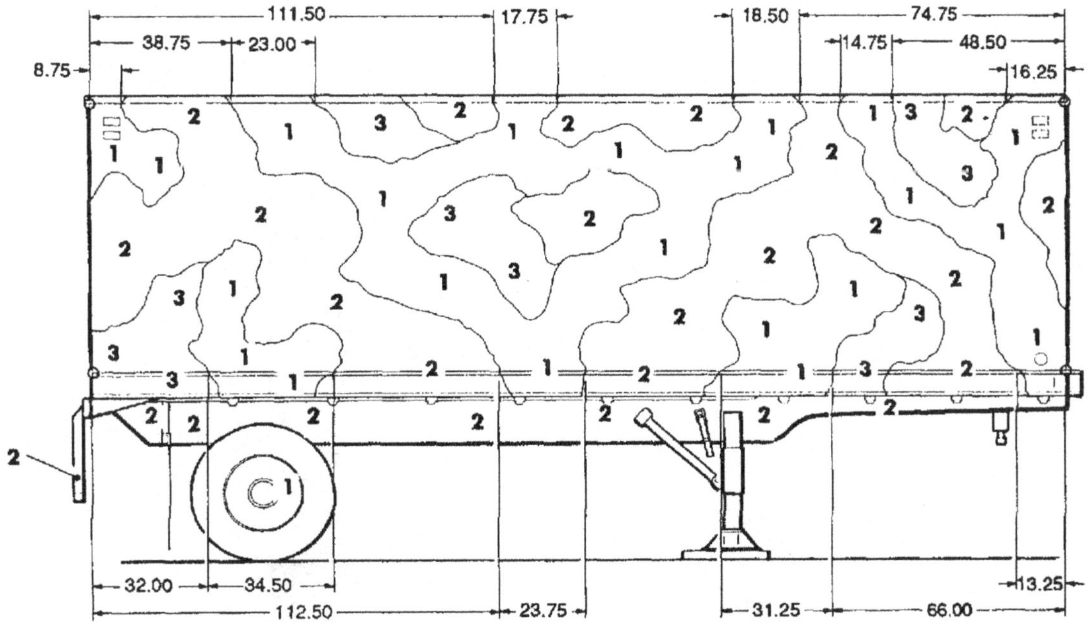

RIGHT SIDE VIEW

Figure 133. Semitrailer, van: cargo, 6-ton, 2-wheel, M119A1. (1 of 2)

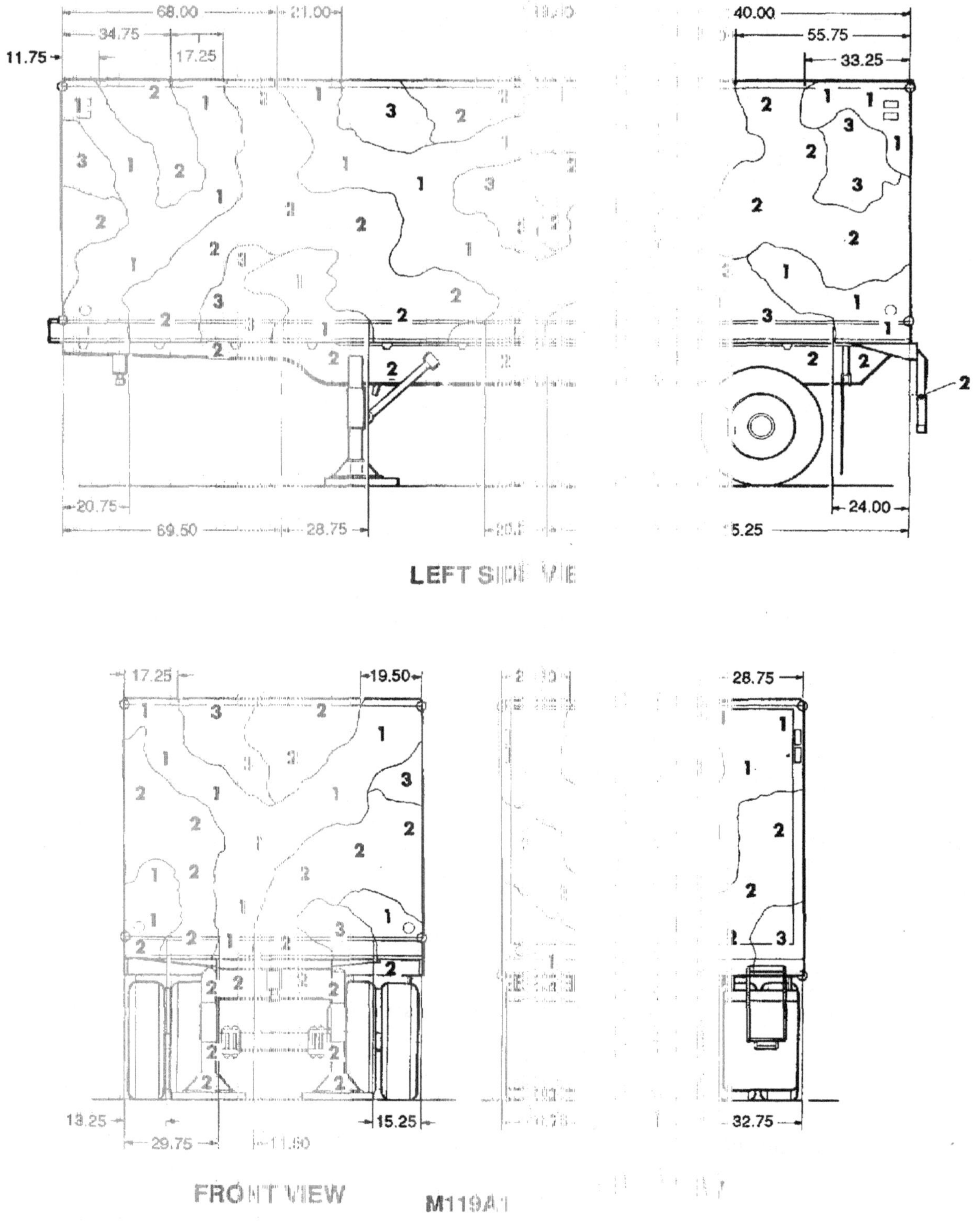

Figure 133. Semitrailer, van: cargo, 6-ton, 2[illegible]1. (2 of 2)

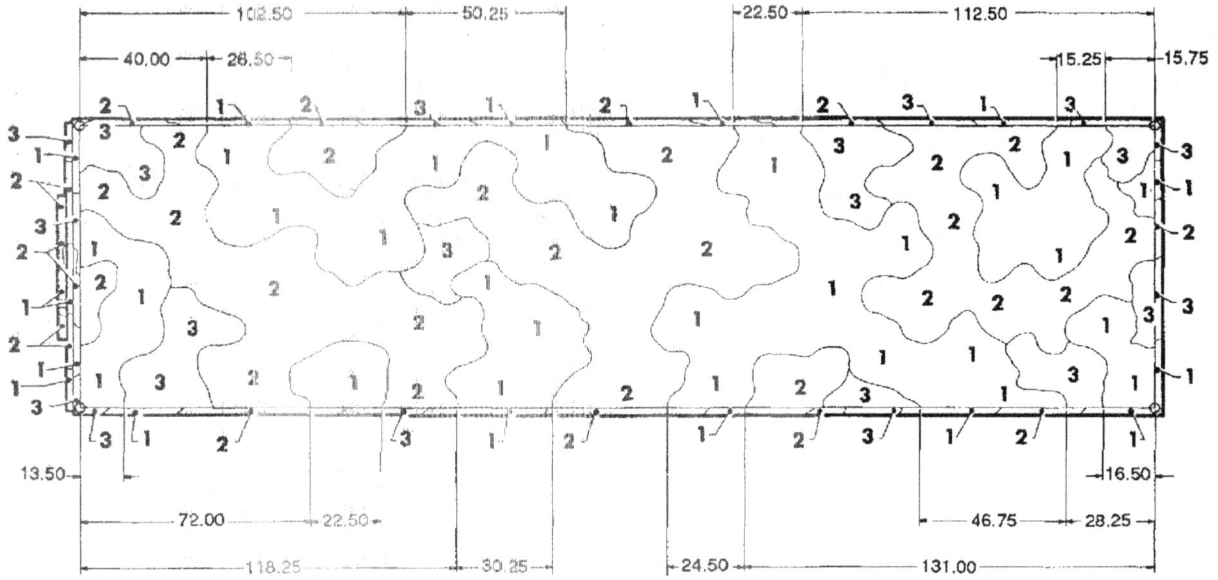

TOP VIEW

M127A2C

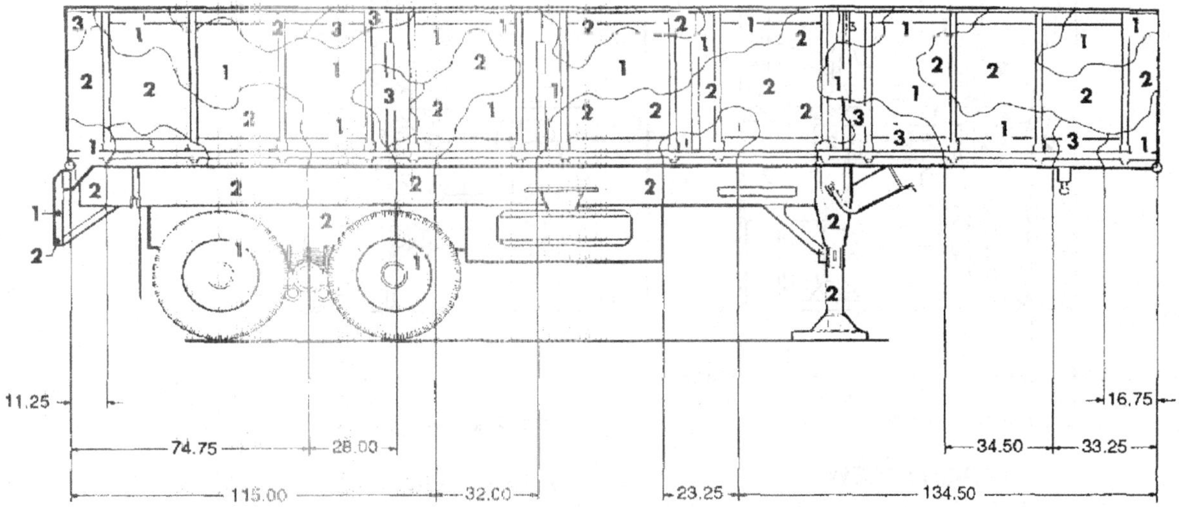

RIGHT SIDE VIEW

Figure 134. Semitrailer,, van: cargo, 12-ton, 4-wheel, M127A2C. (1 of 2)

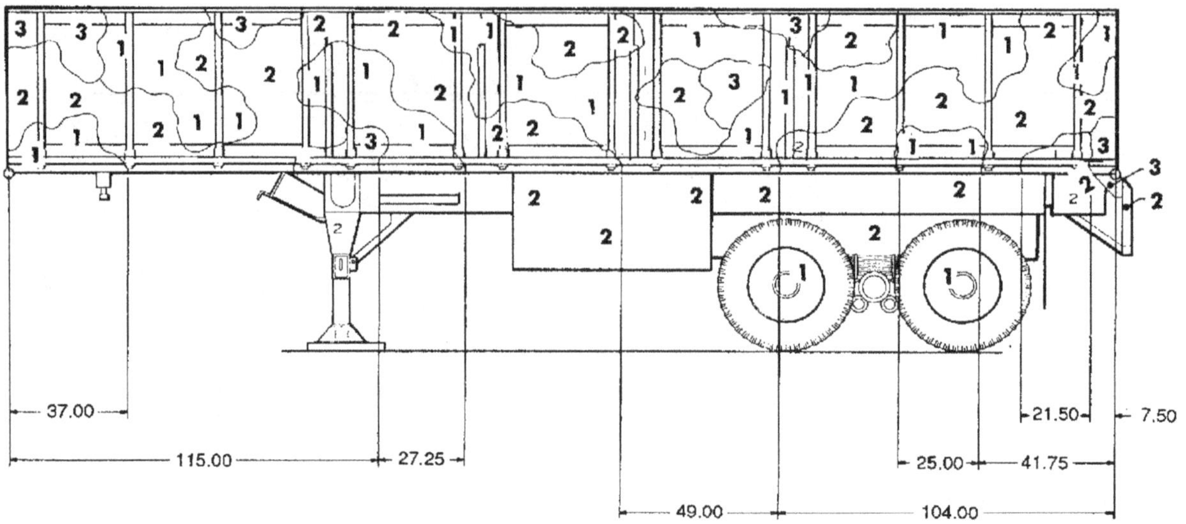

LEFT SIDE VIEW

M127A2C

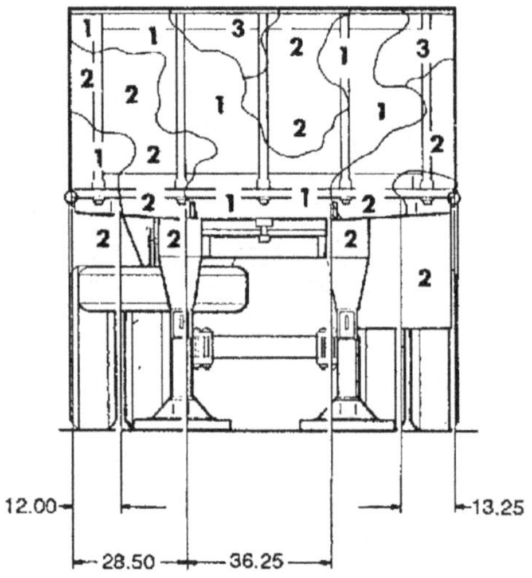

FRONT VIEW

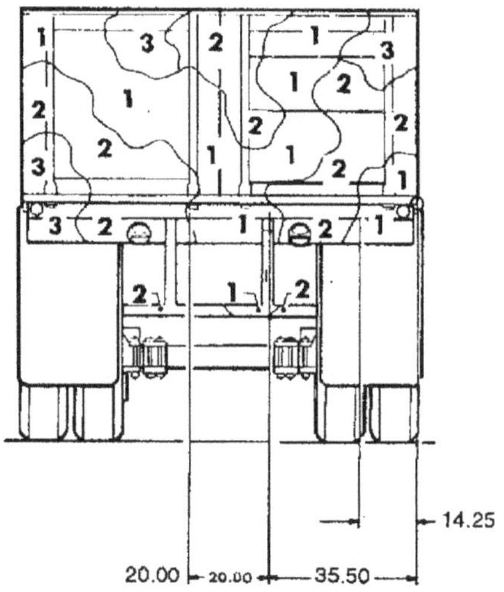

REAR VIEW

Figure 134. Semitrailer, van: cargo, 12-ton, 4-wheel, M127A2C. (2 of 2)

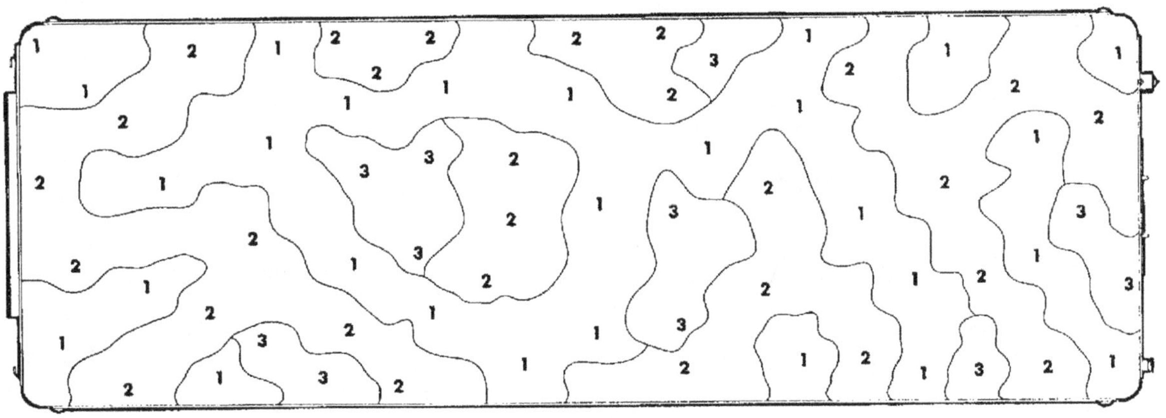

TOP VIEW

M146

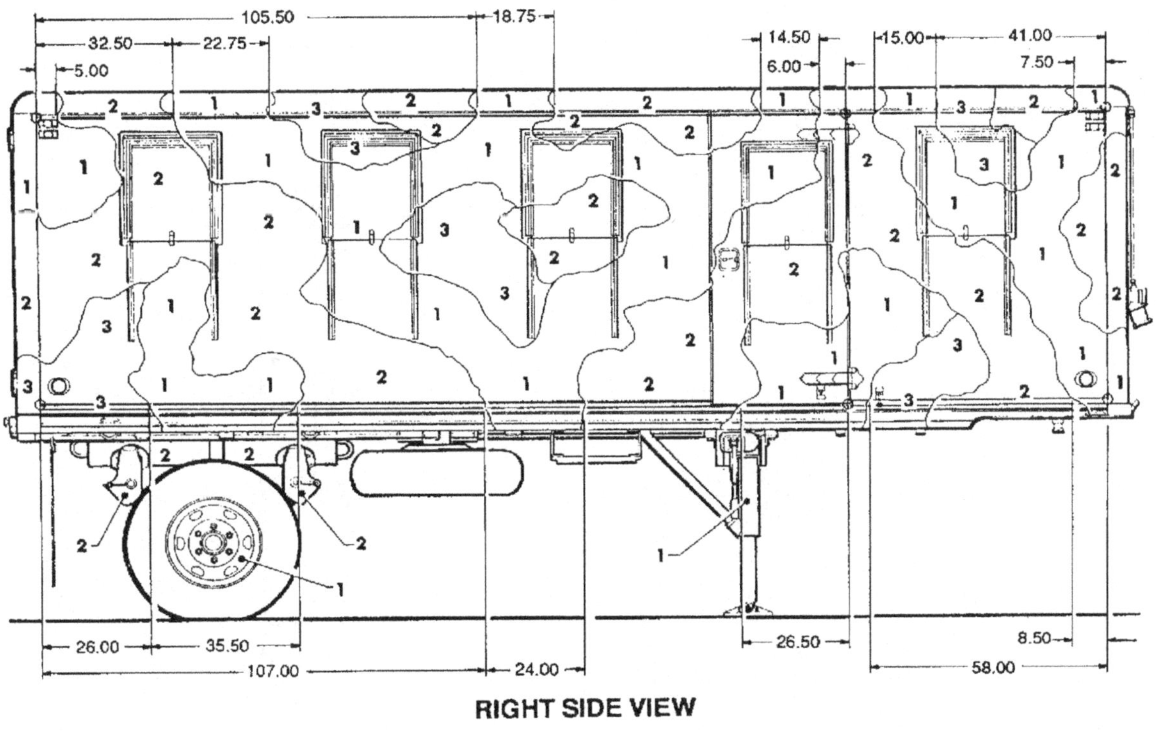

RIGHT SIDE VIEW

Figure 135. Semitrailer, van: shop, 6-ton, 2-wheel, M146. (1 of 2)

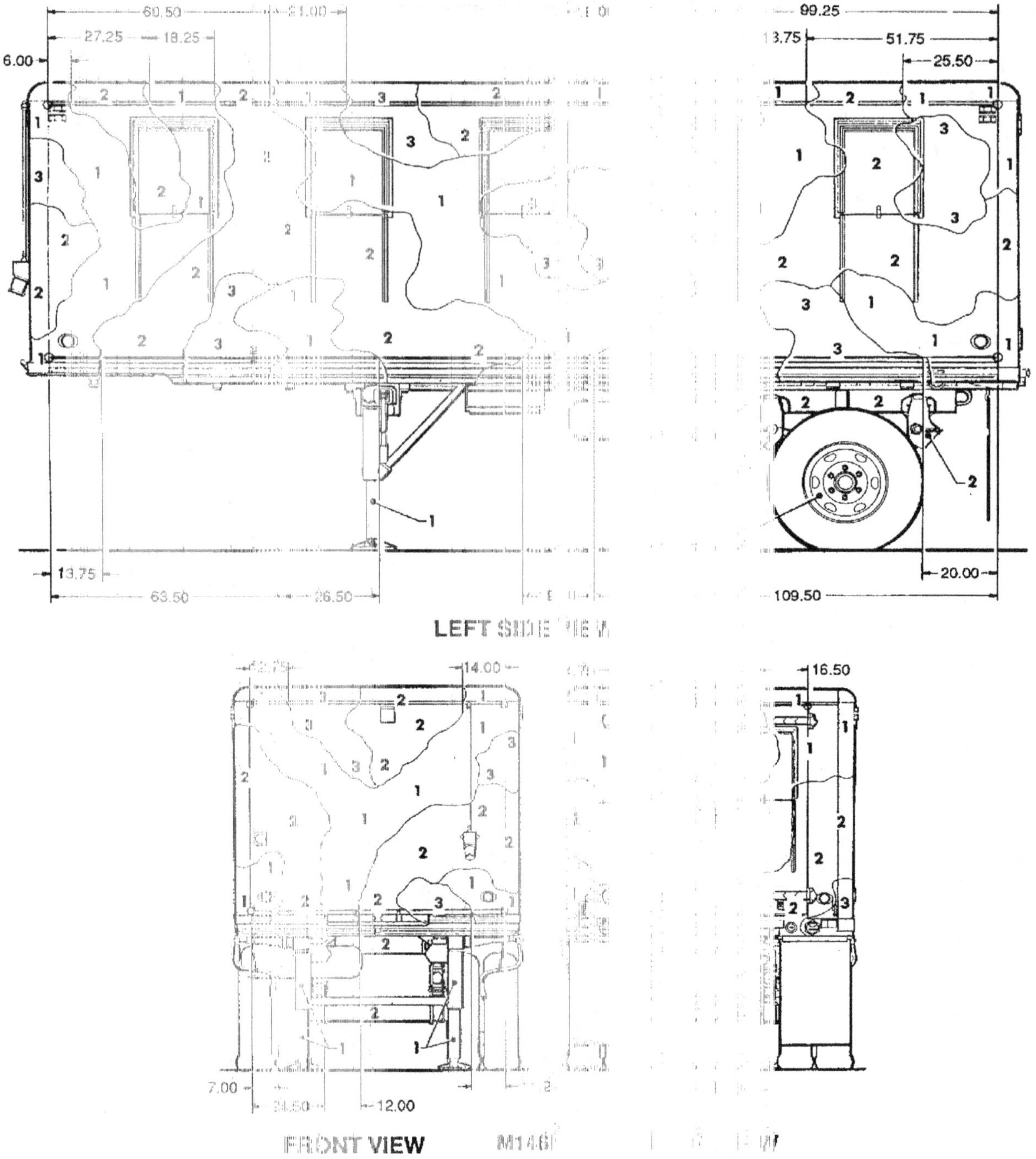

Figure 135. Semitrailer, van: shop, [illegible] (2 of 2)

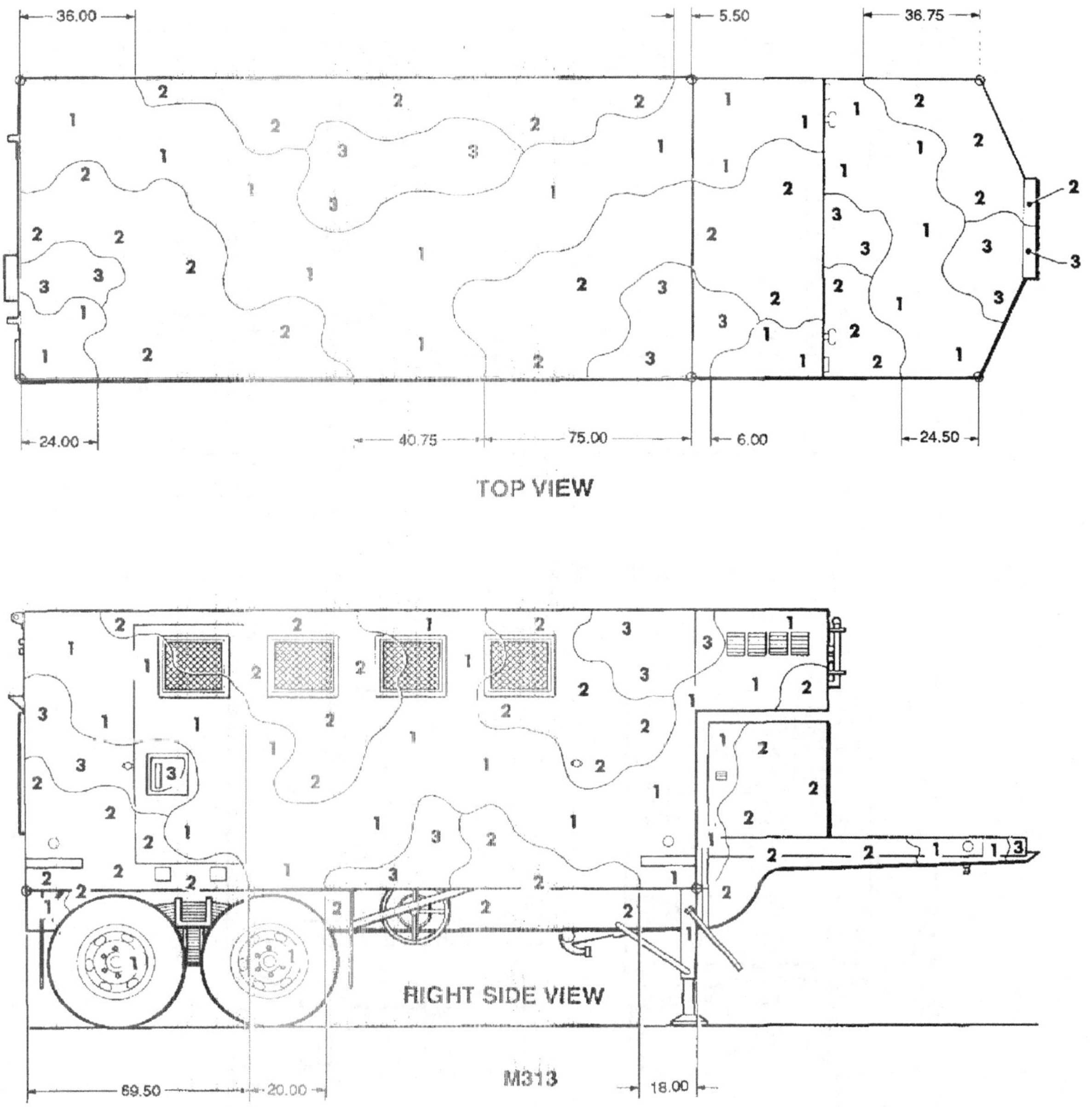

Figure 136. Semitrailer, van: expansible, 6-ton, 4-wheel, M313. (1 of 2)

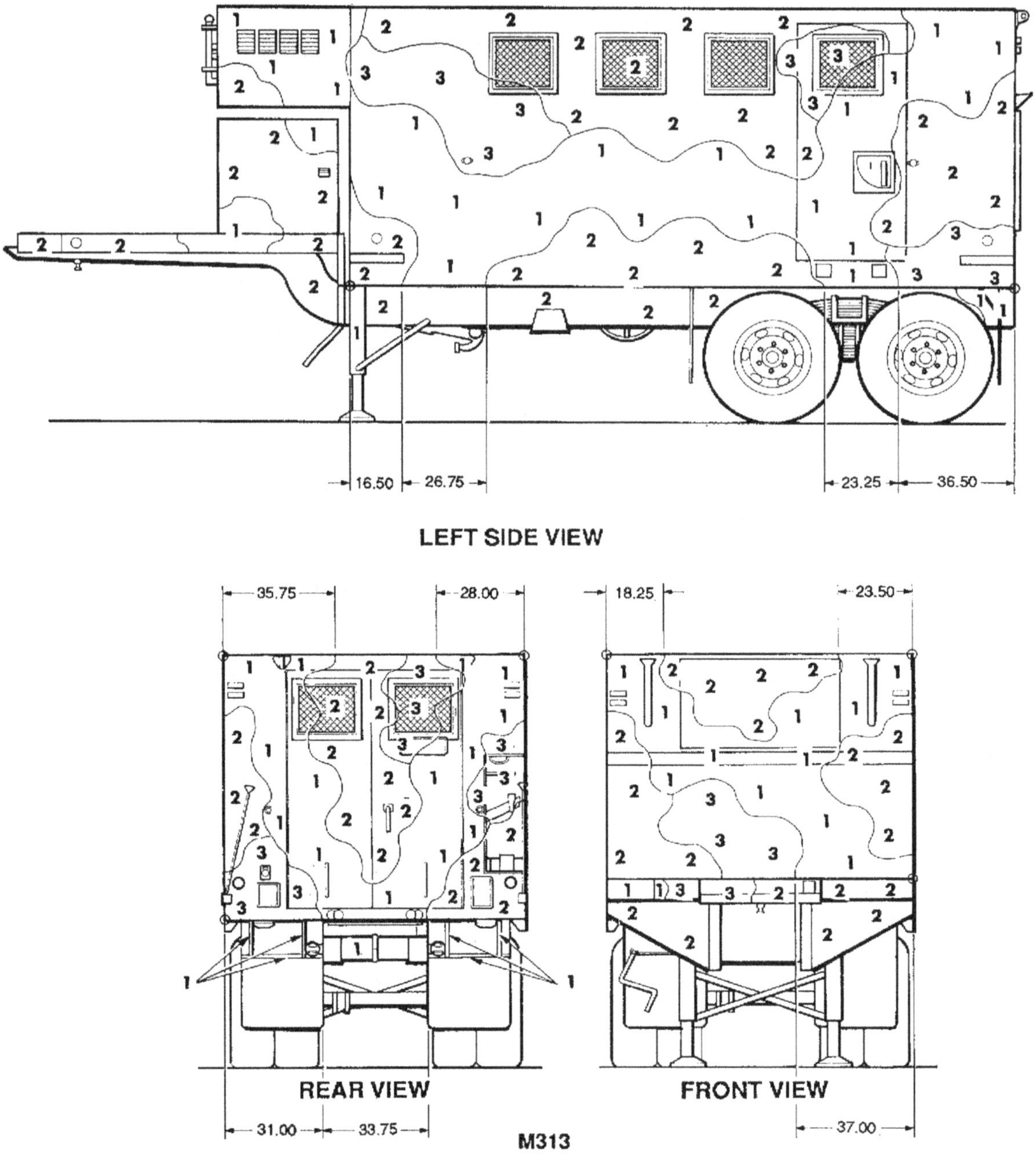

Figure 136. Semitrailer, van: expansible, 6-ton, 4-wheel, M313. (2 of 2)

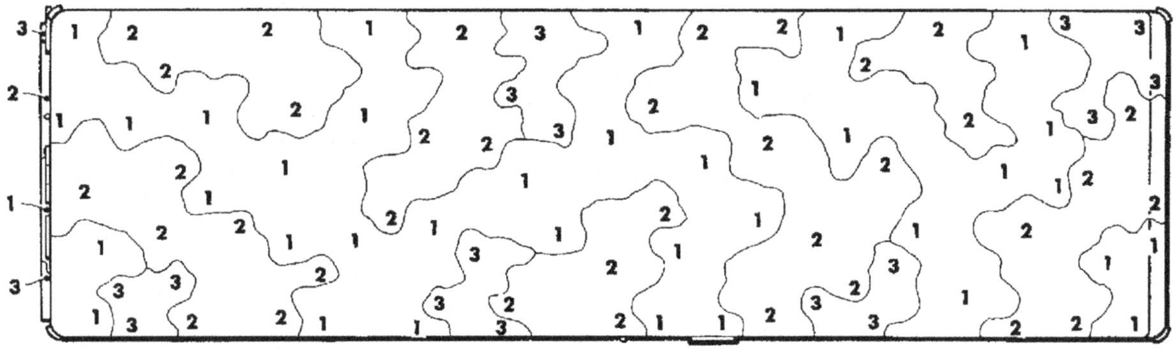

TOP VIEW

M348A2

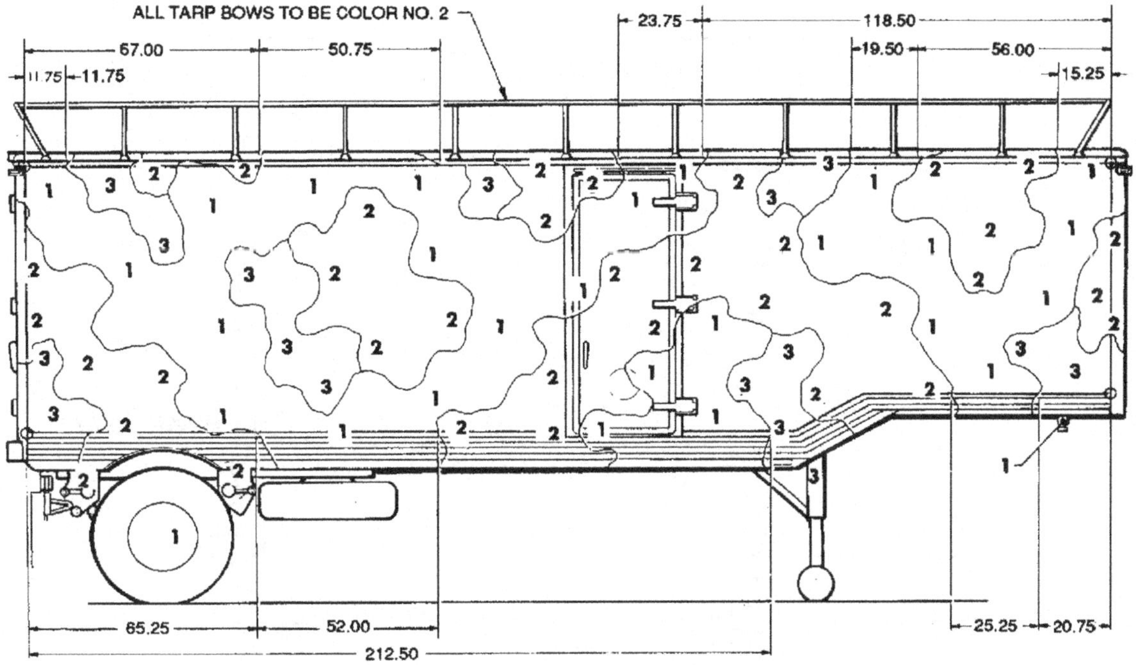

RIGHT SIDE VIEW

Figure 137. Semitrailer, van: composite, 6-ton, 2-wheel, M348A2. (1 of 2)

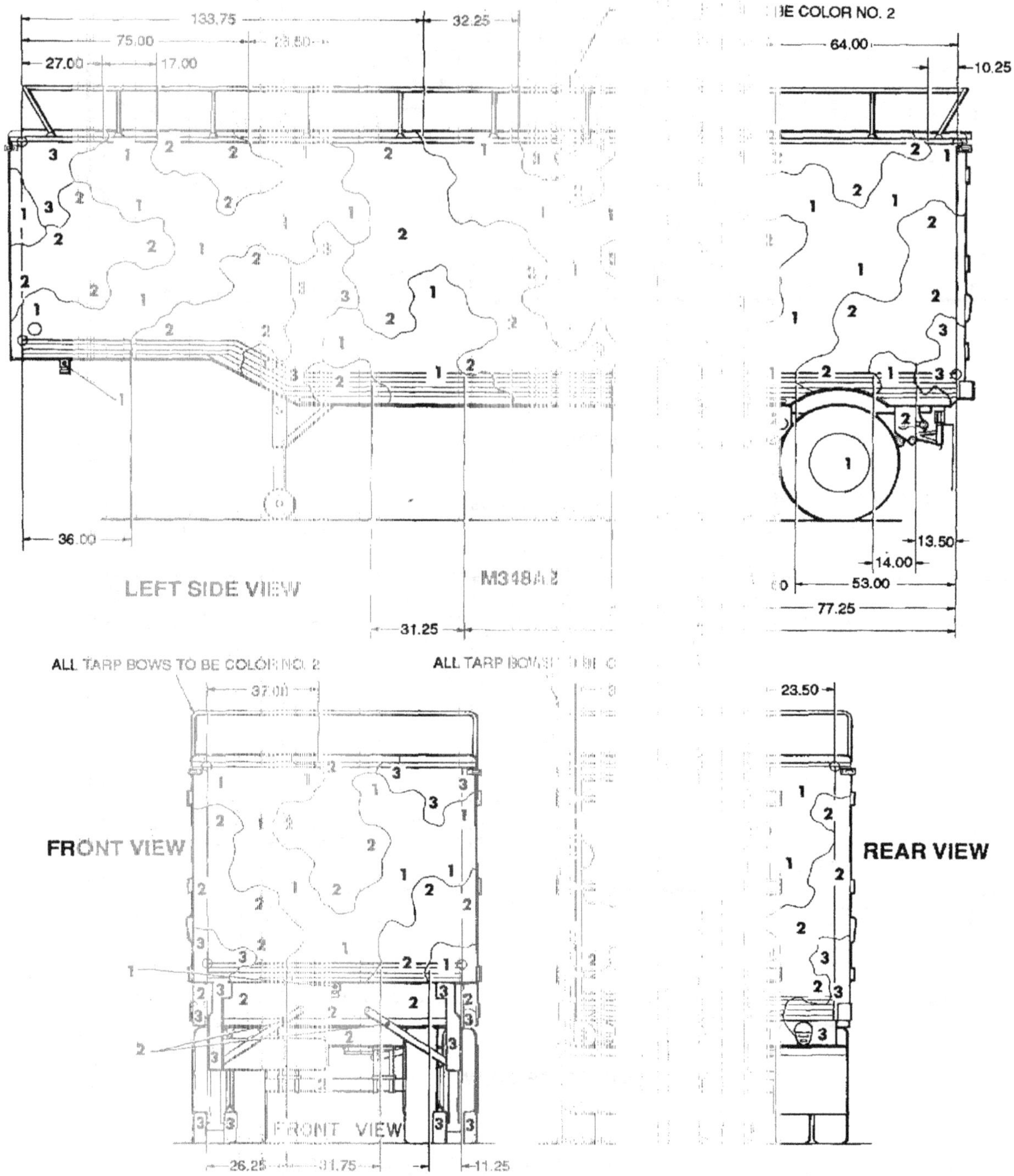

Figure 137. Semitrailer, van: composite, 5 ton, ... 48A2. (2 of 2)

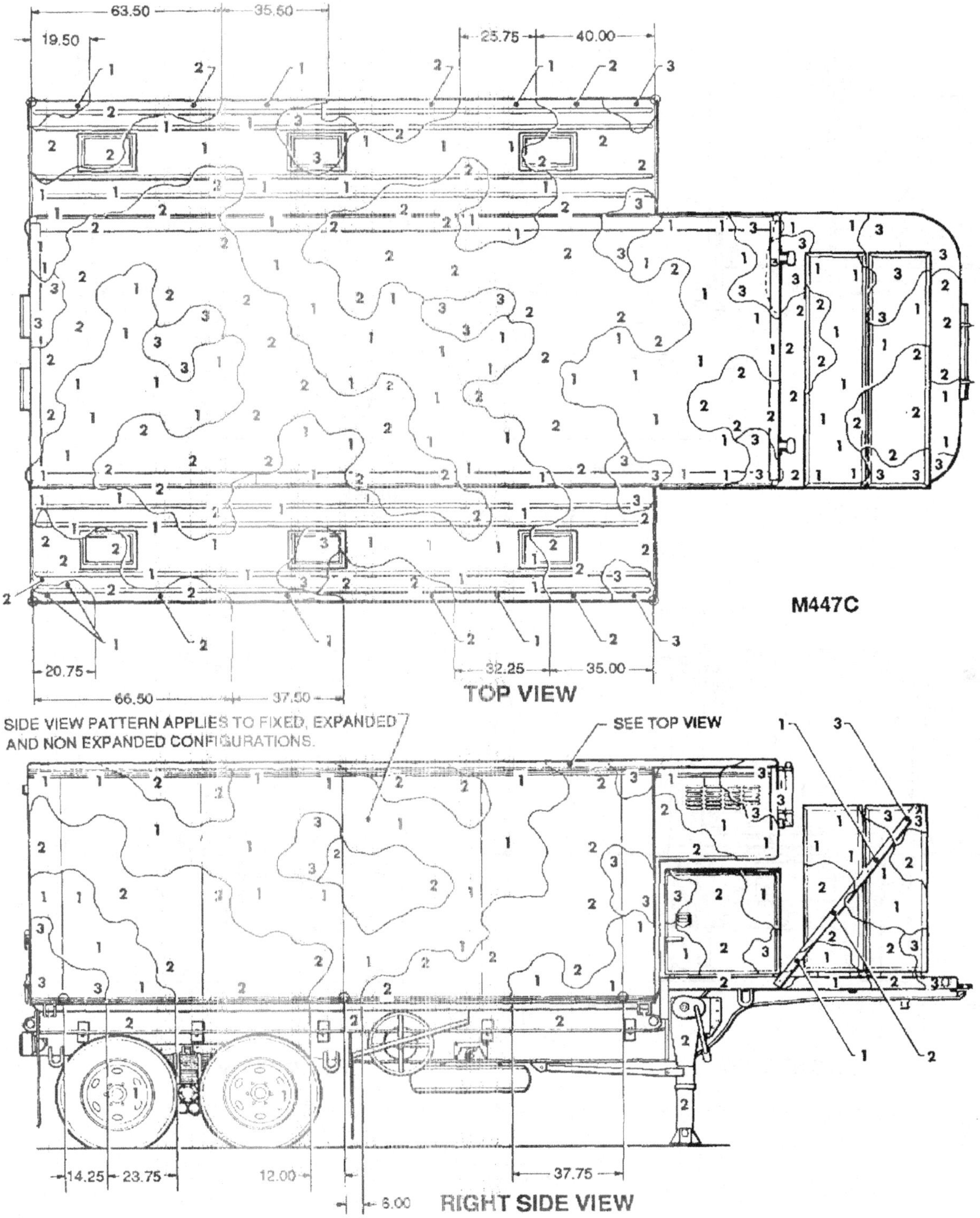

Figure 138. Semitrailer, van: expansible, 6-ton, 4-wheel, M447, M447C, M749. (1 of 2)

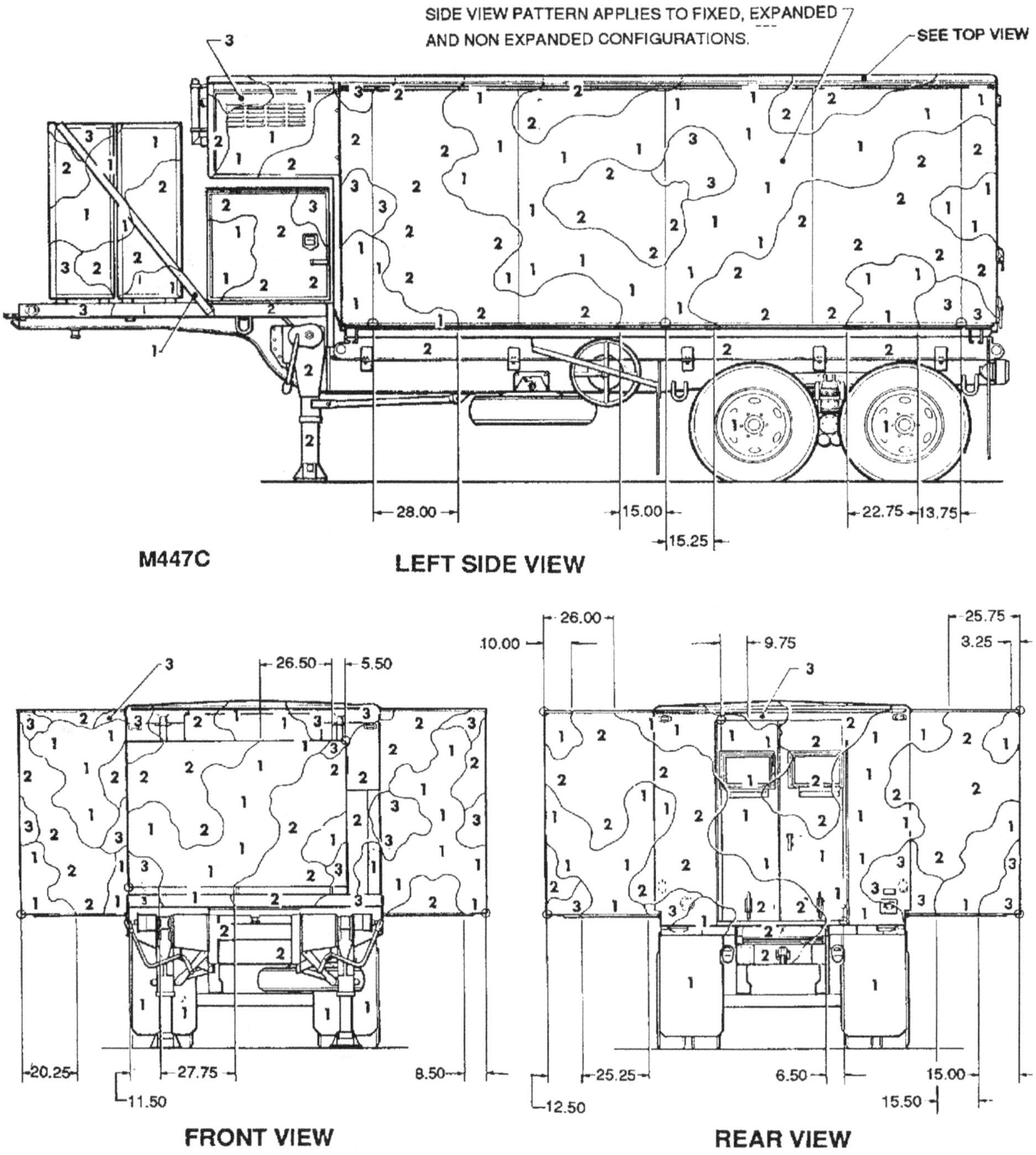

Figure 138. Semitrailer, van: expansible, 6-ton, 4-wheel, M447, M447C, M749. (2 of2)

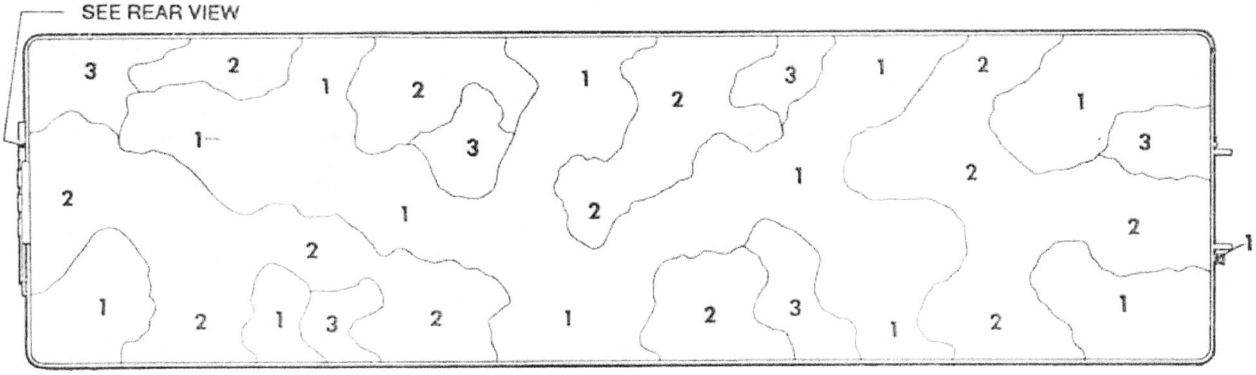

TOP VIEW

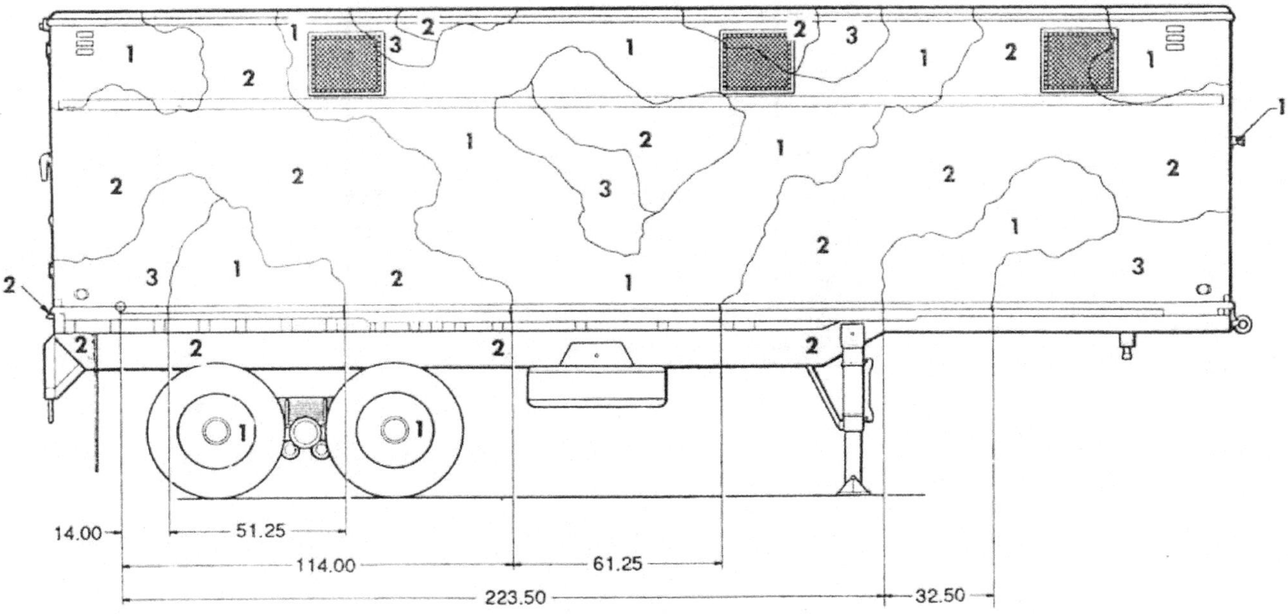

RIGHT SIDE VIEW

M129A2

Figure 139. Semitrailer, van: supply, 12-ton, 4-wheel, M129A2C. (1 of 2)

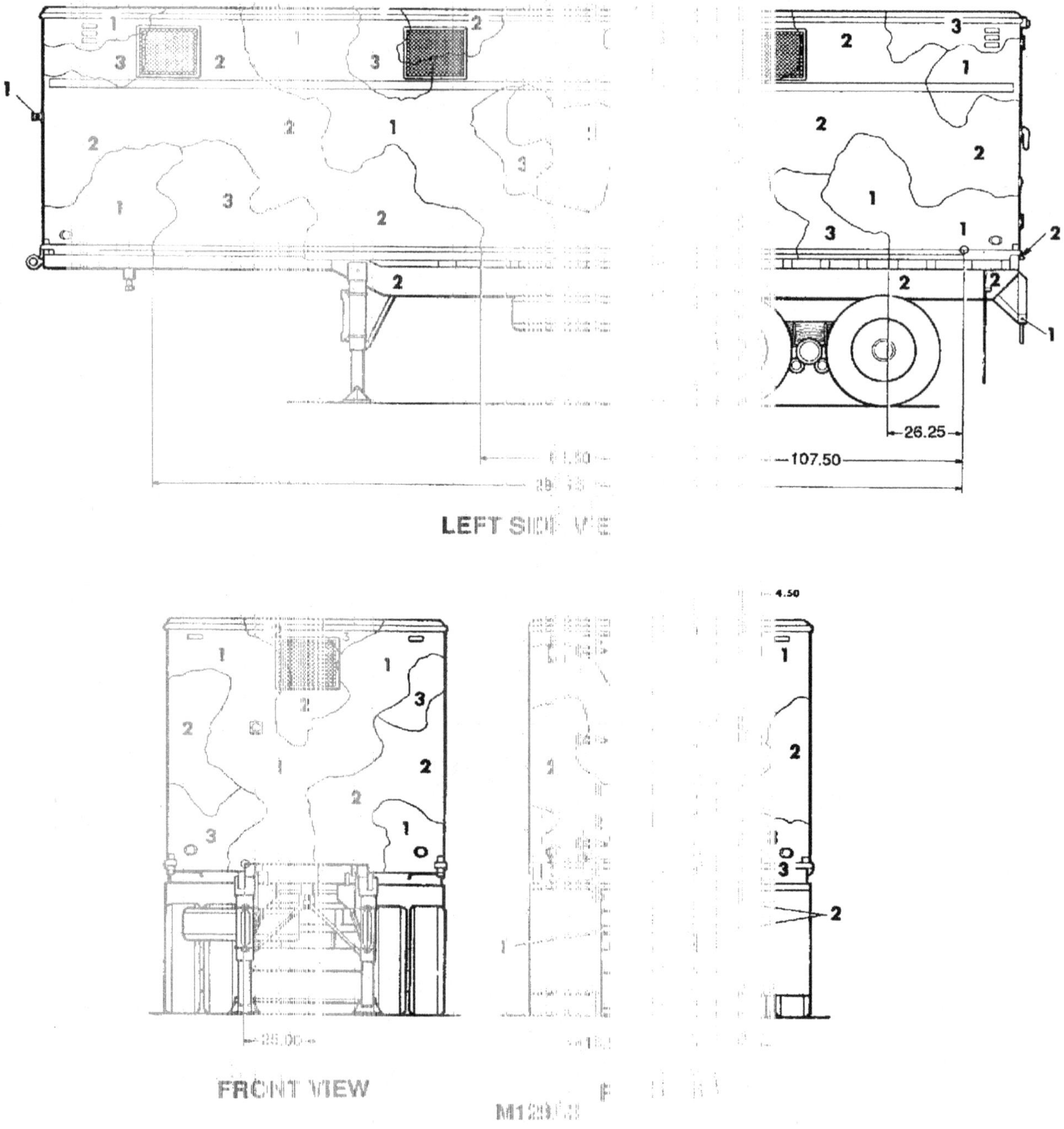

Figure 139. Semitrailer, van: supply, 12 ton, 4 [illegible]. (2 of2)

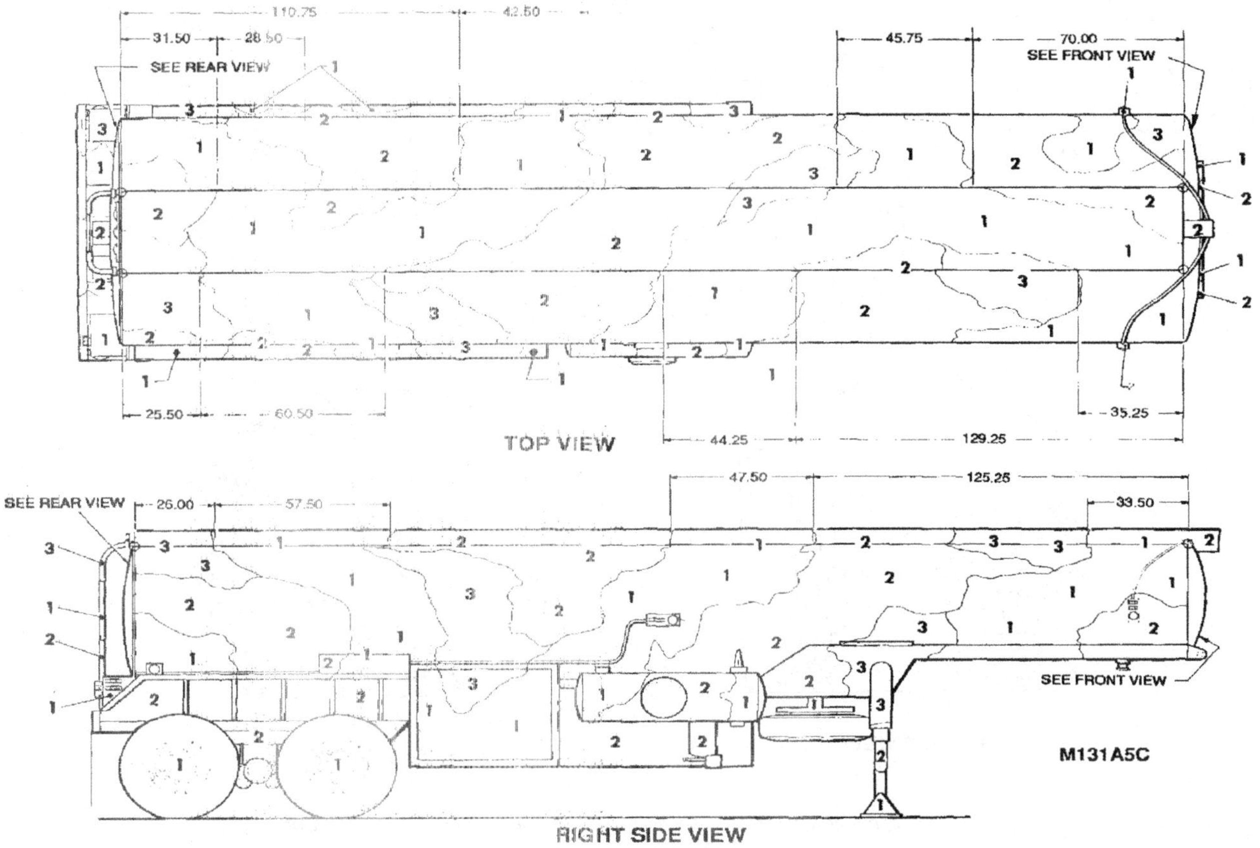

Figure 140. Semitrailer, tanker: fuel servicing, 5000 gallon, 4-wheel, M131ASC. (I of 2)

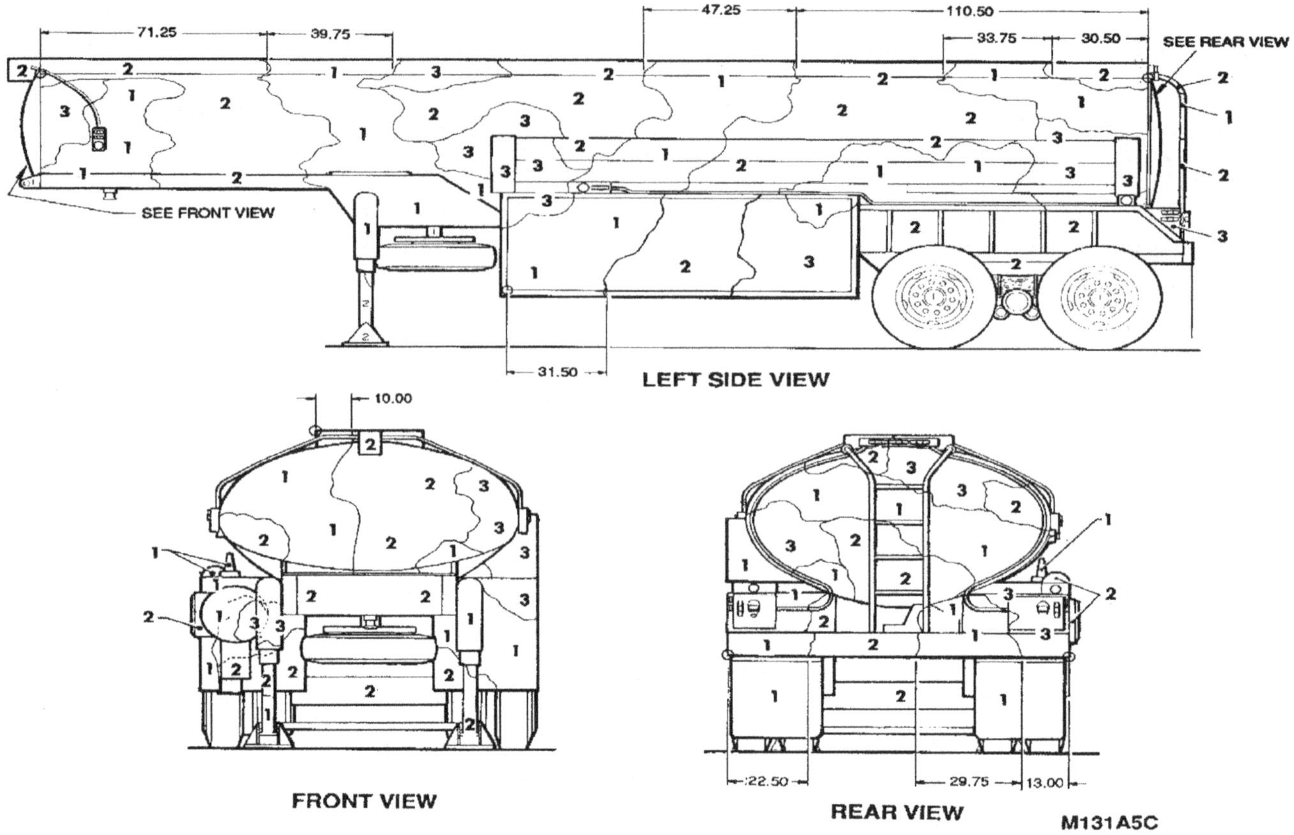

Figure 140. Semitrailer, tanker: fuel servicing, 5000 gallon, 4-wheel, M131A5C. (2 of2)

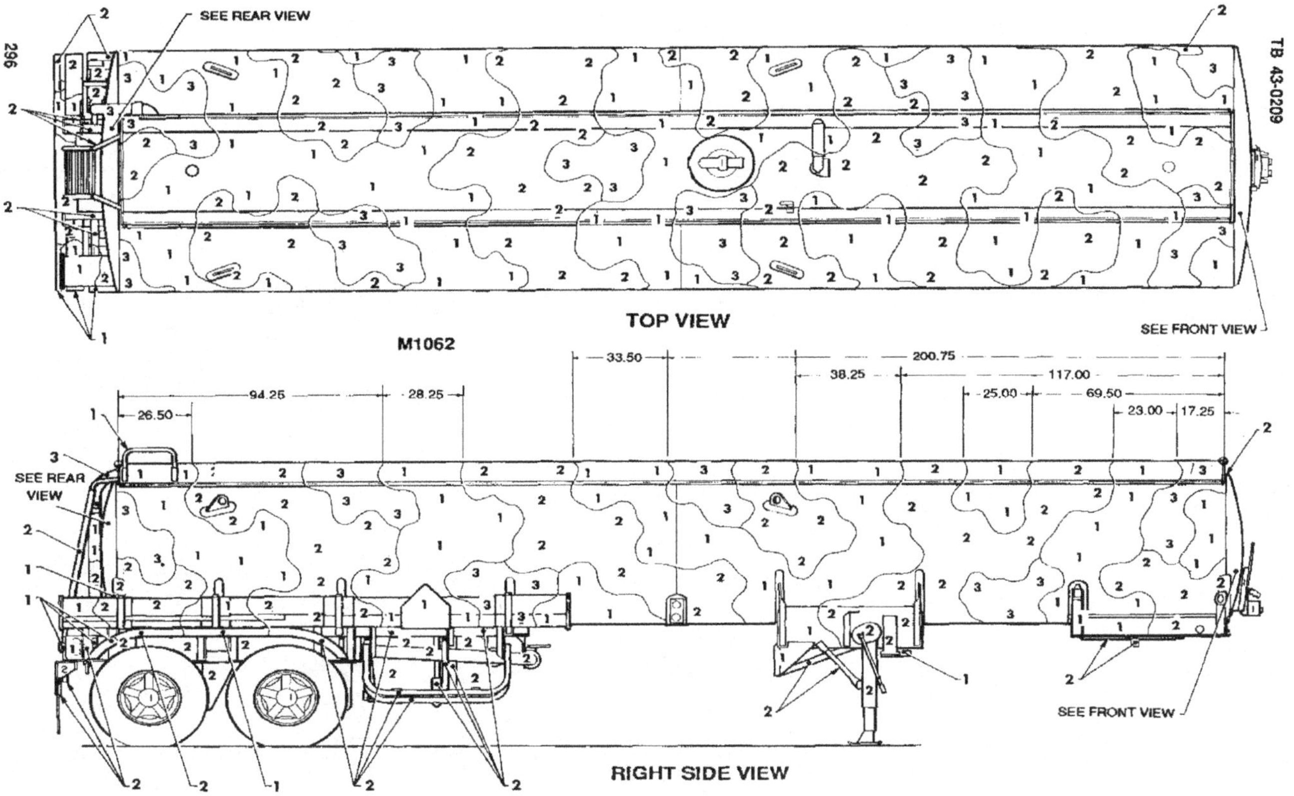

Figure 141. Semitrailer, tanker: 7500 gallon, M1062. (I of2)

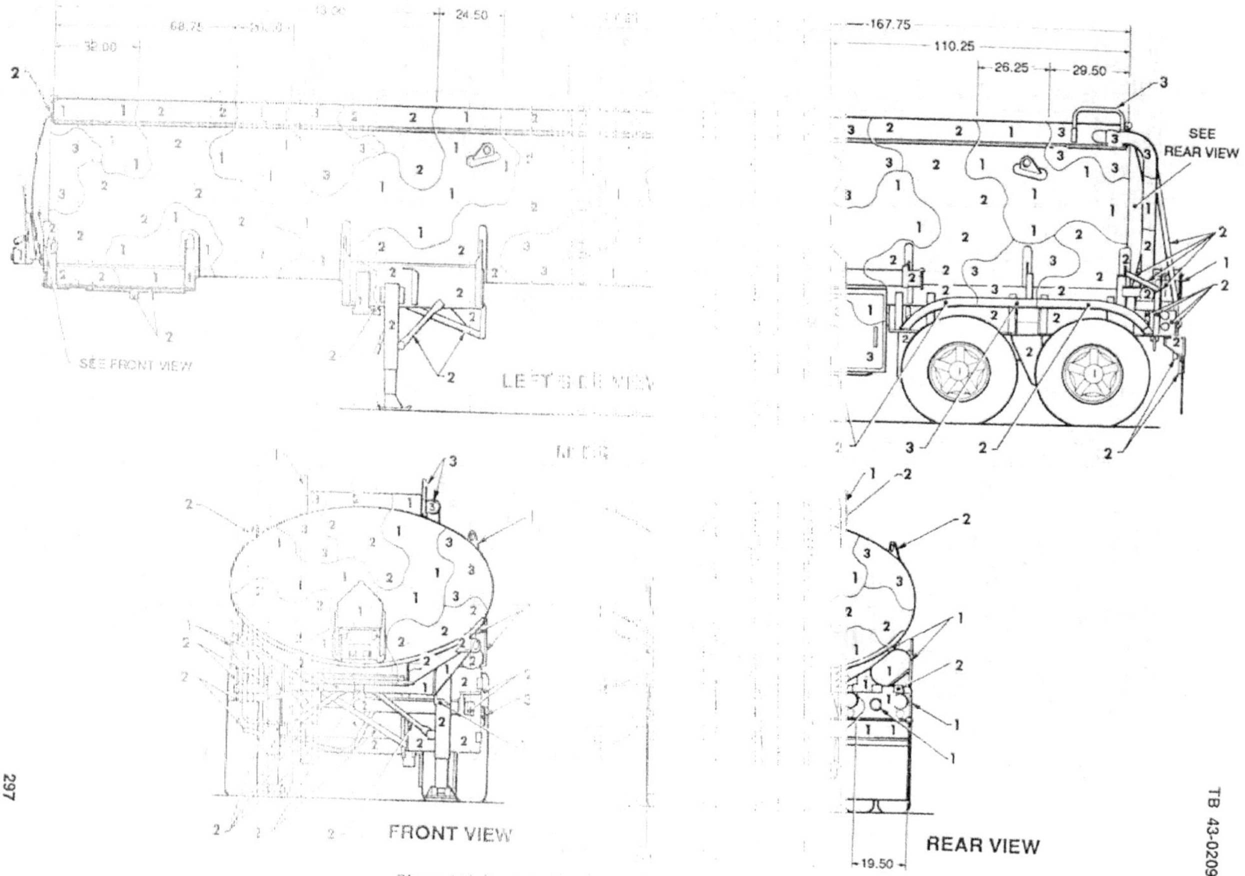

Figure 141. Semitrailer, [illegible]

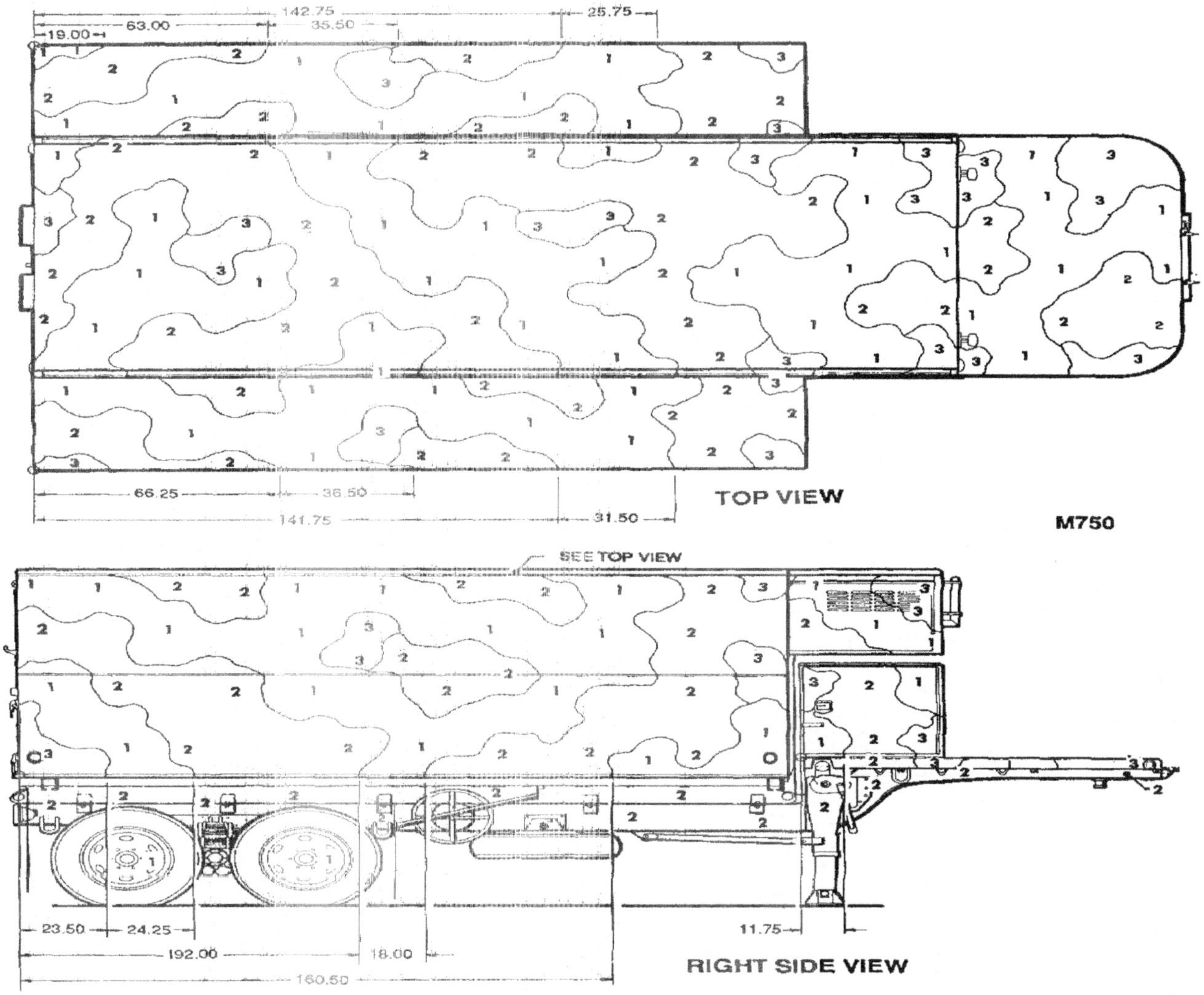

Figure 142. Semitrailer, van: expansible parts storage, 6-ton, 4-wheel, M750. (1 of 2)

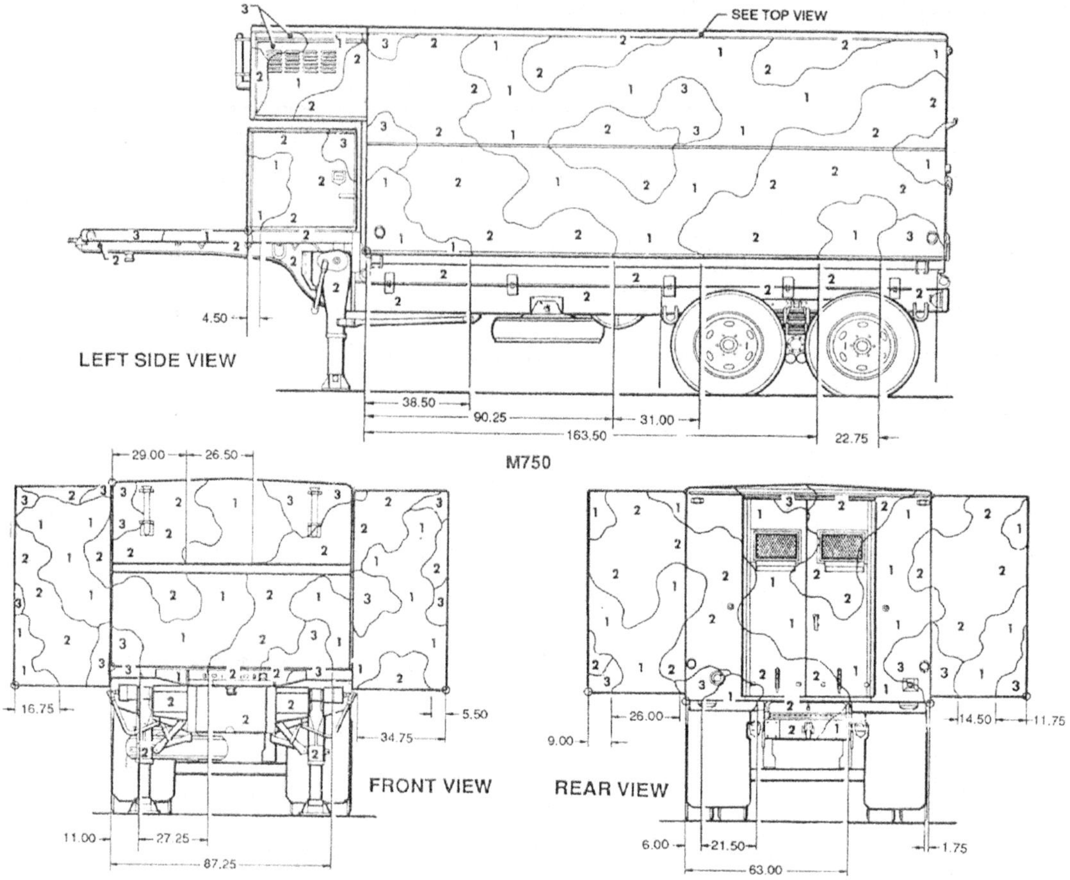

Figure 142. Semitrailer, van: expansible parts storage, 6-ton, 4-wheel, M750. (2 of 2)

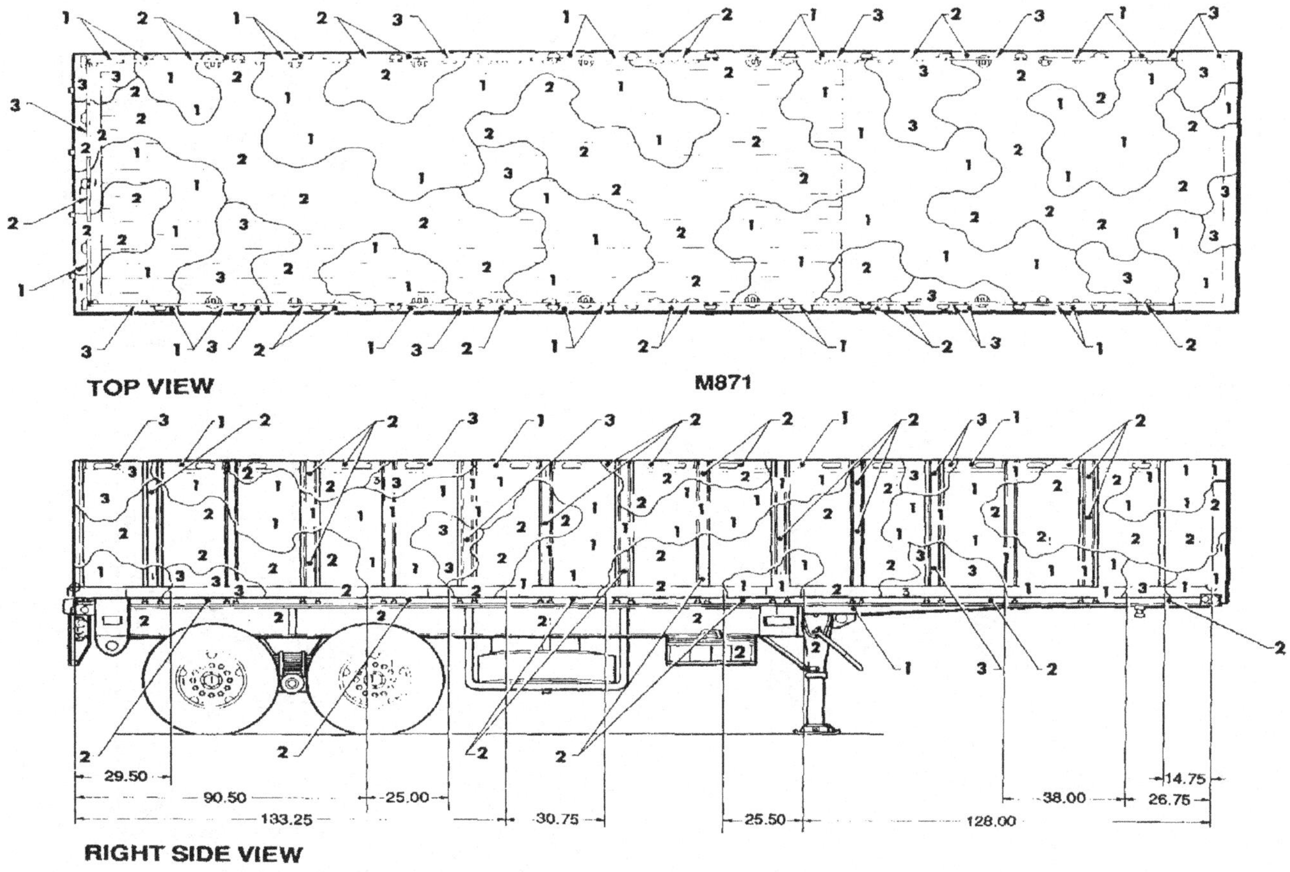

Figure 143. Semitrailer, van: flatbed breakbulk container transporter, 22-1/2-ton, M871. (1 of2)

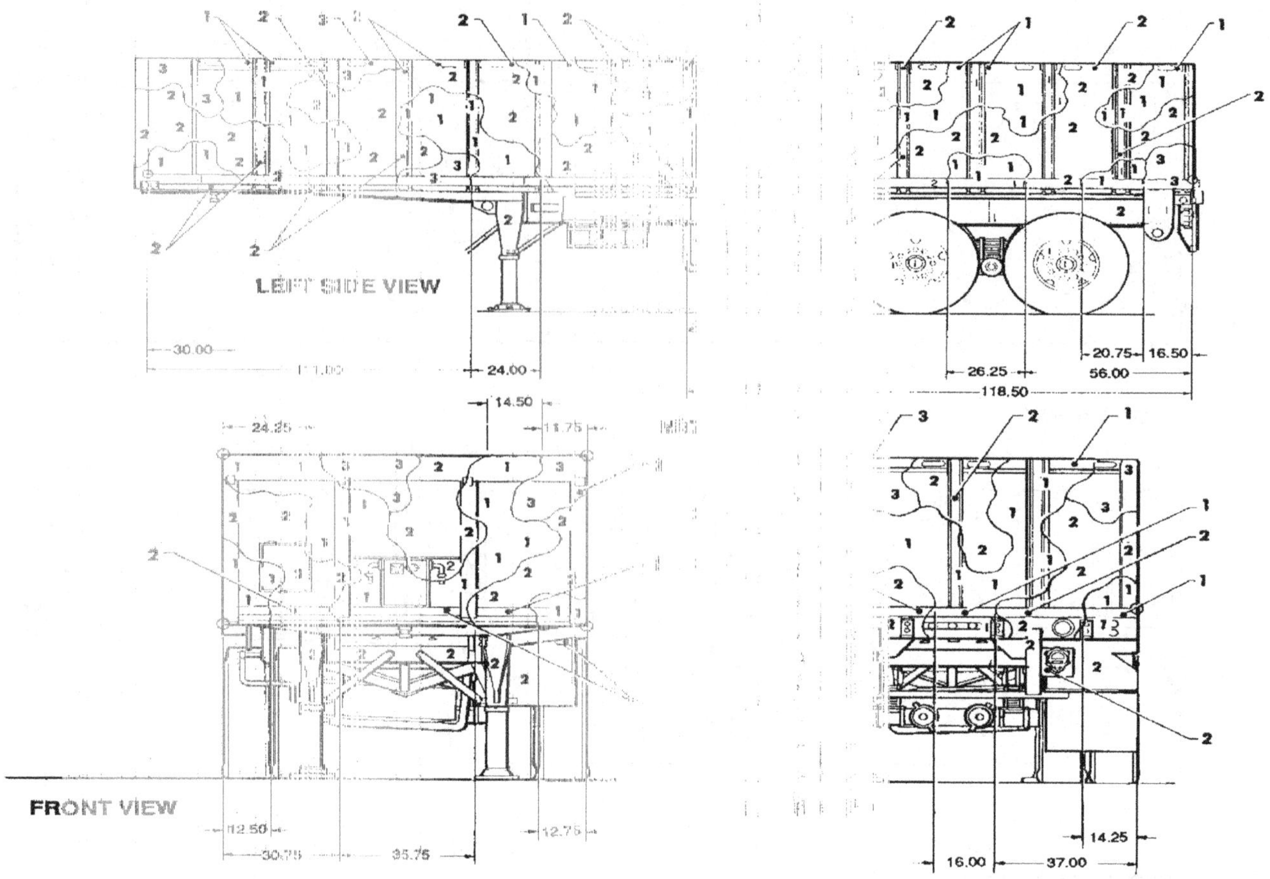

Figure 143. Semitrailer, van :flatbed breakbulk/container [illegible]n, M871. (2 of 2)

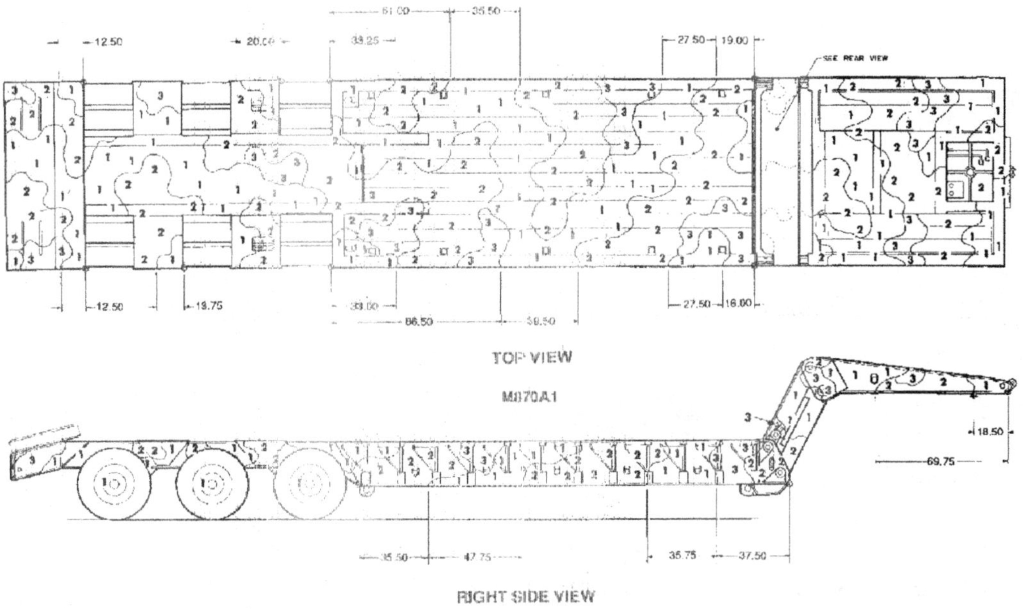

Figure 144. Semitrailer, low bed: 22-1/2-ton, 6-wheel, M870AI. (1 of 2)

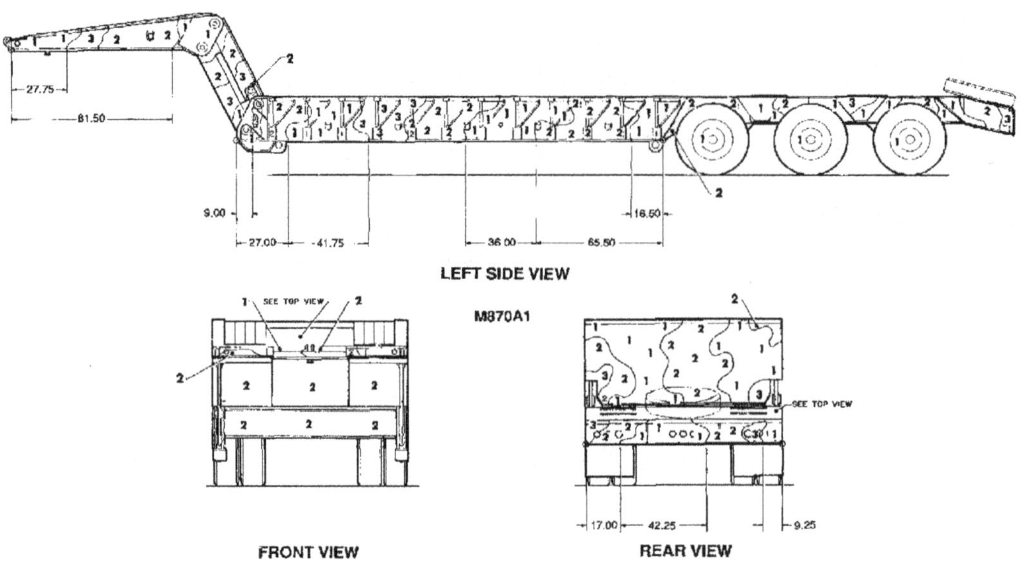

Figure 144. Semitrailer, low bed: 22-1/2-ton, 6-wheel, M870AI. (2 of 2)

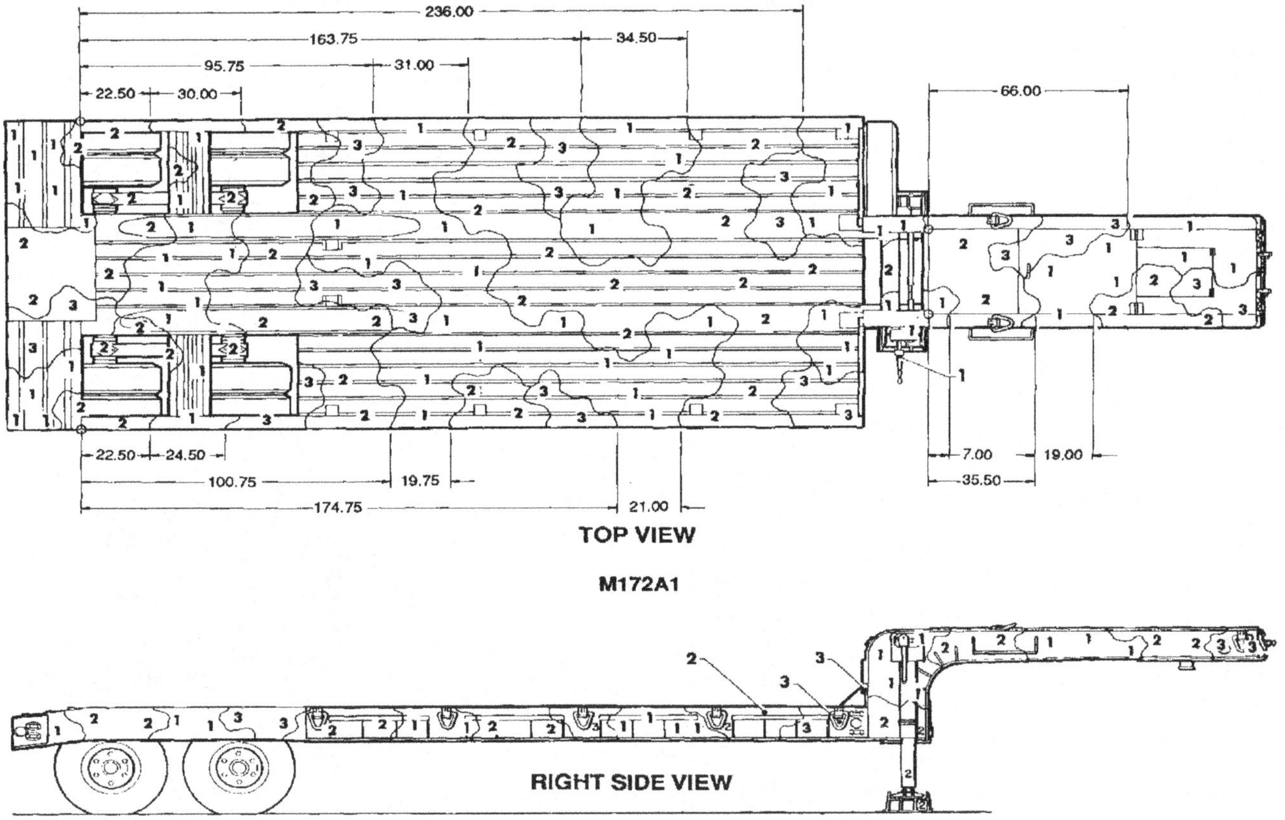

Figure 145. Semitrailer, low bed: 25-ton, 4-wheel, MI 72A1. (I of 2)

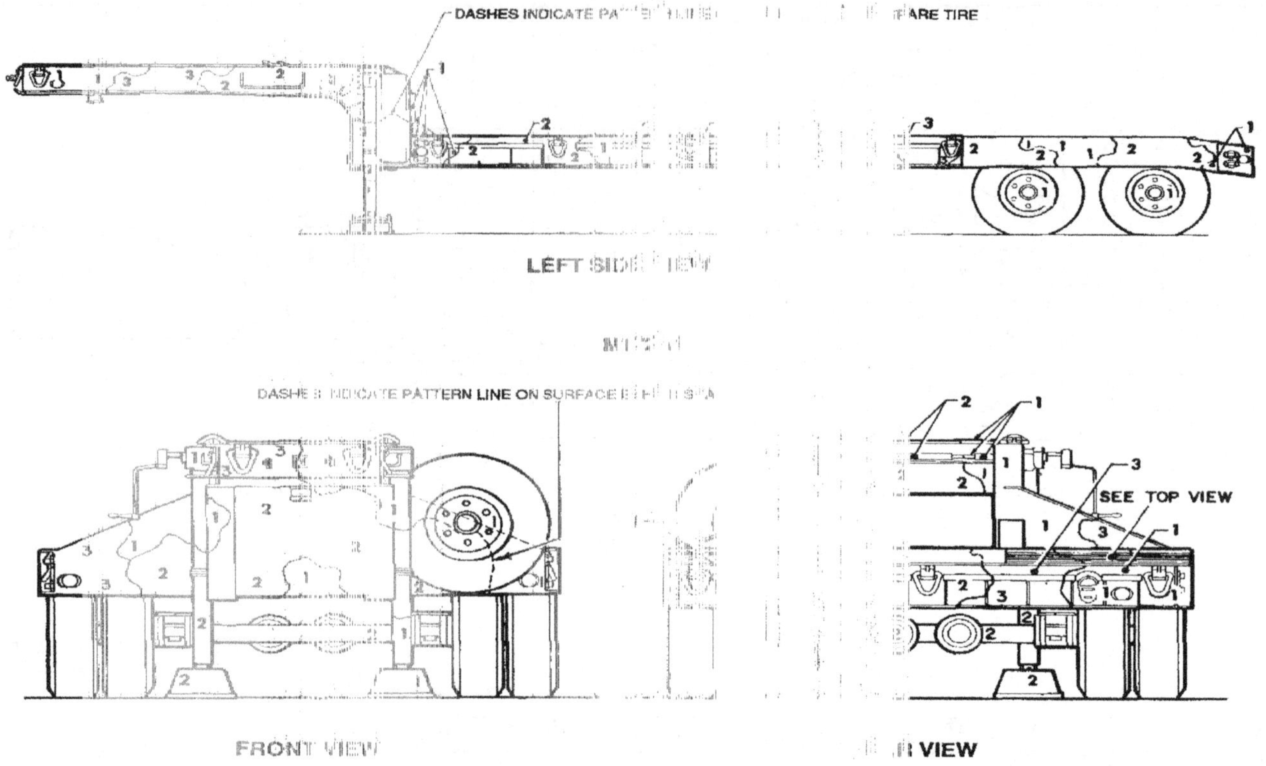

Figure 145. Semitrailer, low bed: 15-ton, [illegible]

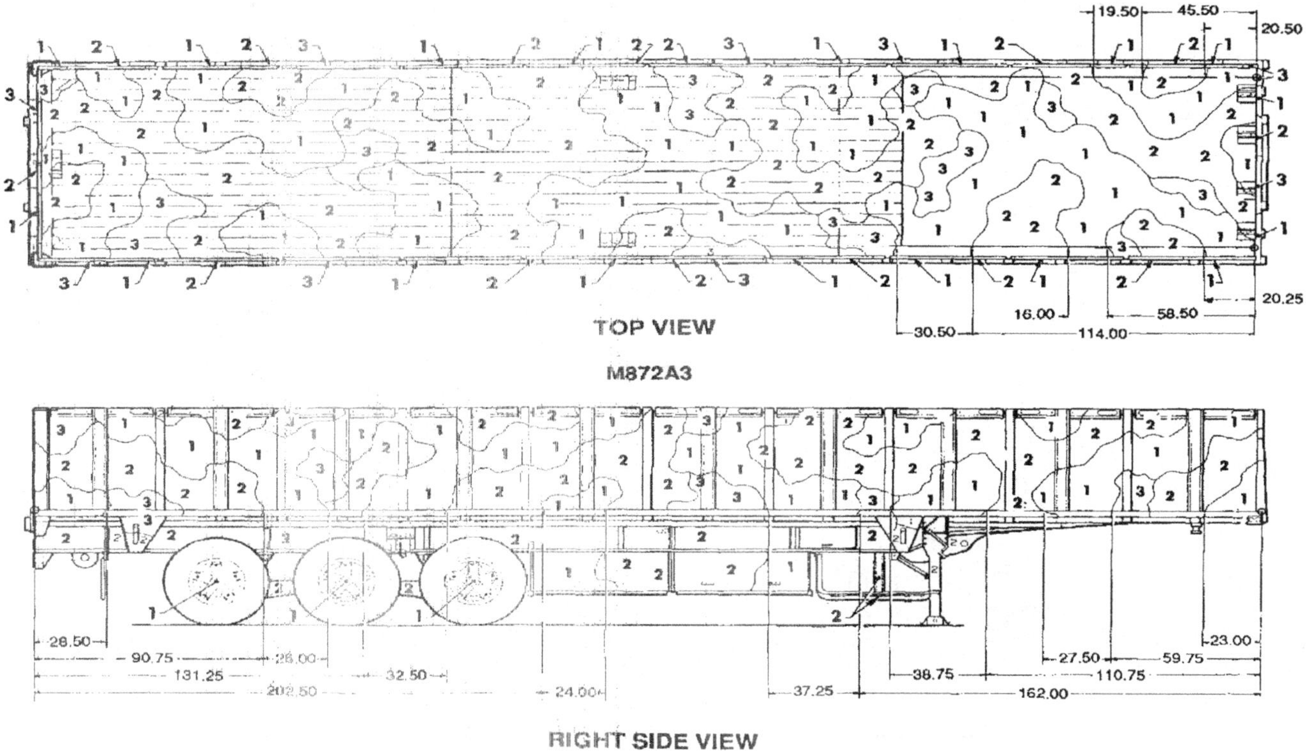

Figure 146. Semitrailer, linehaul: breakbulk container, 34-ton, M872A3. (I of 2)

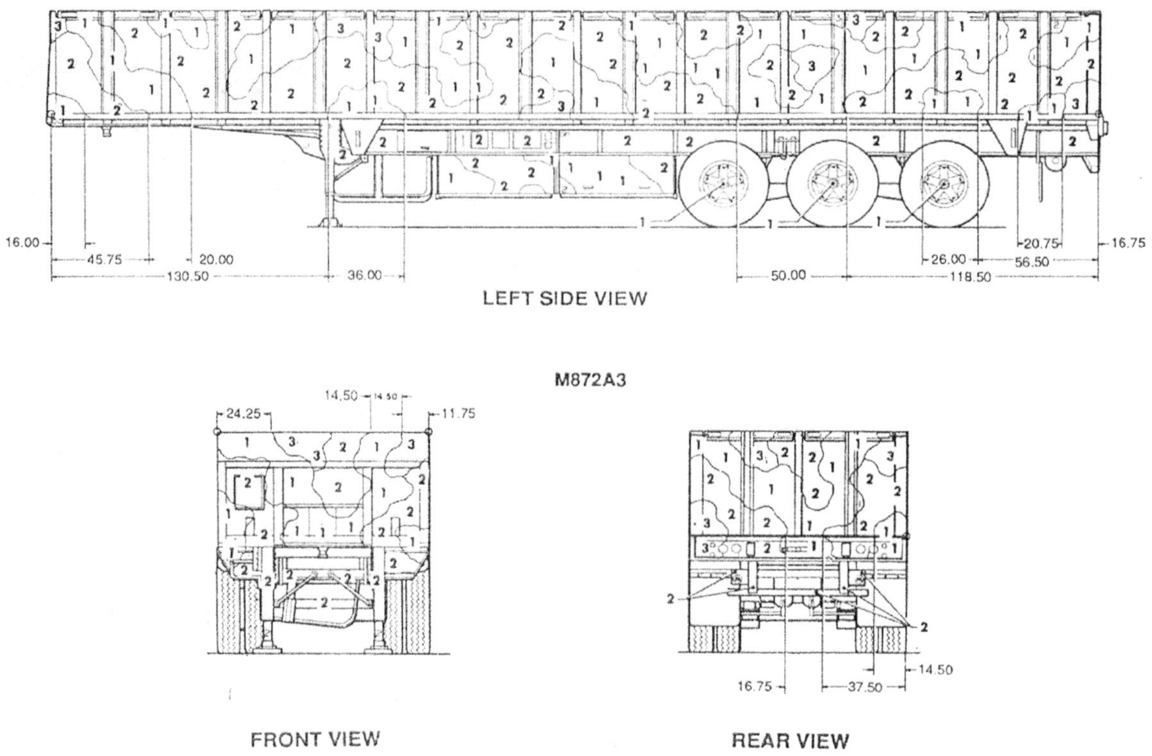

Figure 146. Semitrailer, linehaul: breakbulk container, 34-ton, M872A3. (2 of 2)

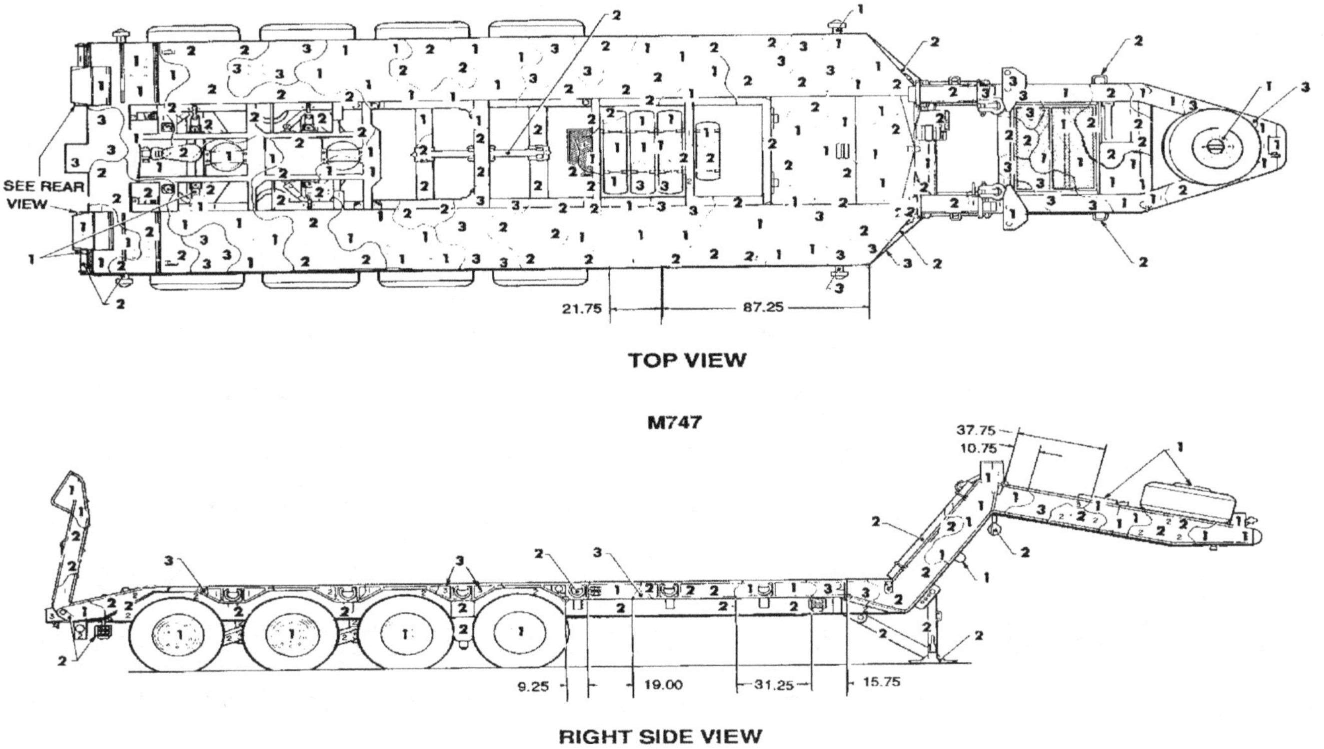

Figure 147. Semitrailer, low bed: heavy equipment transporter, 60-ton, M747. (1 of 2)

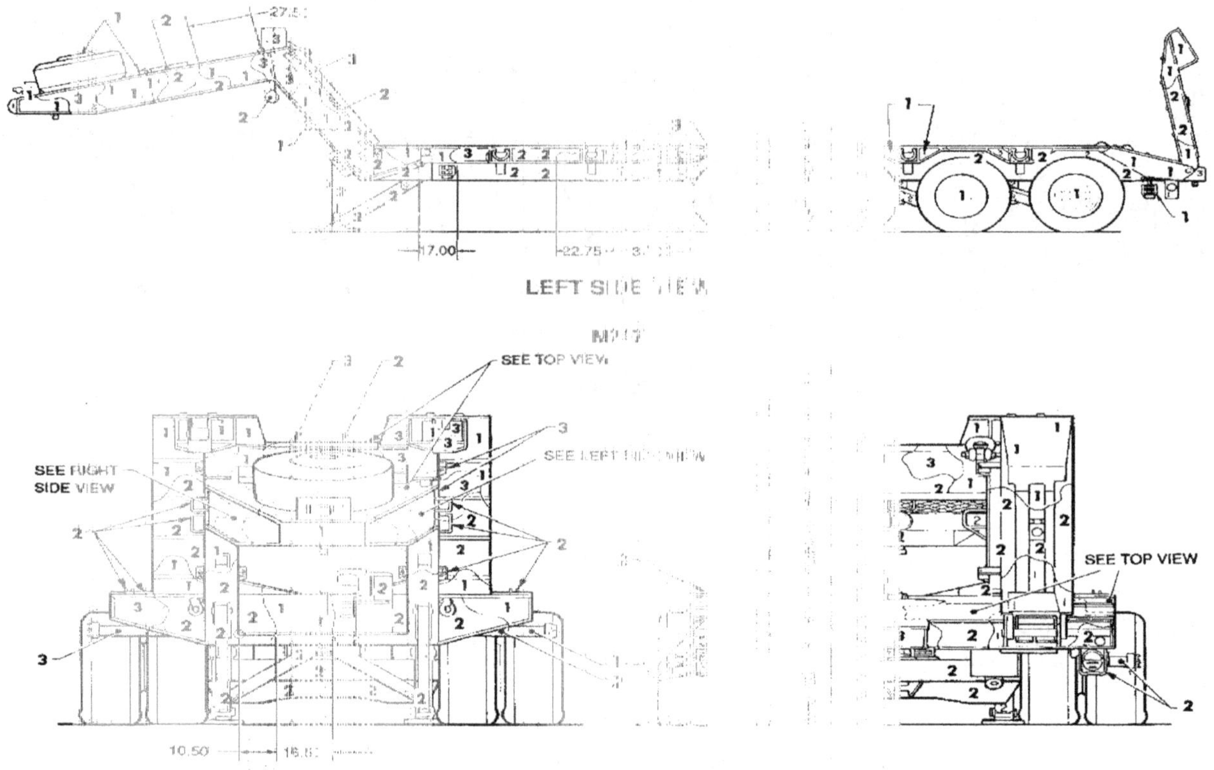

Figure 147. Semitrailer, low bed: heavy equipment [illegible] 47. (2 of 2)

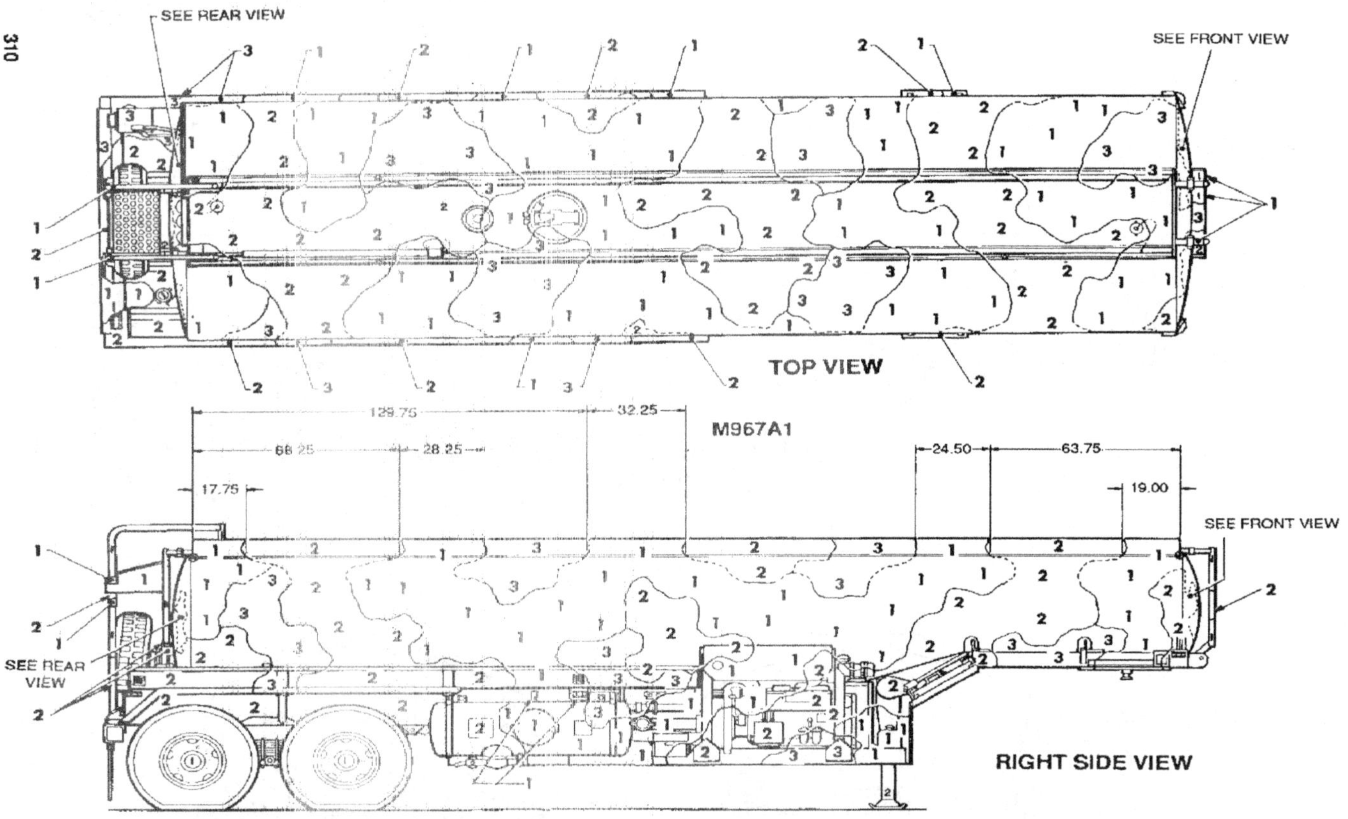

Figure 148. Semitrailer, tanker: 5000 gallon, 4-wheel, M967A1, M969A1 and M970AI. (I of 2)

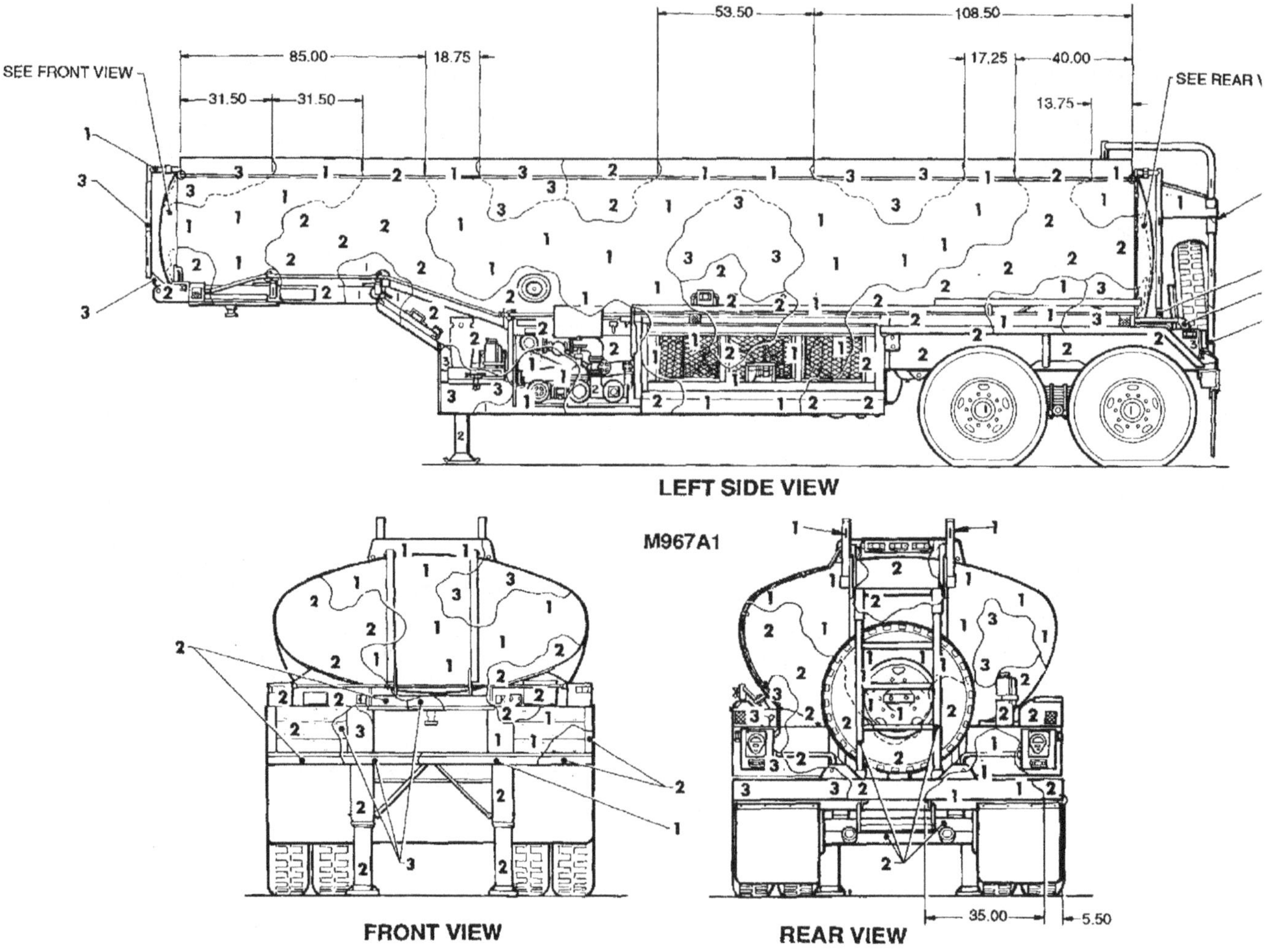

Figure 148. Semitrailer, tanker: 5000 gallon, 4-wheel, M967A1, M969A1 and M970A1. (2 of 2)

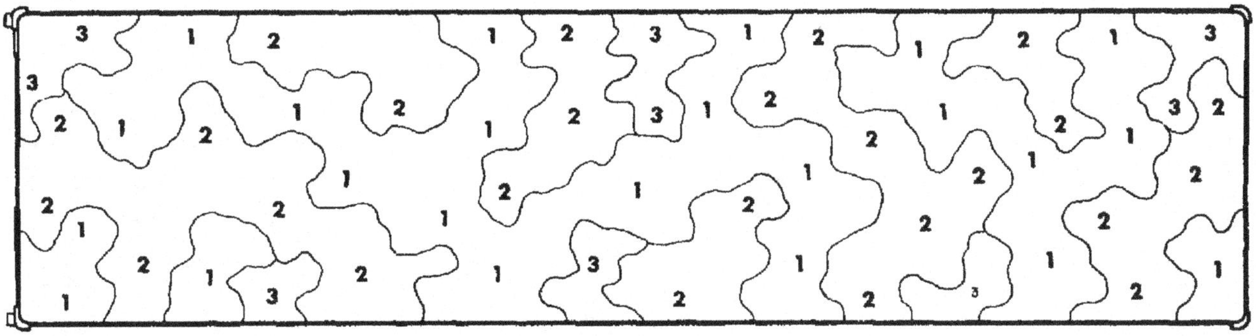

TOP VIEW

M1032

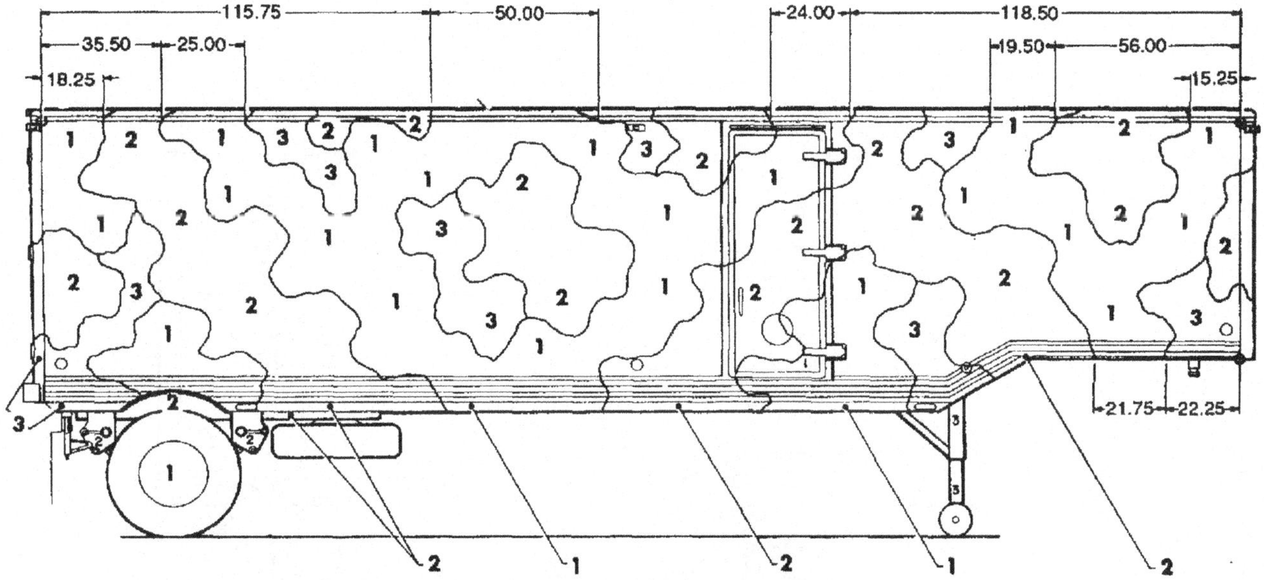

RIGHT SIDE VIEW

Figure 149. Semitrailer, van: small repair parts transporter,, M1032 (also M373A2C). (1 of 2)

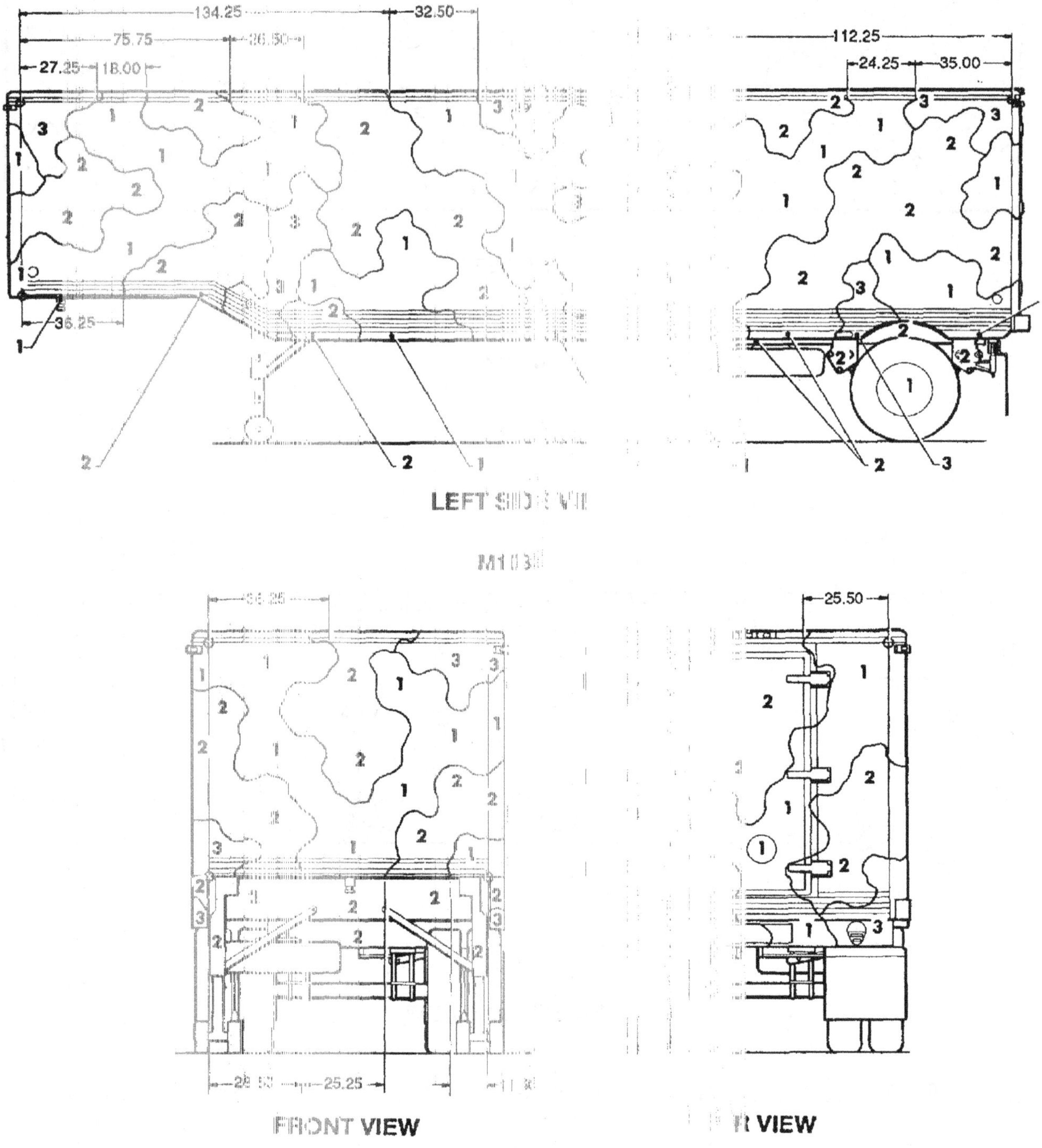

Figure 149. Semitrailer, van: small repair parts … (also M373A2C). (2 of 2)

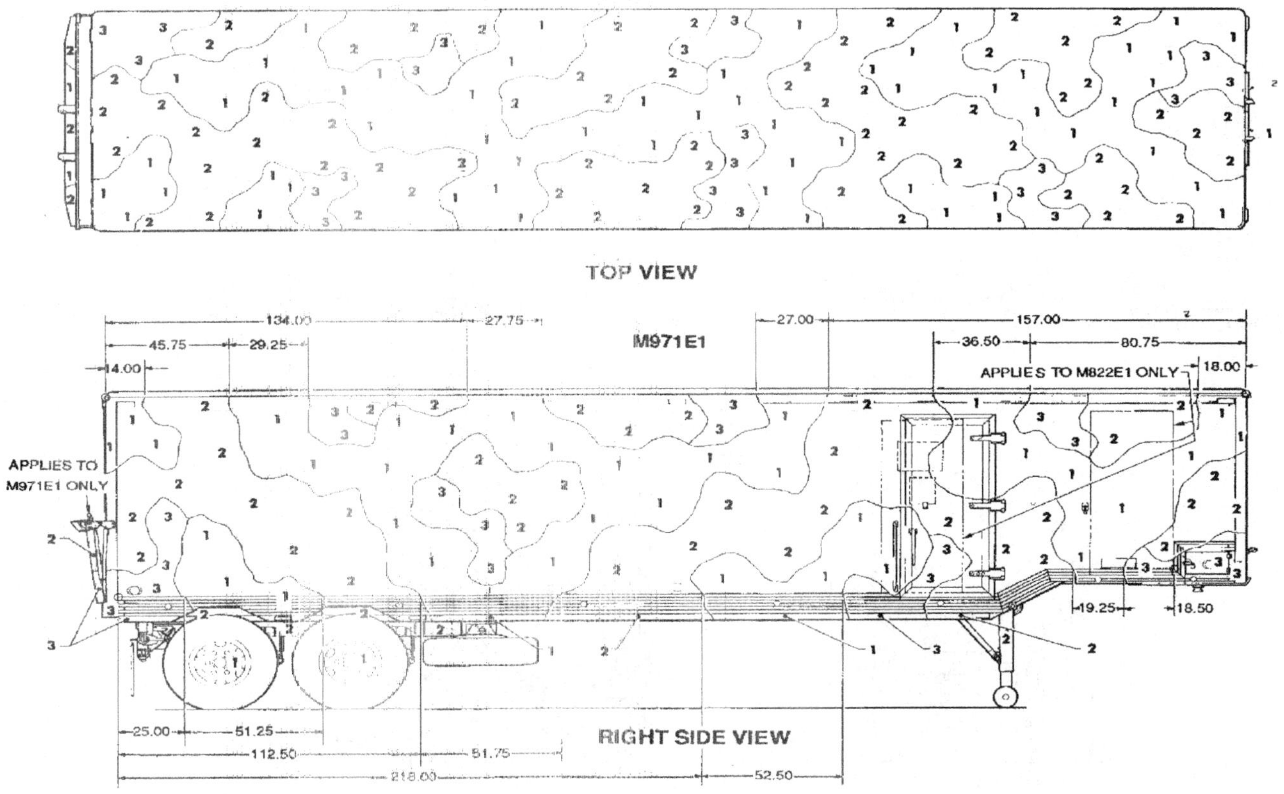

Figure 150. Semitrailer, van: basic, 10-ton, 4-wheel, M971E1, M822E1, M991E2 and M995E2. (1 of 2)

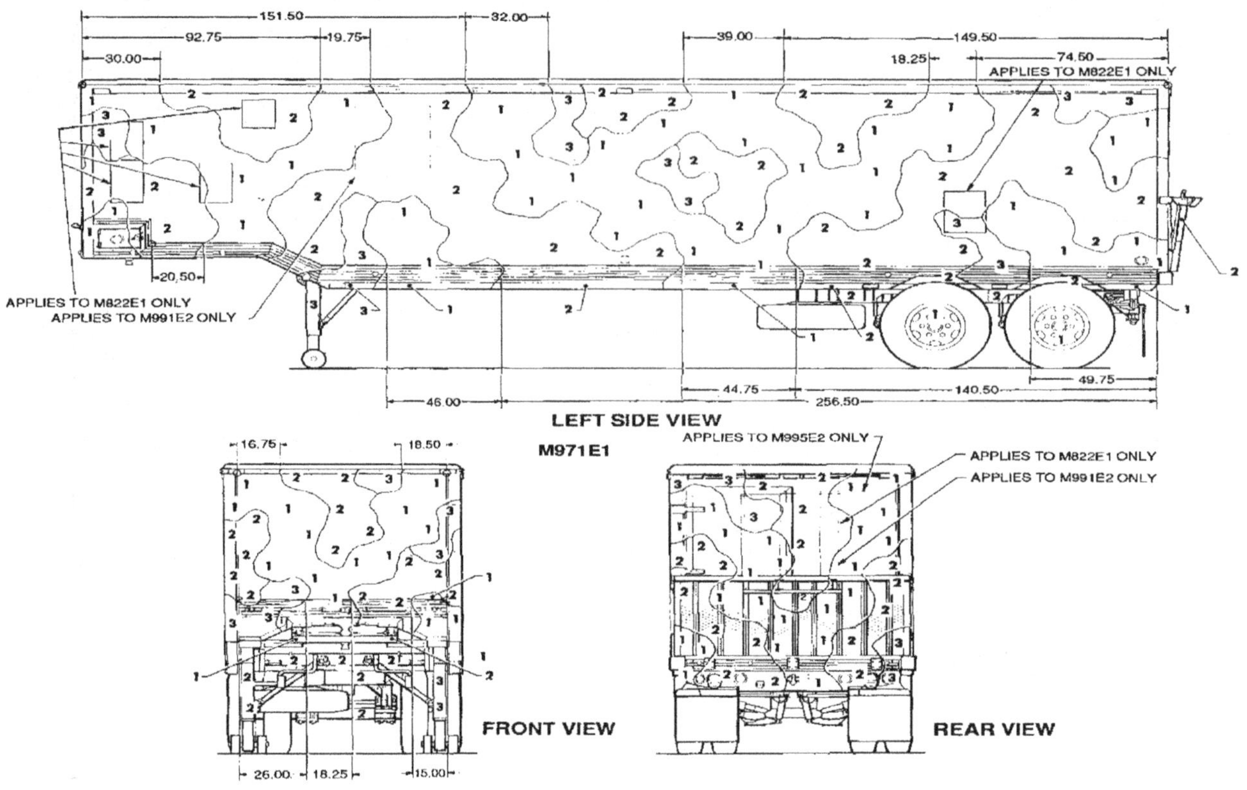

Figure 150. Semitrailer, van: basic, 10-ton, 4-wheel, M971E1, M822E1, M991E2 and M995E2. (2 of 2)

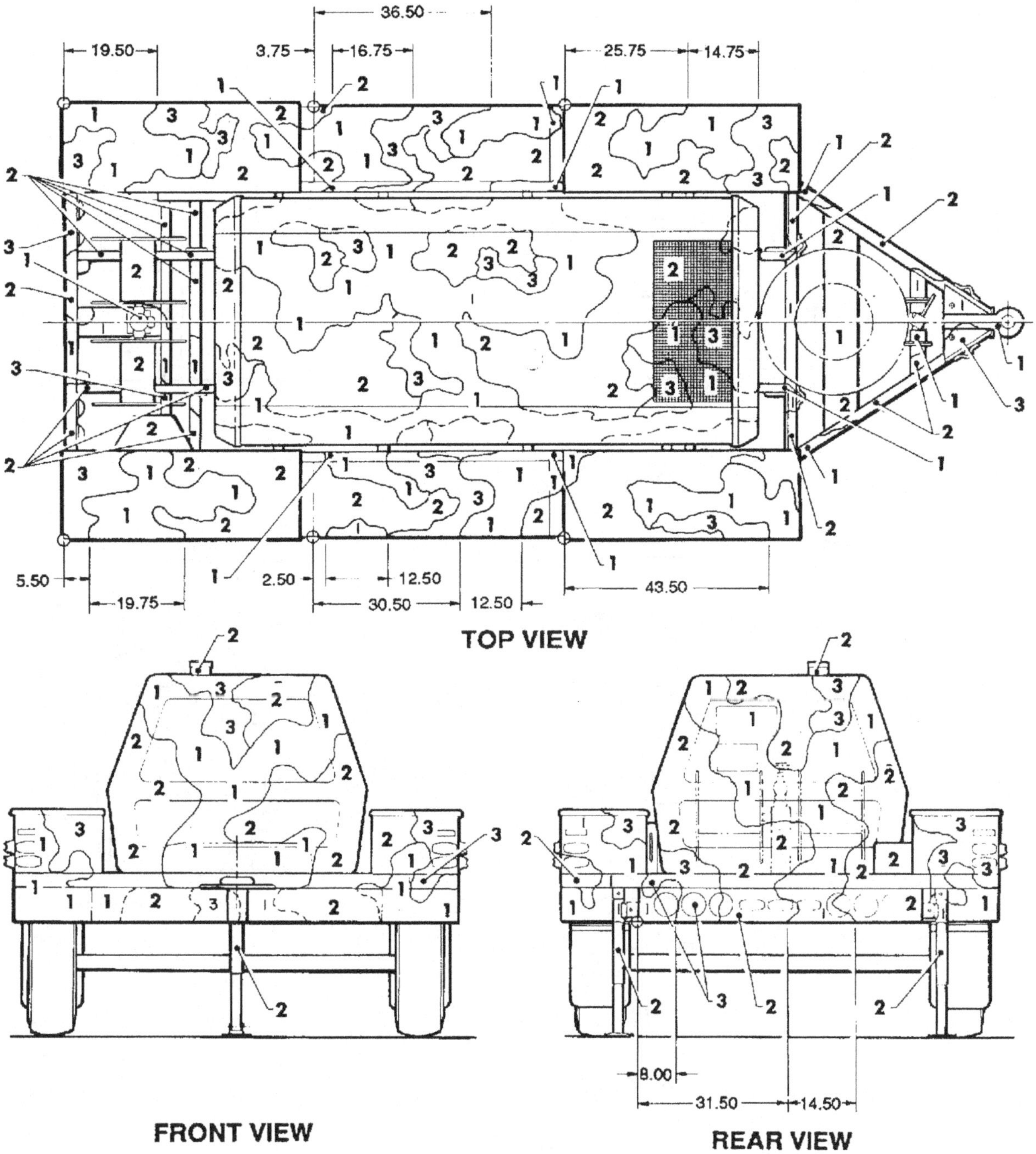

Figure 151. Air compressor: 250-CFM. (1 of 2)

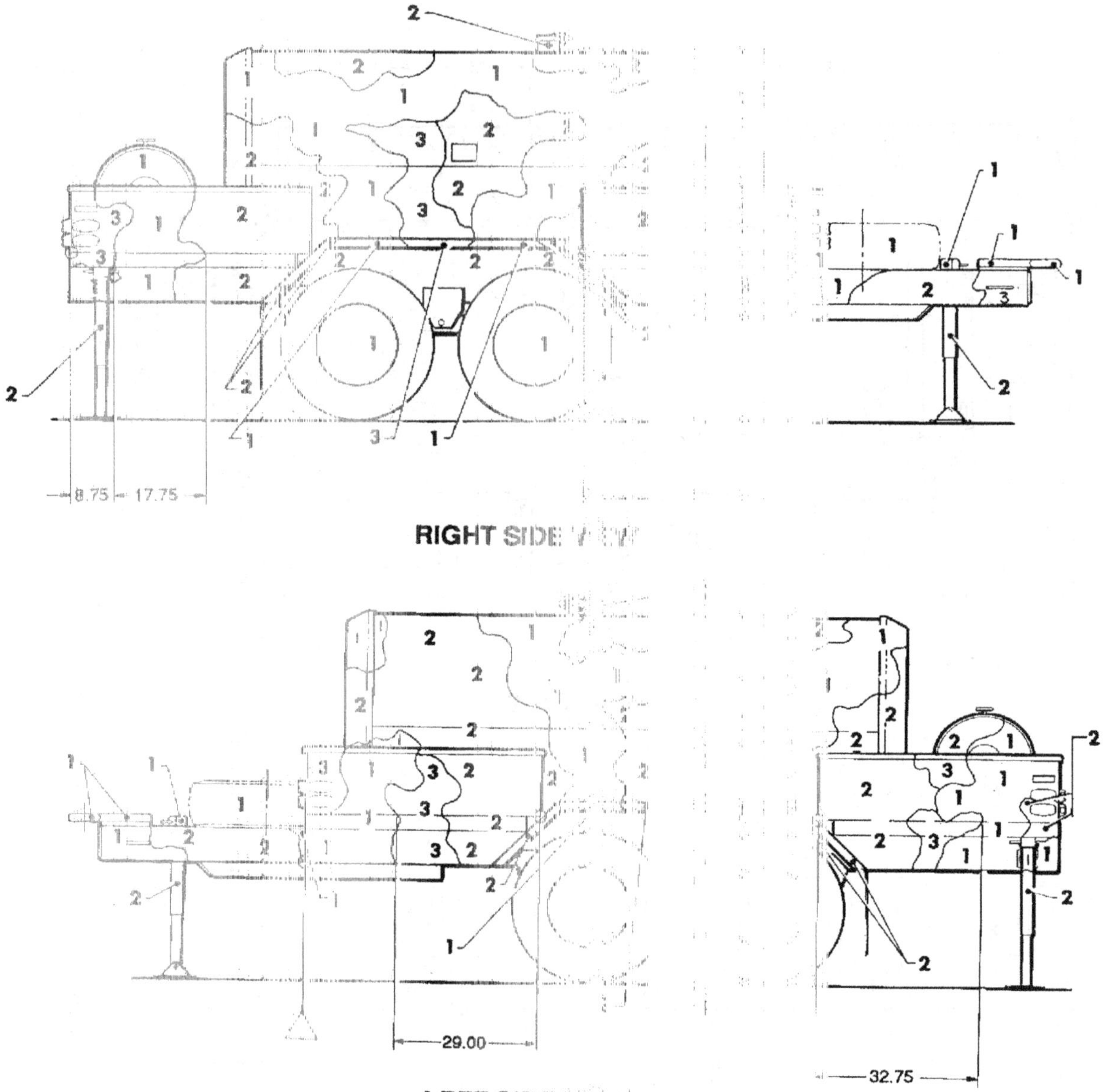

Figure 151. Air compressor 25 ... 2)

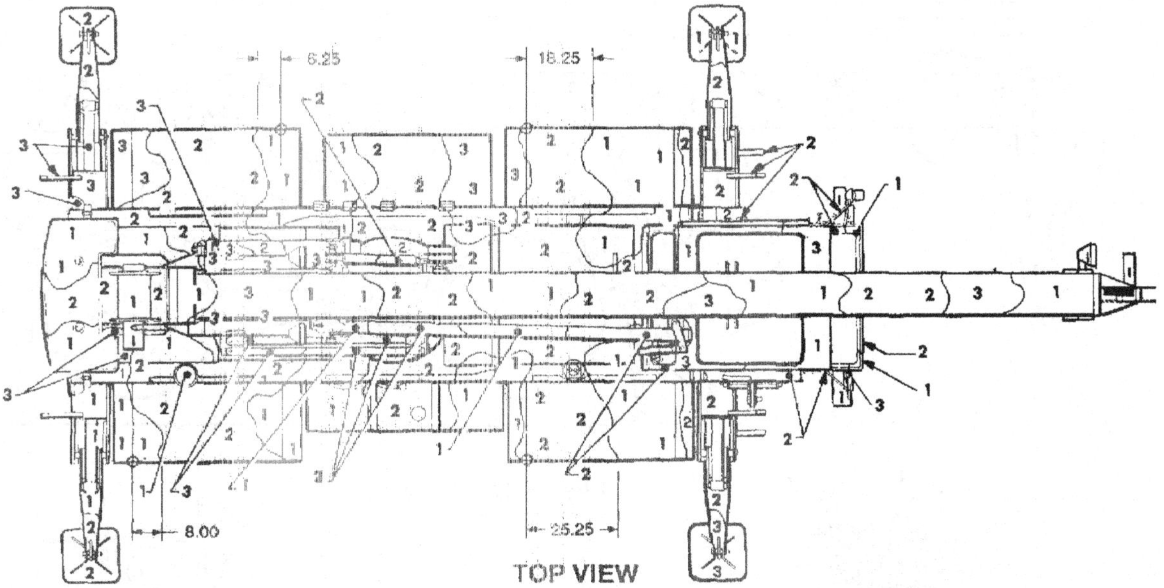

TOP VIEW

R050

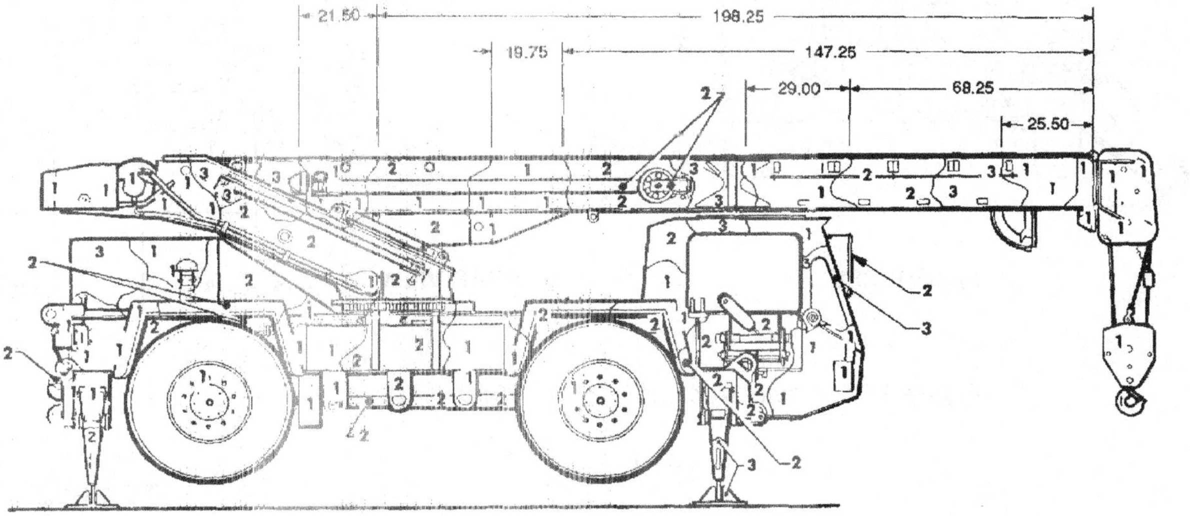

RIGHT SIDE VIEW

Figure 152. Crane, truck: 7-1/2-ton, R050. (1 of 2)

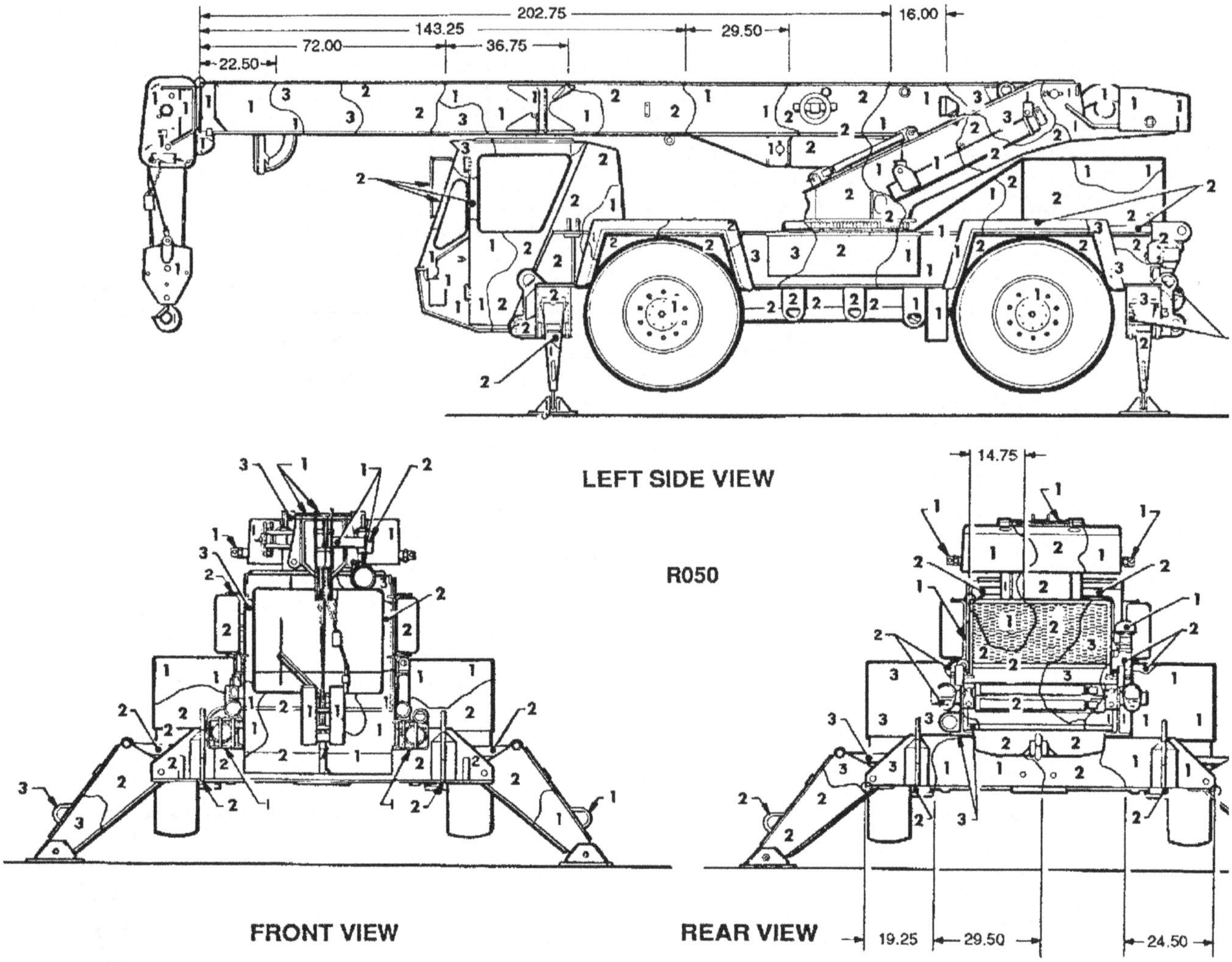

Figure 152. Crane, truck: 7-1/2-ton, R050. (2 of 2)

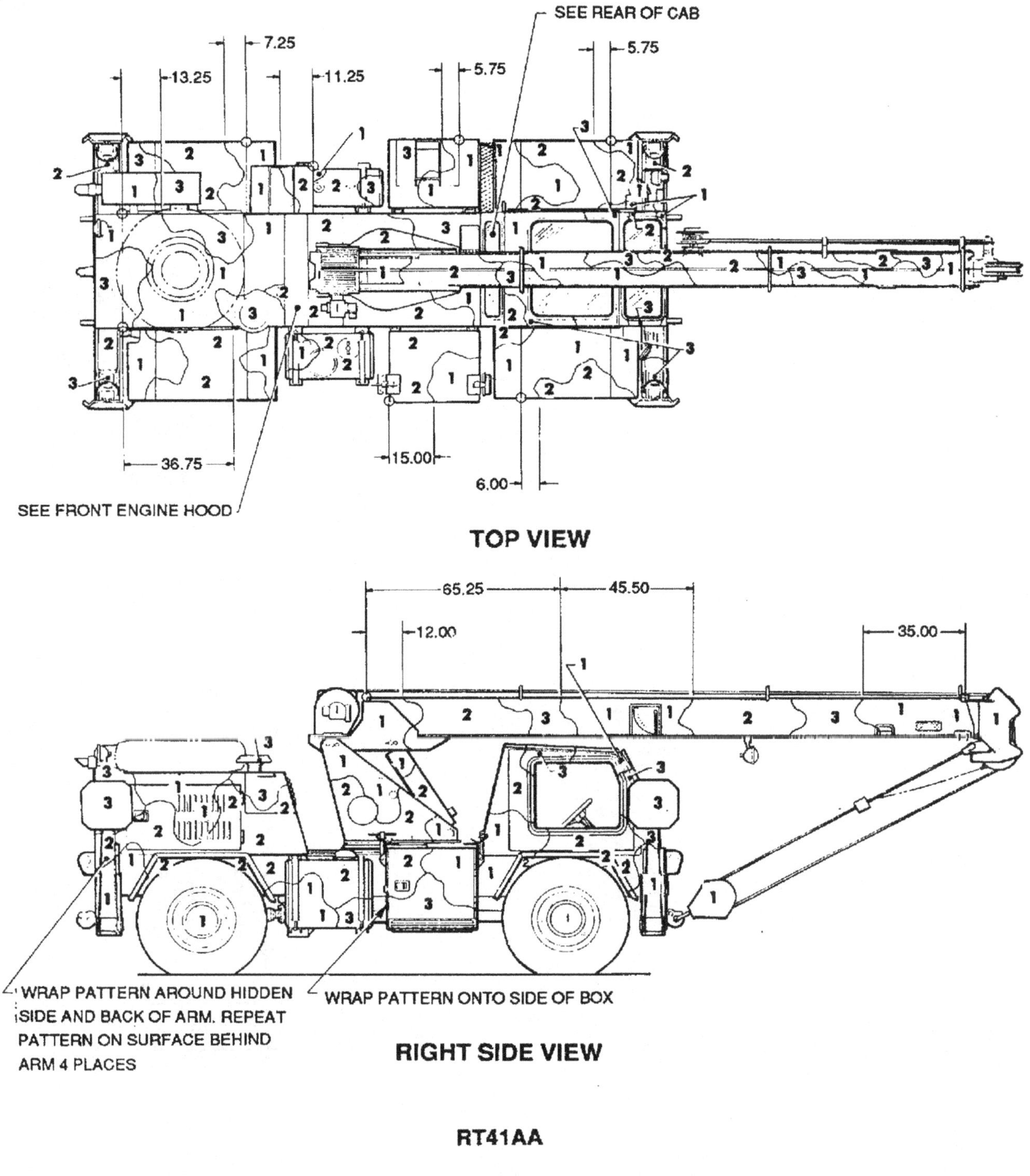

Figure 153. Crane, self-propelled: aircraft maintenance and positioning, RT41AA. (1 of 2)

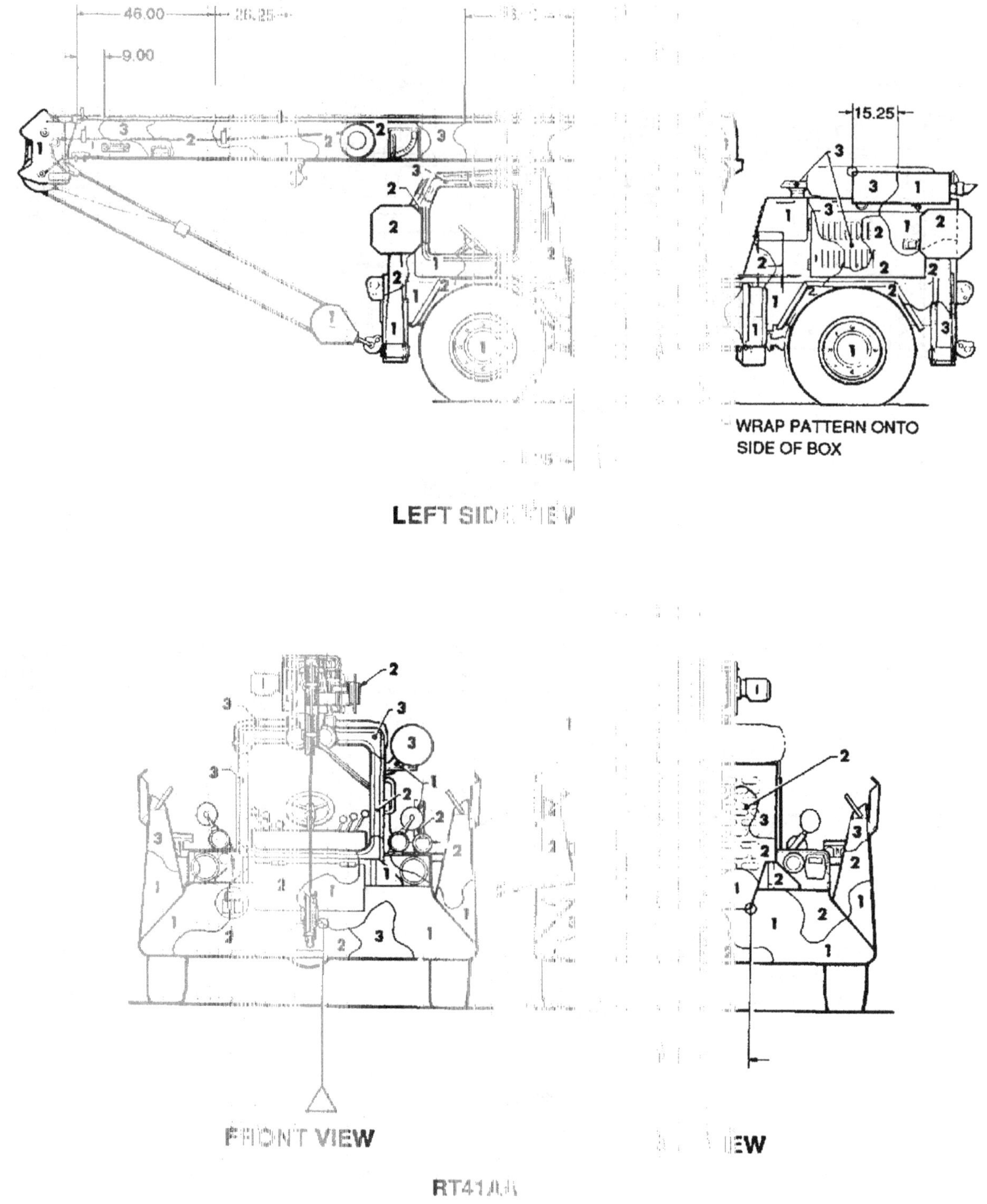

Figure 153. Crane, self-propelled: aircraft maintenance and positioning, RT41AA. (2 of 2)

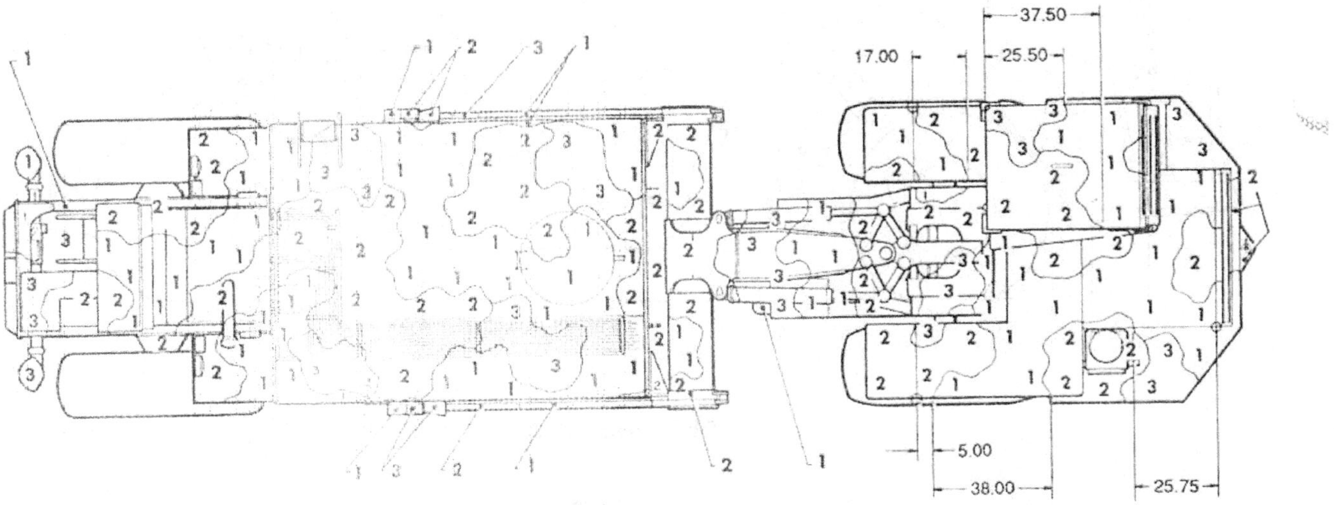

TOP VIEW

613BSS

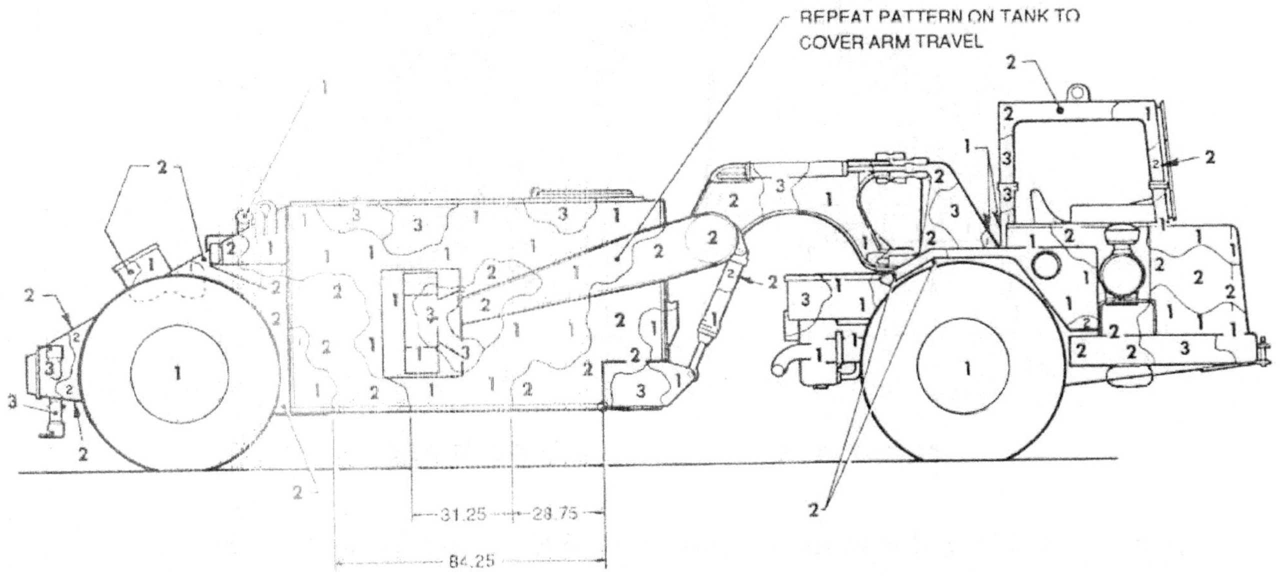

RIGHT SIDE VIEW

Figure 154. Distributor, water, self-propelled 2500 gallon minimum, 613B. (1 of 2)

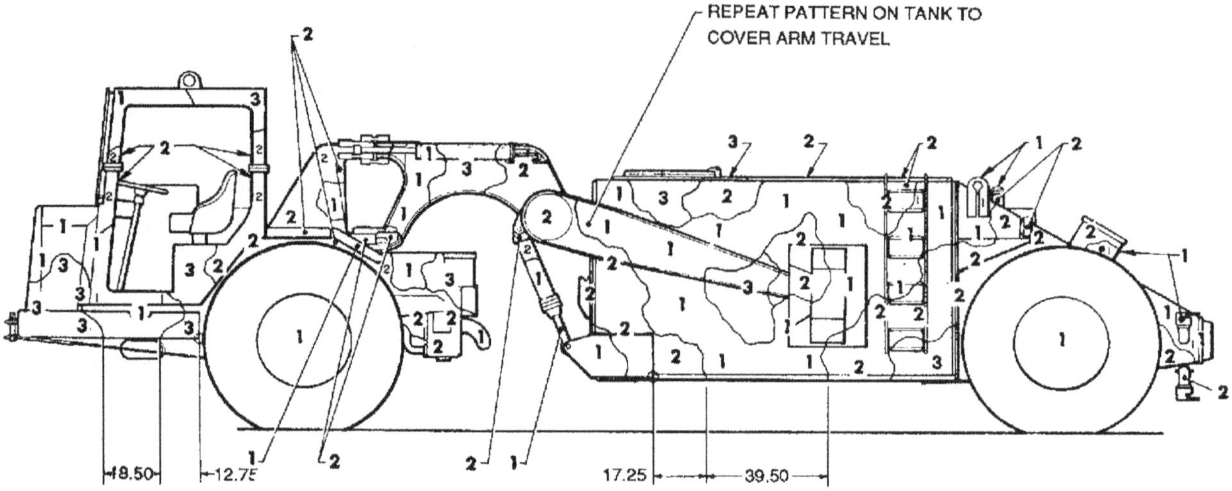

LEFT SIDE VIEW

613BSS

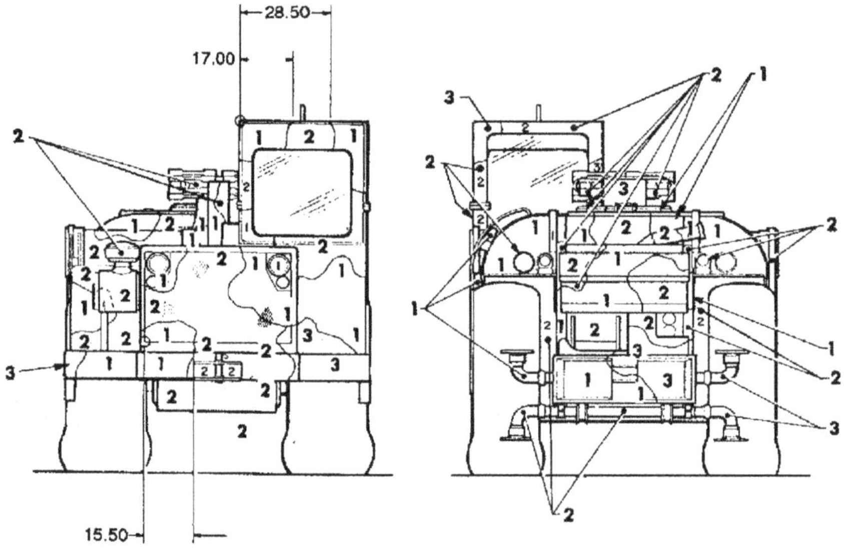

FRONT VIEW **REAR VIEW**

Figure 154. Distributor, water, self-propelled 2500 gallon minimum, 613B. (2 of 2)

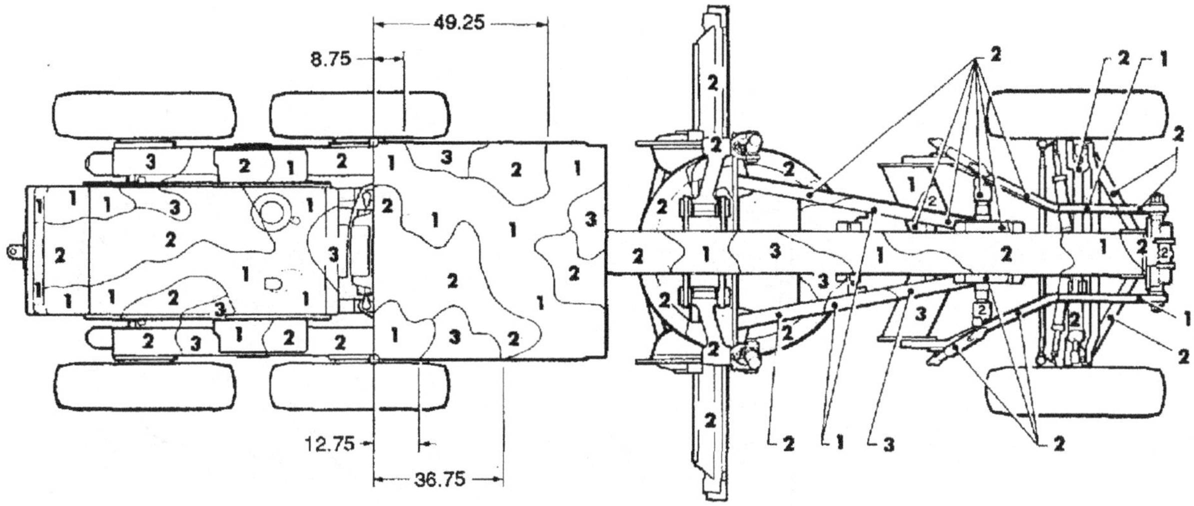

TOP VIEW

130G

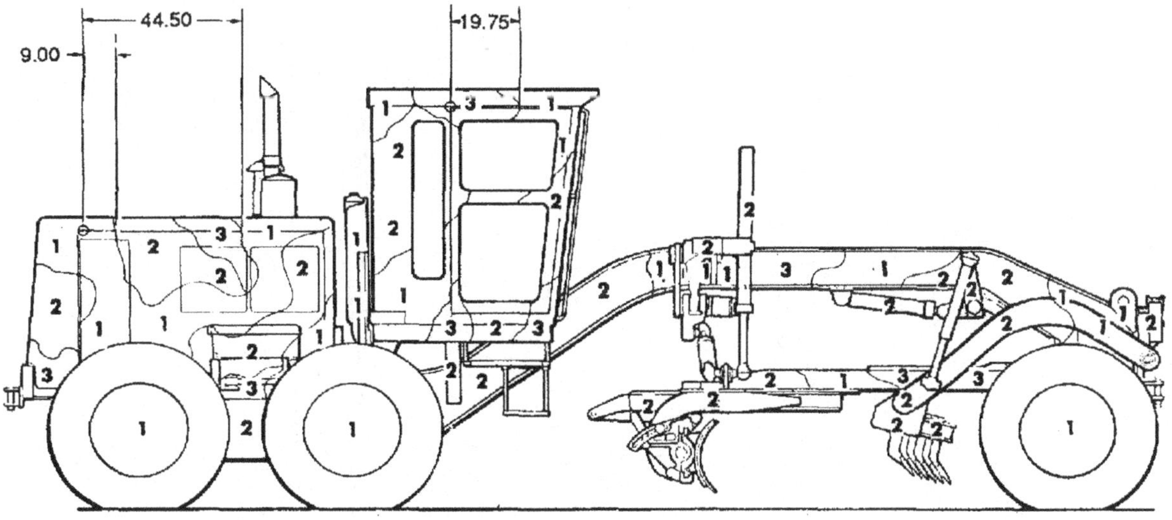

RIGHT SIDE VIEW

Figure 155. Roadgrader, heavy, 130G and 130GS. (1 of 2)

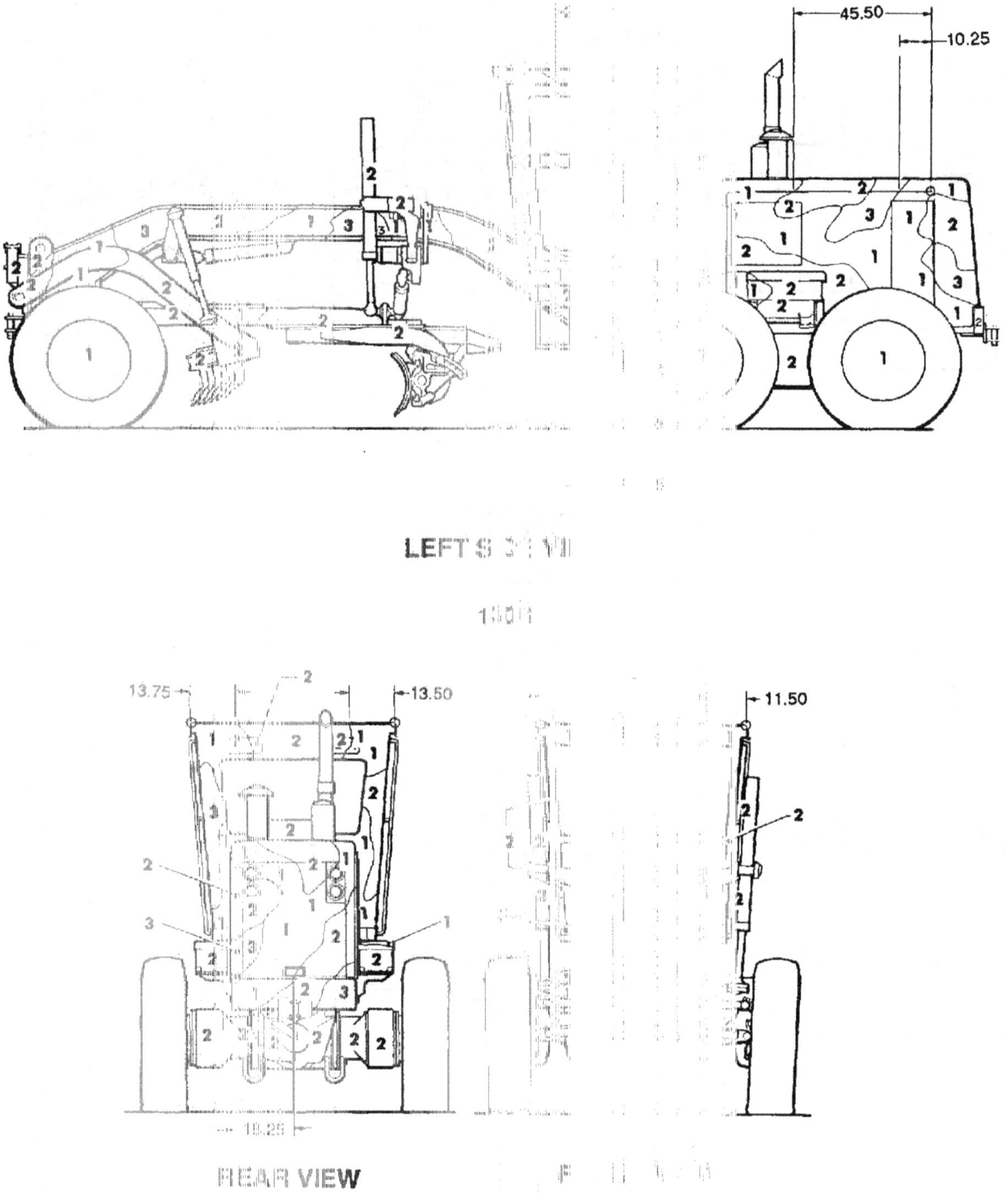

Figure 155. Roadgrader, heavy, 130G ... (2 of 2)

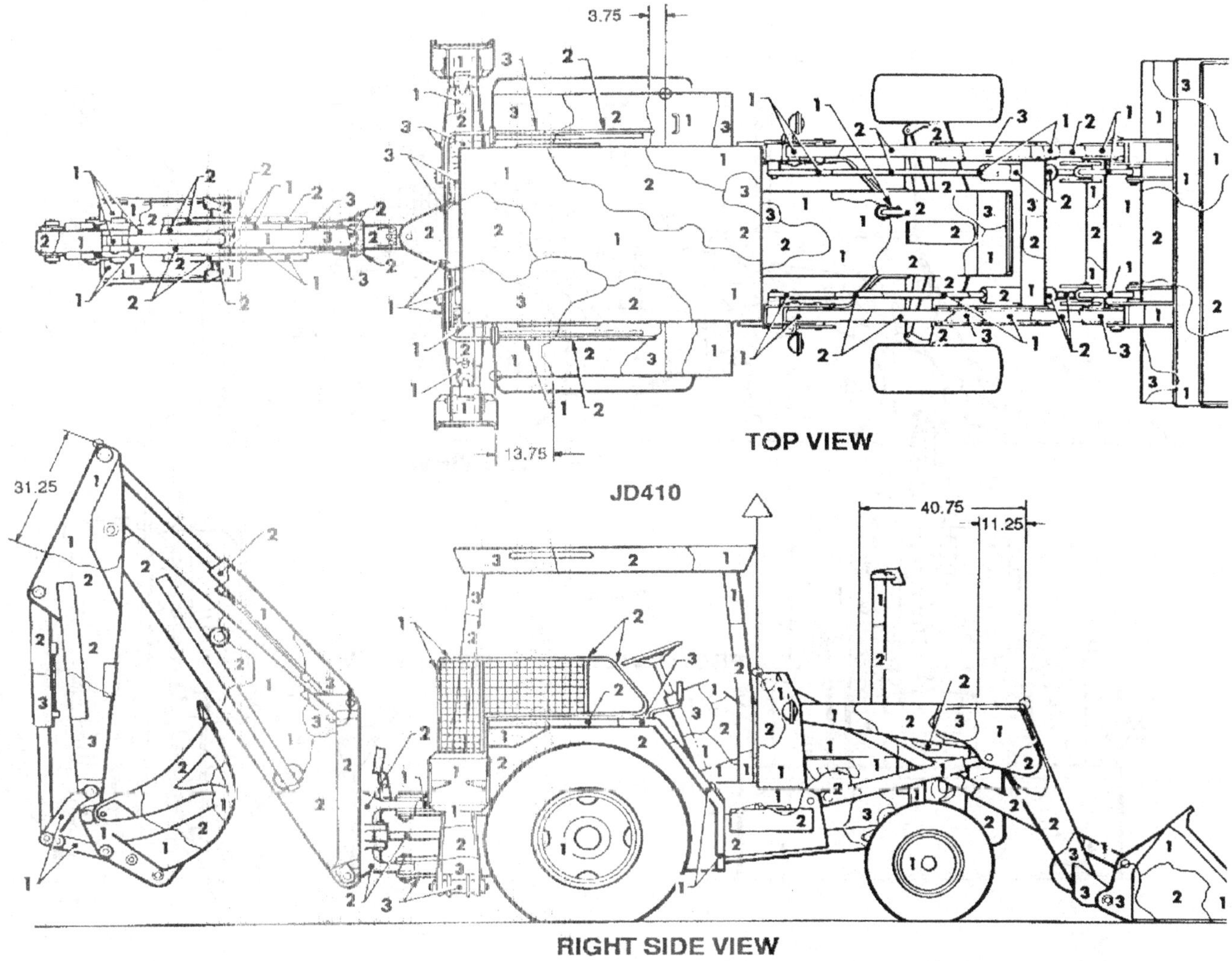

Figure 156. Tractor, wheeled, DED, w/backhoe, w/front loader, JD-140. (1 of 2)

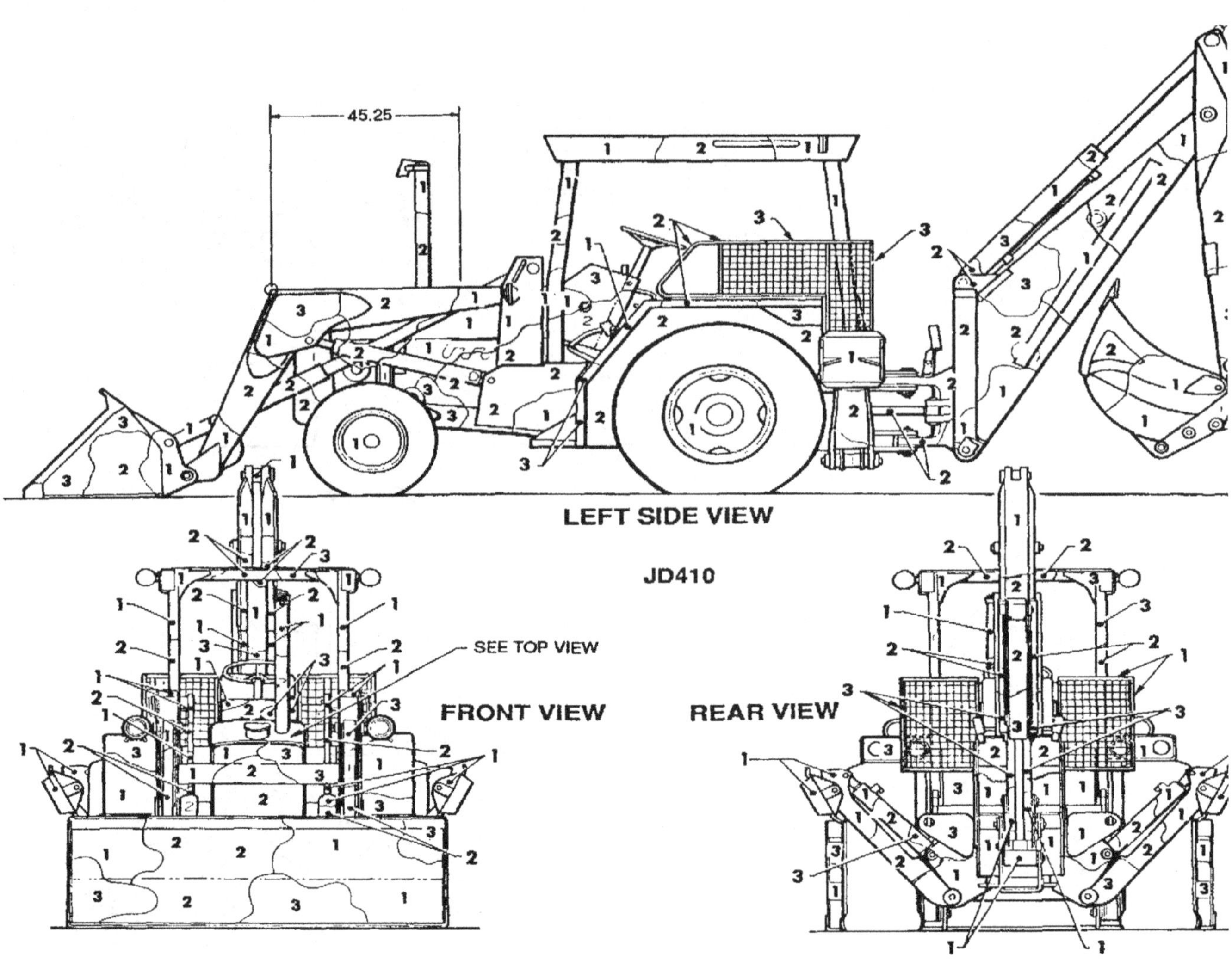

Figure 156. Tractor, wheeled, DED, w/backhoe, w/front loader, JD-140. (2 of 2)

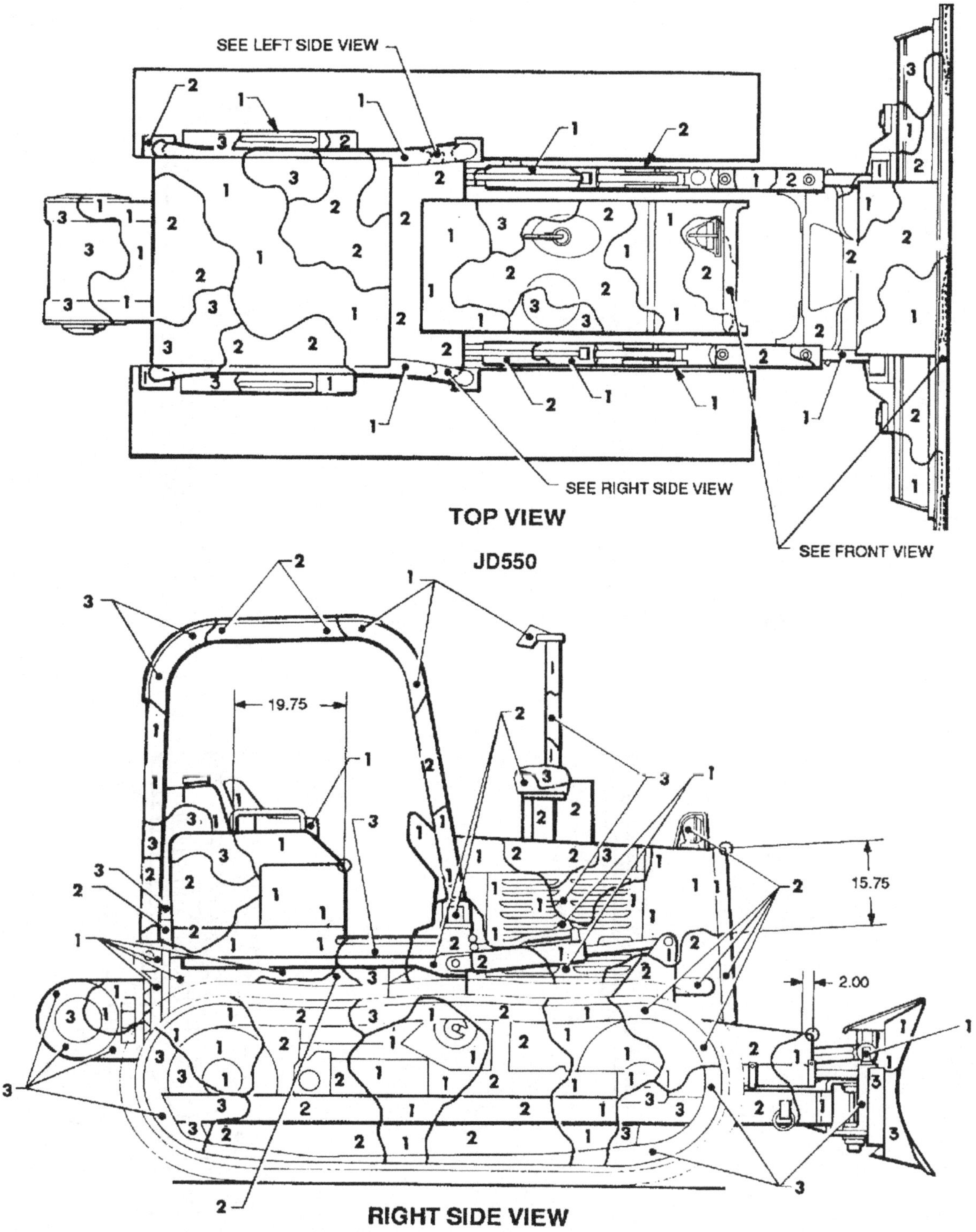

Figure 157. Tractor, full tracked, low speed, DED, air dropable and helicopter transportable, JD550. (1 of 2)

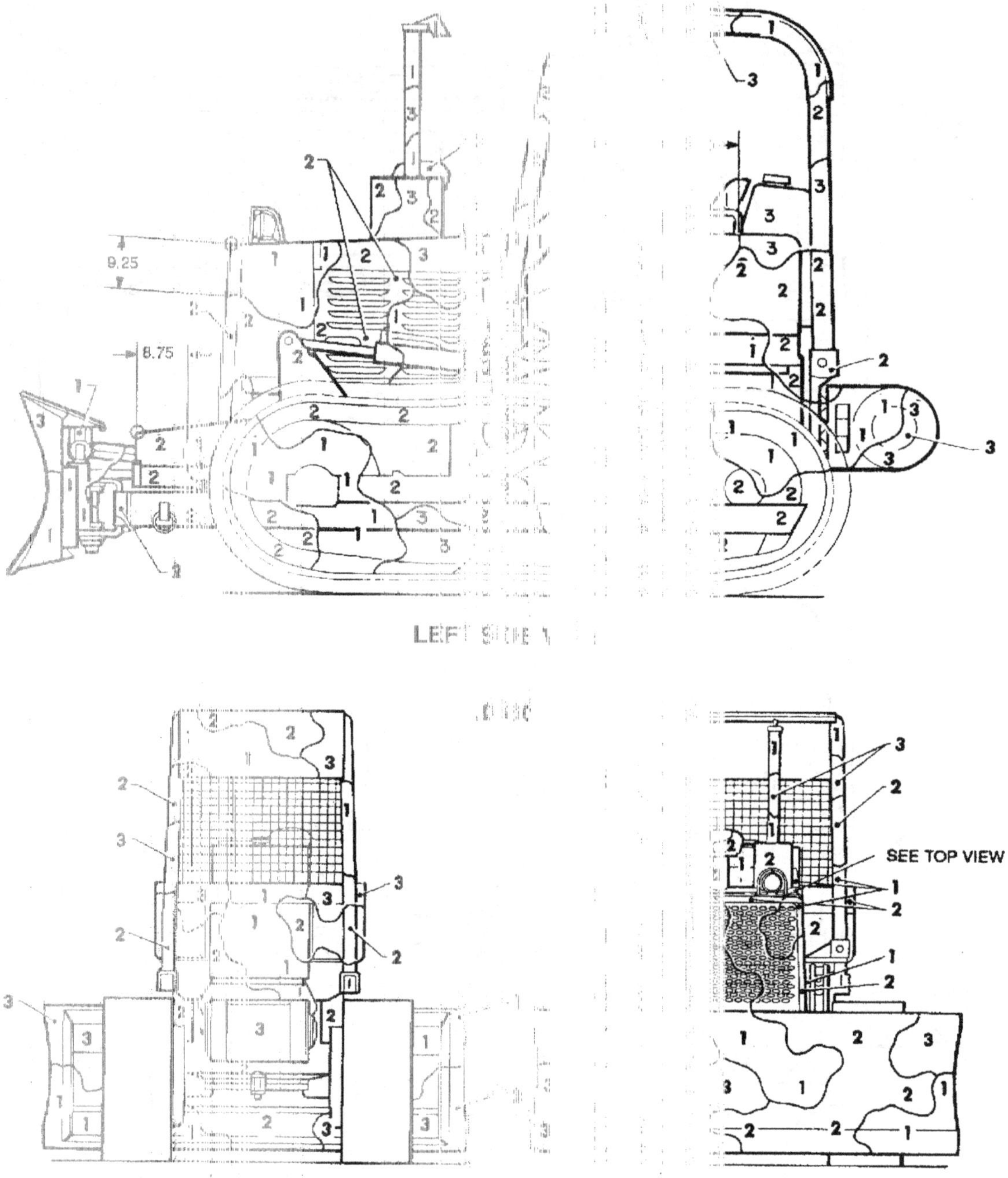

Figure 157. Tractor, full tracked, low speed, DED, [illegible] [illegible]opter transportable, JD550. (2 of [illegible])

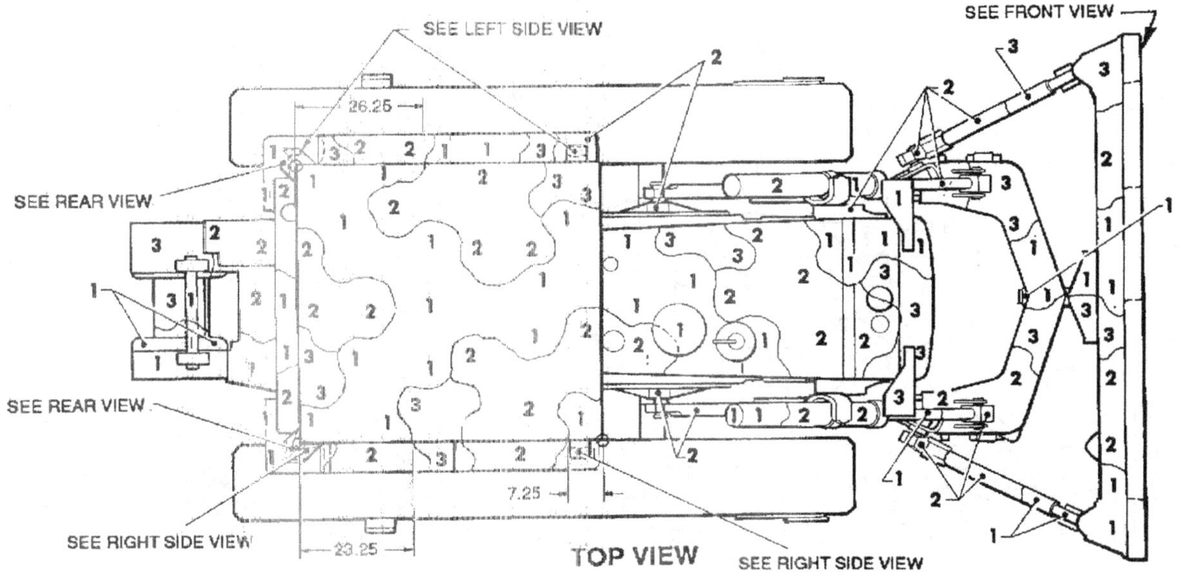

TOP VIEW

D5BS

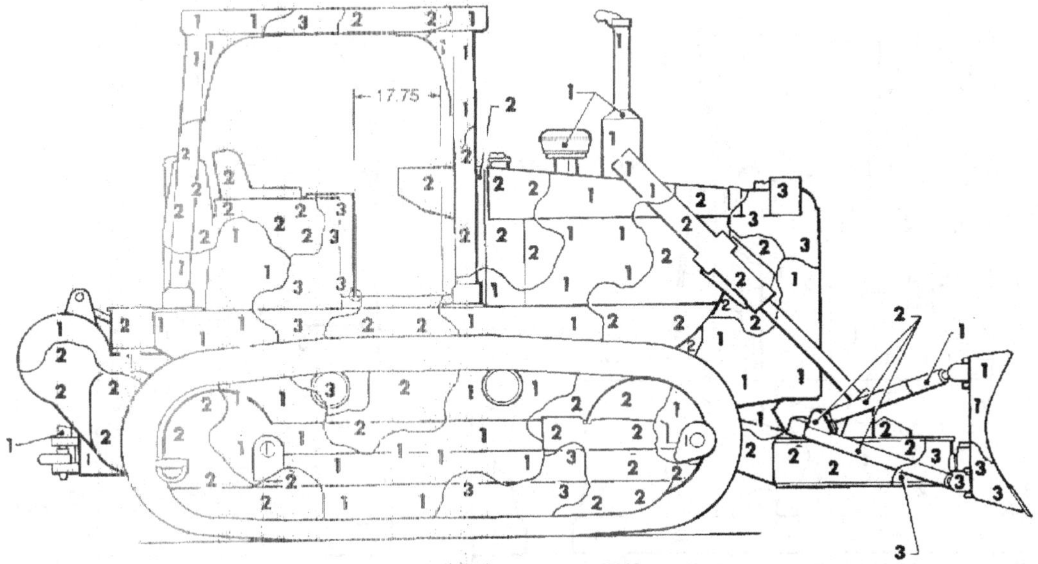

RIGHT SIDE VIEW

Figure 158. Tractor, full tracked, low speed, DED, sectionalized and nonsectionalized, D5B. (1 of 2)

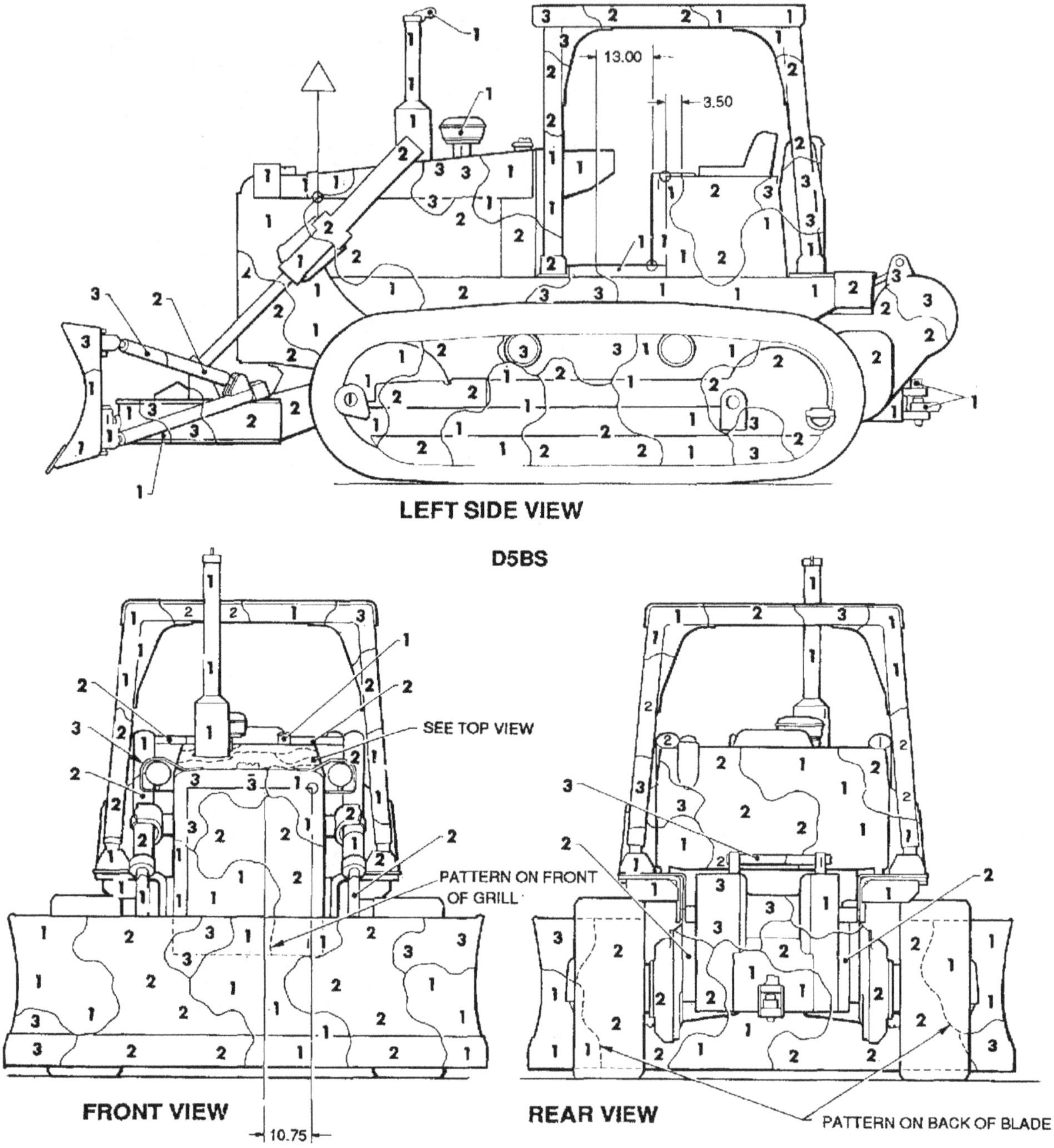

Figure 158. Tractor, full tracked, low speed, DED, sectionalized and nonsectionalized, D5B. (2 of 2)

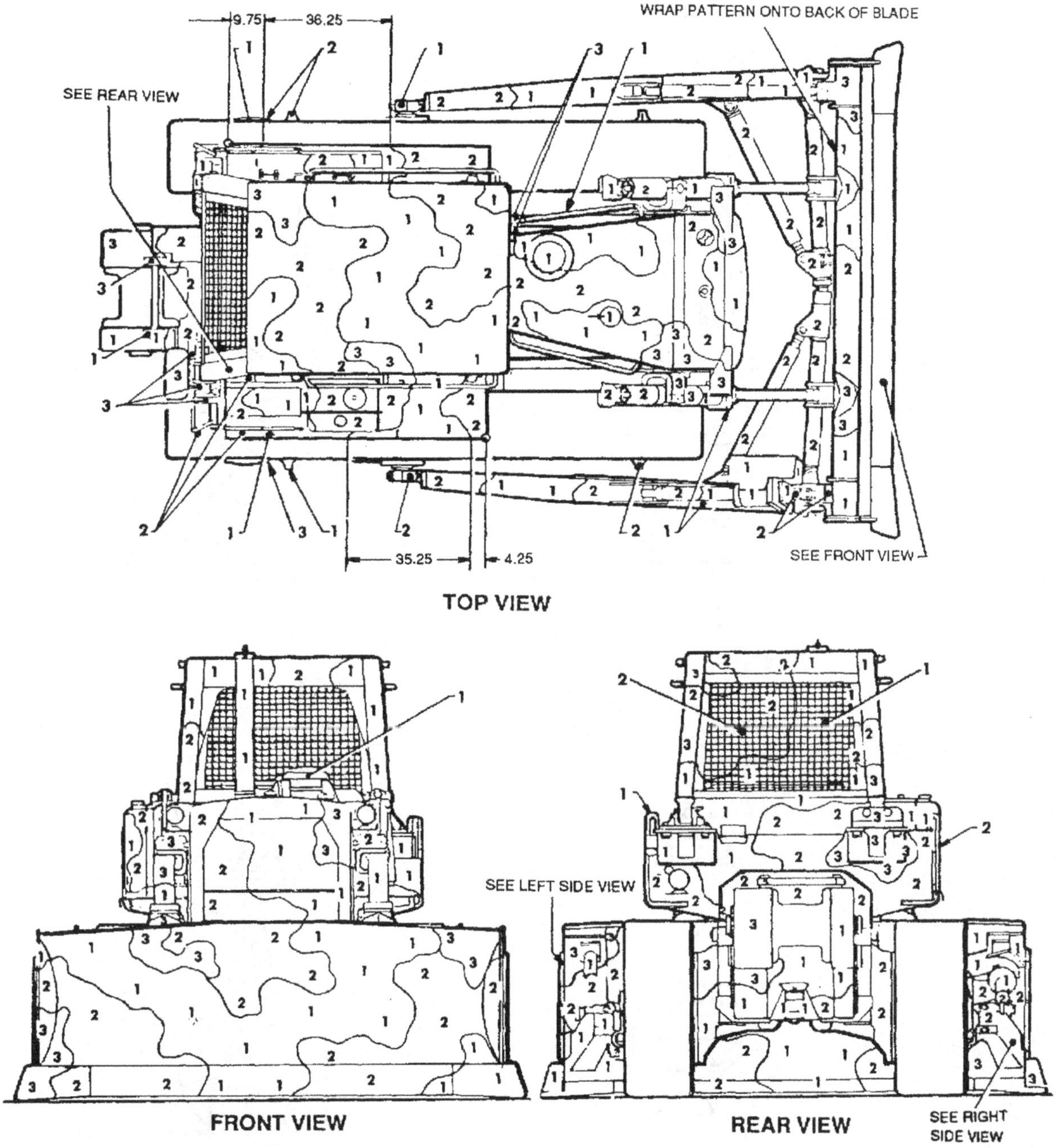

Figure 159. Tractor, full tracked, low speed, DED, medium drawbar pull (T-9), D7G. (1 of 2)

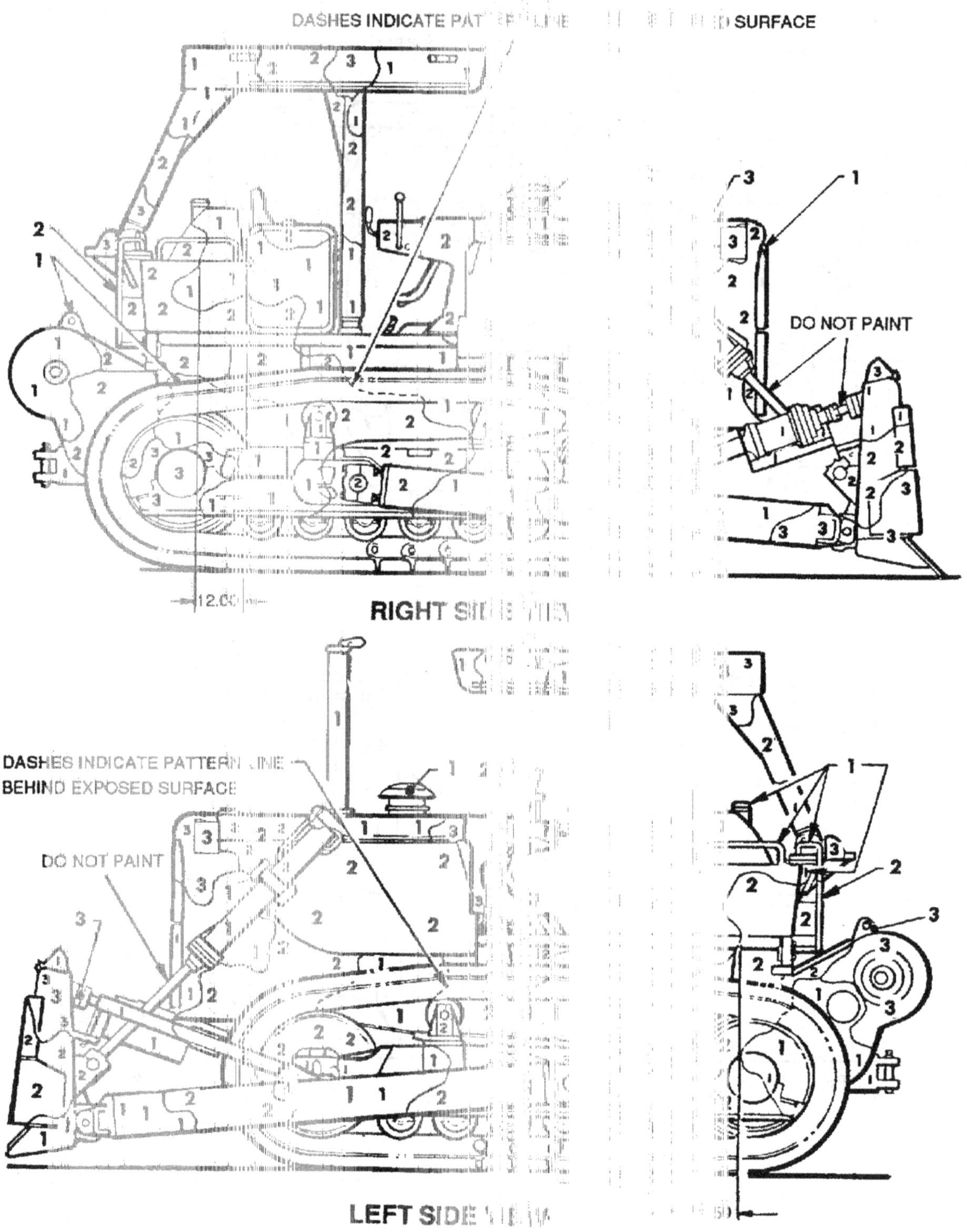

Figure 159. Tractor, full tracked, low speed, DE ... -9), D7G. (2 of 2)

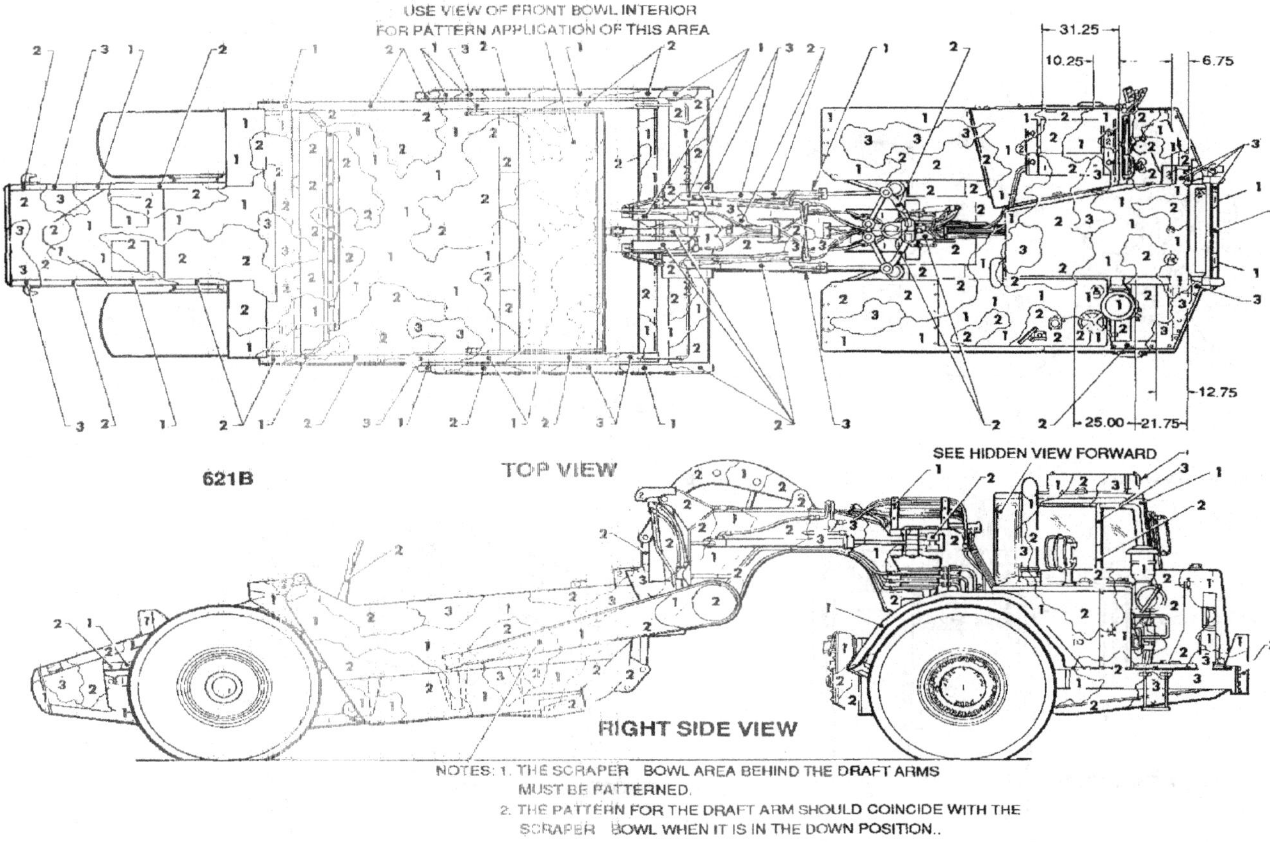

Figure 160. Scraper, open bowl, self-propelled, 621B. (1 of 2)

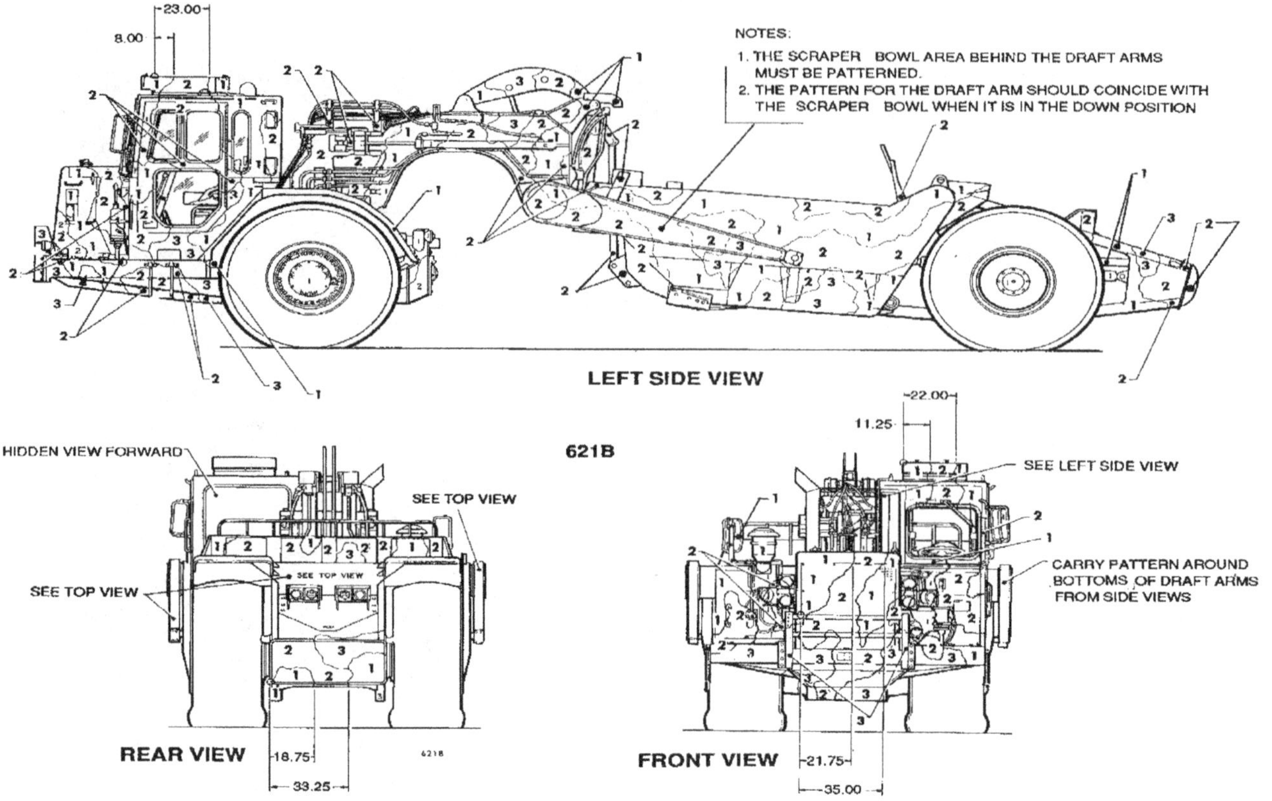

Figure 160. Scraper, open bowl, self-propelled, 621B. (2 of 2)

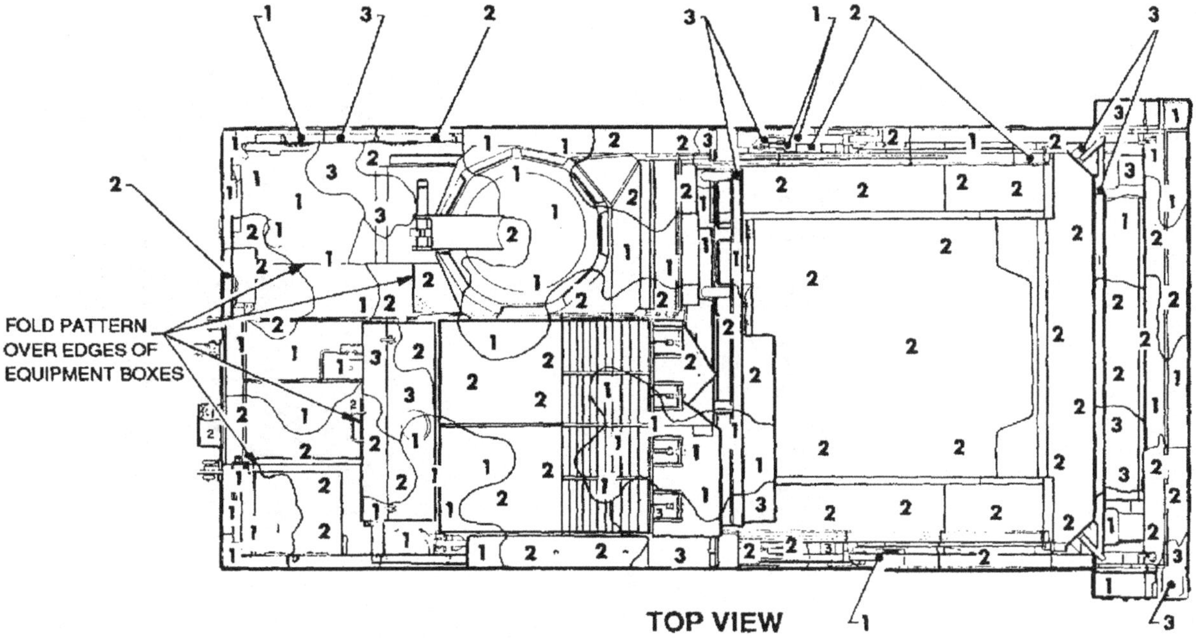

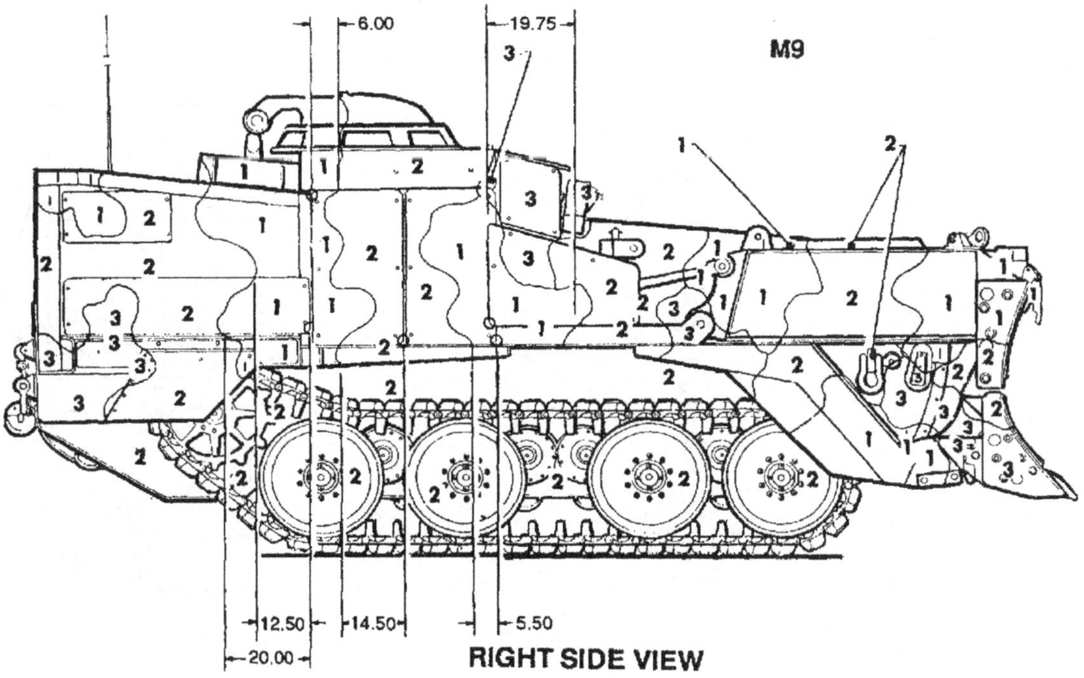

Figure 161. Earthmover, armored, combat, M9. (1 of 2)

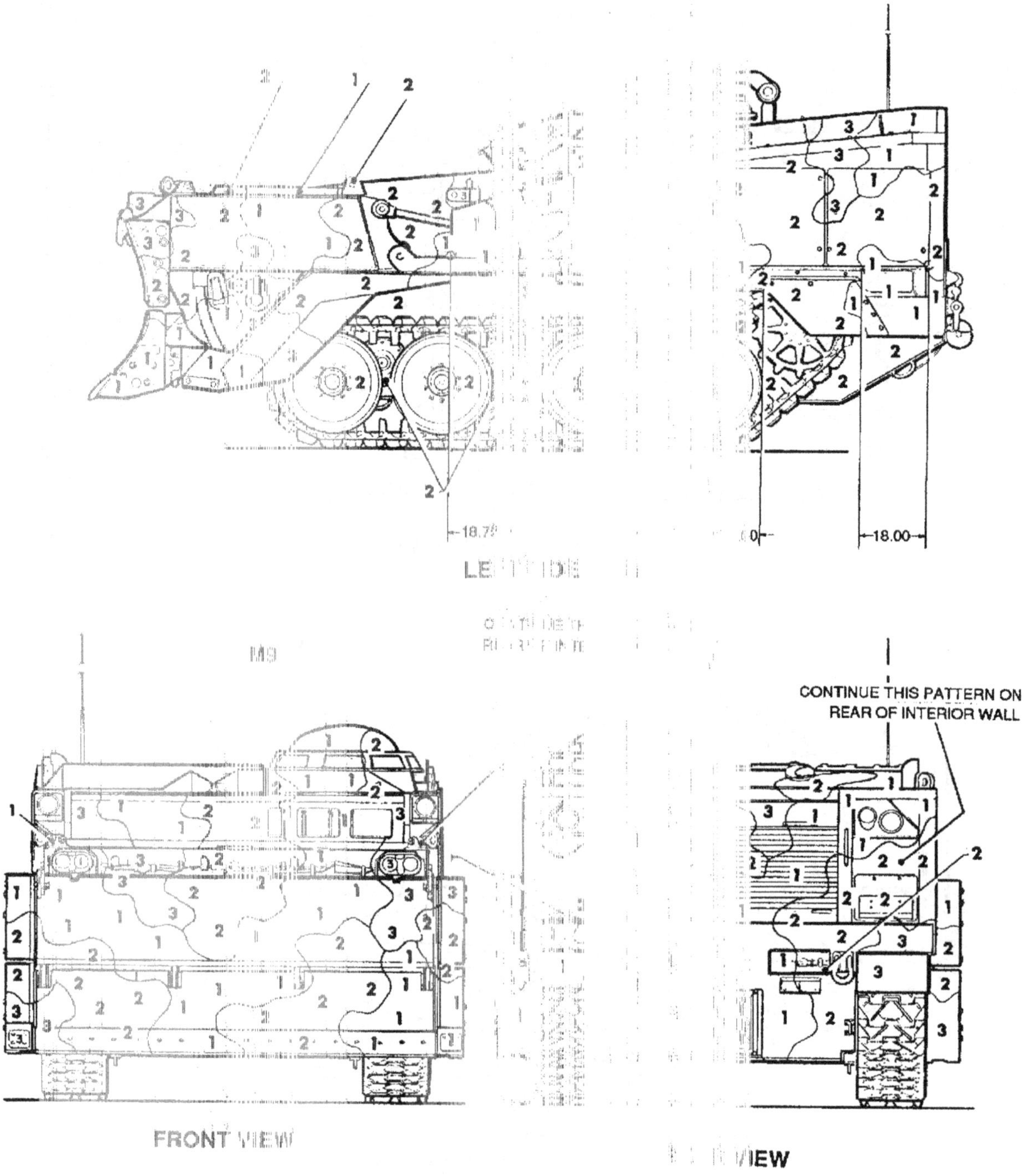

Figure 161. Earthmover,

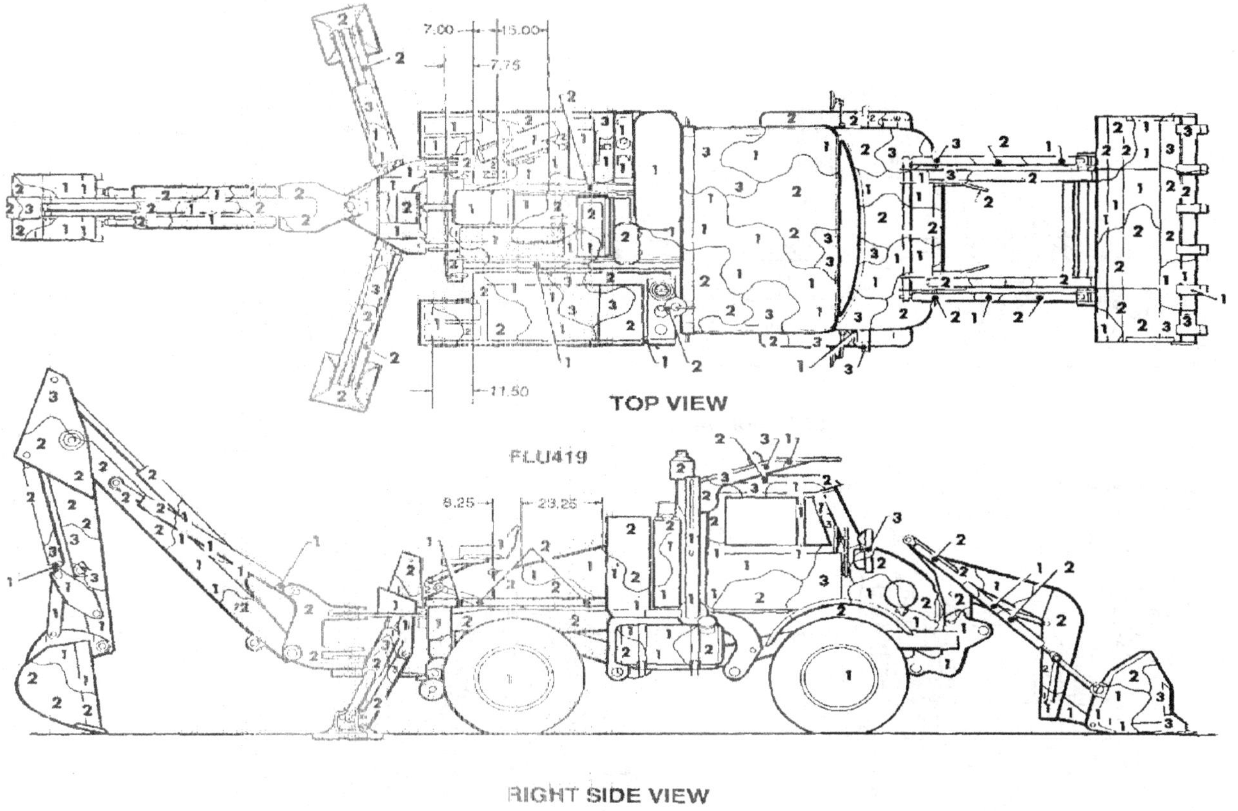

Figure 162. Small emplacement excavator, FLU419. (1 of 2)

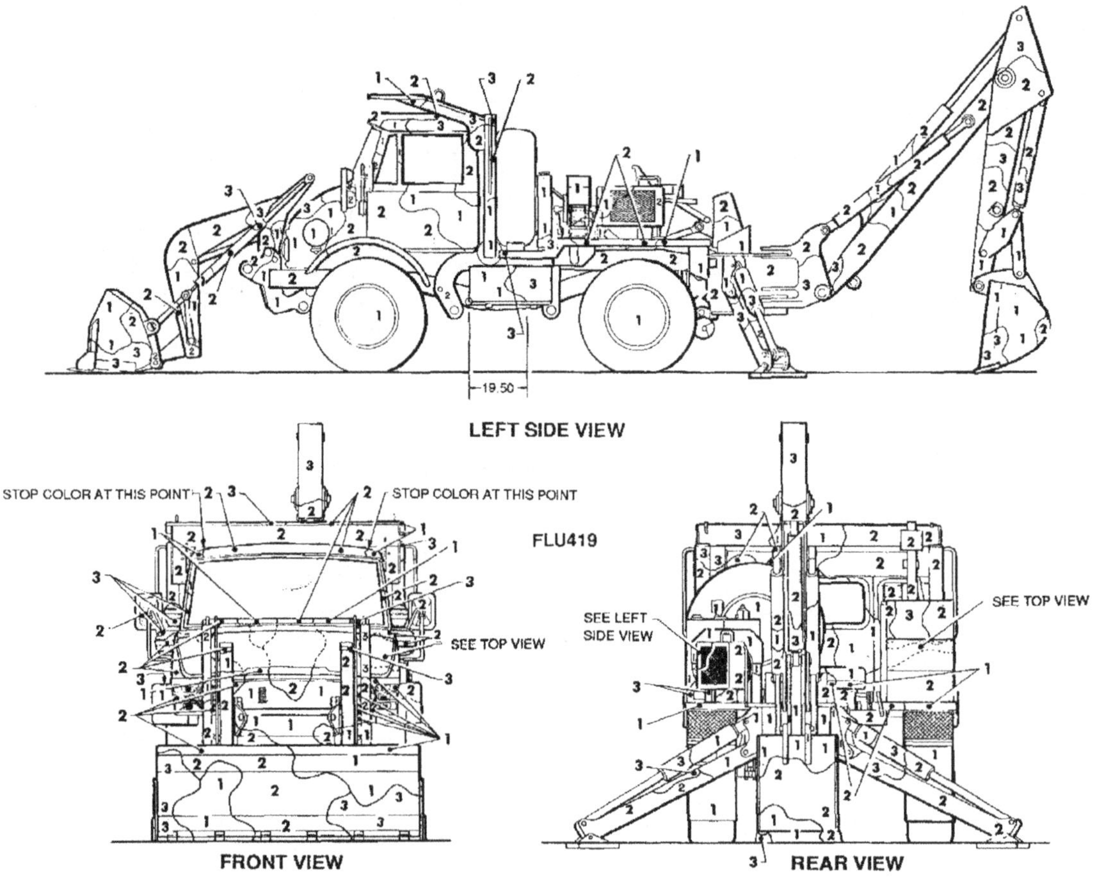

Figure 162. Small emplacement excavator, FLU419. (2 of 2)

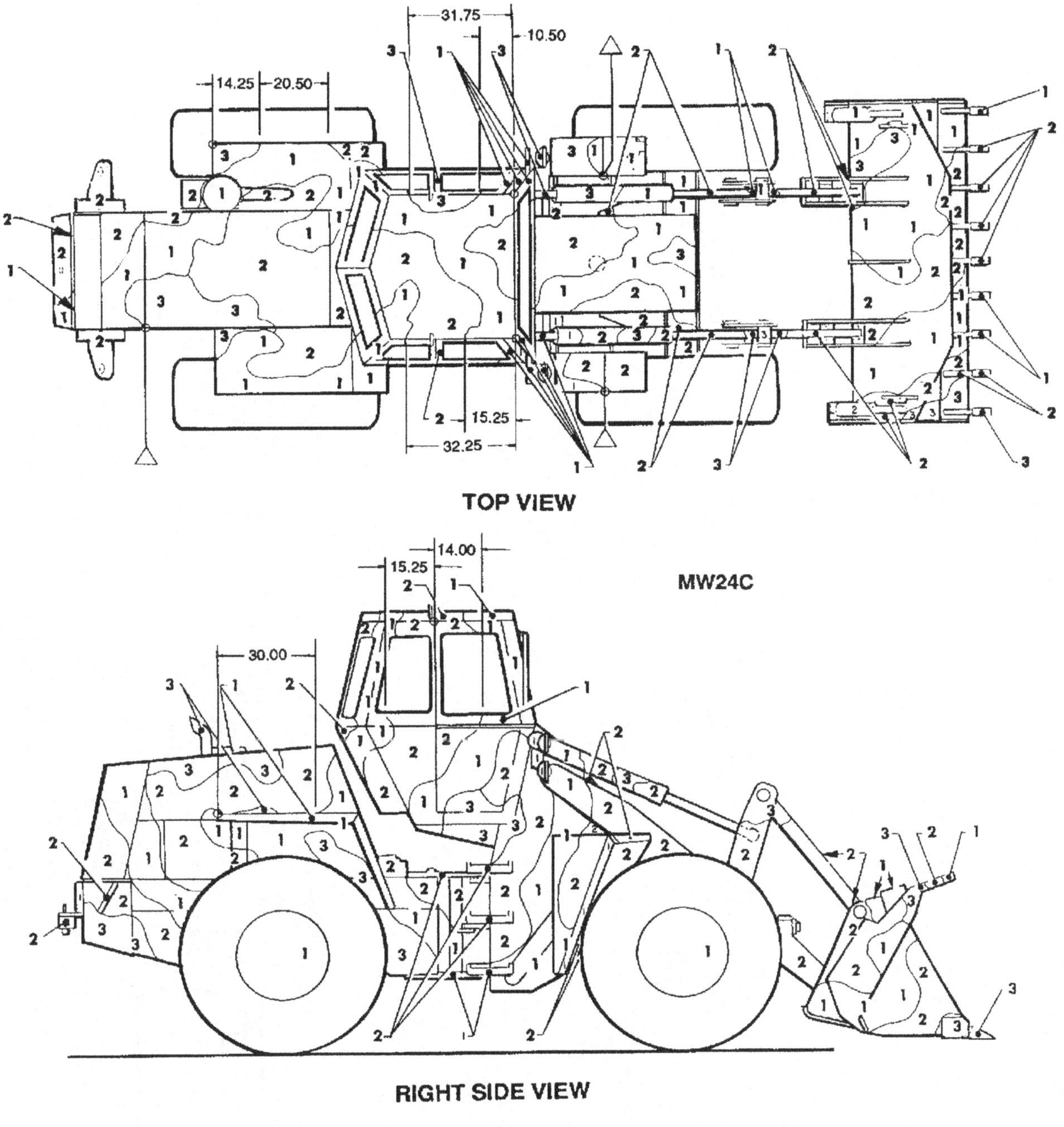

Figure 163. Scoop loader, 2-1/2 CY, MW24C. (1 of 2)

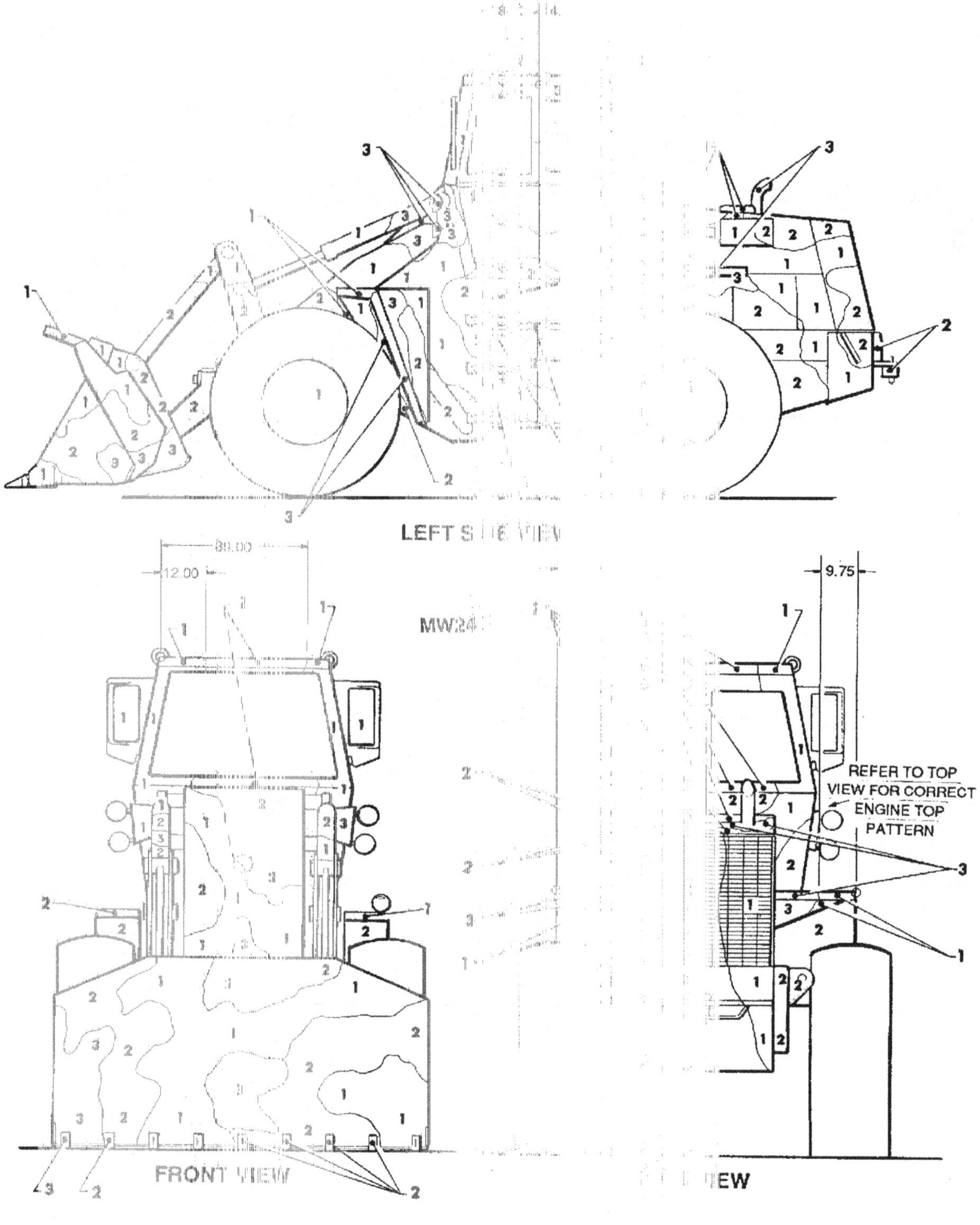

Figure 163. Scoop loader, 2

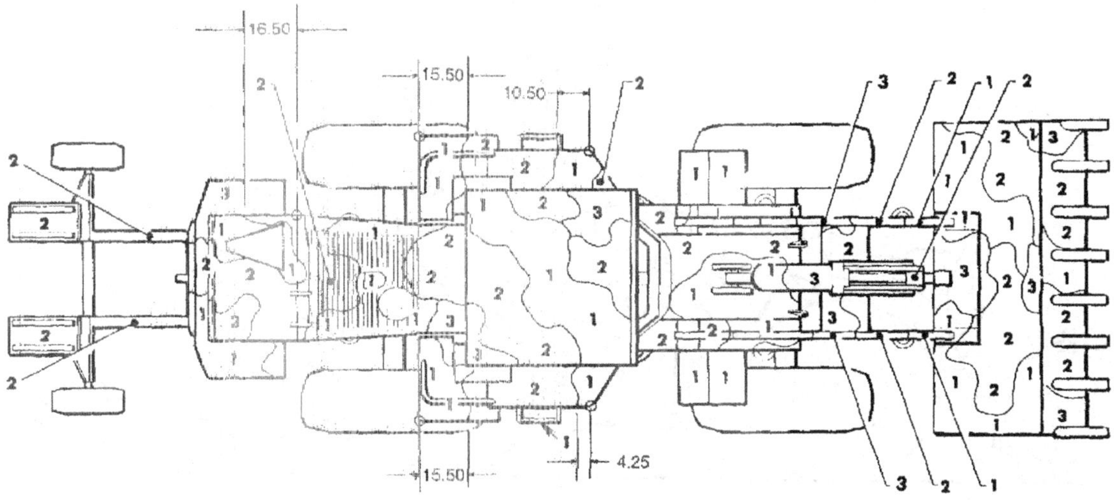

TOP VIEW

950BS 950BNS

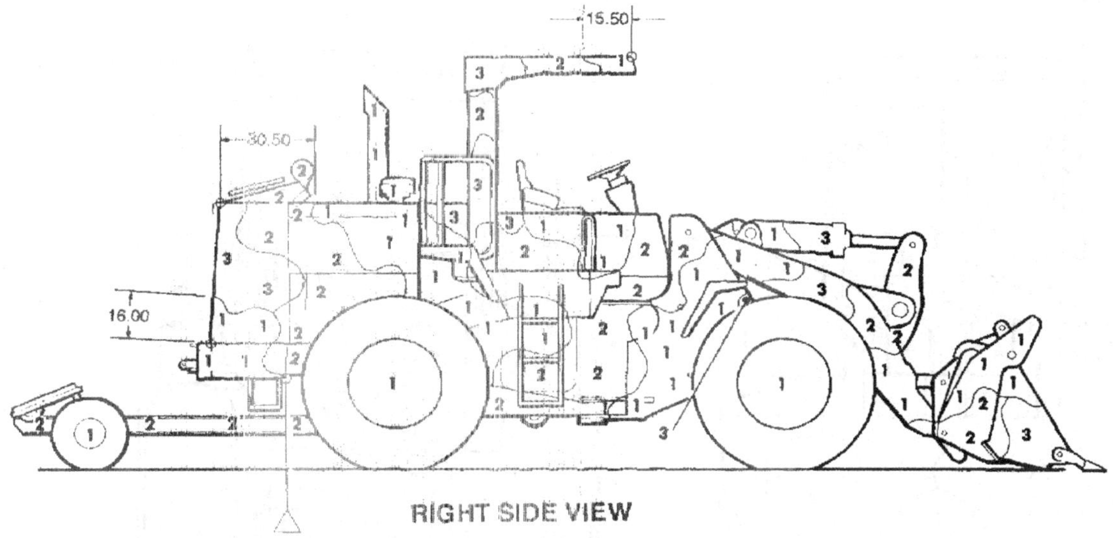

RIGHT SIDE VIEW

Figure 164. Scoop loader, 950BS and 950BNS. (1 of 2)

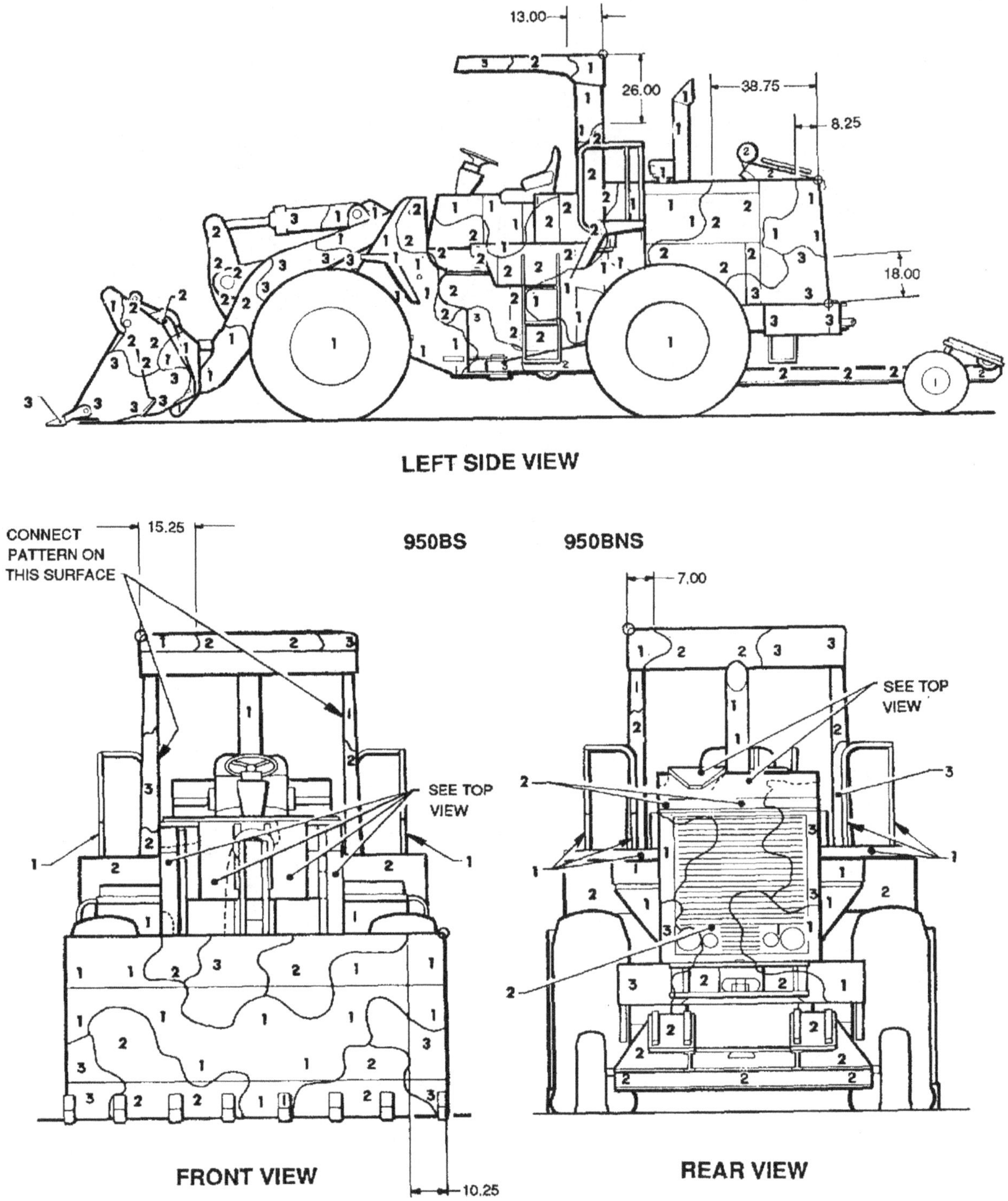

Figure 164. Scoop loader, 950BS and 950BNS. (2 of 2)

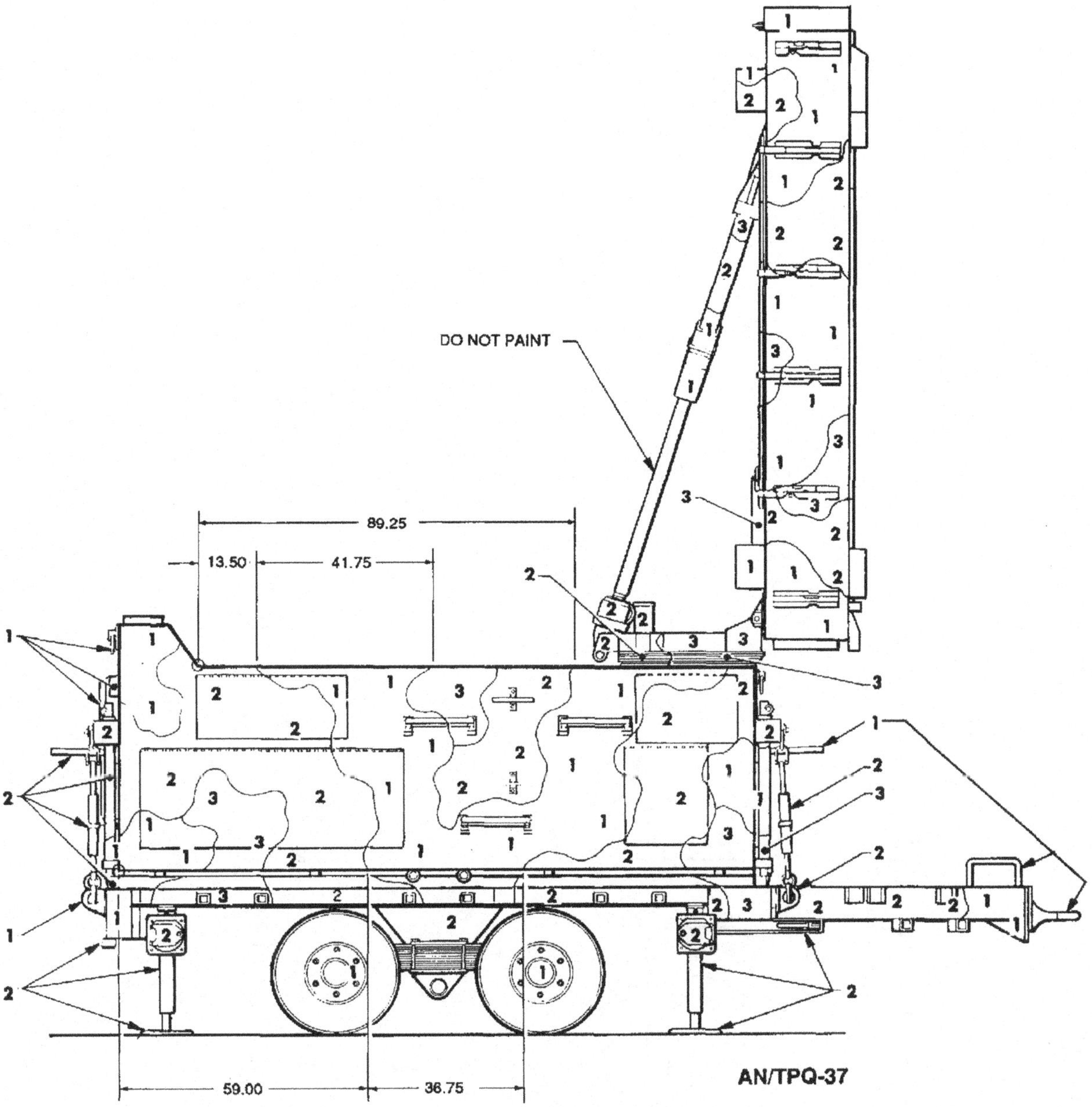

RIGHT SIDE VIEW

Figure 165. Trailer, chassis: antenna transceiver group, AN/TPQ-37. (1 of 5)

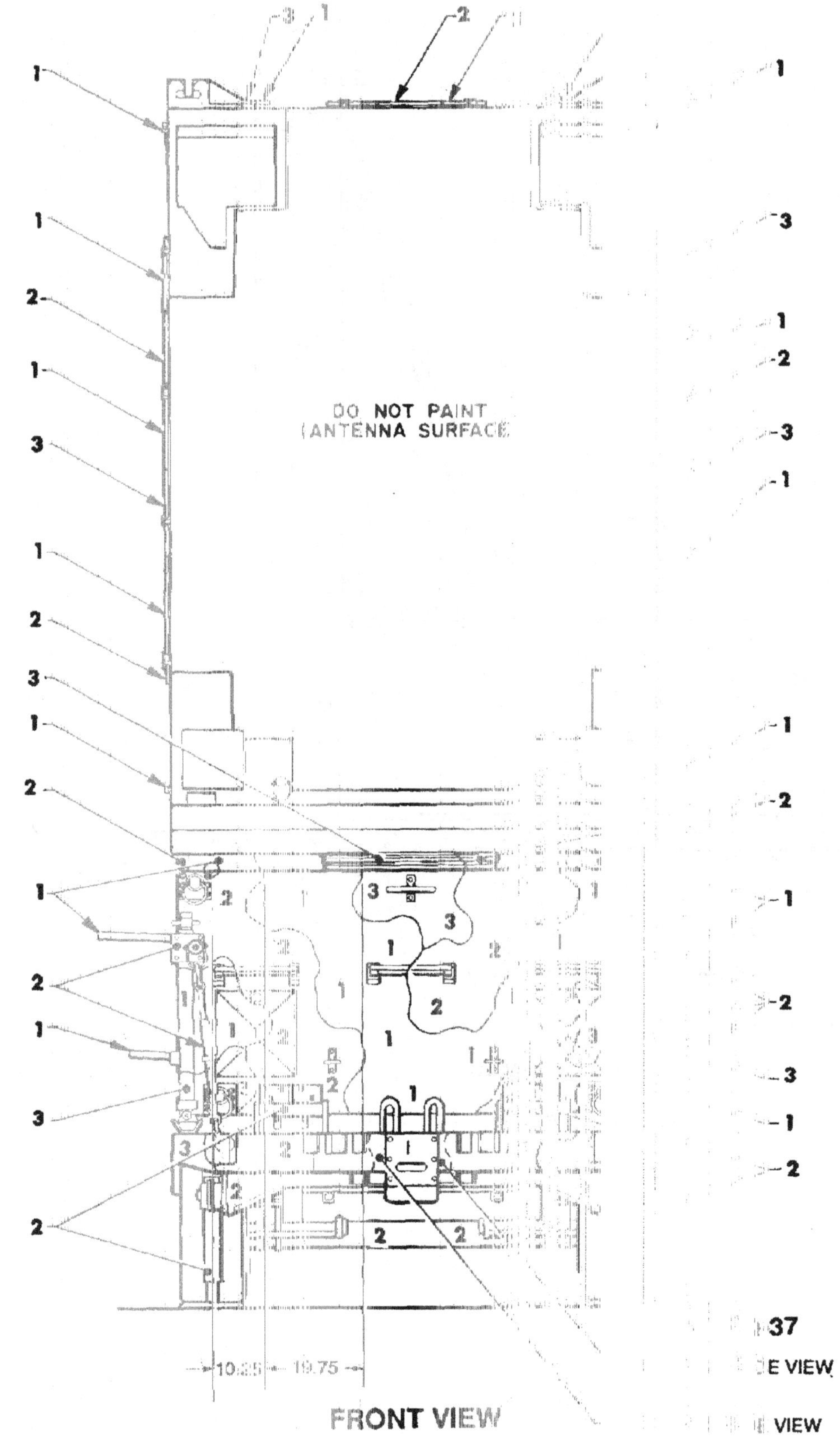

Figure 165. Trailer, chassis: antenna transceiver group, [illegible] 2 of 5)

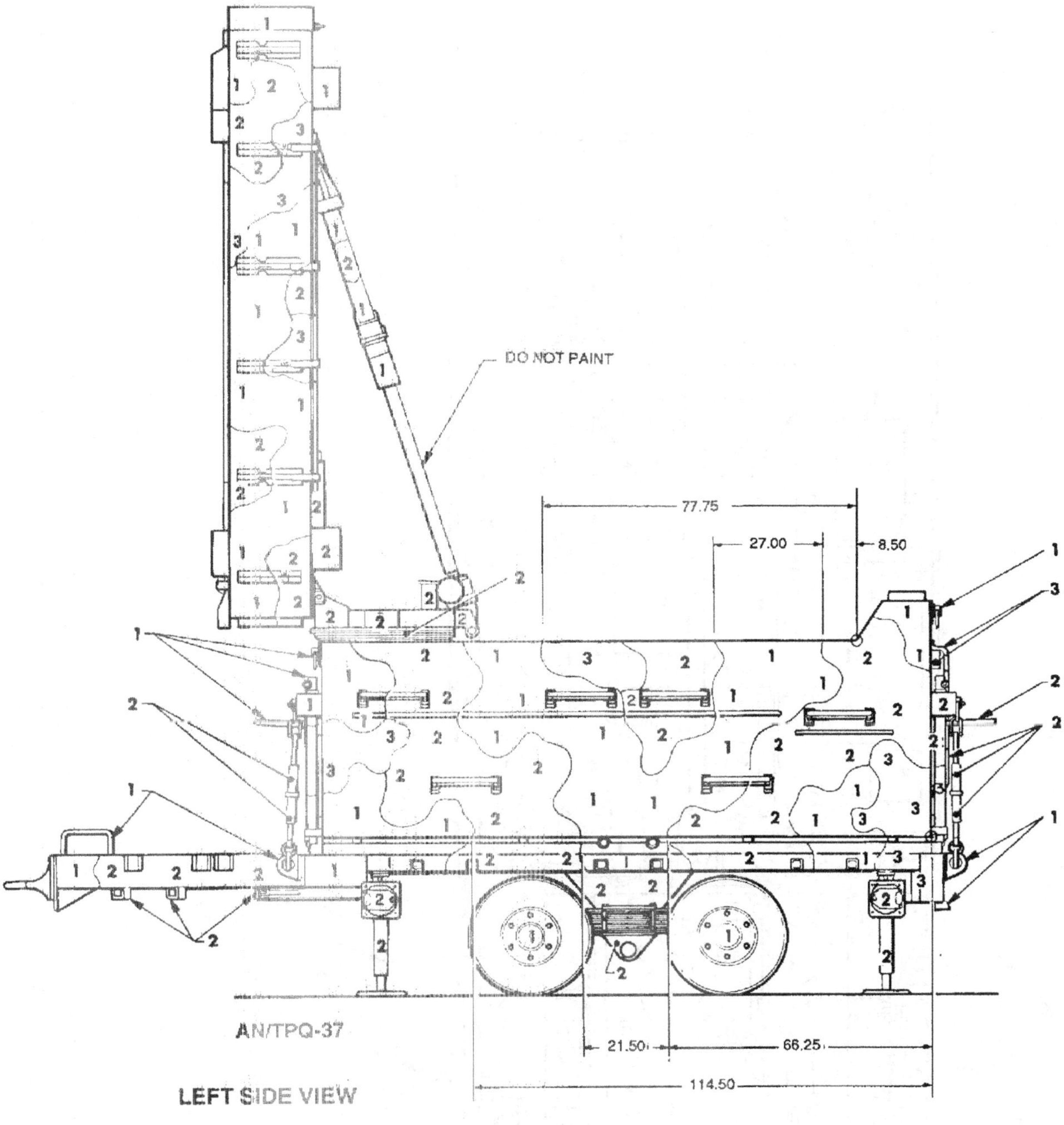

Figure 165. Trailer, chassis: antenna transceiver group, AN/TPQ-37. (3 of 5)

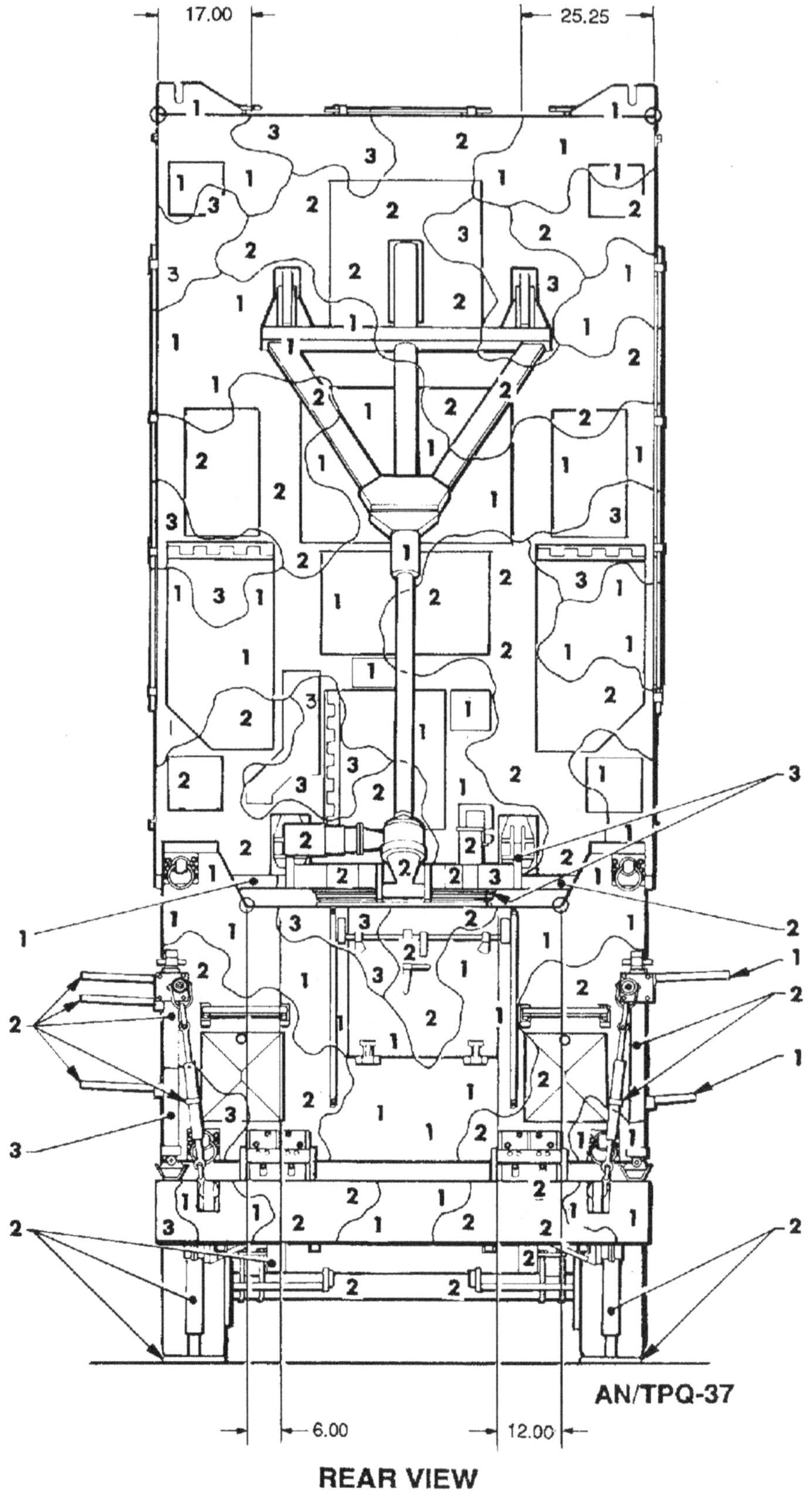

Figure 165. Trailer, chassis: antenna transceiver group, ANITPQ-37. (4 of 5)

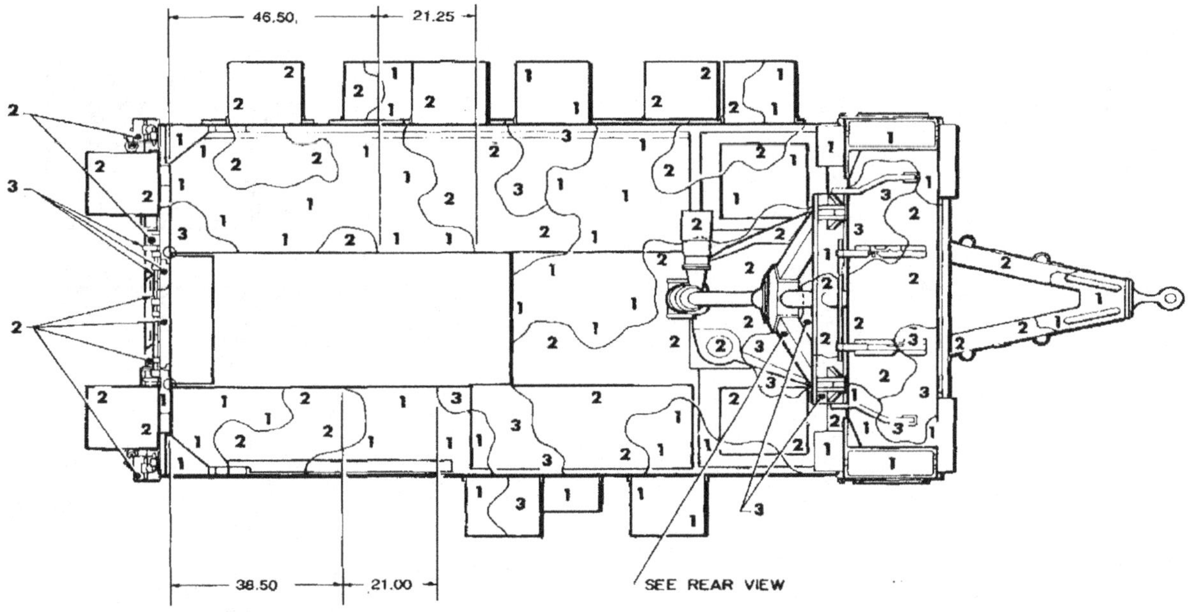

AN/TPQ-37

TOP VIEW

Figure 165. Trailer, chassis: antenna transceiver group, AN/TPQ-37. (5 of 5)

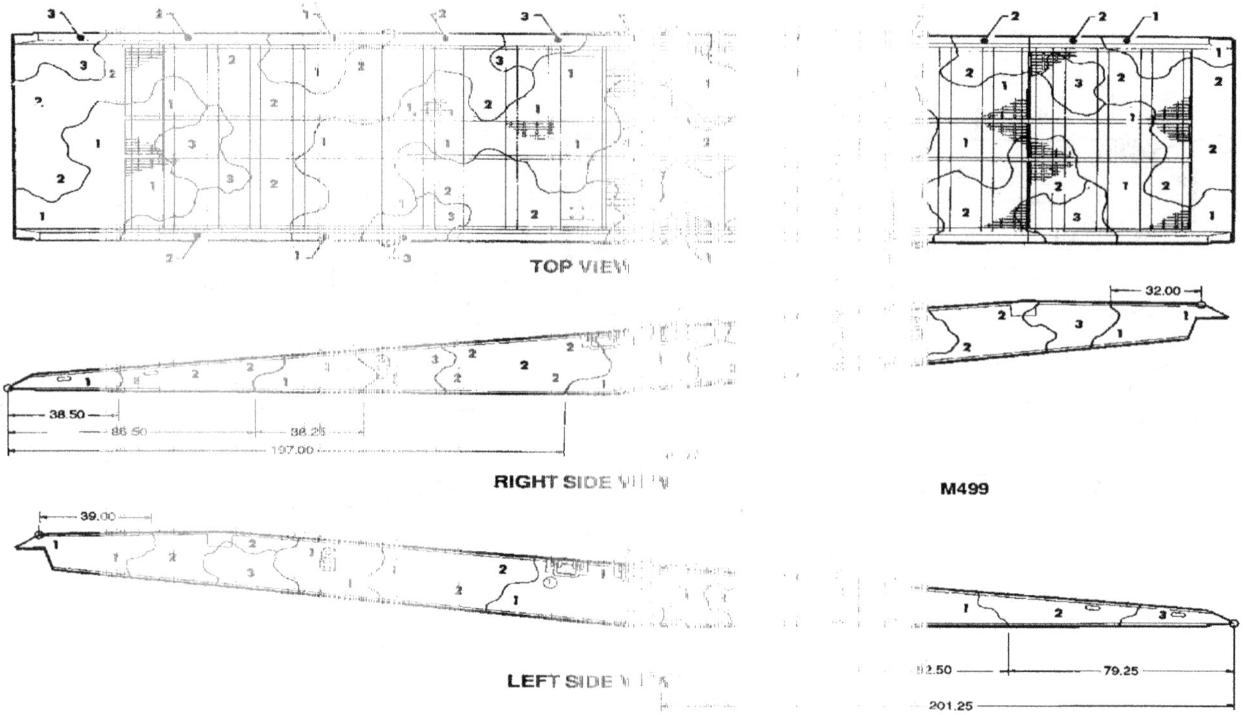
TOP VIEW
RIGHT SIDE
LEFT SIDE
M499
32.00
38.50
86.50
38.25
197.00
39.00
79.25
201.25

Figure 166. [illegible]

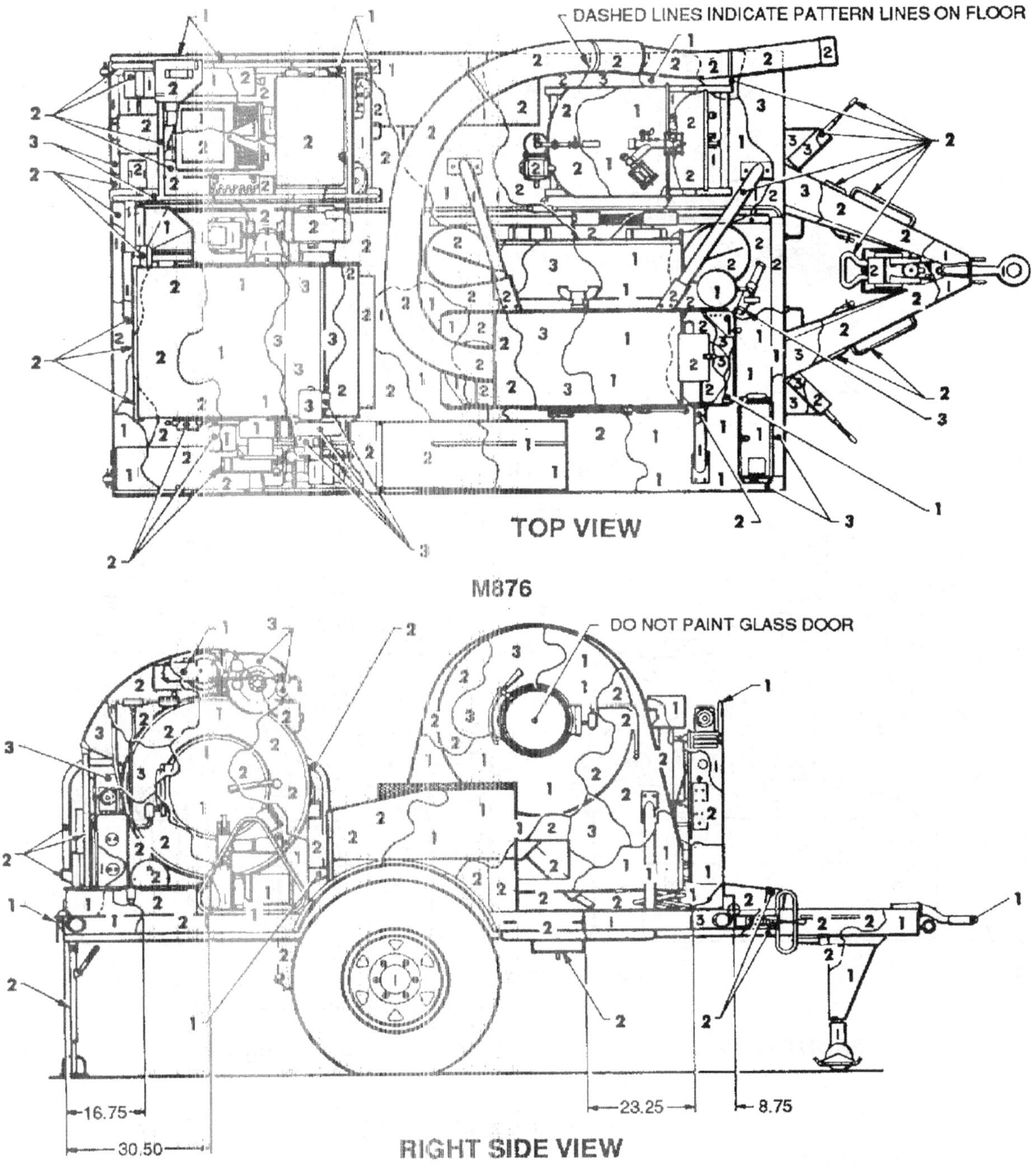

Figure 167. Laundry Unit, M536. (1 of 2)

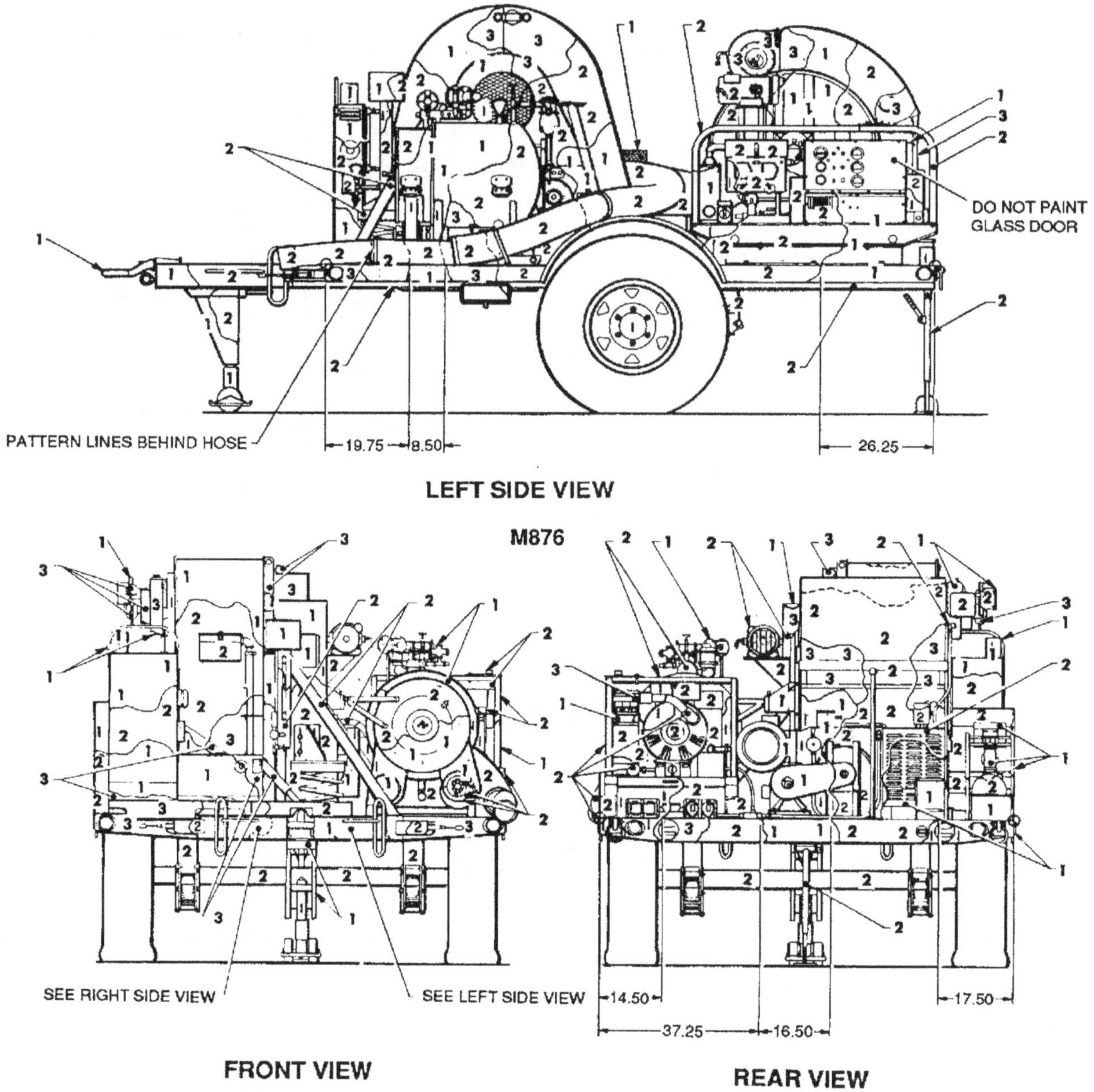

Figure 167. Laundry Unit, M536. (2 of 2)

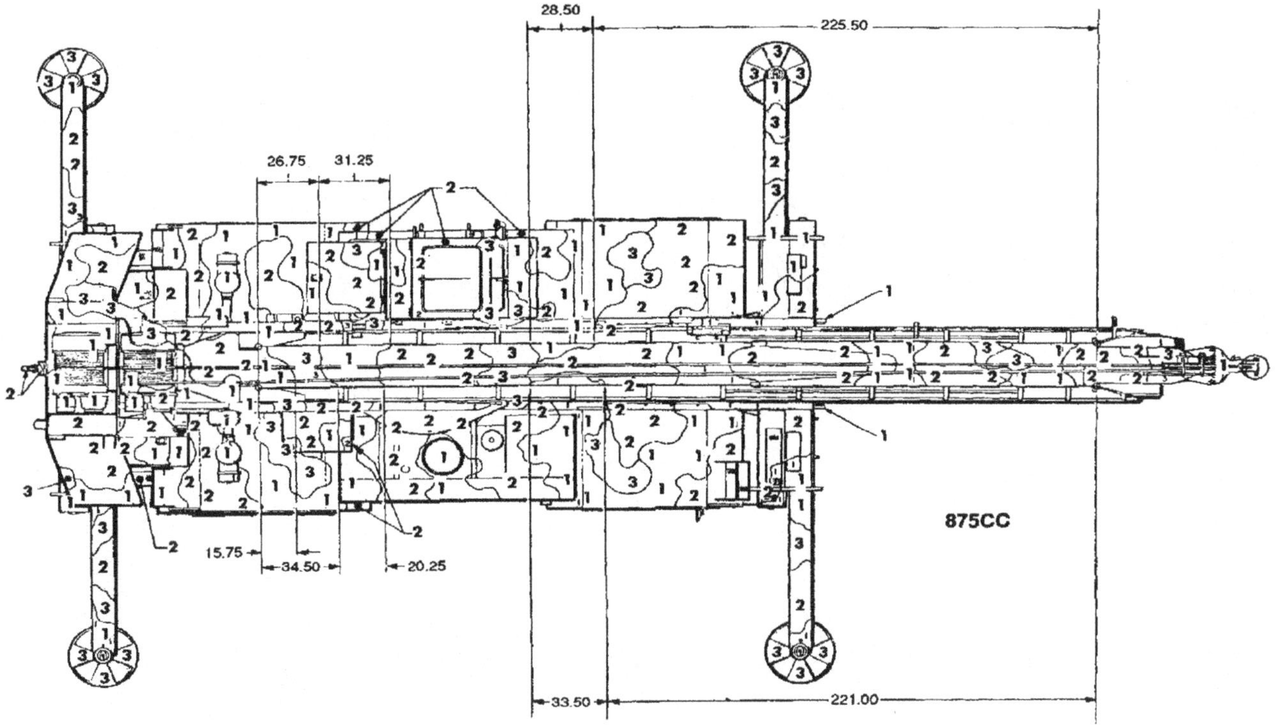

TOP VIEW

Figure 168. Crane, container: rough terrain, 875CC. (1 of 4)

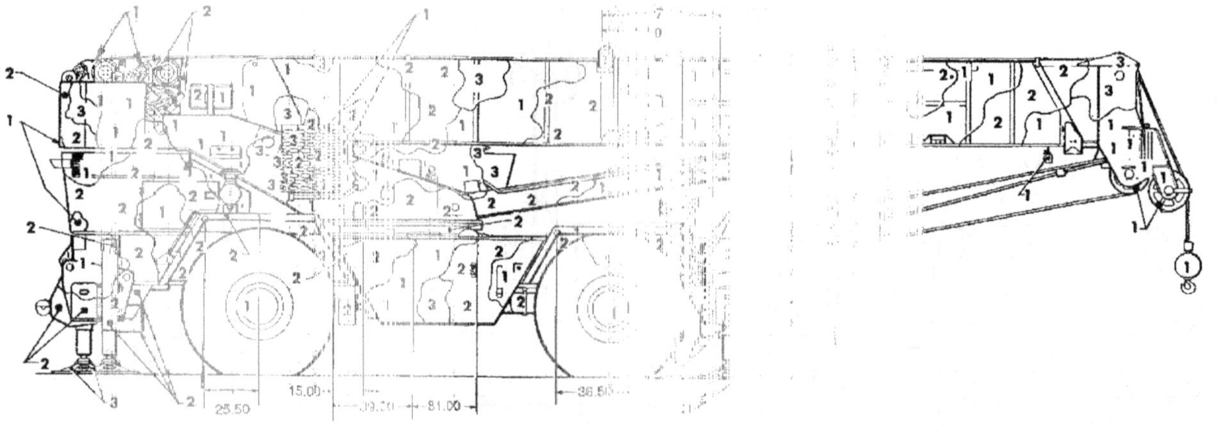

RIGHT SIDE VIEW

875CC

Figure 168. Crane, container: rough terrain, 875CC

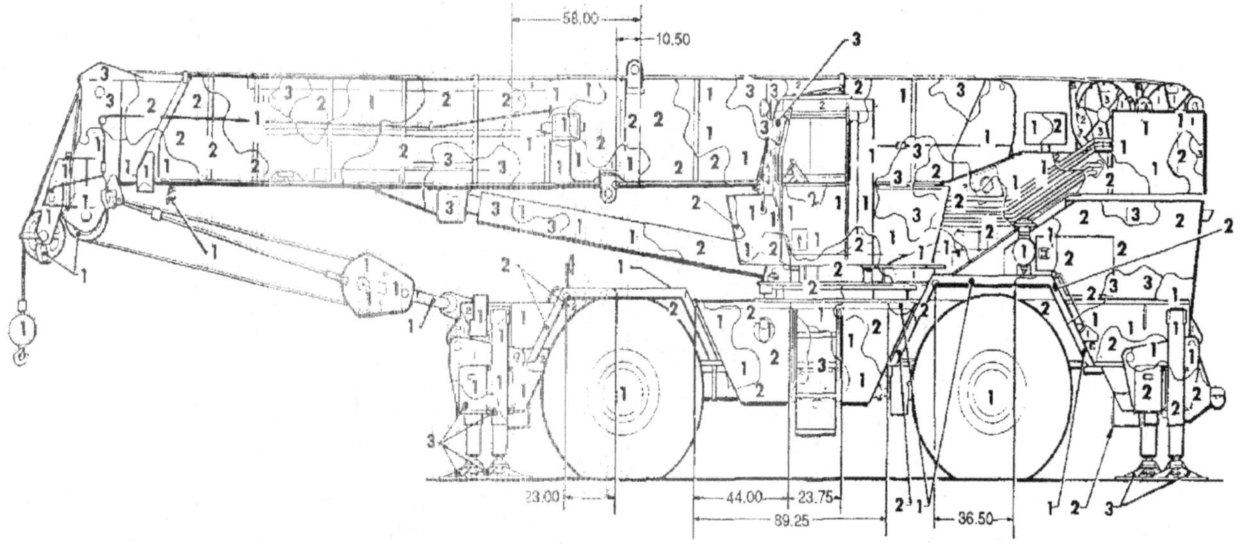

LEFT SIDE VIEW

875CC

Figure 168. Crane, container: rough terrain, 875CC. (3 of4)

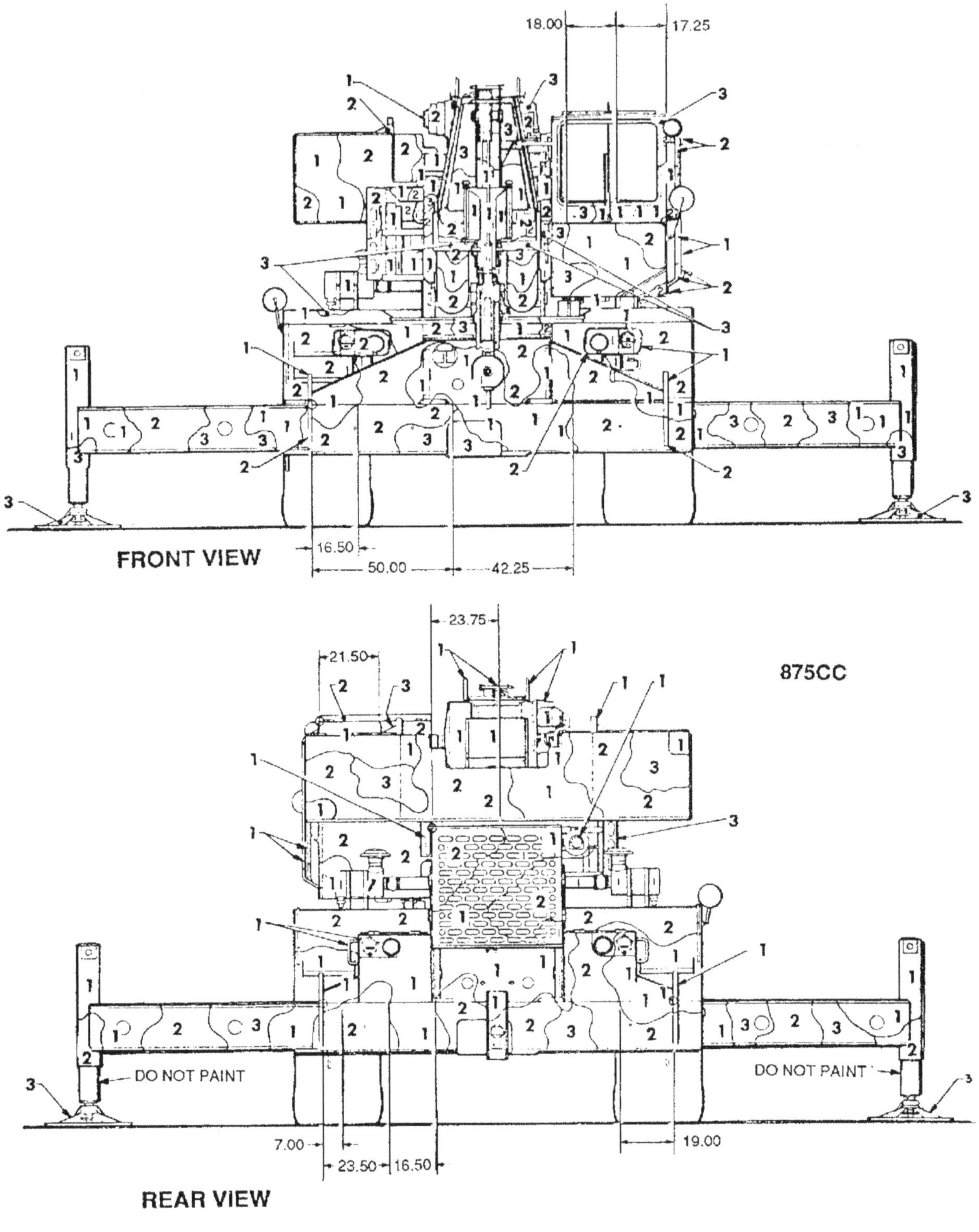

Figure 168. Crane, container: rough terrain, 875CC. (4 of 4)

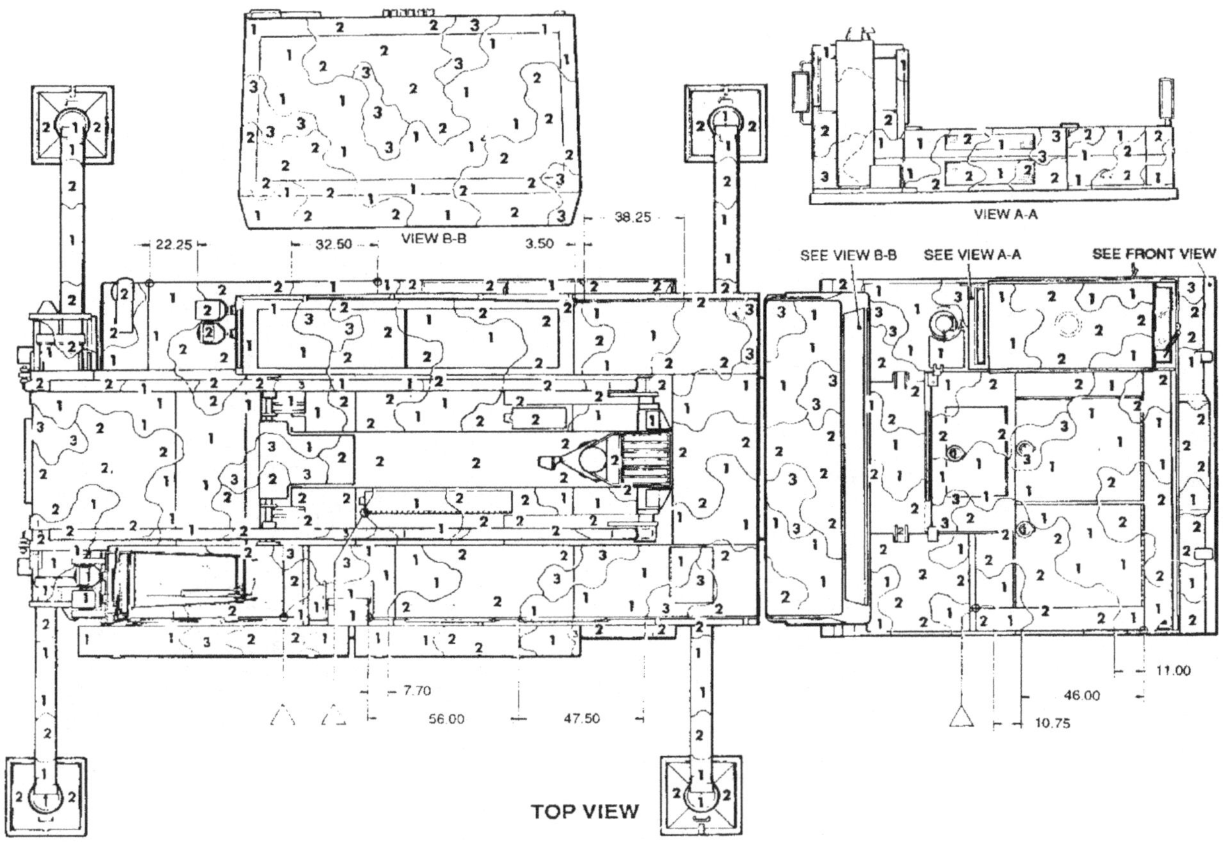

Figure 169. Crane, truck, 140-ton. (1 of 4)

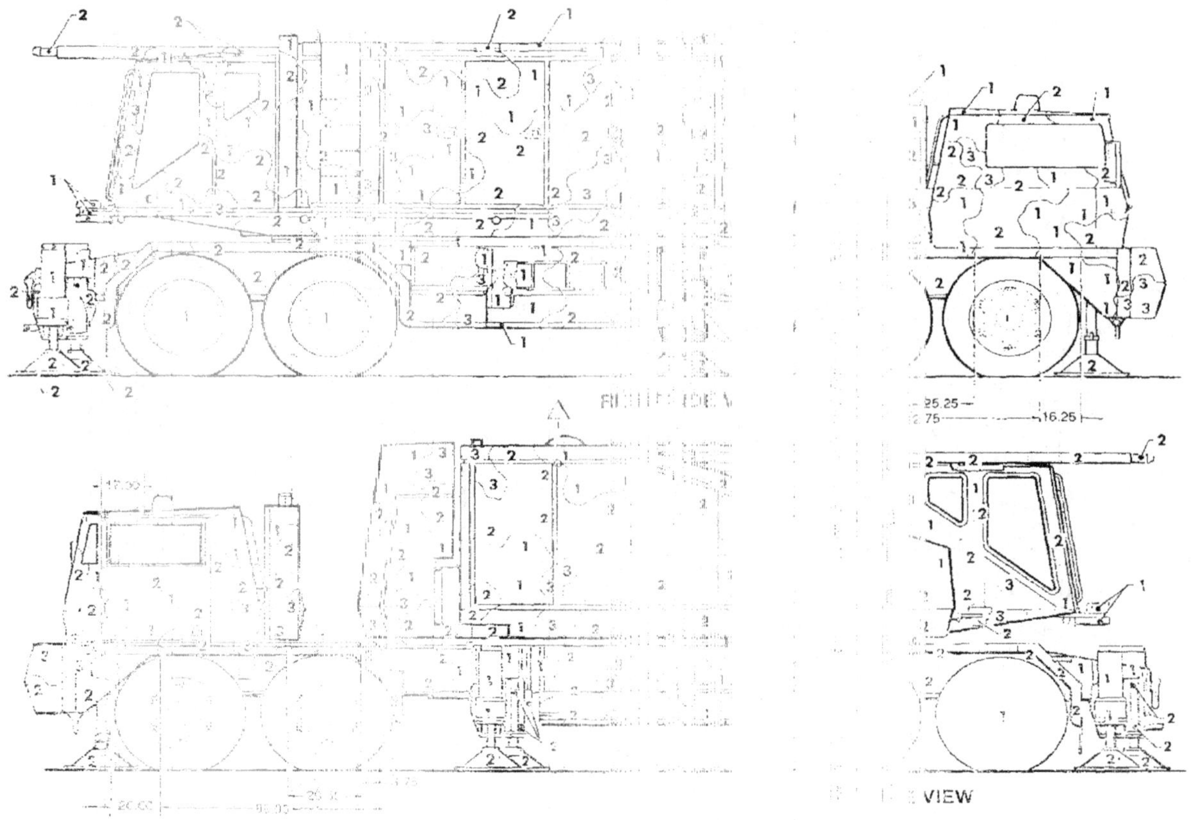

Figure 169. Crane, truck, [illegible]

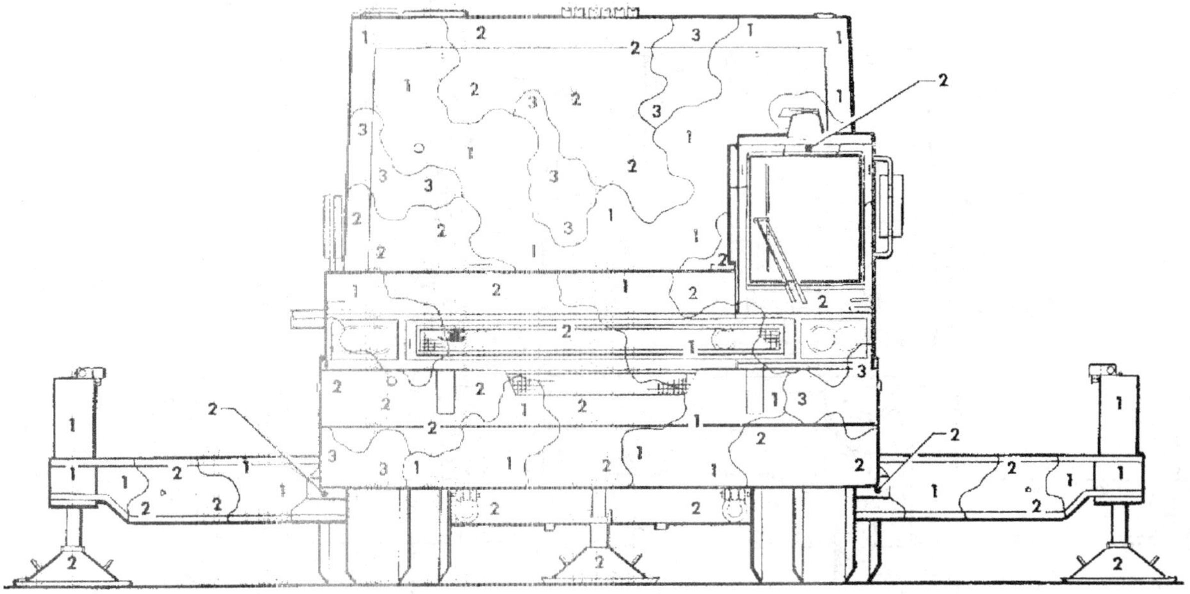

Figure 169. Crane, truck, 140-ton. (3 of 4)

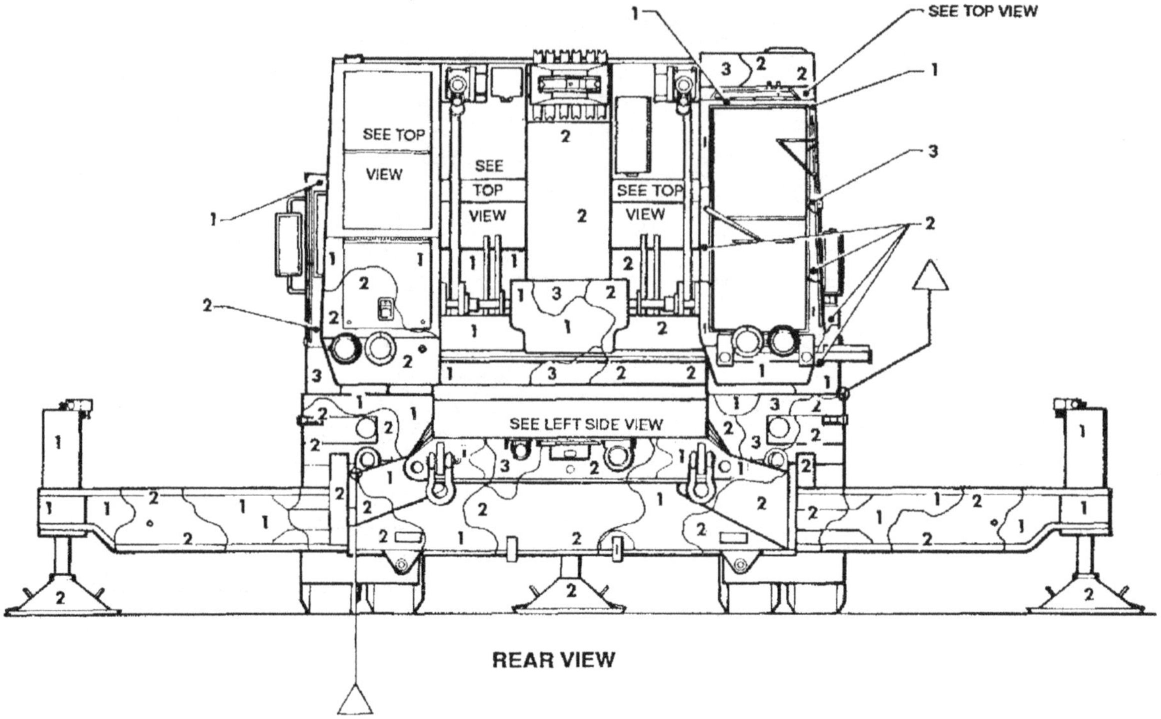

Figure 169. Crane, truck, 140-ton. (4 of 4)

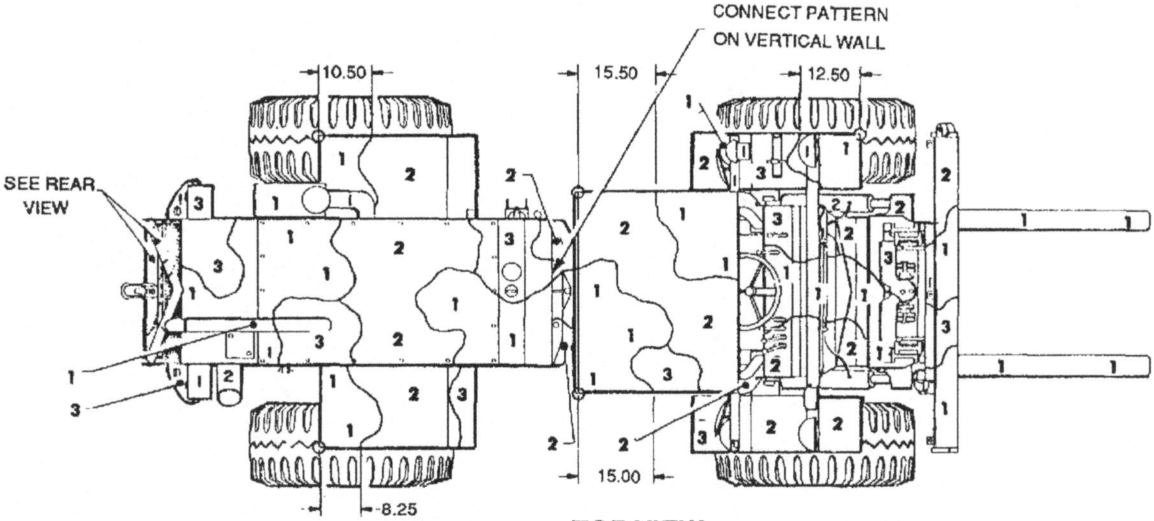

TOP VIEW

M4K

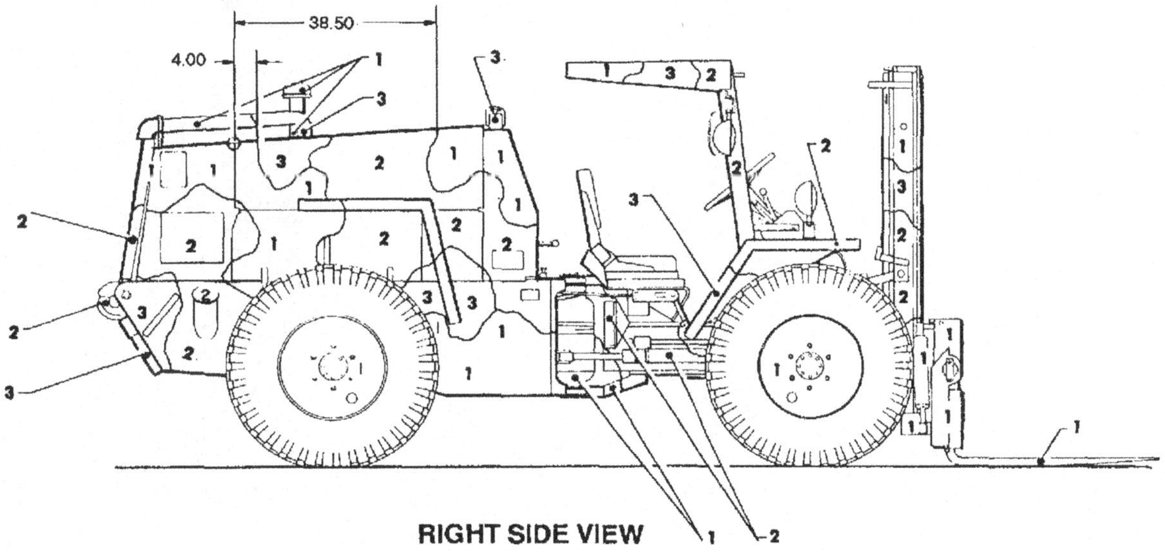

RIGHT SIDE VIEW

Figure 170. Truck, forklift, DED, pneumatic tire, articulated frame steer, 4000 lb. Capacity rough terrain, M4K. (1 of 2)

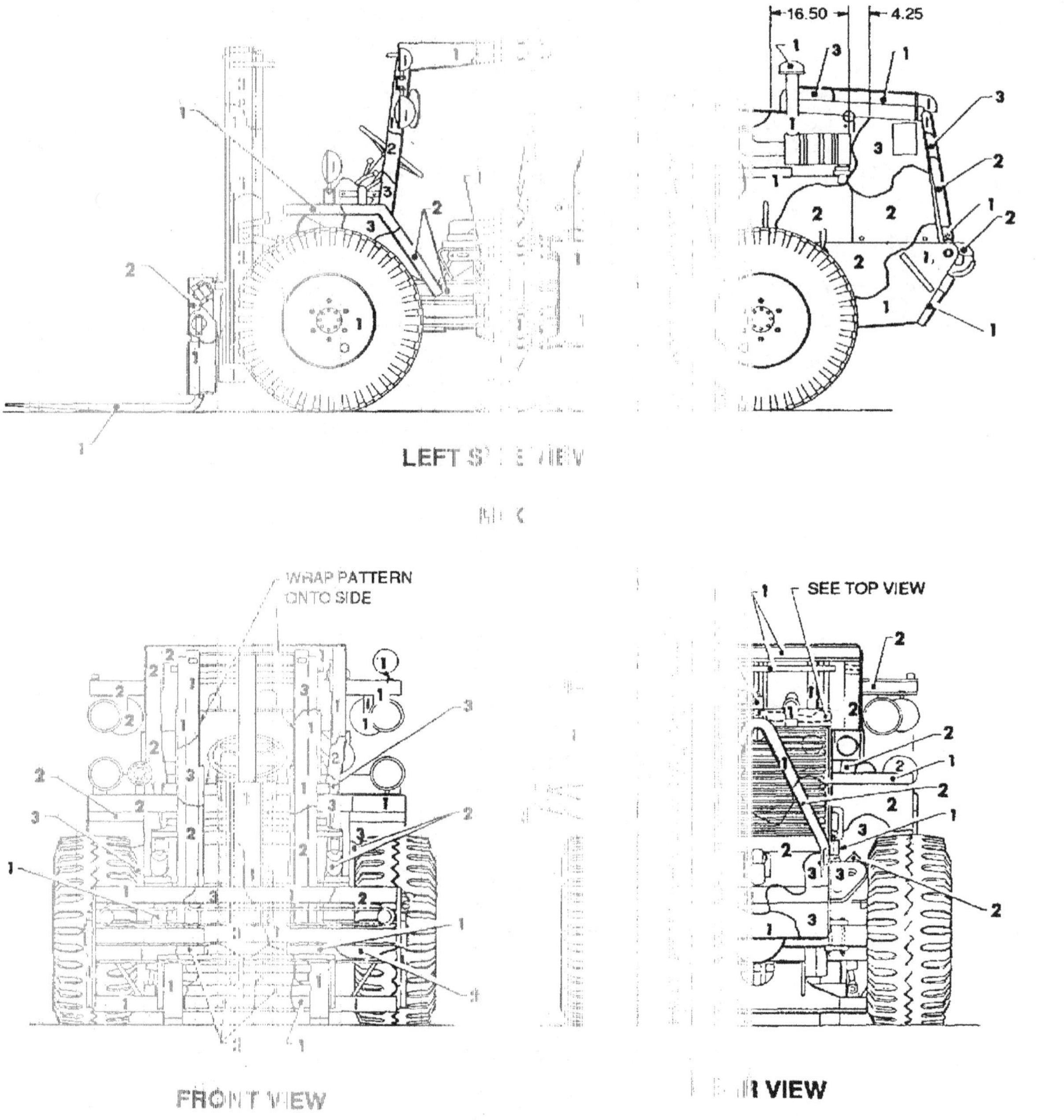

Figure 170. Truck, forklift, DED pneumatic tire, articulated [illegible] capacity rough terrain, M4K. (2 of 2)

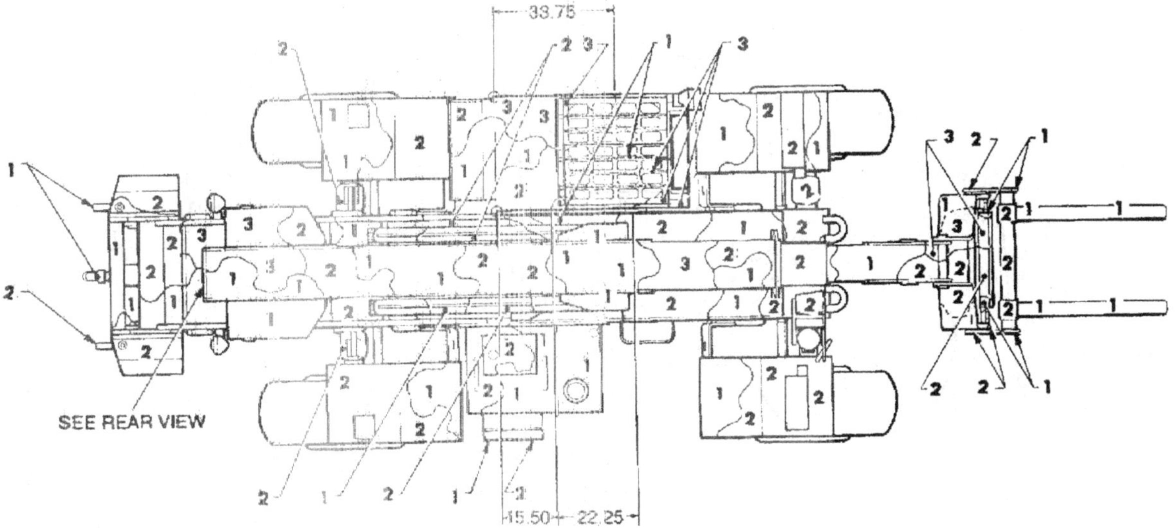

TOP VIEW

M489

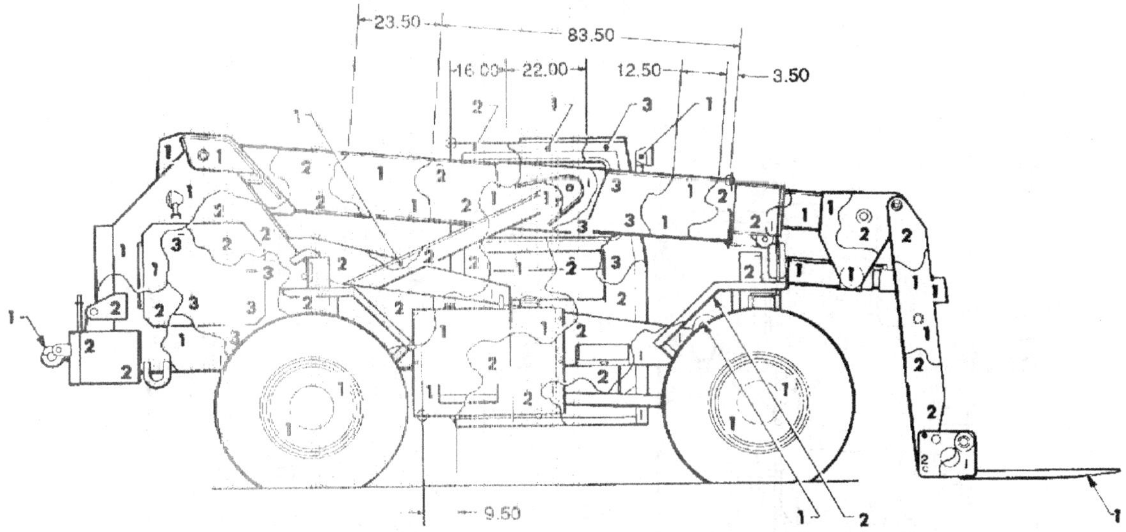

RIGHT SIDE VIEW

Figure 171. Truck, forklift, 6K cap, variable reach, rough terrain, M489. (1 of 2)

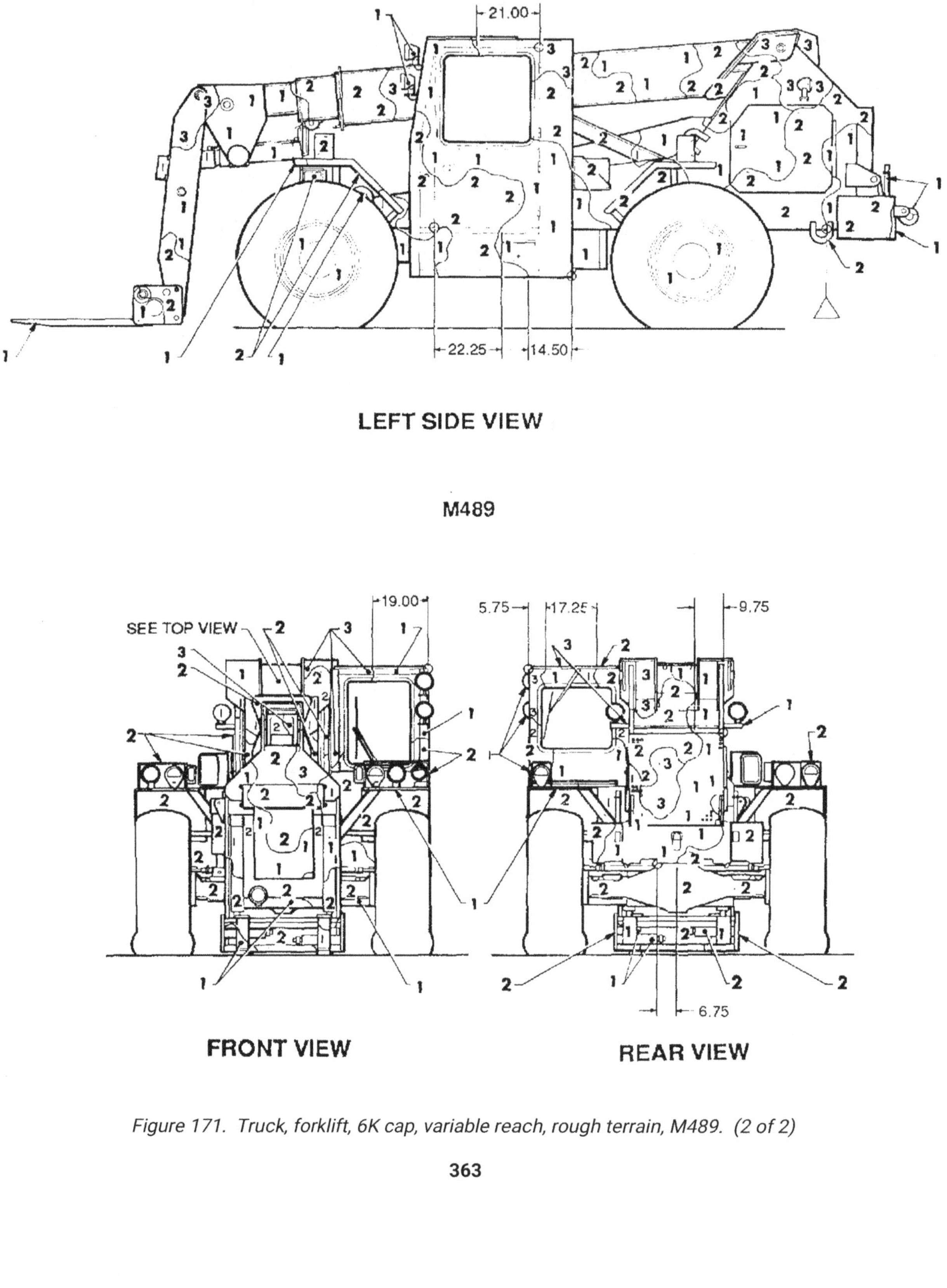

Figure 171. Truck, forklift, 6K cap, variable reach, rough terrain, M489. (2 of 2)

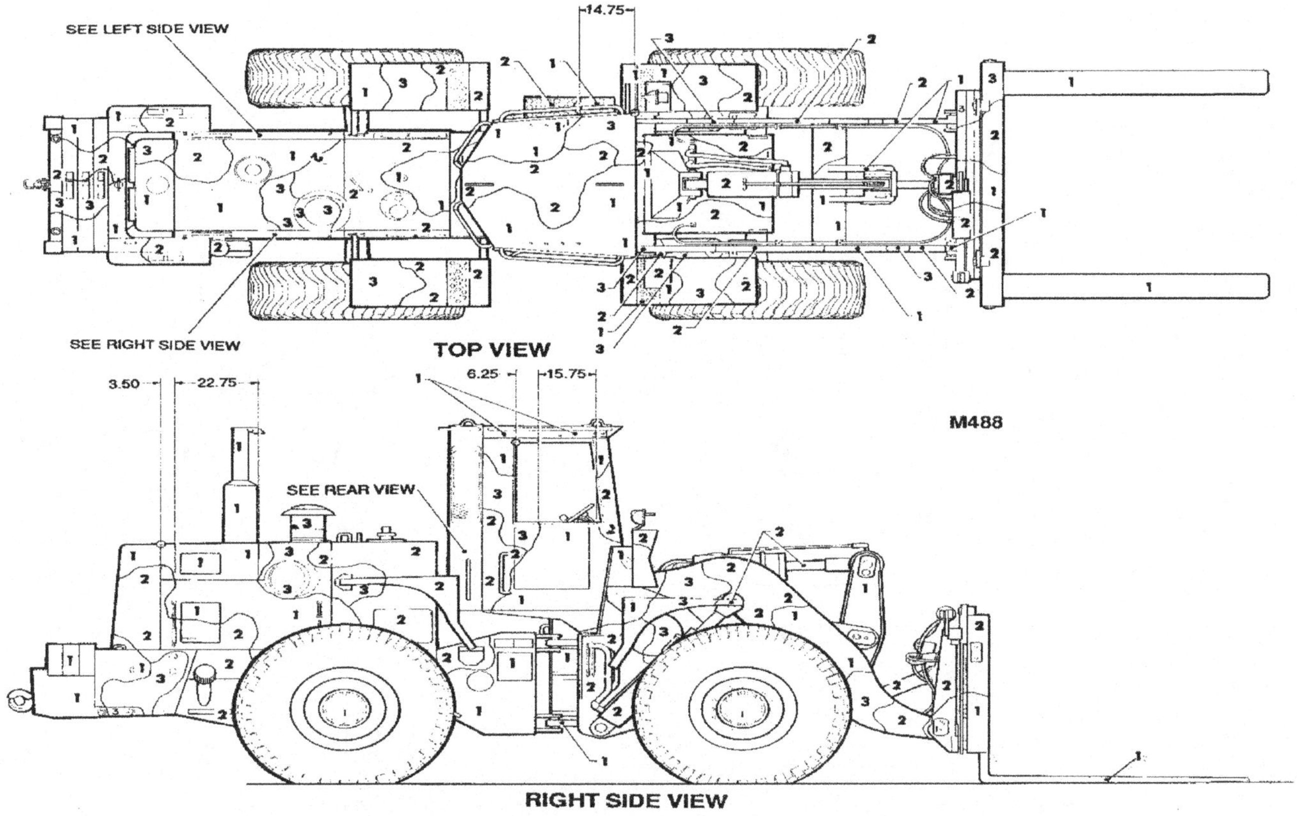

Figure 172. Loader, F/L DED, RT, M10, A488. (1 of 2)

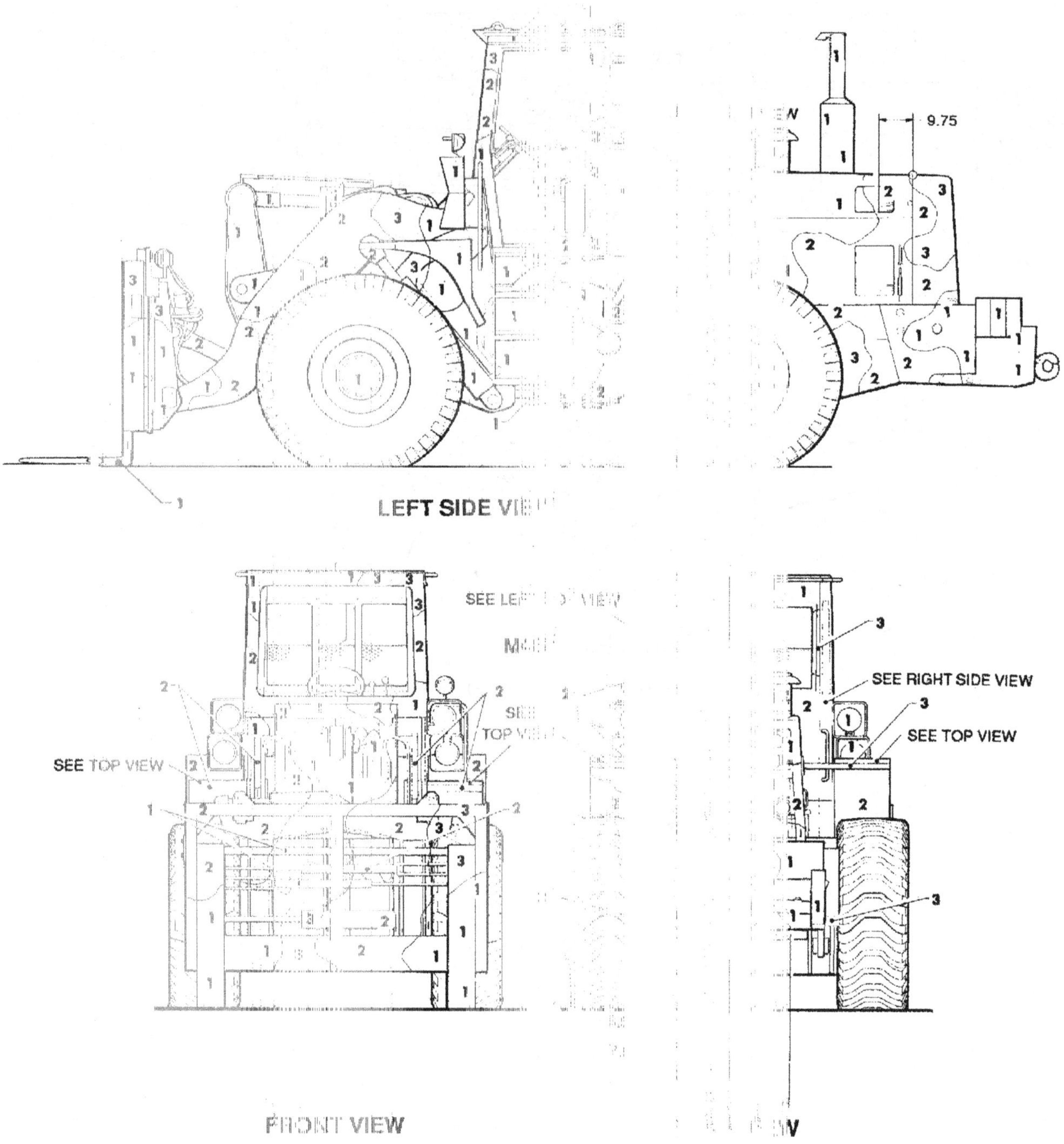

Figure 172. Loader, F/L, DE

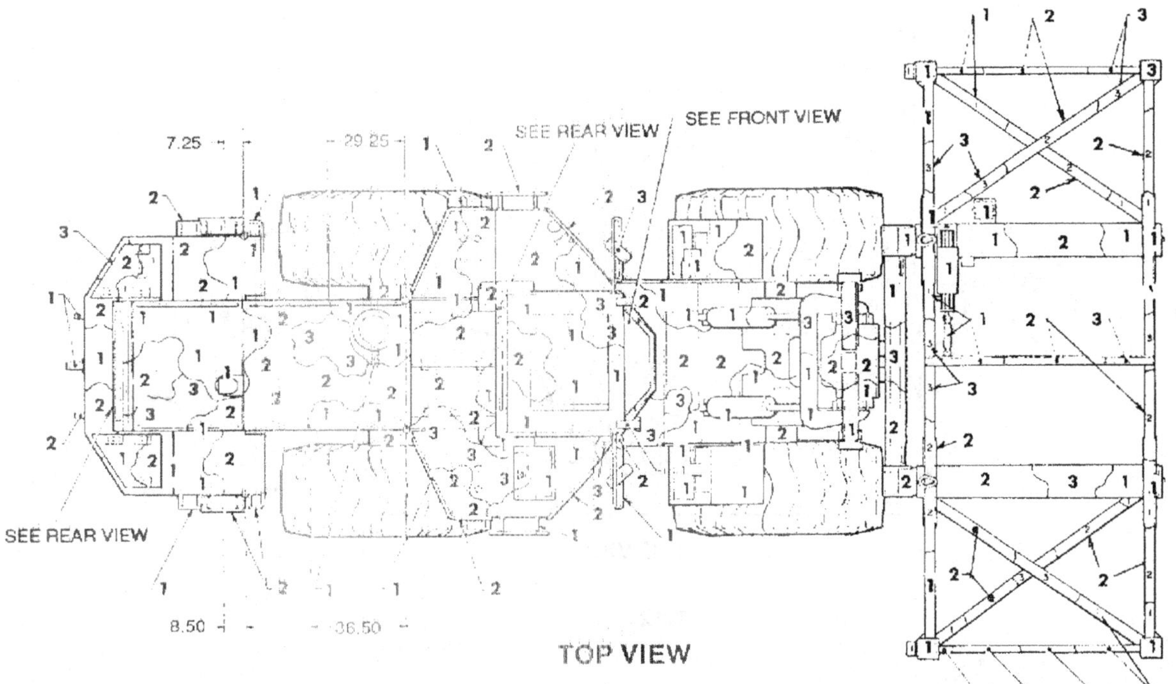

M412

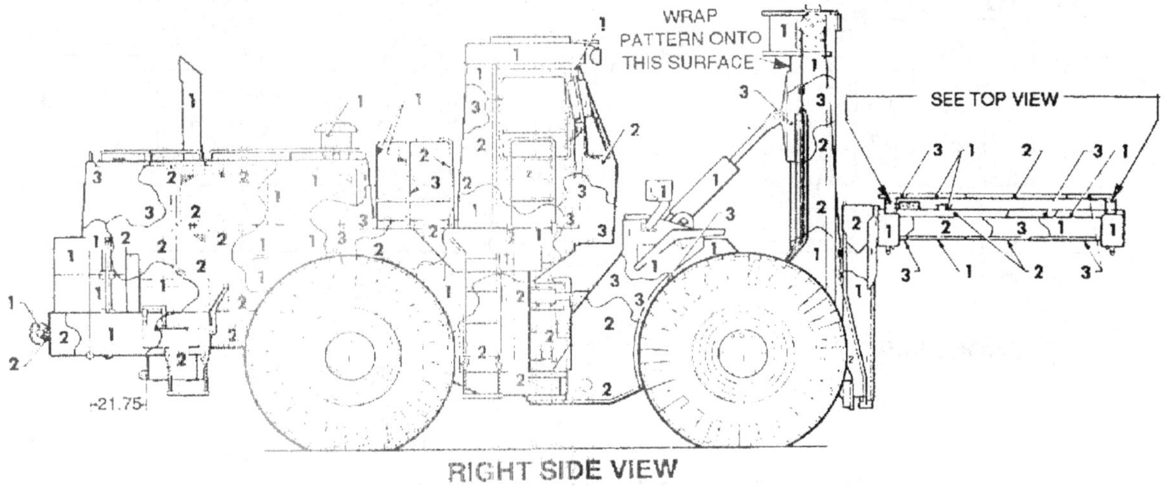

Figure 173. Loader, RT container handler, M412. (1 of 2)

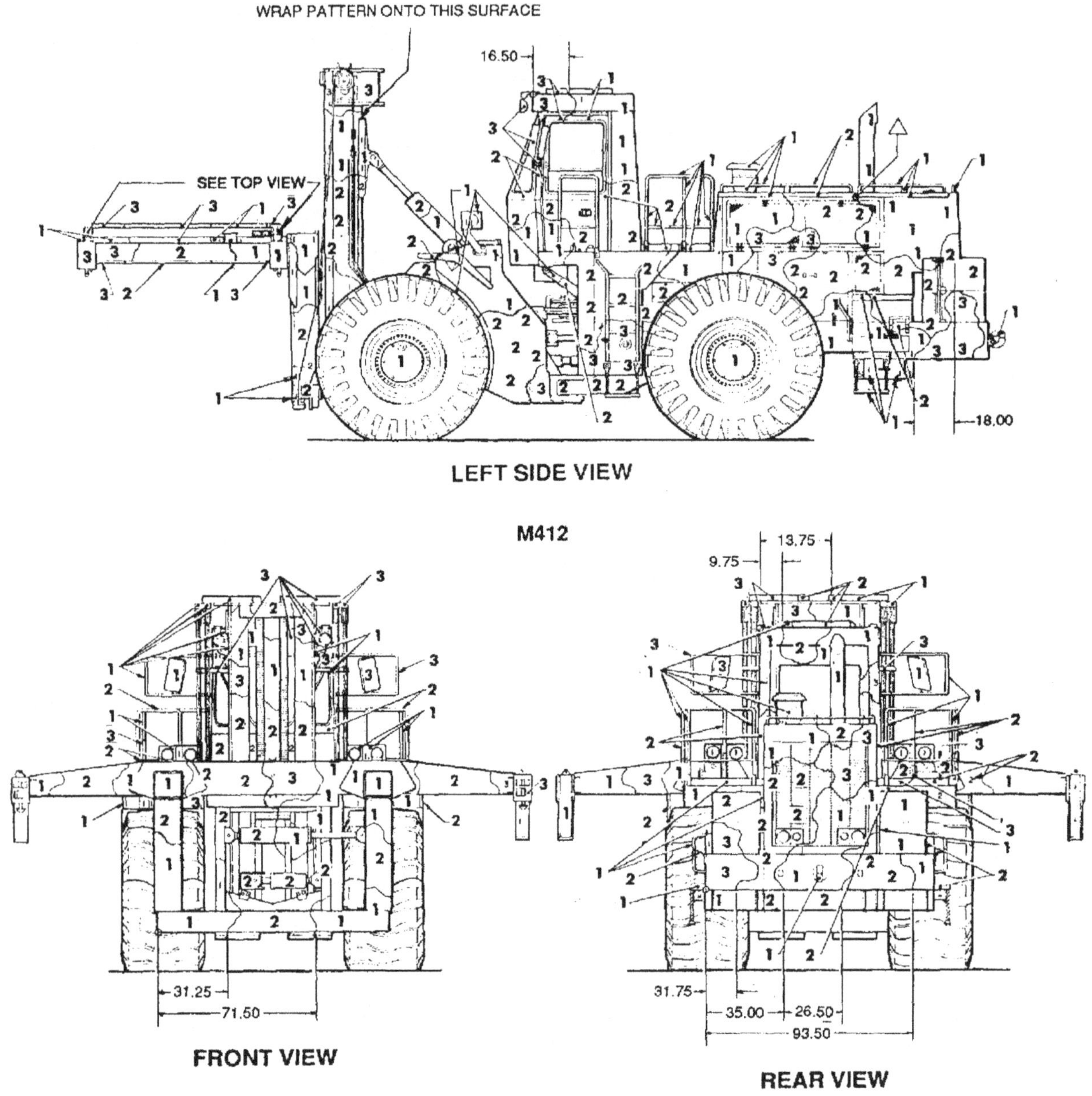

Figure 173. Loader, RT container handler, M412. (2 of 2)

SECTION VII I. COLORS FOR VEHICLES, CONSTRUCTION EQUIPMENT AND MATERIALS HANDLING EQUIPMENT

ITEM NO	EQUIPMENT	COLOR	PLACEMENT
1.	Vehicles, construction equipment, and materials handling equipment from new production, depot reconditioning, or depot stocks; exceptions listed below.	Three-color camouflage pattern using MIL-C-46168 or MIL-C-53039	Exterior surface unless otherwise directed.
		Lusterless Green 383 (34094)	Interior surface unless otherwise directed.
		Walkway Compound (MIL-W-5044)	Under topcoat over walking areas (catwalks, walkways, platforms, cabs, fenders, frames, guards, foot-operated controls, mounting steps, ladders, top bumpers, slope and deck plates, turret floors, crew compartments, ramps, van floors, etc), and interior surfaces.
2.	Firefighting vehicles, auxiliary vehicles assigned exclusively for firefighting or protection against fires and airfield crash rescue vehicles.	Gloss Red 11136 or Gloss Yellow 13670 when in nontactical use; in hot climates, Gloss White 17875 may be used	Exterior and interior surfaces.
3.	Repair and utility vehicles including attachments used for highway construction or maintenance (excluding construction equipment assigned to tactical units).	Gloss Yellow 13538	Exterior and interior surfaces.
4.	a. Materials handling equipment, including aircraft towing, and fuel and oil dispensing vehicles used in nontactical areas.	Gloss Yellow 13538	Exterior and interior surfaces.
	b. Materials handling equipment, including fuel and oil dispensing equipment used in tactical areas.	Three-color camouflage pattern using MIL-C-46168 or MIL-C-53039	Exterior surfaces.

Colors For Vehicles, Construction Equipment [illegible]ment--Continued

ITEM NO	EQUIPMENT	[illegible]	PLACEMENT
		Luste[illegible]	Interior surfaces.
5.	Garbage and refuse collection trucks in nontactical areas.	Gloss W[illegible] Black [illegible]	Exterior and interior surfaces; top of hood handles, and control levers, and stepping areas of running board. Stepping area may be covered with black non-slip walkway compound MIL-W-5044, type IV.
6.	Commercial designed vehicles in administrative use. (Bodies on vehicles need not be painted if constructed of aluminum, stainless steel or other corrosive resistant material, and when ordinarily not painted in common commercial practice.)	Gloss [illegible] where [illegible] special [illegible] to spe[illegible] colors [illegible] camou[illegible] 4616[illegible]	Exterior and interior surfaces.
7.	Vehicles and equipment used in connection with approved research and development test projects.	Any col[illegible] by the [illegible] activity	Exterior and interior surfaces.
8.	Calibration vehicles in CONUS.	In overse[illegible] vehicl[illegible] vironm[illegible] camo[illegible] 4616[illegible] wise, [illegible] Semi-[illegible] 2407[illegible] confor[illegible] MIL-C-[illegible]	Exterior surfaces.
		Semi-[illegible]	Walls, ceiling, door, fittings, and mounted equipment.
		Luste[illegible]	Floors.

Colors For Vehicles, Construction Equipment and Materials Handling Equipment--Continued

ITEM NO	EQUIPMENT	COLOR	PLACEMENT
9.	Refrigerator vans in nontactical use in hot climates; water tank vehicles in nontactical use in hot climates; van-type vehicles in nontactical use in hot climates assigned to centers or schools for the purpose of training personnel in the operation of equipment installed therein; van-type vehicles and shelters in nontactical use in hot climates in which installed equipment would be adversely affected; or personnel would not be able to accomplish assigned technical functions due to excessive heat from solar radiation.	Gloss White 17875 or solar heat-reflecting Semi-gloss Olive Drab Enamel 24087 conforming to MIL-C-46168 or MIL-C-53039, or solar heat reflecting Lusterless Olive Drab Enamel conforming to MIL-E-46117.	Applicable exterior surfaces, upon authorization by the theater commander.
10.	a. Tracked combat vehicles and special purpose armored hull-type vehicles. Exceptions: b and c below.	Three-color camouflage pattern using MIL-C-46168 or MIL-C-53039	Exterior and surfaces which become exterior surfaces during use (i.e. doors, hatches).
		White conforming to MIL-C-22750	Interior.
		Walkway Compound MIL-W-5044, Type IV	Walkways, floors, ramps.
	b. M113 personnel carrier family of vehicles, and the M2 and M3 Bradley Fighting Vehicle.	Three-color camouflage pattern using MIL-C-46168 or MIL-C-53039	Exterior surfaces and surfaces (hatches, ramps, etc) which become exterior surfaces during operational use.
		Seafoam Green 24533 Epoxy MIL-C-22750	Interior surfaces.
		Walkway compound MIL-W-5044	Floors (under MIL-C-22750 topcoat).

Colors For Vehicles, Construction Equipment and Materials Handling Equipment--Continued

ITEM NO	EQUIPMENT	COLOR	PLACEMENT
	c. Vehicles and construction equipment in nontactical use, in Army units not subject to the Army camouflage policy, not already camouflaged with three-color MIL-C-46168 or MIL-C-53039 and not identified in items 2, 3, 4a, 4b, 5, 6, 7, 8, 9, 12, 13, 14, and 16.	Lusterless Green 383, 34094, MIL-C-46168 or MIL-C- 53039	Exterior and interior surfaces unless otherwise directed.
11.	Interior surfaces of van-type bodies equipped with interior lighting in which personnel are required to perform certain detail operations.	Seaform Green Epoxy 24533, MIL-C-22750	Interior surfaces.
		Walkway Compound MIL-W-5044	Floors, under MIL-C-22750 topcoat.
12.	Commercial designed vehicles assigned to Armed Forces Policy Units.	Gloss Black 17038	Exterior and interior surfaces.
13.	Commercial designed vehicles for military police traffic accident prevention.	Gloss White 17875 when deemed necessary by the responsible commander	Exterior and interior surfaces.
14.	Vehicles used for intelligence, criminal investigation, and similar purposes, requiring concealment of the true identity of the activity involved.	Painted and marked as deemed appropriate by the responsible commander	Exterior and interior surfaces. **NOTE** **Proper identification of such vehicles will be marked in an inconspicuous location as prescribed in applicable technical publications and AR 381-143.**
15.	Ground support equipment in Army units subject to the Army camouflage policy.	Three-color camouflage pattern using MIL-C-46168 or MIL-C-53039	Exterior surfaces.

Colors For Vehicles, Construction Equipment and Materials Handling Equipment,--Continued

ITEM NO	EQUIPMENT	COLOR	PLACEMENT
16.	Ambulances in tactical use.	Three-color camouflage pattern using MIL-C-46168 or MIL-C-53039	Exterior surfaces, and surfaces (doors, etc) which become exterior surfaces during operational use.
		Seafoam Green 24533, MIL-C-22750	Interior surfaces.
		Lusterless Black 37030, MIL-C-53039 or MIL-C-46168	Interior of cab when vehicle is commerical design procured with standard color options.
17.	National symbol (5-pointed star).	Star will be a size to fit into a 3-inch diameter circle in Lusterless Black 37030, MIL-C-46168 or MIL-C-53039	Star will be applied at time of manufacture or by depots only when material is specifically designated for issue to US Army units.
	a. Tactical and combat vehicles and related equipment, including support and special purpose vehicles in CON US and overseas.	Lusterless Black 37030, MIL-C-46168 or MIL-C-53039	One each, centered on front and rear, on any verticle or near vertical surface such as bumper or tailgate, where it will not be obscured by canvas items, gasoline cans, pioneer tools or other objects. Normally, stars will not be applied to ambulan-; ces or other medical dedicated vehicles, or to ground support equipment at Army airfields in CONUS.
	b. Mounted equipment.	Lusterless Black 37030, MIL-C-46168 or MIL-C-53039	Wherever there is suitable surface on the vehicle.

Colors For Vehicles, Construction Equipment and Materiel Handling Equipment--Continued

ITEM NO	EQUIPMENT	COLOR	PLACEMENT
18.	Unit identification (abbreviations representing the complete identification of the unit to which the vehicle or equipment is assigned). Specific abbreviations to be used and methods of application are prescribed in Section III, paragraph 3.	Lusterle[illegible] 4616[illegible] MIL-C-5[illegible]	a. Markings will be uniform Gothic style letters the numerals to be the largest size practical for use in available space. but not to exceed 4 inches in height b. Normally markings will appear on both front and rear of each item of equipment, usually on the bumpers when so equipped. Where a more suitable surface is available, such surface may be used provided the location is not in conflict with the location of other markings. Where a suitable surface is not available on the front or rear of the equipment, use an appropriate location on the sides of the item. c. Markings will be applied to vehicles and equipment by the organization to which material is assigned. Such markings will be maintained on vehicles and equipment at all times, but will be removed when th[lie item is permanently transferred from the operating unit. d. When directed by the responsible Commander for security purposes, the first two categories of markings (see a and b of NOTE below) will be thoroughly removed from all vehicles and equipment being shipped with units from CONUS to overseas commands. To insure effective removal, the original figures must b; physically erased rather than painted over.

Colors For Vehicles, Construction Equipment and Materials Handling Equipment--Continued

NOTE

Unit identification markings are divided into four elements arranged from left to right, as follows:

a. Major command, organization, or activity. The major headquarters having jurisdiction over the vehicle or equipment, normally not lower than a division, brigade, group, or major subordinate.

b. Intermediate organization or activity. The next lower headquarters having jurisdication over the vehicles or equipment, normally the headquarters above the unit to which the vehicle or equipment is assigned. This Includes regiments, separate battalions, and separate companies or detachments not assigned to an Intermediate headquarters.

c. Unit or activity. The lowest unit or activity to which the vehicle is assigned, normally a company type unit. This space may be used to Identify the type of separate company or detachment already Identified above.

d. Vehicle or equipment number, the sequence number of the vehicle or equipment In the normal order of march within the unit to which It Is assigned.

ITEM NO	EQUIPMENT	COLOR	PLACEMENT
19.	Agency identification and registration number:	Colors specified on three-color camouflage drawings, but contrasting with color patch painted over	
	a. Registration number (Tactical wheeled vehicles).	Colors specified on three-color camouflage drawings, but contrasting with color patch painted over	Any appropriate interior area which is visible from outside a locked or secured item.
		Lusterless Black 37030 MIL-C-46168 or MIL-C-53039	
	b. (Non combat) Combat Vehicles		Exterior: Both sides and rear in uniform Gothic letters, no larger than 4 inches high.
	c. Agency identification; e.g., "US ARMY", will be painted on those vehicles, construction equipment and materials handling equipment in nontactical use, in Army units not subject to the Army camouflage policy, and vehicles and equipment as identified in items 2, 3, 4a, 5, 6, 7, 8, 9, 10c, 12 and 13.	Lusterless Black 37037, MIL-C-46168 or MIL -C-53039	Markings will be uniform Gothic style letters, the letters to be the largest size practical for use in available space, but not to exceed 4 inches in height. Normally, agency identification will appear on both sides of each item of equipment.
	d. Mounted equipment.	Lusterless Black 37030, MIL-C-46168 or MIL-C-53039	May be applied when necessary.

Colors For Vehicles, Construction Equipment and Materials Handling Equipment--Continued

ITEM NO	EQUIPMENT	COLOR	PLACEMENT
	e. Registration number assigned in accordance with AR-708-1 will be carried by vehicles construction equipment, and materials handling equipment that carry agency identification marking (e.g., US Army), in accordance with criteria set forth in b above.	Lusterless Black 37037, MIL-C-46168 or MIL-C-53039	Exterior; Both sides and rear in uniform Gothic letters, no larger than 4 inches high.
20.	General officer vehicle identification.		
	a. Plate.	Gloss enamel, color in accordance with AR 840-10 and MIL-P-14631	Right end of the front bumper. If plate interferes with lights or functional components of the vehicle, mount in front center. Remove or cover when general officer is not riding in vehicle.
	b. Flag.	See AR 840-10	
21.	Weight classification for bridge crossing.	Black numerals on a yellow circular background 9 inches in diameter. When towing another vehicle, the weight classification number of the combination will be shown with the letter "C" in red above the number	On the front of applicable self-propelled vehicles. For design of plate and mounting, see Paragraph 13.
	a. All vehicles with a gross weight of over 3 tons; all trailers with a rated payload of 1/2 tons and over.	On vehicles painted MIL-C-46168 or MIL-C-53039 lusterless colors in accordance with camouflage policy, numbers 3 inches high will be painted in contrasting colors directly over camouflage pattern On vehicles painted in gloss colors, use Gloss Yellow 13538 or Gloss Red 11136	

Colors For Vehicles, Construction Equipment and Materials Handling Equipment--Continued

ITEM NO	EQUIPMENT	COLOR	PLACEMENT
		On vehicles painted in semi-gloss or lusterless colors use Lusterless Yellow 33538, Gloss Black 17038 and Gloss Red 11136.	
	b. Combination vehicles.	Gloss Black 17038 numerals on a Lusterless Yellow 33538 circular background 6 inches in diameter	The gross weight classification of the prime mover alone and of the towed vehicle alone, will be marked on the right side of the vehicles.
22.	Safety.		
	a. Stripes(vehicles and equipment in nontactical use, that because of size, construction, or function present a possible hazard.)	Alternate Gloss Yellow 13655 and Gloss Black 17038 stripes in accordance with AR 385-30.	Normally, stripes are applied only to the rear, but also may be applied to the front and certain side surfaces when the conditions warrant.
	b. "FLAMMABLE" and "NO SMOKING WITHIN 50 FEET"--vehicles used for bulk transportation of gasoline, fuel oil, or other flammable liquids by Army units subject to the Army camouflage policy.	Brown 383 over green and black. Lusterless Black 37030 or 37038 over brown. The word "FLAMMABLE" will be in 6-inch letters; the words "NO SMOKING WITHIN 50 FEET" will be in 3-inch letters.	On both sides and rear of body, "NO SMOKING WITHIN 50 FEET" should be on a line below "FLAMMABLE".
	"FLAMMABLE" and "NO SMOKING WITHIN 50 FEET"--vehicles used for bulk transportation of gasoline, fuel oil, or other flammable liquids by Army units not subject to the Army camouflage policy, or in nontactical use.	When vehicle is commonly used on public highways in CONUS, Title 49, Sec. 177.823, Code of Federal Regulations, requires 6-inch block letters in Gloss Red 11105 or 11136 for word "FLAMMABLE" and in 3-inch letters "NO SMOKING WITHIN 50 FEET" also in Gloss Red 11105 or 11136. Glosss White 17875 will be used as background extending one inch all around the lettering. If available, reflective markings should be used.	On both sides and rear of body. "NO SMOKING WITHIN 50 FEET" should be on a line below "FLAMMABLE". Vehicles used infrequently on public highways must employ removable or reversible signs that are securely fastened while in use. These safety markings will be applied upon receipt by the using service.

Colors For Vehicles, Construction Equipment and Materials Handling Equipment--Continued

ITEM NO	EQUIPMENT	COLOR	PLACEMENT
	c. "GALS CAP"--Fuel and oil dispensing vehicles and equipment used for servicing aircraft.		Outside of tank near each manhole cover. In addition, type of fuel or oil, by military designation, will be clearly marked.
	d. Placards for vehicles carrying explosive or other dangerous articles.		
	(1) Over public highways.		Placards will be used in accordance with applicable DOT regulations and AR 55-355.
	(2) In training areas.		Placards will be used in accordance with applicable DOT regulations and AR 385-63.
	e. Reflective markings (vehicles and ground support equipment used on airfield flight lines).		As specified in TM 55-1500-204-5/1.
	f. Flags (all vehicles not painted in accordance with items 2 and 9 of this table and e above).		Vehicles will bear a distinctive flag whenever operating on landing areas, runways, taxiways, or peripheral roads at airfields. The flag will be square, and at least 3 feet on each side, and will be divided into 9 equal size squares forming a checkerboard pattern with the center and the corner squares in international orange and the remaining 4 squares in white. A red flag will be mounted on vehicles to indicate danger when considered necessary to caution personnel of a hazardous condition in the area.

Colors For Vehicles, Construction Equipment and Materials Handling Equipment--Continued

ITEM NO	EQUIPMENT	COLOR	PLACEMENT
	g. DA Label 76 (A Good Driver), to promote good driving.		Display in all motor vehicles to be readily visible to the driver. Requisition through normal publication channels.
23.	Red Cross insignia.	Aircraft Red 31136 on Aircraft White 37875, MIL-C-46168 or MIL-C-53039	On tactical ambulances, cross will be approximately 36" to 48" square and will be placed on roof, both sides, rear and front if practical. On other medical dedicated TOE vehicles, a 3" square cross will be placed on front and rear in lieu of the national symbol.
24.	Military police and security guard; tactical type vehicles:		
	a. "MILITARY POLICE"	Gloss Black 17038 letters on Aircraft White 37875 background, MIL-C-46168 or MIL-C-53039.	As specified in applicable DA technical publications. On 1/4-ton military trucks, a horizontal strip with lettering will be placed under the windshield and on a disk covering the spare wheel.
	b. "SECURITY POLICE" (at class II installations and activities).	Same as above, except the words "SECURITY POLICE" will be used in lieu of "MILITARY POLICE"	Same as above.
25.	Rifle and pistol team vehicles (decals of approved design).		On both sides of vehicle.
26.	Calibration service vehicles.		In accordance with applicable DA technical publications.
27.	Convoy markings (as prescribed by FM 55-30).	Serial letter or number may be black on reflective signboard background	

Colors For Vehicles, Construction Equipment and Materials Handling Equipment--Continued

ITEM NO	EQUIPMENT	COLOR	PLACEMENT
28.	Tactical markings (may consist of stripes, geometrical figures, other simple designs, or naming an individual vehicle). The markings will be used to provide a more visible means of identifying the vehicles of tactical units by personnel of those units during tactical operations. They will be of a suitable size to facilitate identification from the ground or from a reasonable distance. They will be designed to be as nearly unintelligible as possible to enemy observation. Under no circumstances will the markings represent the numerical designation or distinctive insignia of any unit. Under actual combat conditions, no written record will be made, nor published system of tactical markings used.	Any appropriate color	As prescribed by commanders of major Army commands; any changes must be approved by these commanders.
29.	Priority signs (The signs will be authorized by the area commander. They are valid only within the area under his jurisdiction. Signs must not be displayed when vehicle is not actually being used for a priority mission and must be removed when no longer required for that mission).		

Colors For Vehicles, Construction Equipment and Materials Handling Equipment--Continued

ITEM NO	EQUIPMENT	COLOR	PLACEMENT
30.	Air recognition panels.	Red fluorescent on one side and yellow fluoresce on the reverse side.	Marked as directed by the appropiate major commander as to the arrangements and conditions under which they will be used. They will be draped on the vehicle in a standard unchanging pattern different from displays prescribed for other recognition purposes, such as front lines, targets, etc.; Size: 6'x2'3", each panel having a tie cord of adequate length In each corner.
31.	Commerical designed vehicle markings. (unless otherwise specified vehicles assigned to TOE or TOD units, or to other organizations and activites for non-administrative use.)		Similiar to markings used on military designed vehicles.

AP[illegible] [illegible]

REF[illegible]

AMC-P-750	Maintenance of Suppl[illegible]
AR 40-5	Preventive Medicine
AR 381-143	(C) Logistic Policies [illegible]
AR-385-30	Safety Color Code Ma[illegible]
AR 385-63	Policies and Proced[illegible] [illegible]r Training, Target Practice [illegible]
AR 58-1	Management, Acqui[illegible] [illegible]t ve Use Motor Vehicles.
AR 310-50	Authorized Abbrevia[illegible] subscription for[illegible]
AR 420-47	Solid and Hazardou[illegible]
AR 708-1	Cataloging and Sup[illegible]
AR 750-I	Army Materiel Maint[illegible]
AR 750-4	Maintenance of Sup[illegible] [illegible]ateriel Maintenance an[illegible]
AR 840- 0	Flags, Guidons, Stre[illegible] [illegible]bile and Aircraft Plates
ASTM D846	Xylene, Standard Sp[illegible]
CFR 1910	OSHA Safety and He[illegible]
DOD-D-15328D	Primer (Wash), Pretr[illegible] [illegible] Metals) (Metric)
DODI 4145.19-R.1	Hazardous Material[illegible] [illegible]ria
DOT AC 70/7460-1	Obstruction, Markin[illegible]
Federal Specification L-S-300	Sheeting and Tape, R[illegible] [illegible] Adhesive Backing
Fed Std 505	Colors
FM 5-20	Camouflage
FM 5-36	Route Reconnaissan[illegible] subscription [illegible]
FM21-30	Symbols
FM 55-30	Army Motor Transpo[illegible]
MIL-A-8625	Anodic Coatings, for [illegible] [illegible]oys
MIL-C-450	Coating-Compound, [illegible] [illegible]ck (for Ammunition)
MIL-C-554l	Chemical Conversio[illegible] [illegible]oys
MIL-C-8514	Coating Compound, [illegible] [illegible]cid
MIL-C- 10578	Corrosion Removing [illegible] [illegible]pound (Phosphoric Acid [illegible]
MIL-C-22750	Coating, Epoxy Poly[illegible]
MIL-C-4l168	Coating, Aliphatic P[illegible] [illegible]esistant
MIL-C-53039	Coating, Aliphatic P[illegible] [illegible]ent, Chemical Agent [illegible]
MIL C-5307 i2	Chemical Agent Res[illegible] [illegible]n Application Procedures and Quality Cont[illegible]
MIL -C-85570	Cleaning Compound[illegible]

MIL-F-14072	Finishes for Ground Electronic Equipment
MIL-L-11195	Lacquer, Lusterless, Hot Spray
MIL-L-12277	Lacquer, Automotive, Hot Spray
MIL-L-46159	Lacquer, Acrylic, Low Reflective, Olive Drab
MIL-L-52043	Lacquer, Semigloss, Cellulose Nitrate
MIL-L-52909	Lacquer, Acrylic, Camouflage, Lusterless
MIL-M-3171	Magnesium Alloy, Processes for Pretreatment and Prevention of Corrosion On
MIL-M-13231 (ER)	Marking of Electronic Items
MIL-M-43719	Marking Materials and Markers, Adhesive, Elastomeric, Pigmented Legends
MIL-N-15178	Solvent, Naphtha
MIL-P-11414	Primer Coating, Lacquer, Rust Inhibiting
MIL-P-14105	Paint, Heat Resisting (for Steel Surfaces)
MIL-P-14458	Paint, Rubber, Red Fuming Nitric Acid Resistant
MIL-P-14631	Plate, Automobile, Distinguishing, General Officer
MIL-P-15931	Paint, Antifouling, Vinyl (Formulas No121 and No 129)
MIL-P-22636	Primer Coating, for Red Fuming Nitric Acid Resistant Paint
MIL-P-23377	Primer Coating, Epoxy Polyamide, Chemical and Solvent Resistant
MIL-P-24411	Paint, Epoxy Polyamide, General Specification for
MIL-P-52192	Primer Coating, Epoxy
MIL-P-52905	Paint, Arctic Camouflage, Removable
MIL-P-52977	Primer Coating, Alkyd, Anticorrosive, Lead and Chromate Free
MIL-P-52995	Primer Coating, Synthetic, Corrosion Inhibiting, Lacquer Resisting, Lead and Chromate Free
MIL-P-53022	Primer, Epoxy Coating, Corrosion Inhibiting, Lead and Chromate Free
MIL-P-53030	Primer Coating, Epoxy, Water Reducible, Lead and Chromate Free
MIL-P-53032	Primer Coating, Water Reducible, Epoxy Ester-Latex Type, Lead and Chromate Free
MIL-P-85582	Primer Coating: Epoxy, VOC Compliant, Chemical and Solvent Resistant
MIL-S-5002	Surface Treatments and Inorganic Coatings for Metal Surfaces of Weapons Systems
MIL-S-11030	Sealing Compound, Non-Curing, Polysulfide Base
MIL-S-11031	Sealing Compound, Adhesive, Curing, Polysulfide Base
MIL-STD-171	Finishing of Metal and Wood Surfaces
MIL-STD-193	Paint Procedures and Marking for Vehicles, Construction Equipment, and Material Handling Equipment
MIL-STD-642	Identification Marking of Combat and Tactical Transport Vehicles
MIL-STD-1201	Alcohol Denatured and Ethyl, Technical
MIL-STD-1473	Standard General Requirements for Color and Marking of Army Materiel
MIL-T-704J	Treatment and Painting of Materiel
MIL-T-6095	Thinner, Cellulose-Nitrate-Dope, Blush Retarding
MIL-T-19544	Thinner, Aircraft Lacquer
MIL-T-81772	Thinner, Aircraft Coating

MIL-STD-642	Identification Marking of Combat and Tactical Transport Vehicles
MIL-STD-1201	Alcohol, Denatured and Ethyl, Technical
MIL-STD-1473	Standard General Requirements for Color and Marking of Army Materiel
MIL-T-704J	Treatment and Painting of Materiel
MIL-T-6095	Thinner, Cellulose-Nitrate-Dope, Blush Retarding
MIL-T-19544	Thinner, Aircraft Lacquer
MIL-T-81772	Thinner, Aircraft Coating
MIL-V-173	Varnish, Moisture and Fungus Resistant for Treatment of Communications, Electronic, and Assorted Equipment
MIL-V-13811	Varnish, Waterproofing, Electrical, Ignition
MIL-W-5044	Walkway Compound, Nonslip and Walkway Matting, Nonslip
MIL-W-13518	Wood Preservative, Tetrachlorophenol and Pentachlorophenol, Surface Sealing Compound
O-E-760	Ethyl Alcohol (Ethanol), Denatured Alcohol, Proprietary Solvents, and Special Industrial Solvents
Quadripartite Standardization Agreement 248	Identification of Medical Materiel to Meet Urgent Needs
TB MED 502	Occupational and Environmental Health Respiratory Protection Program
TB M ED 514	Occupational and Environmental Guidance for Painting Operations
TB 43-0147	Color, Marking and Camouflage Patterns Used on Military Equipment
TB 43-0213	Corrosion Prevention and Control
TM 43-0139	Painting Instructions for Army Materiel
TG 141	Industrial Hygiene Sampling Instructions
TM 5-200	Camouflage Materials
TT-C-490	Cleaning Method and Pretreatmernt of Ferrous Surfaces for Organic Coatings
TT-C-520	Coating Compound, Bituminous, Solvent Type, Underbody (for Motor Vehicles)
TT-E-485	Enamel, Semigloss, Rust Inhibiting
TT-E-489	Enamel, Alkyd, Gloss (for Exterior and Interior Surfaces)
TT-E-516	Enamel, Lusterless, Quick Drying Styrenated Alkyd Type
TT-E-522	Enamel, Phenolic, Outside
TT-E-527	Enamel, Alkyd, Lusterless
TT-E-529	Enamel, Alkyd, Semigloss
TT-E-1593	Enamel, Silicone Alkyd, CopolymE.r, Gloss (for Exterior and Interior Use)
TT-F-325	Filler, Engraving, Stamped Markin I
TT-L-58	Lacquer, Spraying, Clear and Pigmented (General Interior Use)
TT-L-215	Linseed Oil, Raw (for Use in Organic Coatings)
TT-P-28	Paint, Aluminum Heat Resisting (1200 Degrees F.)
TT-P-98	Paint, Stencil, Flat
TT-P-636	Primer Coating, Alkyd, Wood and Ferrous Metal
TT-P-645	Primer, Paint, Zinc Chromate, Alkyd Type
TT-P-659	Primer Coating and Surfacer, Synthetic, Tints and White (for Metal and Wood Surfaces)

TT-P-664	Primer, Coating, Synthetic, Rust Inhibiting, Lacquer Resisting
TT-P-1757	Primer Coating, Zinc Chromate, Low Moisture Sensitivity
TT-R-25114	Remover, Paint and Varnish
TT-S-300	Shellac, Cut
TT-S-720	Stain (Wood, Nongrain Raising, Solvent-Dye Type)
TT-T-266	Thinner, Dope and Lacquer (Cellulose Nitrate)
TT-T-291	Thinner, Paint, Mineral Spirits, Regular and Odorless
TT-T-306	Thinner, Synthetic Resin, Enamels
TT-T-548	Toluene, Technical
TT-T-801	Turpentine, Gum Spirits, Steam Distilled, Sulphate Wood, and Destructively Distilled
TT-V-51	Varnish, Asphalt
TT-V-121	Varnish, Spar, Water Resisting
TT-W-571	Wood Preservation, Treating Practices
TT-X-916	Xylene, Technical
US/GE	Standard Camouflage Pattern Painting
USAEHA	Water Quality Information Paper No 13 Pretreatment Regulations

APPENDIX B

NSN TABLES

B-1. SCOPE

This appendix contains tables of NSN's for various colors [illegible] es not cover all coatings; however, most Chemical Agent Resistant Coatings (CARC) are cover[illegible] primers, Primer (Wash) Pretreatment DOD-D-15328, and Thinner MIL-T-81772. Table B-1 sho[illegible] ble covers.

Table B-1 TABLE TIT[illegible]

TABLE NUMBER	[illegible]
B-2	Coating, Aliphatic Polyurethane, [illegible]
B-3	Coating, Aliphatic Polyurethane, [illegible]
B-4	Coating, Aliphatic Polyurethane, [illegible]
B-5	Epoxy Coating (Interior), MIL-C-[illegible]
B-6	Non-CARC Coatings (MIL-C-8328[illegible] 05
B-7	Primers (MIL-P-23377, MIL-P-52[illegible] 82
B-8	Primer (Wash) Pretreatment (For[illegible] etric, DOD-D-15328
B-9	Thinner, MIL-T-81772

[illegible] COATING, ALIPHATIC POLYURETHANE [illegible] ANT (CARC) 2-COMPONENT (TOP[illegible] MIL-[illegible]

COLOR	COLOR NUMBER	N[illegible]	SIZE
Green 383*	34094	[illegible]	1 1/4 Qt Kit
Green 383*	34094	[illegible]	1 1/4 Gal Kit
Green 383*	34094	[illegible]	5 Gal Kit
Green 383*	34094	[illegible]	55 Gal Drum
Brown 383*	30051	[illegible]	1 1/4 Qt Kit
Brown 383*	30051	[illegible]	1 1/4 Gal Kit
Brown 383*	30051	[illegible]	5 Gal Kit
Brown 383*	30051	[illegible]	55 Gal Drum
Dark Green	34082	[illegible]	1 1/4 Qt Kit
Dark Green	34082	[illegible]	1 1/4 Gal Kit
Dark Green	34082	[illegible]	5 Gal Kit
Dark Green	34082	[illegible]	55 Gal Drum
Field Drab	33105	[illegible]	1 1/4 Qt Kit
Field Drab	33105	[illegible]	1 1/4 Gal Kit
Field Drab	33105	[illegible]	5 Gal Kit
Field Drab	33105	[illegible]	55 Gal Drum
Earth Yellow	33245	[illegible]	1 1/4 Qt Kit
Earth Yellow	33245	[illegible]	1 1/4 Gal Kit
Earth Yellow	33245	[illegible]	5 Gal Kit
Earth Yellow	33245	[illegible]	55 Gal Drum
Sand	33303	[illegible]	1 1/4 Qt Kit

Table B-2 (Continued)

COLOR	COLOR NUMBER	NSN	SIZE
Sand	33303	8010-01-130-3347	1 1/4 Gal Kit
Sand	33303	8010-01-131-6259	5 Gal Kit
Sand	33303	8010-01-169-5960**	55 Gal Drum
Tan 686	33446	8010-01-260-0910	1 Qt Kit
Tan 686	33446	8010-01-260-0909	1 Gal Kit
Tan 686	33446	8010-01-260-0908	5 Gal Kit
Black*	37030	8010-01-141-2419	1 1/4 Qt Kit
Black*	37037	8010-01-131-6254	1 1/4 Gal Kit
Black*	37037	8010-01-131-6261	5 Gal Kit
Black*	37030	8010-01-169-5961**	55 Gal Drum
Aircraft Green	34031	8010-01-141-2420	1 1/4 Qt Kit
Aircraft Green	34031	8010-01-131-6255	1 1/4 Gal Kit
Aircraft Green	34031	8010-01-131-6262	5 Gal Kit
Aircraft Green	34031	8010-01-129-9348**	55 Gal Drum
Olive Drab	34088	8010-01-146-2650	1 1/4 Qt Kit
Olive Drab	34088	8010-01-055-2319	1 1/4 Gal kit
Olive Drab	34088	8010-01-144-9875	5 Gal Kit
Aircraft Gray	36300	8010-01-144-9882	1 1/4 Qt Kit
Aircraft Gray	36300	8010-01-127-8908	1 1/4 Gal Kit
Aircraft Gray	36300	8010-01-144-9876	5 Gal Kit
Aircraft White	37875	8010-01-144-9883	1 1/4 Qt Kit
Aircraft White	37875	8010-01-144-9872	1 1/4 Gal Kit
Aircraft White	37875	8010-01-144-9877	5 Gal Kit
Aircraft Red	31136	8010-01-144-9884	1 1/4 Qt Kit
Aircraft Red	31136	8010-01-144-9873	1 1/4 Gal Kit
Aircraft Red	31136	8010-01-144-9878	5 Gal Kit
Aircraft Black	37038	8010-01-144-9885	1 1/4 Qt Kit
Aircraft Black	37038	8010-01-146-2646	1 1/4 Gal Kit
Aircraft Black	37038	8010-01-144-9879	5 Gal Kit
Interior Aircraft Black (w/Glass Beads)	37031	8010-01-144-9886	1 1/4 Qt Kit
Interior Aircraft Black (w/Glass Beads)	37031	8010-01-146-2647	1 1/4 Gal Kit
Interior Aircraft Black (w/Glass Beads)	37031	8010-01-146-4376	5 Gal Kit
Insignia Blue	35044	8010-01-144-9887	1 1/4 Qt Kit
Insignia Blue	35044	8010-01-146-2648	1 1/4 Gal Kit
Insignia Blue	35044	8010-01-144-9880	5 Gal Kit
Interior Aircraft Gray	36231	8010-01-170-7583	1 1/4 Qt Kit
Interior Aircraft Gray	36231	8010-01-146-2649	1 1/4 Gal kit
Interior Aircraft Gray	36231	8010-01-170-0132	5 Gal Kit
Aircraft Yellow	33538	8010-01-247-8885	1 1/4 Qt Kit
Aircraft Yellow	33538	8010-01-235-8059	1 1/4 Gal Kit
Aircraft Yellow	33538	8010-01-235-5079	5 Gal Kit
Dark Sandstone	33510	8010-01-260-7480	1 1/4 Qt Kit
Dark Sandstone	33510	8010-01-260-7479	1 1/4 Gal Kit
Dark Sandstone	33510	8010-01-260-7478	5 Gal Kit

* Color for Three-color camouflage system

** This NSN for Component A only; Component B for all 55 Gal Drum Sizes has NSN 8010-01-132-0205. If four 55 Gal Drums of Component A are ordered, one 55 Gal Drum of Component B will also be shipped. (MIL-C-46168 is mixed 4 parts Component A with 1 part Component B).

Table B-3
COATING, ALIPHATIC POLYURETHANE,
CHEMICAL AGENT RESISTANT (CARC)
2-COMPONENT (TOPCOAT)
MIL-C-46168 Type IV

COLOR	COLOR NUMBER	NSN	SIZE
Green 383*	34094	8010-01-260-7481	1 1/4 Qt Kit
Green 383*	34094	8010-01-260-0911	1 1/4 Gal Kit
Green 383*	34094	8010-01-260-0912	5 Gal Kit
Brown 383*	30051	8010-01-260-7482	1 1/4 Qt Kit
Brown 383*	30051	8010-01-260-0916	1 1/4 Gal Kit
Brown 383*	30051	8010-01-260-0917	5 Gal Kit
Field Drab	33105	8010-01-260-0918	1 1/4 Qt Kit
Field Drab	33105	8010-01-260-0919	1 1/4 Gal Kit
Field Drab	33105	8010-01-260-0920	5 Gal Kit
Sand	33303	8010-01-260-0921	1 1/4 Qt Kit
Sand	33303	8010-01-260-0922	1 1/4 Gal Kit
Sand	33303	8010-01-260-7483	5 Gal Kit
Black*	37030	8010-01-260-0913	1 1/4 Qt Kit
Black*	37030	8010-01-260-0914	1 1/4 Gal Kit
Black*	37030	8010-01-260-0915	5 Gal Kit

Table B-4
COATING, ALIPHATIC POLYURETHANE,
CHEMICAL AGENT RESISTANT (CARC)
SINGLE COMPONENT (TOPCOAT)
MIL-C-53039

COLOR	COLOR NUMBER	NSN	SIZE
Green 383*	34094	8010-01-229-7546	1 Qt Can
Green 383*	34094	8010-01-229-9561	1 Gal Can
Green 383*	34094	8010-01-229-7547	5 Gal Can
Green 383*	34094	8010-01-232-8514	55 Gal Drum
Brown 383*	30051	8010-01-229-7543	1 Qt Can
Brown 383*	30051	8010-01-229-7544	1 Gal Kit
Brown 383*	30051	8010-01-229-7545	5 Gal Kit
Brown 383*	30051	8010-01-233-0060	55 Gal Drum
Black*	37030	8010-01-229-7540	1 Qt Can
Black*	37030	8010-01-229-7541	1 Gal Can
Black *	37030	8010-01-229-7542	5 Gal Can
Black*	37030	8010-01-233-1568	55 Gal Drum
Sand	33303	8010-01-234-2934	1 Qt Can
Sand	33303	8010-01-234-2935	1 Gal Can
Sand	33303	8010-01-234-2936	5 Gal Can
Sand	33303	8010-01-234-2937	55 Gal Drum
Aircraft Green	34031	8010-01-246-0717	1 Qt Can
Aircraft Green	34031	8010-01-246-0718	1 Gal Can
Aircraft Green	34031	8010-01-246-0719	5 Gal Can
Aircraft Green	34031	8010-01-246-0255	55 Gal Drum

*Color for Three-color camouflage system

Table B-5
EPOXY COATING (INTERIOR)
MIL-C-22750

COLOR	COLOR NUMBER	NSN	SIZE
Gray	16743	8010-00-082-2438	2 Qt Kit
Gray	36099	8010-01-053-2655	2 Qt Kit
Gray, Lt	36495	8010-01-117-7685	2 Qt Kit
Gray, Lt	36495	8010-01-117-7686	2 Gal Kit
Gray	36320	8010-01-117-7687	2 Qt Kit
Gray	36320	8010-01-117-7688	2 Gal Kit
Gray	36375	8010-01-117-7689	2 Qt Kit
Gray	36375	8010-01-117-7690	2 Gal Kit
Gull Gray	16440	8010-01-053-2646	2 Qt Kit
Gray	36440	8010-01-082-2435	2 Gal Kit
Gray	36231	8010-00-148-7091*	1 Gal Kit
Gray	26492	8010-01-023-0341	1 Gal Kit
Gray	26492	8010-01-028-8577*	2 Gal Kit
Gray	26622	8010-00-112-0725*	1 Gal Kit
Gray	26081	8010-00-082-2436*	2 Gal Kit
Gray	16081	8010-00-082-2437	2 Gal Kit
Gray	36231	8010-01-082-3059*	1 Pt Kit
Gray	36231	8010-01-082-3060	1 Qt Kit
Orange Yellow	13538	8010-01-053-2650	2 Qt Kit
Yellow	13538	8010-00-148-3166*	1 Gal Kit
Yellow	23538	8010-01-156-1421	1 Gal Kit
Yellow	13538	8010-01-106-0110	2 Gal Kit
Insignia White	17925	8010-01-053-2647	2 Qt Kit
White	17925	8010-01-154-2334	2 Pt Kit
White	17925	8010-00-082-2439	2 Gal Kit
White	27875	8010-01-106-0114*	2 Gal Kit
White	17925	8010-00-082-2440	10 Gal Kit
Insignia Blue	15044	8010-01-053-2648	2 Qt Kit
Insignia Blue	35044	8010-01-053-2657*	2 Qt Kit
Blue	35237	8010-01-118-9981	2 Qt Kit
Blue	35237	8010-01-117-7691	2 Gal Kit
Blue	35299	8010-01-106-0113*	2 Gal Kit
Maroon	10049	8010-01-053-2653*	2 Qt Kit
Insignia Red	11136	8010-01-053-2649	2 Qt Kit
Red	31302	8010-01-106-0112*	2 Gal Kit
Orange-Red	12199	8010-01-030-3507	2 Qt Kit
International Orange	12197	8010-01-053-2652	2 Qt Kit
International Orange	12197	8010-00-948-6733	2 Gal Kit
Olive Drab	34088	8010-00-756-2411	2 Pt Kit
Olive Drab	24084	8010-00-117-9992	1 Gal Kit
Olive Drab	24084	8010-01-024-9589	2 Gal Kit
Olive Drab	34088	8010-00-438-4909	2 Gal Kit
Clear	Full Gloss	8010-00-959-4661	2 Pt Kit
Aluminum	Full Gloss	8010-00-559-5583*	1 Pt Kit

*Not listed on AMDF; must be specially ordered.

EPOXY COA[illegible]

MIL-C-22[illegible]

COLOR	COLOR NUMBER	[illegible]	SIZE
Black	37038	[illegible]	2 Qt Kit
Black	37038	[illegible]	2 Gal Kit
Dark Green	14062	[illegible]	2 Qt Kit
Green	24052	[illegible]	[illegible] Gal Kit
Lt Green	14187	[illegible]	2 Qt Kit
Seafoam Green	24533	[illegible]	[illegible]Qt Kit
Seafoam Green	24533	[illegible]	[illegible]Gal Kit
Seafoam Green	24533	[illegible]	[illegible] Gal Kit

* Not listed on AMDF; must be specially ordered.

NON-CA[illegible]

COATING	MILSPEC	COLOR	[illegible]	NSN	SIZE
Coating, Urethane,	MIL-C-83286	Yellow		8010-00-181-8297	2 Qt Kit
Aliphatic Isocynate For	MIL-C-83286	Yellow		[illegible]10-00-181-8292	2 Gal Kit
Aerospace Applications	MIL-C-83286	Yellow		[illegible]10-00-181-8302	2 Qt Kit
(Contains Lead)	MIL-C-83286	Yellow		8010-00-181-8300	2 Gal Kit
Paint, Heat Resistant (For	MIL-P-14105	Green		[illegible]10-01-235-2693	1 Qt Can
Use on Surfaces Exceeding	MIL-P-14105	Green		[illegible]10-01-235-4164	1 Gal Can
400°F (204°C)	MIL-P-14105	Brown		8010-01-235-2694	1 Qt Can
MIL-P-14105	Brown	30051		8010-01-235-2695	1 Gal Can
MIL-P-14105	Black	37030		8010-01-235-4165	1 Qt Can
MIL-P-14105	Black	37030		8010-01-235-4166	1 Gal Can
Paint, Arctic Camouflage, Removable (White)	MIL-P-82905	White		[illegible]10-01-129-5444	1 Gal Can
Paint, Alkyd, Lusterless	Forest				
(Spot Painting Alkyd)	Green			[illegible]10-00-598-5464	1 Gal
Brown				8010-00-597-7862	1 Gal
Sand				8010-00-598-5460	1 Gal
Black				8010-00-297-0546	1 Qt
Black				8010-00-297-0547	1 Gal
Black				8010-00-664-7656	5 Gal
Black				8010-00-914-9147	1 Pt

TABLE B-7

PRIMERS

PRIMER	MILSPEC	TYPE	COLOR	COLOR NUMBER	NSN	SIZE
Primer Coating Epoxy-Polyamine Chemical and Solvent Resistant (Contains Chromate for use on Non-ferrous Surfaces)	MIL-P-2377	I	Deep Yellow		8010-00-229-4813	1 Pt Kit
	MIL-P23377	I	Deep Yellow		8010-00-142-9279	1 QT Kit
	MIL-P-23377	I	Deep Yellow		8010-00-935-7080	2 QT Kit
	MIL-P-23377	I	Dark Green	34052	8010-00-082-2450	2 Gal Kit
	MIL-P-23377	II	Dark Green	34052	8010-01-048-6539	2 Gal Kit
	MIL-P-23377	I	Deep Yellow		8010-00-082-2477	10 Gal Kit
Primer Coating, Epoxy (contains Lead and Chromate) (Rust Inhibiting for use on Ferrous Surfaces)	MIL-P-52192				8010-00-264-8866	1 1/4 Qt Kit
	MIL-P-52192				8010-00-082-1714	1 1/4 Gal Kit
Primer, Epoxy Coating (Corrosion Inhibiting-for use on Ferrous and Non-Ferrous Surfaces)	MIL-P-53022		White		8010-01-193-0516	1 1/4 Qt Kit
	MIL-P-53022		White		8010-01-193-0517	1 1/4 Gal Kit
	MIL-P-53022		White		8010-01-187-9820	5 Gal Kit
Primer Coating, Epoxy, Water Reducible (for use on Ferrous and Non-Ferrous Surfaces)	MIL-P-53030		Reddish Brown		8010-01-193-0519	1 1/4 Qt Kit
	MIL-P-53030		Reddish Brown		8010-01-193-0520	1 1/4 Gal Kit
	MIL-P-53030		Reddish Brown		8010-01-193-0521	5 Gal Kit
Primer Coating: Epoxy VOC Compliant Chemical and Solvent Resistant (Lead-Free, Water Reducible)	MIL-P-85582		Light Green		8010-01-218-0856	1 Qt Kit
	MIL-P-85582		Light Green		8010-01-218-0354	1 Gal Kit

Table B-8

PRIMER (Wash)
PRETREATMENT
(Formula No. 117 For Metals)
METRIC, DOD-D-15328

NSN	SIZE
8030-00-580-7076	1.25 Qt Kit
8030-00-281-2726	1 Gal Kit
8030-00-165-8577	5 Gal Kit

Table B-9
THINNER, MIL-T-81772

TYPE	NSN	SIZE
I (Synthetic Resin Enamel)	8010-00-181-8080	1 Gal Can
I (Aliphatic)	8010-00-181-8079	5 Gal Can
I (Aliphatic)	8010-00-280-1751	55 Gal Drum
II (Epoxy)	8010-01-212-1704	1 Gal Drum
II (Epoxy)	8010-01-212-1704	5 Gal Can
II (Epoxy)	8010-01-168-0684	55 Gal Drum

APPENDIX C

VEHICLE CAMOUFLAGE PATTERNS
WHEELED VEHICLES

Model No	Nomenclature	Figure No
M718A1	Truck, ambulance: 1/4-ton, M718A1	33
M151A2	Truck: utility: 1-1/4-ton, M151A2	34
M561	Vehicle, support: 1-1/4-ton M561	35
M880	Truck, Cargo: 1-1/4-ton, M880	36
	Shop equipment, contact maintenance truck mounted	37
M996	Ambulance, 2-litter, hard top, 4 x 4, M996	38
M997	Ambulance, 4-litter, hard top, 4 x 4, M997	39
M996, M1025 M1026, w/w M1044, M1045, and M1046	Truck, utility: armament carrier, armored 1-1/4-ton, 4 x 4, M996, M1025, M1026, w/winch, M1044, M1045 and M1046	40
M1035, M1038	Truck, utility: cargo/troop carrier, 1-1/4-ton, 4 x 4, M1035, M1038	41
M1037, M1042	Truck, utility: 5250 shelter carrier, 4 X 4, M1037, M1042	42
M1008	Truck, cargo: 1-1/4-ton, 4 x 4, M1008	43- 43a
M1009	Truck, utility :3/4-ton, 4 x 4, M1009	44- 44a
M1010	Truck, ambulance: 1-1/4-ton, 4 x 4, M1010	45- 45a
M1031	Truck, chassis: type D, 1-1/4-ton, M1031	46
M35A2, M35A2C, V18A	Truck, cargo/personnel: 6 x 6, 2-1/2-ton	47
M35A2	Truck, distribution, fuel/water: 6 X 6, 2-1/2-ton	48
M36A2, M36A2C	Truck, cargo: 2-1/2-ton SLWB, M36A2 and M36A2C	49
M49A1C, M49A2C	Truck, tanker, fuel: 2-1/2-ton, wo/winch M49A1 C, M49A2C	50
M50, M50A2, M50A3	Truck, water distribution: 2-1/2-ton, M50 M50A2 and M50A3	51
M109A3,M185A3	Truck, shop van: 2-1/2-ton, 6 x 6, M109A3 & M185A3	52
M44, M275A2	Truck, tractor: 2-1/2-ton, M44, M275A2	53
M44, M342A2	Truck, dump: 2-1/2-ton, M44, M342A2	54
M764 w/w	Truck, maintenance, earth boring machine and pole setter: 2-1/2-ton, 6 x 6, M764 w/winch	55
M51A1, M51A2	Truck, dump: 5-ton, w/winch, M51A1,M51A2	56
M52A1, M52A2 and M52A2	Truck, tractor: 5-ton, w/winch, M52A1	57
M813, M131A1	Truck, cargo: 5-ton, w/winch, M8' 3 and M131A1	58
M814, M55A2	Truck, cargo: 5-ton, 6 x 6, M814 and M55A2	59
M816, M543A2	Truck, wrecker: 5-ton, M816 and M543A2	60
M817, w/w	Truck, dump: 5-ton, M817, w/winch	61
M818	Truck, tractor: 5-ton, 6 x 6, M818	62
M819, w/w	Truck, tractor wrecker: 5-ton, M819 w/winch	63
M553	Truck, wrecker: 10-ton, 4 x 4, M553	64
M820A2	Truck, van, expansible: 5-ton, M820A2	65
M920	Truck, tractor: 5-ton M920	66
M923A1, M925A1	Truck, cargo: 5-ton, M923A1 and M925A1	67

Model No	Nomenclature	Figure No
M60A3	Tank, mine-clearing: robot-line-charge, roller and clams, M60A3	103
M60ATTS	Tank, combat, Full Tracked: 105-mm gun, M60ATTS	104
M88A1	Tank, recovery vehicle, Full Tracked: medium, M88A1	105
M109A2	Howitzer medium, self-propelled: 155-mm, M109A2.	106
M110A2.	Howitzer, self-propelled: 8 inch M110A2	107
M113A3	Carrier, personnel: M113A3	108
M113A2	Carrier, personnel, Full Tracked, armored: M113A2	109
M981	Carrier, Personnel, fire support, M981	110
M548	Carrier, cargo, Full Tracked: M548	111
M551A1	Armored, reconnaissance/airborne vehicles Full Tracked M551A1	112
M577A2	Carrier, command post, Full Tracked, armored, M577A2	113
M578	Recovery vehicle, armored: M578	114
M728	Tank, combat, engineers vehicle, Full Tracked, M728	115
M730A1 E1	Carrier, guided missile equipment M730A1 E1	116
M901 A1	APC, improved tow vehicle, M901A1	117
M973	Carrier, cargo, tracked, 1-1/2-ton, M973	118
M992	Carrier ammunition, tracked: M992	119
M1015	Carrier: electronic warfare, tracked: M1015	120
M1059	Carrier, smoke generator M1059	121

TRAILERS

Model No	Nomenclature	Figure No
M101A1	Trailer, cargo: 3/4-ton, 2-wheel, M101A1	122
M105A2	Trailer, cargo: 1 1/4 ton, two wheel, M105A2	123
M107A2	Trailer, tank: water, 1-1/4-ton, two wheel, M107A2	124
M149A2	Trailer, tank: water, 400-gallon, 1-1/2-ton, 2-wheel, M 149A2	125
M310	Trailer, cable reel: M310	126
M332	Trailer, ammunition: 1-1/2-ton, 2-wheel M332	127
M345	Trailer, flatbed: 1 0-ton, M345	128
M416	Trailer, cargo, 1/4-ton, 2-wheel, general purpose, M416	129
M796	Trailer, bolster: 4-ton, 4-wheel, general purpose, M796	130
M989	Trailer, ammunition: 11-ton, heavy, expanded mobility, M989	131
MILVAN	Container, cargo: general, MILVA 1	132
M119A1	Semitrailer, van: cargo, 6-ton, 2-wheel M119A1	133
M127A2C	Semitrailer, van: cargo, 12-ton, 4-wheel M127A2C	134

Model No	Nomenclature	Figure No
M146	Semitrailer, van: shop, 6-ton, 2-wheel M146	135
M313	Semitrailer, van: expansible, 6-ton, 4-wheel M313	136
M348A2	Semitrailer, van: composite, 6-ton, 2-wheel, M348A2	137
M447, M447C and M749	Semitrailer, van: expansible, 6-ton, 4-wheel M447, M447C, M749	138
M129A2C	Semitrailer, van: supply, 12-ton, 4-wheel M129A2C	139
M131A5C	Semitrailer, tanker: fuel servicing, 5000-gallon, 4-wheel, M131A5C	140
M1062	Semitrailer, tanker: 7500 gallon, MI 062	141
M750	Semitrailer van: expansible parts storage, 6-ton, 4-wheel, M750	142
M871	Semitrailer, van,: flatbed, breakbulk container transporter, 22-1/2-ton M871	143
M870A1	Semitrailer, low bed: 22-1/2-ton, 6-wheel, M870A1	144
M172A1	Semitrailer, low bed: 25-ton, 4-wheel, M172A1	145
M872A3	Semitrailer, linehaul, breakbulk/container, 34-ton, M872A3	146
M747	Semitrailer, low-bed: heavy equipment transporter, 60-ton, M747	147
M967A1, M969A1 M970A1	Semitrailer,tank: 5000-gallon, 4-wheel, M967A1, M969A1, and M970A1	148
M1032, also M373A2C	Semitrailer, van: small repair parts transporter, M1032 (also M373AC)	149
M971 El, M822E1 M991E2, M995E2	Semitrailer, van: basic, 10-ton, 4-.vheel, M971 El, M822E1, M991 E2 and .:995E2	150
M098	Air compressor: 250-CFM	151
R050	Crane, truck: 7-1/2-ton, R050	152
RT41AA	Crane, self-propelled: aircraft maintenance and positioning RT41AA	153
613B	Distributor, water, self-propelled 2500 gallon minimum, 613B	154
130G, 130GS	Roadgrader, heavy, 130G and 130GS	155
JD-140	Tractor, wheeled DED, w/backhoe, w/front loader JD-140	156
JD550	Tractor, Full Tracked, low speed DED air dropable and helicopter transportable, JD550	157
D5B	Tractor, Full Tracked, low speed, DED, sectionalized and nonsectionalized,D5B	158
D7G (T-9)	Tractor, Full Tracked, low speed DED, medium drawbar pull, T-9 D7G	159
621B	Scraper, open bowl, self-propelle, , 621B	160
M9	Earthmover, armored, combat, h1	161
FLU 419	419 Small emplacement excavat, r, FLU 419	162
MW24C (M064)	Scoop loader, 2-12 CY, M064, M', V24C	163
950BS, 950BSN	Scoop loader, 950BS and 950Bns	164

Model No	Nomenclature	Figure No
AN/TPQ-37	Trailer Chassis: antenna transceiver group	
ANITPQ-37		165
M499	Ramp, M499	166
M536	Laundry, Unit, M536	167
875CC	Crane, container: rough terrain 875CC	168
	Crane, truck, 140-ton	169
M4K	Truck, forklift, DED, pneumatic tire articulated frame steer, 4,000 lb capacity rough terrain M4K	170
6K,M489	Truck, forklift, 6K cap, variable reach, rough terrain M489	171
M10,M488	Loader, F/L DED RT, M10, M488	172
M412	Loader, RT container handler M412	173

APPENDIX D

Special CARC-Painting Requirements

1 The following surfaces have to be kept free of CARC overspray ... sible:

- restraint systems (seat belts)
- electrical connections and terminals (plug connectors)
- radio and infrared wiring
- voltage regulator box (wheeled vehicles)
- turning signal sending unit including contacts
- piston rods
- fitting and sealing areas
- glass
- door-windshield and access door sealing
- steering wheel
- plastic knobs and levers (interior of vehicle)
- radiator fins
- duct personnel heater, exhaust system

2. The following flexed materials are not to be painted, ... former remaining paint is acceptable):

- Wheeled vehicles:
- boot steering knuckle
- brake hose rubber type
- water hoses (on radiator, engine and personnel heating)
- air direction tubes on heating unit
- intake hoses air filter turbocharger
- wash and wiping system reservoir tubes
- air-heating tubes (interior of vehicle)
- actuating pedal pads (brake clutch etc)
- boot power takeoff
- boots on air supply connections
- air intake hoses
- splash-guard
- hydraulic tubes
- tires
- cone belts

b Tracked vehicles
- access covers of receptacles for external power supply -
- rubber parts on rifle mounting brackets
- rubber rolling surfaces on roadwheels, sprocketwheels and track-pads

c Inaccessible areas
- At areas where protection is impossible due to inaccessibility over-painting of parts mentioned before is allowed.

3 On parts exposed to a temperature exceeding 204' C (400' F) neither CARC nor CARC overspray is permissible:
- turbocharger
- exhaust manifold
- fuel fired personnel heaters
- exhaust grill doors MI-tank

By Order of the Secretary of the Army:

CARL E. VUONO
General, United States Army
Chief of Staff

Official:

THOMAS F. SIKORA
Brigadier General, United States Army
The Adjutant General

Distribution:

To be distributed IAW DA Form 12-34C-E (Block No. 0528) Color, Marking, and Painting of Vehicles, Constr, MHE

*·U.S. GOVERNMENT PRINTING OFFICE: 1995-395-927/30744

THE METRIC SYSTEM AND EQUIVALENTS

LINEAR MEASURE

1 Centimeter = 10 Millimeters = 0.01 Meters = 0.3937 Inches

1 Meter = 100 Centimeters = 1,000 Millimeters = 39.37 Inches

1 Kilometer = 1,000 Meters = 0.621 Miles

SQUARE MEASURE

1 Sq Centimeter = 100 Sq Millimeters = 0.155 Sq Inches

1 Sq Meter = 10,000 Sq Centimeters = 10.76 Sq Feet

1 Sq Kilometer = 1,000,000 Sq Meters = 0.386 Sq Miles

CUBIC MEASURE

1 Cu Centimeter = 1,000 Cu Millimeters = 0.06 Cu Inches

1 Cu Meter = 1,000,000 Cu Centimeters = 35.31 Cu Feet

LIQUID MEASURE

1 Milliliter = 0.001 Liters = 0.0338 Fluid Ounces

1 Liter = 1,000 Milliliters = 33.82 Fluid Ounces

TEMPERATURE

5/9 (°F -32) = °C

212° Fahrenheit is equivalent to 100° Celsius

90° Fahrenheit is equivalent to 32.2° Celsius

32° Fahrenheit is equivalent to 0° Celsius

9/5 C° +32 = F°

WEIGHTS

1 Gram = 0.001 Kilograms = 1,000 Milligrams = 0.035 Ounces

1 Kilogram = 1,000 Grams = 2.2 Lb.

1 Metric Ton = 1,000 Kilograms = 1 Megagram = 1.1 Short Tons

APPROXIMATE CONVERSION FACTORS

TO CHANGE	TO	MULTIPLY BY
Inches	Centimeters	2.540
Feet	Meters	0.305
Yards	Meters	0.914
Miles	Kilometers	1.609
Square Inches	Square Centimeters	6.451
Square Feet	Square Meters	0.093
Square Yards	Square Meters	0.836
Square Miles	Square Kilometers	2.590
Acres	Square Hectometers	0.405
Cubic Feet	Cubic Meters	0.028
Cubic Yards	Cubic Meters	0.765
Fluid Ounces	Milliliters	29.573
Pints	Liters	0.473
Quarts	Liters	0.946
Gallons	Liters	3.785
Ounces	Grams	28.349
Pounds	Kilograms	0.454
Short Tons	Metric Tons	0.907
Pound-Feet	Newton-Meters	1.356
Pounds Per Square Inch	Kilopascals	6.895
Miles Per Gallon	Kilometers Per Liter	0.425
Miles Per Hour	Kilometers Per Hour	1.609

TO CHANGE	TO	MULTIPLY BY
Centimeters	Inches	0.394
Meters	Feet	3.280
Meters	Yards	1.094
Kilometers	Miles	0.621
Square Centimeters	Square Inches	0.155
Square Meters	Square Feet	10.764
Square Meters	Square Yards	1.196
Square Kilometers	Square Miles	0.386
Square Hectometers	Acres	2.471
Cubic Meters	Cubic Feet	35.315
Cubic Meters	Cubic Yards	1.308
Milliliters	Fluid Ounces	0.034
Liters	Pints	2.113
Liters	Quarts	1.057
Liters	Gallons	0.264
Grams	Ounces	0.035
Kilograms	Pounds	2.205
Metric Tons	Short Tons	1.102
Newton-Meters	Pound-Feet	0.738
Kilopascals	Pounds Per Square Inch	0.145
Kilometers Per Liter	Miles Per Gallon	2.354
Kilometers Per Hour	Miles Per Hour	0.621

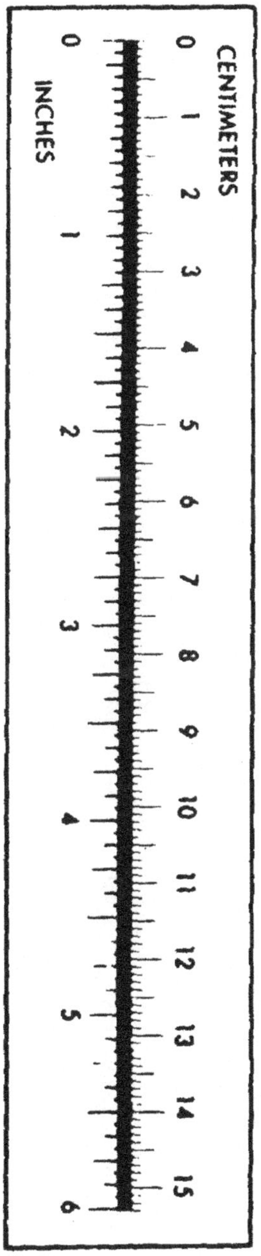

PIN: 015[illegible]001

Made in the USA
Monee, IL
05 November 2026

67387156R00227